D1633985

£28 50

P mL

This book is

16 APR

BUILDING SYSTEMS, INDUSTRIALIZATION, AND ARCHITECTURE

BUILDING SYSTEMS, INDUSTRIALIZATION, AND ARCHITECTURE

Barry Russell

JOHN WILEY & SONS

London · New York · Sydney · Toronto

British Library Cataloguing in Publication Data:

Russell, Barry
  Building systems, industrialization
  and architecture.
  1. Buildings, Prefabricated
  I. Title
  721   NA4140   80-41692

ISBN 0 471 27952 8

Typeset by Photo-Graphics, Yarcombe, Devon
and printed by Page Bros, Norwich, England

for
Jean
Jane
Jil
Sarah

CONTENTS

PREFACE

The notion of writing a book on industrialization and building systems first surfaced during my third encounter with such ideas in Britain in 1962. I became alarmed at the way in which building systems were being promoted and implemented since, often, this seemed to proceed with little understanding or regard for the body of knowledge that had already been built up in Britain and elsewhere. Further, this renewed interest frequently seemed to be pursued with little regard for any experience of building or concern with architecture itself. My view of all this was sharpened by having been exposed to the pragmatic and direct approach taken by the design and construction process in North America during a period of working in Canada.

I was subsequently invited to teach at Portsmouth School of Architecture by Jack Ogden, then Acting Head, and I am grateful to him for creating the impetus for me to develop my ideas with a succession of students. My thanks are also due to Professor Geoffrey Broadbent, Head of the Portsmouth School of Architecture, who created a context for me to pursue the ideas and who has given continuous support and encouragement to this project. Needless to say we have been exchanging views on the topic of this book throughout this period.

My thoughts have been sharpened by teaching a variety of courses and my thanks go to all those students who have been involved. Similarly, a wide range of colleagues have, unwittingly or otherwise, provided a stimulus to the project over the years and I am grateful for all these encounters, and in particular to Alan Balfour, Laurie Fricker and Michael Thompson. Information, comment, illustrations and permissions to use material has involved a wide range of people and organizations: these are credited in the text, bibliography or list of sources of illustrations. The responsibility for the form this has been fashioned into is, of course, entirely mine.

Although many of the photographs are my own I have received invaluable help from a number of photographers including Brian Barfoot, Karen Goodwin, Nigel Grundy, and the Polytechnic Photographic department. The majority of the typing (and retyping) was ably carried out by Edna Eglinton, Sharon Lipman, Marylyn Parker and Jil Russell. I am grateful to my editor at John Wiley, Celia Bird, and to her predecessor Michael Coombs. I thank all these and the following who have assisted me in various ways:

Ball State University, College of Architecture and Planning, Muncie, Indiana, USA; Professor Juan Pablo Bonta;. Alan J. Brookes; Arthur Corney; Ole Dolva; Barrie Evans; Leslie Fairweather; Mr and Mrs A. Fisher (owners of a Lustron home); Norman Foster; Dan Gagen; Professor D. Hermansen; Karl Koch; Uwe Kohler; John LeGood; Jeff Merriman; James Paddock; Portsmouth Polytechnic Library (David Francis, Ian Mayfield, and Daphne May) Administrative and Technical staff; Andrew Rabeneck; Helge Saatvedt; Barrie Seaman; Walter Segal; Professor A.W. Skempton; Glen Smith (and National Homes, Indiana); Nick Woolfenden; John Wright.

I owe a great debt to my wife Jean, who has provided continuous support, criticism, and encouragement, and to my three daughters Jane, Jil and Sarah who have all assisted in various ways. It was my wife who said that without the family I would have been able to finish the book in half the time: however, without them I would probably never have started such an undertaking let alone finished it.

CONVENTIONS USED IN TEXT

References:
In the text these are given in the form:
Morgan (1961), (Morgan 1961), or Giedion (1969, 1948). The latter form gives the text referred to followed by the date of first publication, which is normally of particular interest.

Dates:
The life dates of individuals are given in the form:
Palladio (1508-80)

For buildings it is normally the completion date that is given:
Leicester Engineering Building (1963).
Although on occasion the design and construction period is given so that this may be related to other work.

INTRODUCTION

'If everything is founded on sound efficiency, this
efficiency itself, or rather its utility, will form its own
aesthetic law. A building must be beautiful when seen
from outside if it reflects all these qualities...'

Bruno Taut
Modern Architecture 1929

'The point is that material effectiveness, practicality,
does not exist in any absolute sense, but only in the
measure and form projected by a cultural order.'

Marshall Sahlins
Culture and Practical Reason 1978, 1976

A variety of images is conjured in the mind by the phrase building
system. For some it denotes the rigours of many a vast concrete
housing complex or metallic school, whilst for others it will invoke the
subtle fusion of craftsmanship, order, social structure and philosophy
evident in the traditional Japanese house. Yet again it may call to mind
the beauties of dimensional harmony sought by Palladio or Le
Corbusier. Or it may recall the significance of Ronan Point, the high
rise system built block in London, where a minor gas explosion drew
attention to the subtle intertwining of a philosophy with political and
technical decision making to cause a part of the building to
'progressively collapse'. All buildings, of course, are systems with
ranges of parts and rules for putting them together, from the
speculative builder's house, with its many prefabricated elements,
including the brick, to the Gothic cathedrals with their complex
geometry, symbolism and building methods.

How is it then possible that the notion of a building system has come
to mean something quite specific both to architects and to the public at
large? By what means has it had such a powerful influence on the day
to day conduct of putting up buildings? Part of the answer lies in the
relationship between the ideals of the modern movement in
architecture and industrialization. Whether called prefabrication,
building systems, system building, or industrialized building, the
name has usually carried assumptions which favour, encourage and
reinforce the necessity of applying the methods of industrialized mass
production to the building process. The development and validity of
this approach is a central thread in this study which sets out to offer
some cohesive explanation for the repeated attempts to create building
systems. Some of the possibilities suggested by the growing body of
knowledge surrounding general systems theory itself are examined but
only in so far as they offer an appropriate context for the ideas being

discussed, and it will be argued that building systems have often developed in contradiction to the central notions surrounding systems theory. This phenomenon is not confined to architecture but can also be seen in systems engineering: indeed it is more to this latter field that architecture appears to have looked for its models. The body of knowledge that has grown up around the systems idea as propounded by von Bertalanffy (1968) and others, appears to offer a more adequate model in that it concerns itself with the relation of both physical and non-physical systems. The field of studies referred to as sociotechnical is a reflection of a growing concern felt about the relation of man to his technologies. However, it is only in recent years that the implications of this have begun to be explored more thoroughly for architecture. An examination of some of the issues related to this question can be seen in a special issue of *Architectural Design* edited by Landau in 1969 and significantly entitled 'Despite popular demand, AD is thinking about architecture and planning'. These ideas were subsequently pursued in a sequel in 1972 entitled 'Complexity'. In this Landau (1972) drew attention to the shift in conceptual understanding necessary in viewing problems as complex rather than as simple, and to the importance of the growth of those approaches which sought to take a holistic view rather than one that was atomistic. This suggested serious questions concerning the reduction or simplification of difficult social and architectural problems. Such simplifications appear in the way the industrialized building systems idea has usually been taken up by politicians and architects. It has frequently been used as a 'technology fix' in the crudest sense, and such a notion still seems to hold considerable attractions for politicians offering programmes of reform. The results of this can be seen in the highly industrialized countries, but the most serious implications affect the less industrialized countries who attempt to import or develop building systems which are heavily dependent upon sophisticated technologies.

In the first *General Systems Yearbook* published in 1956, von Bertalanffy, who claims to have introduced the term General System Theory, drew attention to the importance of the problem of organized complexity. The difference of approach demanded by complex problems was not best dealt with by the analytical methods of scientific enquiry which in the modern world have been applied to any and everything regardless of suitability. Stafford Beer (1963) characterized it in this way: 'We are not trained to think in systemic terms because our scientific approach has been analytical to the ultimate degree. We took things apart, historically, and described the atomic bits. We did an experiment, historically, on these bits — deliberately holding invariant the behaviour of other bits with which in fact they were systemically interacting in real life.' The significance of the systems notion lay, not in the systematic building block approach of systems engineering or systems analysis but in the way it draws attention to the importance of wholes. Angyal, in 1941, pointed out that it was this holistic viewpoint which made the idea so useful: 'In aggregates, it is significant that the parts are added; in a system it is significant that the parts are arranged . . . the system cannot be derived from the parts; the system is an independent framework in which the parts are placed.' This is a potent idea for designers who should readily understand that it is the relationship between the parts that is significant, and not merely the effectiveness of those parts. However, many of those involved in the creation of building systems have concentrated on the efficiency of the parts at the expense of the whole. We frequently find, for example, that good components have been produced, but that joint problems remain unresolved. Or good subsystems produced, but poor buildings result. Or, the most frequent occurrence, an interesting technology has been developed, but little or no consideration of the administrative, marketing and management problems has taken place. Similarly little attention has been given to post-occupancy evaluation.

The study examines the development of industrialized building systems in Britain, and relates this to developments in Europe and North America. In a work of this nature, a decision has to be made to present the material chronologically or thematically. The structure adopted is, in principle, chronological. However, this has not been pursued dogmatically since to do so would unnecessarily fragment material which should be seen as related. The approach, therefore, is one of overlapping continuity which enables a particular technique or topic to be examined without necessarily confining it to a particular point in time or to a particular place. The brief chapter on proportional systems is an example of this as is that on timber-frame methods. The two separate chapters dealing with housing and school building systems in Britain discuss events that ran roughly parallel in time but which developed along distinct paths. While the importance chronologically of the first consortium of local authorities to develop a building system, CLASP, has meant that it is discussed, not in the chapter on the consortia, but in a chapter prior to those on both housing and schools. For different reasons another schools consortium, The Metropolitan Consortium of Education (MACE), is examined after the work of Ezra Ehrenkrantz. A general discussion of the outcome, benefits, disadvantages and problems arising from the attempt to implement system ideas occurs towards the end of the book as does a broader commentary on those ideas. The approach adopted then examines theoretical, methodological and practical interactions in a historical context to provide both explanation and evaluation of the repeated attempts to industrialize building processes. It attempts to bring together the concepts, whether written or drawn, the propositions in the form of single buildings which offer a model, and the application of these ideas in the day to day business of making buildings.

Existing literature either offers a polemic in favour of industrialization, unmitigated hymns of praise to the possibilities of modern technology, or it consists of a series of catalogues of available or once available systems. This study aims to relate thoughts and ideas to practical actions and to do this in a social context. It is for this reason that the book includes, for example, early propositions concerning industrialization as well as specific case histories.

The overall pattern adopted can be divided into three parts. The first two chapters draw attention to the fact that systems of one sort or another have occurred whenever there is building and discuss the first impact of industrialization. The second part, which constitutes the major section of the work, examines the various responses to industrialization by architects and designers from the turn of the century. Developing theory and technique is followed in Britain, Europe and North America up to the enormous interest in recent years in system approaches, culminating with a series of examples which discuss the initial impact of environmental engineering and energy issues. These have paved the way for a more inclusive view of what level of system should be of concern to designers. The third part examines some of the assumptions, successes and failures of building systems, particularly in Britain. The book concludes by setting the development of building systems against a broader framework and by raising questions about the nature of such notions in an architectural and social context.

In this study I have attempted to show not only the broad sweep of ideas but specific solutions adopted by building systems. It will invariably be the case that initial ideas will have been modified after their first use, sometimes even discarded completely. This has led some architects to comment on the value of showing specific technical solutions. It is my view that the historical development of ideas, their use, acceptance and rejection can only be fully understood in relation to the specific application in practice of those ideas. In pursuing information I met, on occasions, what I recognize as the system builder's lament: 'Yes, we did do it like that then but, of course, we have improved the solution and do not do it that way now'.

The politics of system credibility demand that whatever is current practice be defended as the very best course of action, even if past procedures are open to comment. If pursued, the logic of this stance has a very convenient outcome for those involved. It is that there is little point in showing or discussing what was done at a particular point in time, however much critical feedback exists, because it has all now changed for the better. The current practices have then to be as equally well defended as were the earlier ones, until they in their turn are succeeded. The ebb and flow of publicity about a given system approach is also relevant here. Most building systems are launched with coverage in the architectural press, sometimes this extending to general press coverage. Bold claims are made, optimism runs high. Often that is all that is heard but more frequently there is a gradual accumulation of positive and negative feedback. Usually this diminishes as well and, although the system may continue to survive in one form or another, little more is published about its activities. The building system designer's complaint that his later successes and improvements have not been taken note of by the architectural world or the public at large can often be seen as a result of this process. For the fact is that after the euphoria of the early heady days of the birth of a building system, even those involved with sustaining it fail to make positive efforts to demonstrate the success or failure of subsequent development. In discussion with those associated with building systems I have found that they will often point out significant improvements made to initial methods, but usually no attempt has been made to publicize this. More commonly they will indicate that, although a particular system was not successful, many of the principles it introduced have been generally absorbed by the building industry.

The much publicized failure of many building systems has encouraged reticence on the part of many of those involved and it has often been left to public inquiries or investigative reporting by the press to uncover the problems encountered. In the present situation I am, therefore, very appreciative of all those who have given time to answering questions and providing illustrations. In piecing together the account offered here, I am aware that there is still much work to be done. My hope is that it will offer a useful commentary on this somewhat uncharted aspect of architectural endeavour.

The visions of a new society with a new architecture sustained the modern movement for many years and the impact of industrialization was one of its central concerns. During the recent debate surrounding architecture's modern movement, much that it stood for has been rejected. The process of industrialization, however, continues apace, and architects still have to find humane solutions which make sensible use of whatever technologies are available. Wherever this happens systems of one sort or another will be found.

1

'A ROSE BY ANY OTHER NAME'

SYSTEMS IN TRADITIONAL BUILDING

'What's in a name? That which we call a rose
By any other name would smell as sweet'

Juliet in *Romeo and Juliet*
William Shakespeare

Whilst this work is concerned with the more recent events in building systems there is, of course, a long history of buildings being organized around sets of principles, implicit or explicit. The Oxford English Dictionary defines a system as a 'complex whole, set of connected things or parts, organised body of material or immaterial things'. Indeed any building might be said to conform to this definition in that, however primitive, it must comprise a set of connected things or parts and is certainly an organization of both physical bits and pieces together with non-physical considerations such as the sociological, cultural, psychological and managerial principles. The fact that a building is there at all suggests that it is some sort of a system.

Nevertheless at different periods differing aspects of building have been emphasized within this general assumption and specific approaches made more explicit than others. In more recent times it is this degree of explicitness which has enabled some specific approaches to have bestowed upon them the title of a system. Moreover the strong tradition of Cartesian rationalism and scientific endeavour has given greater and greater credibility to this notion of explicitness. However, the notion of explicitness, 'distinctly expressing all that is meant; leaving nothing merely implied' (usage 1613) and 'free from folds or intricacies' (usage 1697), can be seen, in this sense, as being in opposition to what great building has always been about.

Architecture has always relied to a lesser or greater degree on a complex net of material or immaterial folds and intricacies. Certainly a drive to be distinct in meaning lays an impossible intellectual burden on architects since the many subtle and ambiguous meanings carried by buildings are usually less than perfectly understood by their creators.

However, it is clear that there are many things one can be explicit about in these terms. What architects have chosen to be explicit about during the recent history of the modern movement has been closely tied to the growth of scientific method and industrialization. It is this area that forms the background for this work.

Whilst the notion of explicitness has become an exceedingly powerful and, I believe often destructive, force in modern architecture, it should not be assumed that for a building to be a whole system all things must be made explicit. This particular trap and its consequences will be discussed in a later section. That designers and builders in the past have always used some sort of a system is self-evident: this section illustrates two examples of such traditional building systems. The first, and probably the most well known example of a whole system in building, is that of the Japanese house. Well documented, particularly in Engel's beautiful book (1964), it stands as a model of the harmony that can be achieved in building when the philosophy, the religion, the family structure, the visual quality, the methods of building and the materials used are all interrelated by evolutionary tradition. Further, it has acted as a powerful stimulus to the modern building systems movement. Unfortunately it is only the visual qualities in isolation that have been sought and reproduced, together with the attention to mathematical order it displays. This latter aspect alone has had a profound effect upon thinking on the dimensions of building components and modular co-ordination and, as Engel is at pains to point out, has been very much misrepresented.

The second example is of a different sort, that of the use of proportional systems. For Panofsky, discussing the theory of human proportion, the meaning is 'a system of establishing the mathematical relations between the various members of a living creature, in particular of human beings, in so far as these beings are thought of as subjects of an artistic representation' (Panofsky 1970). The search for proportional systems which relate this concern for the human with the architecture that man produces is a recurring one: however as will be shown, this approach has suffered a heavy onslaught from the time of the industrial revolution.

THE JAPANESE HOUSE

'While Western civilisation with its enormous technical achievements in building long ago succeeded in making life within the house independent of climate changes, in the Buddhist world nature has never been considered as something to be fought against, conquered, and mastered.'

Heinrich Engel
The Japanese House 1964

The Japanese house, and garden, has had a profound influence upon architects in the twentieth century. Its visual purity found ready acceptance amongst those reacting to the excesses of the late nineteenth century and the associated displays of symbolic opulence. The realization that the houses were based on some dimensional order gave those architects grappling with the implications of machine production an argument that was to become powerful as it developed. The fusion of pure forms, purged of all decoration, with a coherent dimensional system has been a powerful force in the modern movement and in the development of the idea of a building system. Invariably it seems, it is the visual quality that has primarily attracted the interest of designers who have often attempted to apply it directly to western society. However this ignores the fact that the order and visual qualities possessed by these buildings was the result of very specific sociocultural pressures:

> '... this unique physical order in building was due to the environmental pressure of insularity and of political isolation that left the Japanese no choice but to rigidly control consumption and application of existing resources and to avoid any essential change that could upset the precious balance established to safeguard a social order in which the nobles and warriors were supreme.' (Engel 1964)

The Buddhist scholars returning from China to Japan during the 7th and 8th centuries began an acceptance of Chinese culture. This brought about administrative centralization and created social order based on the civil code, Taihō-ritsuryō, or the 'code of the great treasure'. This, amongst other things concerned itself with city organization, individual dwellings and a measure system.

From the seventeenth century began Japan's 200 years of isolation during which progress was deliberately arrested rather than stimulated by the military rulers. This was a policy of the conservation of resources including a reduction in population, trade and art. Engel notes that this brought about the 'noted Japanese "economy in matter and spirit"' and that 'the only architectural virtues that could spring from such a state of affairs were those of economising, rationing and standardising, which presupposed an exact and comprehensive planning and a tight control of the entire economy, including building activities' (Engel 1964).

Rudofsky (1965) points out that a Japanese farmer with a five hundred bushel harvest was permitted to live in a 60 foot (18.3m) long house, but with no parlour or roof tiles. He was not allowed to eat rice. Everything was controlled right down to lanterns and wooden hairpins. Such laws existed until 100 years ago, when one Townsend Harris, says Rudofsky, pointed out that 'it would be an endless task to attempt to put down all the acts of a Japanese that are regulated by authority' — clearly a bureaucracy to be reckoned with.

Since 1945 Western Europe has come increasingly to a similar situation. Britain particularly has seen a period of attempts at comprehensive planning of the economy with tighter and tighter control. Initially the impetus for this came from socialist ideologies concerned with a just distribution of wealth. Latterly however, it has gained added philosophical and practical support from the 'whole earth' view with its emphasis on the conservation of global resources including energy. That eastern religion has seen a resurgence in the West during this period is particularly interesting in the light of Japanese history. It is essential to be quite clear that those characteristics of visual clarity and dimensional order which have commended themselves to western designers are but a part of a whole cultural system and one that has undergone rapid changes in recent years. Therefore before we discuss those notions that have been transferred into western architectural language some aspects of the social context will be briefly discussed. One of these is the attitude to the family and family life. Traditionally this has four main characteristics:

1. disregard of the individual
2. the absolutism of the head of the house
3. right of primogeniture (rights of the first born)
4. subordination of the female.

In Japanese terms the house is merely a mirror image of the family. Engel points out that its essence is 'receptive-reflective rather than causative-formative' and in this we can see the two totally different sets of values at work — those of the meditative harmonious relation with nature, characteristic of the Buddhist, versus that of the control and subjugation of nature which had been the hallmark of the western approach. The traditional Japanese house also reflects a series of attitudes to privacy which run throughout Japanese life. Engel seems to suggest that because of the sociocultural pressures outlined above the economy and order apparent in the houses has itself been partly responsible for constructing these attitudes to privacy. The Japanese have developed attitudes to bodily functions — washing, excreting, sleeping, eating — which show less concern with privacy than does the westerner. Indeed the elaborate rituals and beauty associated with all aspects of this life might be said to have developed from the pressures of exposure. After all one does tend to eat in public in a different way to that in which one would eat in private. Most significant of all is the fact that the Japanese language has no word for 'privacy'. The materials used are lightweight, space is minimal and multipurpose.

The relation of the outside and the inside, the house and the garden has also been one that has exerted a strong influence on western designers. To the Japanese a house with no garden is not a home. Indeed Ka-tei, the Chinese pair of ideographs which stand for the concept home, mean respectively house and garden. The siting of houses and the position and relationship of special functions is also given guidance by a set of rules (feng-shui) which have been cast off by westerners as superstitions but which do seem to embody 'a handy rule of thumb, the observance of which would avoid errors in design' (Engel 1964). The rules are built into the cardinal points of the compass and embody the climatic implications of sun, wind and the form of the land, into a practice called geomancy and were interpreted by a geomancer — something between a priest and an architect.

Having discussed briefly the sociocultural background, the family, the house and the garden, its planning and siting, it is appropriate to turn to those aspects more directly relevant to building itself: materials, methods and measures. Engel points out how wood, being an essential to existence after food and clothing, was subject to the same pressures and became a trading article. Certain size preferences had emerged naturally but these pressures led to standard sizes. This then influenced the way in which the builder went about his work. Western architects are fond of citing the Japanese house as an example of simplicity of means and, further, to assume that this is due to a very limited range of materials. However, as Engel demonstrates, this is fallacious since there is in fact considerable variety. What observers are responding to is the unity which such houses possess. This unity, however, is dependent upon a wide range of factors: the cultural effects on space organization, the way in which natural materials are employed, the almost ritualized techniques associated with the crafting of the houses and the way in which the approach lays importance on the necessity for the inherent characteristics of each material to be brought out. House building is thus not singled out as an art or special activity but is seen as part of daily life in which any person can make his own house. The travelling craftsman existed of course but his abilities are more akin to those of the bread maker and the weaver than the architect as we know him.

One of the pieces of folk lore associated with the Japanese house, and known by many architects, is that it is 'modular' (**1.2.1**) and that the dimensions all relate to the Tatami mat. This mat, originally a portable element (**1.2.2**), did service in many ways in Japanese life —

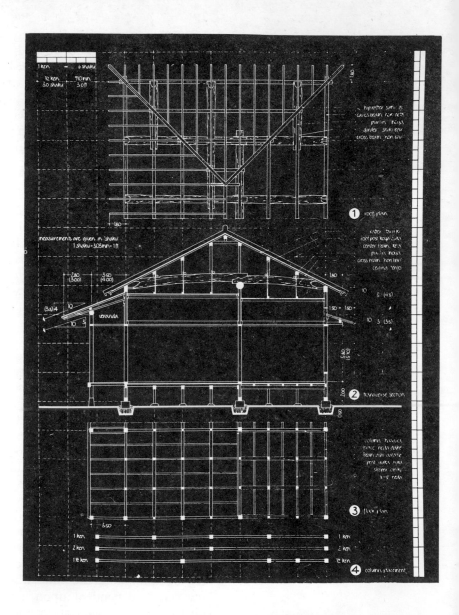

1.2.1 The ordered construction of the Japanese house

to sit on, to sleep on, to talk on, and as a table. However Engel is at pains to point out that, although there is a connection between methods of layout and the tatami it:

> 'has never, not even fictitiously, functioned as a module of any kind in the Japanese house, as is most frequently assumed'. (Engel 1964)

The tatami is made of rice-straw bound together with string to a thickness of two inches or more. It is edge trimmed and the two long sides are bordered on the top surface and edged with black linen an inch in width. Mat making is a separate trade from that of straw matting and Morse in his record of the Japanese house published in

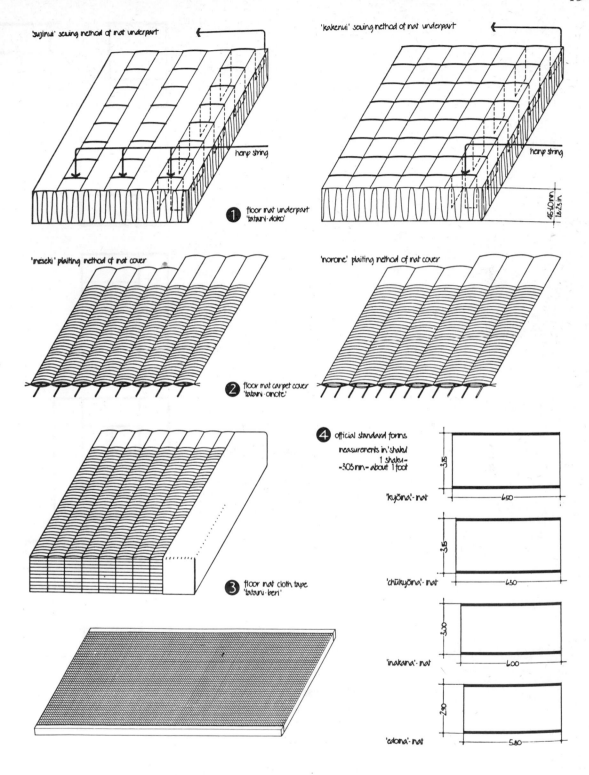

'sujinui' sewing method of mat underpart

'kakenui' sewing method of mat underpart

hemp string

hemp string

45-40 mm
1.6-2.5 in

1 floor mat underpart
'tatami-doko'

'neseki' plaiting method of mat cover

'norone' plaiting method of mat cover

2 floor mat carpet cover
'tatami-omote'

3 floor mat cloth tape
'tatami-beri'

4 official standard forms
measurements in 'shaku'
1 shaku =
=303 mm = about 1 foot

3.15

6.50

'kyōma'-mat

3.15

6.30

'chūkyōma'-mat

3.00

6.00

'inakama'-mat

2.90

5.80

'edoma'-mat

1.2.2 The tatami: sizes and
construction

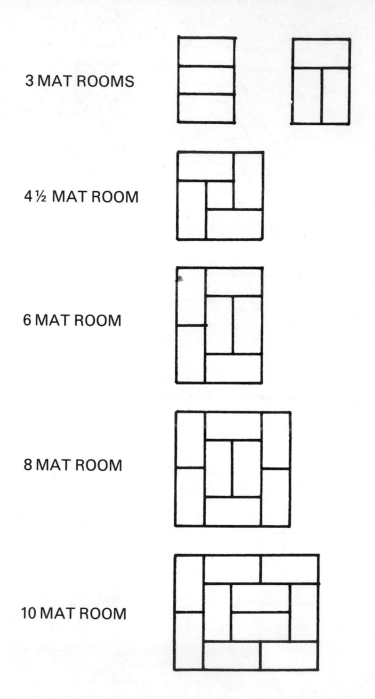

3 MAT ROOMS

4 ½ MAT ROOM

6 MAT ROOM

8 MAT ROOM

10 MAT ROOM

1.2.3 Tatami arrangements

1885 describes how the mat maker could be seen at work in front of his house.

The mat is approximately 910 × 1820 mm (3 × 6 feet), with a thickness of 45–60 mm (1.8–2.4 inches), and room sizes were designed to accommodate a number of mats (**1.2.3**). The three mat room had the mats side by side or two one way and one the other, the four and one-half mat room had the half mat in the centre, with the six and eight mat rooms being the most common and indicating the small

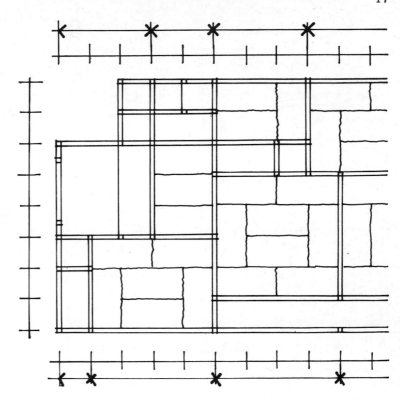

1.2.4 Layout to the 'Maka-ma' method: columns centred on grid lines

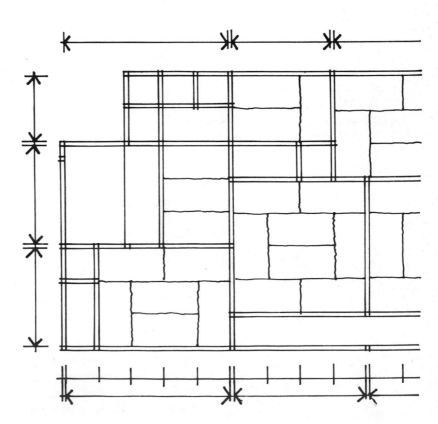

1.2.5 Layout to the 'Kyo-ma' method: column zones between room spaces. A discontinuous tartan grid

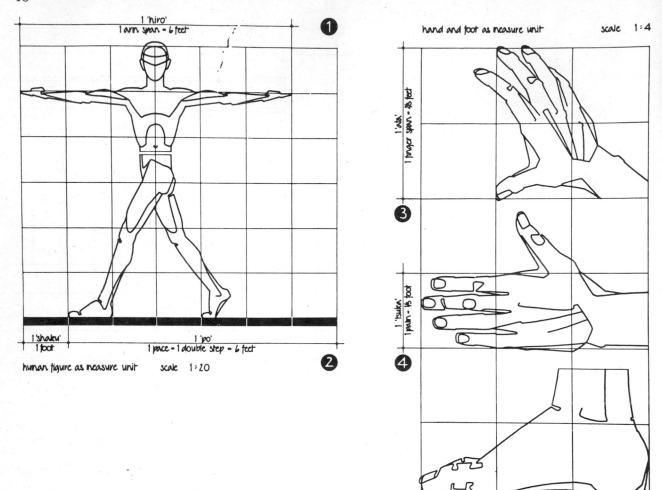

1 1 'hiro'
1 arm span = 6 feet

hand and foot as measure unit   scale 1:4

1 'ata'
1 finger span = 2½ feet

3

1 'tsuka'
1 span = ½ foot

4

1 'shaku'
1 foot

1 'po'
1 pace = 1 double step = 6 feet

2

human figure as measure unit   scale 1:20

1 'shaku'
1 foot = ⅓ step

5

early human standard measures in relationship to foot unit
1 foot = about 275 mm = 10.8 in

1.2.6 Proportional relationships in the human figure

scale of such rooms. The corners of the mats are not allowed to abut and the resulting form is often that of a spiral.

The tatami originally accommodated one man sleeping or two sitting and this then became translated to room sizes as described above. The structural system that developed had its own effect on this essentially human scaled system. It is at this point that it becomes clear that, whilst the tatami may be used for room sizing, it is not used as a module in a building sense. The room sizes are based on mats and the structure is related to it by two basic methods.

The 'maka-ma' uses a consecutive grid (**1.2.4**) with the columns on gridline centres. The 'kyo-ma' uses the mat sizing on a column face to column face basis (**1.2.5**), and with the rooms adjusted to co-ordinate the differences arising from the resulting tartan type grid that is not continuous through the plan. The maka-ma method is using special mat sizes at room perimeter to deal with the thickness problem. Clearly it is not a constructional module as we understand it since it is not uniting planning, structure and construction in an additive way. It is, however, a much more subtle relationship between cultural and human needs and the available techniques.

Since the mat sizes varied in different regions this also caused the constructional measures to alter. The introduction of a standard mat size in one region standardized the open width between columns. However, again the constructional dimensions will be governed by the column to column centre distance, and not the size of the tatami mat. So, either the mats remain standard and the construction adjusts, or alternatively the construction relates to the tatami size used by differing sized mats used in the same room. Architects who have grappled with the so-called 'thickness-problem' in relation to grid planning and partition walls, will recognize these difficulties.

Heights were also related to the room sizes although they were not, as is sometimes thought, related to the tatami size itself. There was a different height for an eight mat room, a six mat room and a three mat room and so on. Further the basic unit of measure is the Japanese foot — the shaku — although there are various sizes for this, depending on the systems used. One shaku equals a third of a step and relates to other body dimensions (**1.2.6**). Engel points out that although the metric system was introduced into Japan in 1891, the shaku is still in use. Whilst it is not accurate to attach the dimensional unit that at first seems apparent in the Japanese house directly to the modular order, there emerges a complete integration of systems from the cultural to the constructional. This causes Engel to describe the Japanese house as 'unique in the history of world architecture'.

A MATTER OF PROPORTION

'What differentiates Palladio's proportions from Alberti's is that they are used in integrated systems that bind plan and elevation, interior and exterior, room and room, giving a sense of the architect's control.'

James S. Ackerman
Palladio 1977, 1966

We saw in the Japanese house how the proportions of the human body were used as a basis for sizing. One palm width is about a third of a foot, 4 inches or 10 centimetres. One finger span, that is from tip of outstretched thumb to top of outstretched finger is about two-thirds of a foot, 8 inches or 20 centimetres. One foot is therefore three times the palm width and also one-third of a step, or about a yard. Six times one foot gave both a man's height and the distance between his outstretched finger tips. Very many elaborate theories have been developed around this basic idea as man in different cultural situations has tried to find harmonious ways of relating the ideas and technologies of his time to the human proportion. Such theories of proportion can become very elaborate indeed, as did Le Corbusier's Modulor (1961), but most of them embody in some way the basic dimensions we have seen used in the Japanese house. Four inches and the dimension of approximately a foot recur in many cultures and systems of proportion, right down to the present. For example the preferred sizes set out in the first British Standard which tried to get to grips with the problem of component dimensions, BS 4011:1966 were 300, 100, 50 and 25 millimetres. Both Bemis (1936) in the United States in the thirties, and the Modular Society in postwar Britain called for 100 millimetres (10 centimetres or 4 inches) to be the basic component and planning module.

The three classical orders of the Greeks, Doric, Ionic and Corinthian, were also related to the human proportion. The Doric order with its height six times its base, was said to reflect the fact that man's foot is equal to one-sixth of his height and the simple straightforward capital was said to show his muscular strength. The Ionic order is said to embody the proportions of the female figure, and was thus eight times the height of the base diameter, the volutes of the capital representing the hair style of the Greek woman. The last and most elaborate is the Corinthian order. Various stories lay behind this, the most favoured being that of the lovely sick girl who died. Her best loved possessions were collected and placed in a basket on the top of her tomb, surmounted by a roof tile. An acanthus plant grew up next spring around the basket and, reaching the underside of the roof tile, bent and curved its leaves around. A passing architect saw this and subsequently used it as the basis for the columns of a temple he was asked to build in Corinth. Whatever the validity of such stories there is no doubt that the Greeks developed a sophisticated approach to proportion and its visual effects.

There is evidence that the Ancient Egyptians used the technique of squaring up a design — possibly as an aid to sculpting, whilst according to Harvey (1972) in England, from the end of the tenth century to the beginning of the twelfth, 'the key to planning was essentially the use of squared paper for design sketches'. A variety of attempts have been made to draw out the proportional bases upon which the mediaeval cathedrals were designed but Harvey points out that these must be treated cautiously since any overall system would have been modified by eye. Furthermore, the fact that lines may be centred on structural elements, outside or inside of any given set of mouldings suggests that it may be possible to create almost any system in retrospect. The mediaeval philosophers placed great importance on number systems and their place in the ordered universe and master masons, even today, will refer to the importance of this 'hidden language' evident in the Cathedrals. For example 1 represented man, 2 woman and 3 courtship. Through this, complex messages could be built up by the way the pattern was deployed in the layout and detail of the building.

Conant's work at Cluny (c 1080) showed how the layout accorded with a modular unit of five Roman feet (11⅝ inches or 295 mm) and with a major grid of 25 Roman feet, the latter being the basis upon which the columns were laid out (Harvey 1972). It is clear that basic figures like the square and the equilateral triangle were used in setting out the cathedrals; however, it is the reasons for the selection of particular dimensions that are probably more interesting. The symbolic basis for the choice of one set of numbers over another thought to be used by master masons and referred to above, is set aside by Morgan in his intriguing study *Canonic Design in English Mediaeval Architecture* (1961) when he says:

> 'such procedures were essentially generated by the need to solve practical problems of building, and any metaphysical connotations associated with them were acquired advantageously in the course of time, and are superfluous to their basic nature and purpose.'

It is interesting that Morgan's work was published at a time when, in Britain, there was a strong practical and 'functional' thrust to architectural thought and when the more spiritual and less tangible aspects were being set aside. A major part of Morgan's work centred around examining the form of the mason's square used in design and setting out the cathedrals. He describes two forms of mason's square; the first he calls the 'general square' the second the 'canonic square'.

The former is based upon the 'general' right triangle of 60 and 30 degrees; the latter is based upon the 'canonic' right triangle with angles of 58° 17' and 31° 43'. Morgan suggests that the two triangles were combined in the one mason's square by the device of tapering the arms. After an analysis of 28 buildings Morgan put forward some very interesting conclusions, two of which deserve mention here:

'(i) From the early thirteenth century until the last quarter of the fourteenth century the bay ratios and section ratios in 'Royal' work are greater than those used by designers working outside the royal sphere. In effect, the 'royal' buildings would give an impression of being taller than their counterparts, although in terms of absolute height there is often no significant difference between them.

(ii) In both 'royal' and 'non-royal' work the use of a tall bay proportion is matched by a high section ratio. Also as the section ratios show tendencies to increase or decrease, such changes are reflected in the bay ratios.' (Morgan 1961)

After the completion of Bristol Presbytery in 1298 the two separate proportional traditions began to come together. The Kings' Master Masons, it seems, used a square based upon the 'Canonic Triangle' which in turn was closely related to the golden number triangle:

'Golden number' triangle:	90° 0'	31° 43' 03"	58° 16' 57"
'Royal Canon' triangle:	90° 0'	31° 23' 0"	58° 37' 0"
Discrepancy:		0° 20' 3"	0° 20' 3"

(Morgan 1961)

The implication of this is that the bays in 'Royal' buildings were given by a triangular construction using the golden number and starting from a base of 23 feet 5 inches (7.137 m). Morgan points out that this dimension was also that of the French perch used in the spacing of the bays at Amiens, Bourges and Reims cathedrals.

An important point emerges from this work. By the use of an understood and specific proportional device the designers of the 'Royal' buildings were able to gain a particular desired effect (that of a soaring loftiness) whilst at the same time keeping their actual heights very similar to that of the 'Provincial' buildings. Here proportion was used as a way of relating the intended meaning of the buildings to the perceptions of the individual.

As Wittkower shows (1962, 1949), Alberti (1404-72) drew heavily upon the classical tradition, emphasizing the perfection of the human body and describing how, with hands and feet extended, it fitted those most perfect of geometrical figures, the circle and the square.

'This simple picture seemed to reveal a deep and fundamental truth about man and the world, and its importance for Renaissance architects can hardly be overestimated.' (Wittkower 1962, 1949)

24

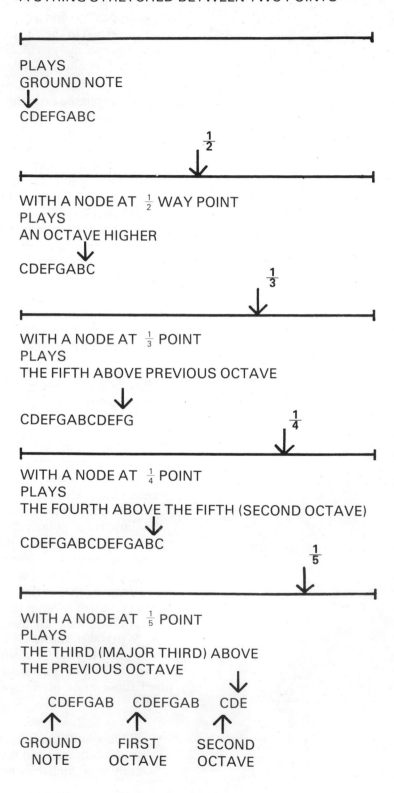

A STRING STRETCHED BETWEEN TWO POINTS

PLAYS
GROUND NOTE
↓
CDEFGABC

$\frac{1}{2}$

WITH A NODE AT $\frac{1}{2}$ WAY POINT
PLAYS
AN OCTAVE HIGHER
↓
CDEFGABC

$\frac{1}{3}$

WITH A NODE AT $\frac{1}{3}$ POINT
PLAYS
THE FIFTH ABOVE PREVIOUS OCTAVE

↓

CDEFGABCDEFG

$\frac{1}{4}$

WITH A NODE AT $\frac{1}{4}$ POINT
PLAYS
THE FOURTH ABOVE THE FIFTH (SECOND OCTAVE)
↓
CDEFGABCDEFGABC

$\frac{1}{5}$

WITH A NODE AT $\frac{1}{5}$ POINT
PLAYS
THE THIRD (MAJOR THIRD) ABOVE
THE PREVIOUS OCTAVE
↓

CDEFGAB   CDEFGAB   CDE
↑   ↑   ↑
GROUND   FIRST   SECOND
NOTE   OCTAVE   OCTAVE

1.3.1 Relation between musical
harmony and mathematics

The Renaissance also continued the classical tradition of the interdependence of music and geometry, which had been found by Pythagoras. This may be demonstrated by taking a single stretched string of a given length, fixed at each end (**1.3.1**). If this string is plucked it will play the ground note. When the string is divided into two by the creation of a still point or node at its centre the note played by the vibrating string will be an octave higher. With the node one-third of the way along, the note played will be a fifth above the previous octave. When the node is one-quarter the way along the string the note played is a fourth (above the fifth), which is again an octave above the note played by the string with the node at halfway point. Subsequent to Pythagoras it was also shown that with the node one-fifth of the way along the string, the note played was a major third above the second octave played by the string divided in two.

That there exists such a relationship between the visual and the musical implied that there was an underlying harmonic order to the natural world and for Renaissance man the existence of such ordering principles meant that man could be brought into a harmonious relationship with the world: the microcosm and the macrocosm could be seen as one, and apparent chaos could be shown to be ordered.

Alberti's 'conformity of ratios and correspondence of all the parts' was to be applied to all buildings, although particularly to churches. Not only was the building seen as a harmonious education device, but contact with the outside was to be limited by positioning the windows high enough so that only sky could be seen. The various versions of the Vitruvian figure given by Wittkower (1962, 1949) from Fra Giocondo's edition of 1511, Leonardo's, that of Francesco di Giorgio to that of Cesariano in 1521, the development can be seen of the idea of man and the perfect figures. Fra Giocondo's figure stands confidently demonstrating, whilst di Giorgio's is more relaxed within the circle and square. Leonardo's Vitruvian figure, circle and square are precisely delineated and demonstrate the argument to be true in two separate positions, thus adding a little more 'objective credibility' (**1.3.2**). Cesariano's figure, on the other hand is very different (**1.3.3**). Fiercely muscular, he looks as if he has just sprung to his extended position, showing his mastery of the world with the added force of an erect penis. Furthermore, the circle and the square are gridded at about 100 mm (or 4 inches approximately) intervals. His larger than life appearance is enhanced by the fact that his height is one grid square more than one would expect on the principle of each square being 100 mm or 4 inches. The six feet height of the earlier proportional men has been increased to six feet plus one-third of a foot.

If a building is thus to be composed of a whole, together with its harmonious parts, it has to be seen as such both inside and out. It was this necessity perhaps that led Leonardo da Vinci to state that:

> 'a building should always be detached on all sides so that its true form may be seen.' (Wittkower 1962, 1949)

In this way the proportional whole of the building can be shown, with each of its elevations carefully related to the other, consistent within itself. This has been one of the basic 'rules' of all monumental buildings that continues to this day — but with one modification or extension. The modern movement in its search for democratic solutions and its search for order through all buildings, has applied this principle of the total object to almost every building type. Thus we have factories, schools, hospitals and offices, all consistent within themselves as freestanding objects in space. Further than this they are more consistent with each other than with the place or environment in which they find themselves.

Palladio (1508–80) emphasized that beauty came from the relationship of the parts and of the parts to the whole, in this way the

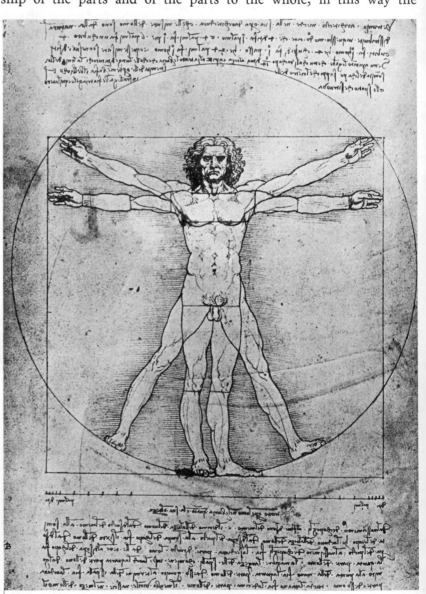

1.3.2 Leonardo da Vinci, *c.*1492. The proportions of the human figure

building would be a complete and entire whole. This must start with the plan (**1.3.4**) and extend right through the building. Alberti had recommended that those ratios of one to one, one to two, one to three, three to four shown by Pythagoras to be the basis of musical harmony were to be used in building (**1.3.1**). For example, the height of the Pantheon in Rome is equal to its diameter and one-half of the latter equals the height of the building from the springing of the dome to the ground. Further, many Italian cities had the standard units of measurement on display in a central place. In Vicenza's market place the public could see the Vicentine foot and the standard sizes for bricks and roof tiles. Palladio's Villa Foscari at Malcontenta (**1.3.5**), commenced some time during the 1550s, has a basic room module of a square of 16 ft 0 in (4.87 m) side. In measuring the villa in 1962 Forssman (1973) and his students demonstrated that

> 'all interior spaces (except for the stairs and the length of the main hall) are integrated according to the proportions 1:2:3:4 or expressed in musical terms, they are harmonised as key tone, fourth, fifth and octave'.

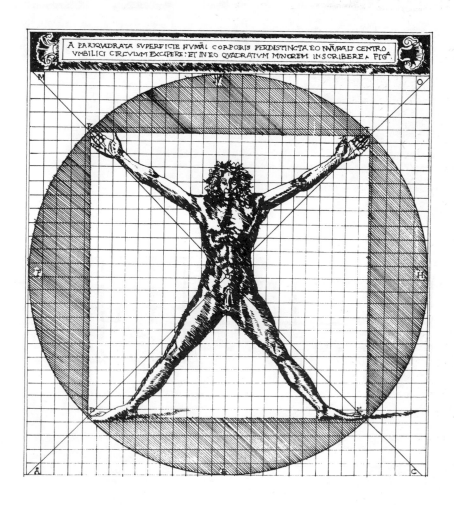

1.3.3 Cesaviano edition of Vitruvius, 1521, Vitruvian man (redrawn)

28

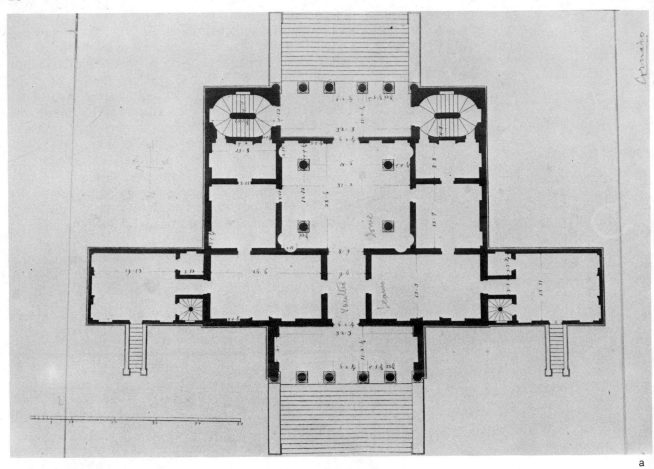

a

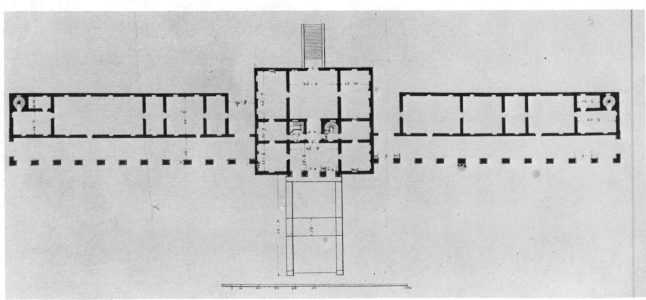

b

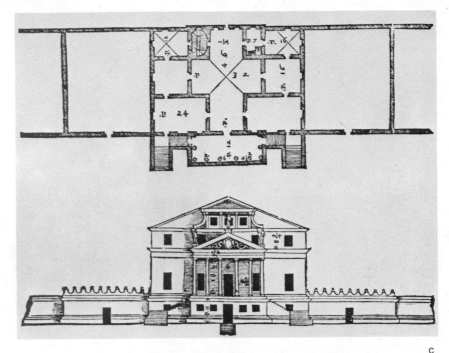

c

d

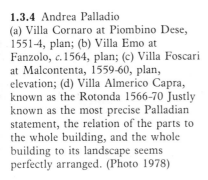

1.3.4 Andrea Palladio
(a) Villa Cornaro at Piombino Dese, 1551-4, plan; (b) Villa Emo at Fanzolo, *c.*1564, plan; (c) Villa Foscari at Malcontenta, 1559-60, plan, elevation; (d) Villa Almerico Capra, known as the Rotonda 1566-70 Justly known as the most precise Palladian statement, the relation of the parts to the whole building, and the whole building to its landscape seems perfectly arranged. (Photo 1978)

The system of proportion also governs the height of rooms which has an arithmetic, geometric or harmonic relationship to the lengths and breadths of the rooms.

As Wittkower points out, Alberti was concerned to show that such harmonies in a building resulted from objective reasoning and not from 'personal fancy'. Here we see the search for some authority, some outside agency to validate the decisions taken. In the Renaissance, man was to be put in accord with his universe and with God; this could best be done by the demonstration that there were laws outside man, 'objective' truths which could be called upon to support a particular approach which in turn embodied a particular set of values. With the advent of machine technology man began to seek

1.3.5 Villa Foscari at Malcontenta, 1559-60. View across Brenta canal. (Photo 1978)

new sets of laws, a new authority for his actions. His search for a connection between the new technology and himself brought about a clash between the traditional proportional systems and the demands of the machine. In this clash the human connections postulated by Vitruvius and Alberti were gradually set aside and authority was sought in the capabilities of the machine.

No longer was the appeal for harmonious whole building, but for architecture to be tailored to what machines could do well. What the early machines could do well was produce a continuous stream of identical objects in a limited range of sizes or types. If this was so, the argument ran, then the ranges of sizes available to architects must be limited to what it is economic for a machine to produce. A limited range of dimensions would thus begin to use ranges of components in related sizes and the resulting buildings would, ergo, symbolize the machine age. A new authority thus developed to validate the dictates of necessity and to provide architects with the considerable credibility necessary for their actions. Through all this the 100 mm (approximately 4 in) module remains, but merely on an additive basis, all suggestions that it be used as part of an overall human based proportional system having been discarded.

2

INDUSTRIALIZED CORNUCOPIA

STANDARDS AND ICONS

'In Handicrafts and manufacture, the workman makes use of a tool, in the factory, the machine makes use of him.'

Karl Marx
Capital 1867 (1946 edition)

Standardization plus Mechanization plus Power

The massive industrialization which took place in Britain during the nineteenth century transformed not only the countryside but the minds of men. The particular proclivities of machine production gave rise to a whole new approach to the making of buildings and, ultimately, to a whole new way of thinking about them. The power of this was, of course, constantly under question but it has taken until now for its full implications to begin to be thoroughly realized.

As Mumford outlines in *The Pentagon of Power,* 'mechanisation had assumed formidable proportions before the seventeenth century' (1970). However, it was the use of mechanical power that made developments in the nineteenth century so far reaching. Mumford again points out that

'Standardisation, prefabrication, and mass production, were all first established in state-organised arsenals, most notably in Venice, centuries before the "industrial revolution" (Mumford 1970),

and Venice could, by prefabrication, build a whole vessel within a month, with a galley being towed along a warehouse lined canal and the arms and equipment passed through the windows in the proper sequence (George 1968). But more than anything else it was the invention of interchangeable part manufacture which had the most profound effects generally and on building. Although the Venetians had mastered mass production, it was its combination with mechanical power and the concept of interchangeability which opened new vistas both for military production and factory production in general. Mechanization swept through almost all human activity and the necessities of war drew forth approaches which were most clearly defined as demonstrations of what was possible.

Prior to Britain's defeat of the combined Franco-Spanish fleet at the battle of Trafalgar on 21st October, 1805, the country had been

a

continuously at war with France since 1793 with a brief break in 1802. In his summary of the battle of Trafalgar, Davies points out that

'The Battle of Trafalgar marked both the beginning and the end of an epoch: the beginning of more than a hundred years of undisputed British Naval supremacy, and the end of the large scale battles fought under sail.' (Davies 1972)

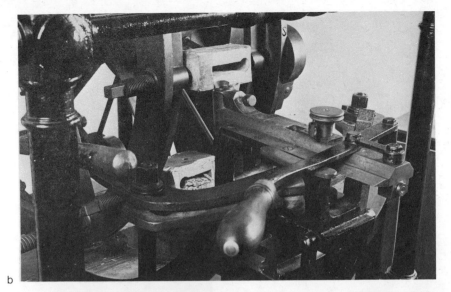

b

c

2.1.1 Portsmouth Blockmill. Marc Isambard Brunel, 1801. Machines now in Southsea Castle Museum, Portsmouth. (a) Shaping engine. The belt drive wheel is partly hidden on the right side. A partly cut shell can be seen in the centre. (b) Detail showing shell and cutting control lever. (c) Shaping engine in the Blockmill showing drive shaft, wheels, and belt. Corner saw rear left. (continued over)

d

The pressures of the long drawn out struggle with the French, culminating in the Battle of Waterloo ten years after Trafalgar, created a favourable climate for improving productivity. The sailing ships upon which strength relied used an enormous number of pulley blocks and therefore when Marc Isambard Brunel proposed the introduction of mechanization for their manufacture, it was rapidly implemented. The machinery (**2.1.1**) was introduced by Marc Isambard Brunel (the father of Isambard Kingdom Brunel) who approached the new Inspector General of Naval Works, Sir Samuel Bentham, in 1801. It is significant that Brunel had previously been Chief Engineer of New York and was an American citizen. His first British patents on blockmaking machinery were taken out whilst he was still in America in 1801. It is ironic that this, the first real example of machine production of standard parts under controlled conditions, was introduced almost simultaneously with the last great battle fought under sail and was intended to improve the production of the very parts used in such battles. Nevertheless the implications of Brunel's achievement are enormous.

The three parts of the block, the sheave (or pulley), the pin and the shell (**2.1.2**) had previously been made by a combination of machine and hand work, with saws and lathes horse-driven. The market demand of some 100,000 blocks each year, clearly suggested a field where rationalization would pay off — a single 74 gun ship, for

e

f

2.1.1 continued (d) Mortizing
machine. The first of its kind and,
once set in motion, mortised two
blocks at once (or a two sheaved
block) without attention. (e) Boring
machine. This bored the squared
piece of wood with holes for the pin
and for the mortise chisel. (f) The
Blockmill at Portsmouth *c.*1910

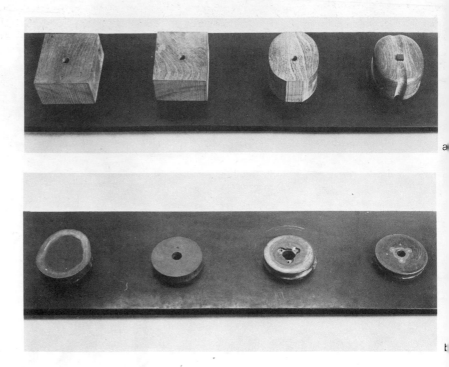

a

b

2.1.2 Stages in the manufacture of a pulley block: (a) the shell; (b) the sheave

example, required 922 blocks. Brunel's machines were installed on two floors of the building and powered by two steam engines by means of line shafting and belt drives. A vertical shaft passed to the ground floor and from this the large sawing machines were driven. The risk of fire was dealt with by some foresight in the provision of a cistern mounted on top of the building with a supply to all parts.

The mechanized blockmaking plant at the Royal Naval Dockyard at Portsmouth first began to manufacture pulley blocks for the Royal Navy in 1803, and by 1805 the production had reached the point where there was no further use for the supplies from the external contractors previously employed. Because of its scale Gilbert (1965) cites the Portsmouth Blockmills as 'the first instance of the use of machine tools for mass production'. Its capacity by 1808 was 130,000 blocks per annum and as he points out, the method of production was extensively written up in the encyclopaedias of the time, indicating that it was considered a relevant innovation. Further the machine tools were the earliest to be made entirely of metal and, although their major use was during the nineteenth century, several machines continued into the middle of the twentieth century with three of them still in use at the time Gilbert wrote his study in 1964.

Giedion documents many of the efforts to mechanize industry in his *Mechanisation takes Command* (1969, 1948), although he dismisses Brunel in four lines. He rightly gives prominence to the developments of Bramah, who, in 1784, had developed the famous bank lock, Linus Yale (1865), the standardization of lock production and the Mechanized Mill of 1783 designed and built by Oliver Evans. Clearly, however, Brunel's combination of standardization, mechanization, power and production runs was a unique step, which probably owed a great deal to what he had witnessed in America. In addition, Maudslay, who built the machinery to Brunel's designs, had previously worked for Bramah. It is probably no accident that the beginnings of industrialization had to do with both the manufacture of components used to wage war and the development of better and better locks on a universal basis. The plethora of property and possessions had to be protected.

Brunel's Blockmill is therefore significant as the first substantial application of machine tools to mass production on a powered basis. It demonstrated that rationalized sets of components produced on a continuous basis under controlled conditions could be made to a constantly high standard with minimum intervention by the human hand. The art and craft of the hand was about to be refashioned into the art and craft of the machine.

The Palace and Paxton

No account of the development of building systems could fail to mention the Crystal Palace of 1851 and its designer, Joseph Paxton. Indeed this building occupies the centre of the stage in any work on the architecture of the past 100 years and there are good reasons why it has become significant. Nevertheless, it has been taken out of its context and used in a very particular way, to fit into a very specific set of arguments concerning modern architecture. For the apologists of building systems it offers a part of the armoury that none can afford to do without. However, when replaced in their context, both Crystal Palace and Paxton become even more interesting — although not quite in the same way that we have been led to believe. There are two questions to which answers will be attempted here: first, what was its context, and second for what purposes was it subsequently used as an example. For the answer to the first question a closer look must be taken at Paxton himself for, as Chadwick observes, by 1851

> 'Paxton had been a builder of glass structures for more than twenty years before the Crystal Palace took shape, a railway speculator for fifteen years and mainly as a result of this a man of some financial standing, a designer of more conventional buildings for over fifty years.' (Chadwick 1961)

From this it is clear that Paxton, a Bedfordshire farmer's son, was a man with a manysided talent who busied himself in many differing fields of activity. Paxton became head gardener to the Duke of Devonshire at Chatsworth in 1826 and this developed his concern with gardens to include forestry, roads and all estate managerial matters. Later, starting in 1834, he travelled widely and entered Parliament in the Victorian tradition of 'profitable philanthropy' (Chadwick 1961).

Paxton was then a gardener, landscape gardener and landscape manager who also designed buildings. This gives rise to a different

40

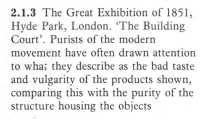

2.1.3 The Great Exhibition of 1851, Hyde Park, London. 'The Building Court'. Purists of the modern movement have often drawn attention to what they describe as the bad taste and vulgarity of the products shown, comparing this with the purity of the structure housing the objects

approach to problems than is normally adopted by those who are primarily building designers, who then become interested in landscape and gardening. The latter will usually see the place as an adjunct to his buildings, referring to it as the space between buildings, whereas the former will see the buildings as part of the overall landscape, the buildings between the spaces. The two different approaches can lead to significantly different results. The model village, Edensor, demonstrates this most clearly, the housing of which shows:

> 'the extraordinary compendium of styles evolved by Paxton between 1838 and 1842.' (Chadwick 1961)

These buildings and some 30 of Paxton's traditional (i.e. non-glass) projects and buildings range from Anglo-Italianate to a vernacular picturesque, and from gothicized revival to a French Renaissance. Chadwick makes the point that Paxton was 'first and above all' a gardener and that his range of styles was the response to the necessity for focus in the Reptonian idealized landscape. To Chadwick this is Paxton's acceptance 'of the conventional architectural styles of the time', but this eclecticism is exactly where Paxton is of interest to us now since it allowed him to respond to each problem and each place in a unique manner. From this, of course, sprang the Crystal Palace itself as well as his other work. And this, I believe, is where the problem has arisen. The structure and mechanical innovations which emerged from Paxton's glass buildings came together in the Palace, a short life exhibition building set in a particularly valued landscape — Hyde Park, London. In plucking it from its cultural context and

a b

2.1.4 Crystal Palace. Sir Joseph Paxton, 1851. (a) Part south elevation showing cladding module of 8ft 0in (2.4 m). (b) Interior showing structural module of 24ft 0in (7.32 m). Also shows the arch sections introduced to span existing trees

attaching symbolic importance to it as the apotheosis of building in the machine age, it has become one of the foundation stones of the modern movement. Giedion gives it pride of place in *Space Time and Architecture* (Giedion (1967, 1941)) as do most architectural histories (**2.1.3**). For the building system designers it is a powerful progenitor since it embodies the main points of their claims:

1. designed to a 24 ft 0 in (7.32 m) structural and 8 ft 0 in (2.4 m) cladding module (**2.1.4**)
2. components prefabricated, mass-produced and standardized (**2.1.5**)
3. dry assembly (**2.1.6**)
4. many components interchangeable
5. rapid erection (39 weeks for 989,884 sq ft (91,960 sq m) of floor space) and demountability
6. light steel structure with a weatherproof lightweight skin, or curtain wall
7. the framework was its own scaffolding
8. the use of mechanized erection techniques, for example the roof glazing wagon (**2.1.7**)

a

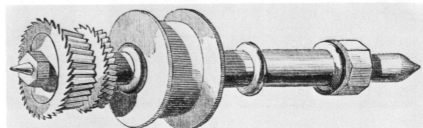

b

c

d

e

2.1.5 Crystal Palace. (a) The sash bar machine. Originally used by Paxton during the construction of the Great Conservatory at Chatsworth in 1838. Power was by means of a steam engine. 307 planks passed through the machine in ten hours, each plank giving three sash bars. (b) Sash bar machine cutters. The saws could be changed to make bars of any form. Paxton introduced a grooved bar to protect the putty. (c) Sash bar finishing machine. (d) Drilling machine. (e) Sash bar painting apparatus. After priming, the bar was placed in a tank (next to machine) of paint, then passed between the four brushes of the machine to remove surplus paint, which ran off into the wooden chute to be re-used

2.1.6 Crystal Palace: (a) frame assembly; (b) column and beam assembly

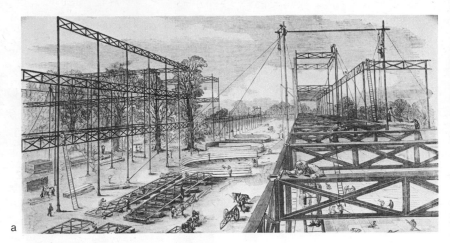

a

b

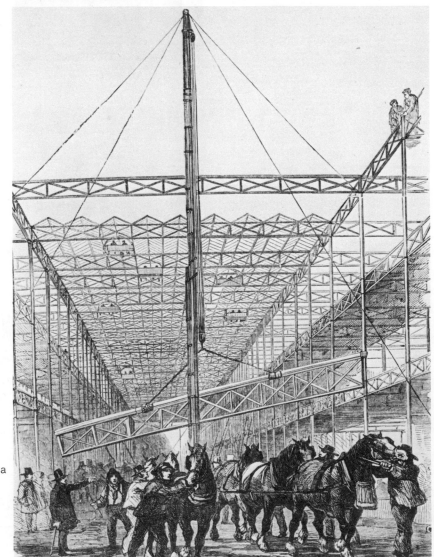

a

2.1.7 Crystal Palace: (a) the use of horsepower to erect components. Notice the 'glazing waggons' at work in the roof; (b *opposite*) roof 'glazing waggon'

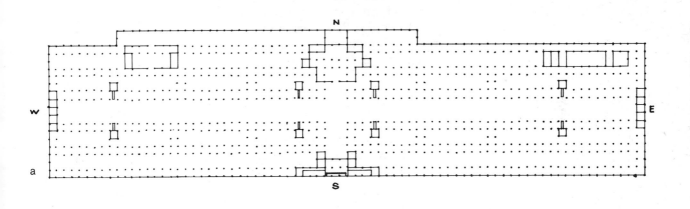

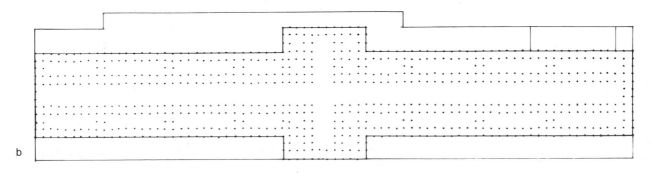

2.1.8 Crystal Palace: (a) ground plan;
(b) upper level plan

9. the designer, engineers and suppliers worked as one organization. Paxton, Fox and Henderson (contractors and engineers) and Chance (glass supplier) between them controlled the companies working on the building.

The building (2.1.8) was 1848 ft (563.3 m) by 408 ft (124.4 m), with an extension on the north side, and was in standard cast iron structural multiples generally of 24 ft (7.32 m) on plan and height. Its tender cost was £150,000 or 'only £79,800 if the materials remained the property of Fox and Henderson after dismantling' (Chadwick 1961). The roof, almost 18 acres (7.29 ha), was to Paxton's patent ridge and furrow system with his special drainage method, the gutters draining into hollow column downpipes. The external cladding, in 8 ft 0 in (2.44 m) bays had two timber columns between the 24 ft 0 in (7.32 m) spaced cast iron columns. The ground floor was clad in vertical timber boarding and horizontal louvres, elsewhere glazed with wooden sashes. However, the picture of a building of few, large standardized components is not a suitable description since the truth was more complex, and more significant. The simple overall form of the building concealed 'an incredible complexity of construction' (Herbert 1978) and both the columns and the external envelope involved ranges of parts forming complicated subsystems.

The Crystal Palace was a great popular success which extended the range of architectural possibilities: possibilities which were both recognized and questioned at the time:

> '. . . an effect of space, and indeed on actual space hitherto unattained a perspective so extended, that the atmospheric effect of the extreme distance is quite novel and peculiar; a general lightness and fairy-like brilliancy never before dreamt of; and above all — to our minds one of the most satisfactory of attributes — an apparent truthfulness and reality of construction beyond all praise. Still, the conviction has grown upon us, that it is not architecture: it is engineering — of the highest merit and excellence — but not architecture. Form is wholly wanting; and the idea of stability or solidity is wanting.' (The Ecclesiologist quoted in Chadwick 1961)

We know now that it has been those very qualities of transparency, space and impermanence that have imbued it with as much significance to modern architecture and system builders as its standardized and rapid construction. Further than this, these attributed qualities have devalued Paxton's holistic approach to problems; after all it was this that caused him to arrive at the design for the Crystal Palace at all. To have suggested that its effect would be to encourage a narrowing of architectural responses would have filled his entrepreneurial mind with horror: to have seen it as a prescription for a new 'total' architecture in the manner of Wachsmann (1961) exactly 110 years later would have at the very least caused him to ask some questions in parliament.

2.1.9 Edensor, Derbyshire, England. Joseph Paxton. Some of the 'extraordinary compendium of styles evolved by Paxton' (Chadwick) in this model village, 1838-42

Of all the references Giedion's has been the most significant since he was concerned to use the Crystal Palace as a cornerstone for the new architecture. Here the suggested implications of the Einsteinian revolution and the changes in painting and sculpture were brought together with the Eiffel Tower and the Crystal Palace as seminal events. Engineering solutions were the new virtues; the stylistic games of architecture were made out to be irrelevant in the industrialized era. Such solutions were given their credentials by this means as morally appropriate, and architects who deviated from this were certainly not seen to be 'modern'. In latter years deviations from the machine aesthetic have attracted cries of heresy within the architectural profession. The building systems have often formalized such cries and

2.1.10 Mentmore, Buckinghamshire, England. Joseph Paxton and G.H. Stokes, 1852-4. (a) Elevation; (b) detail of elevation

firmly set their faces against any digression from their institutionalized style.

How surprised Paxton would have been to realize that he had fathered 'building systems', especially since, in modern terms, his approach is a far truer systems approach to problem-solving and one in which the time, the genius loci of the place and the technical means are all brought into play in the solution. Although Paxton's Crystal Palace is important, much more important is this approach that can select from a range of alternative solutions the one most appropriate (**2.1.9**, **2.1.10**). It was an informed eclecticism of this sort that became obscured by the drive of the polemicists of the modern movement towards a unified machine aesthetic. Perhaps Paxton's broad formal vocabulary can be paralleled by Loudon, the inventor of the term 'Gardenesque'

> 'the characteristic feature of which is the display of the beauty of trees and other plants, individually'. (Loudon 1840 quoted in Chadwick 1961)

and perhaps the suggestion (Harnden 1973) that a use of style such as Paxton's, eclectic but related to the whole, should be termed 'Buildingesque' offers a suitable model for those, like Jencks (1977) searching for a suitable post-modern idiom.

CAST IRON

'There is no creativeness in the mechanical production
of cast-iron grills and ornaments.'

Sigfried Giedion
Mechanisation Takes Command 1969, 1948

The vision of prefabricated building units rolling off the production line is a recurring theme not only in architectural histories since the inception of the industrial revolution, but of many people associated with the building industry as fabricators, manufacturers and suppliers. The concept of a kit of parts which would form some sort of industrialized vernacular cannot be said to have been really achieved anywhere in recent years in spite of the propaganda of architects, politicians and other interested parties. However, with the development of cast iron, just such a vernacular was available and operated successfully over many years: further, those involved were not above visions of their own, these having an uncanny match with utterances of more recent years.

Abraham Darby produced cast iron from a coke blast furnace in 1735 and, according to Bannister (1950), cast iron columns were used to support a chimney stack at Alcobaca, Portugal in 1752. Cast iron railings had been made in 1714 to surround St. Paul's Cathedral (Robertson and Robertson 1977), but the first recorded structural use in England was in St. Anne's Church, Liverpool in 1770 where cast iron columns support the gallery.

Cast iron began to be introduced in textile mills in England in the 1790s as a response to the many serious fires in such buildings. In 1792 work commenced on a six storey mill at Derby designed by William Strutt (1756–1830). The external walls were load bearing, but the internal columns were of cast iron. Timber beams supported brick arches, with brick tiles on top and plastered beneath. Two further buildings were constructed at Milford and Belper employing similar construction between 1792 and 1795. The first all cast iron column and beam design was the five storey flaxmill at Shrewsbury completed in 1797 by Charles Bage with help from Strutt (Johnson and Skempton 1955/7). Strutt subsequently built a series of similar cast iron frame mills at Belper in Derbyshire, the first being North Mill of 1803/4, a five storey structure (**2.2.1**). In 1812 Thomas Rickman, an accountant and Professor of Architecture at Liverpool Academy, worked on the design of a church with Thomas Cragg, owner of the Mersey Iron Foundry. The design was used for St. George's Church Liverpool employing stone as the external walling material but with the interior entirely made of cast iron units. Two further churches

COTTON MANUFACTURE.

PLATE XIV.

Sections of one of Mess.rs Strutt's COTTON MILLS at Belper in Derbyshire.

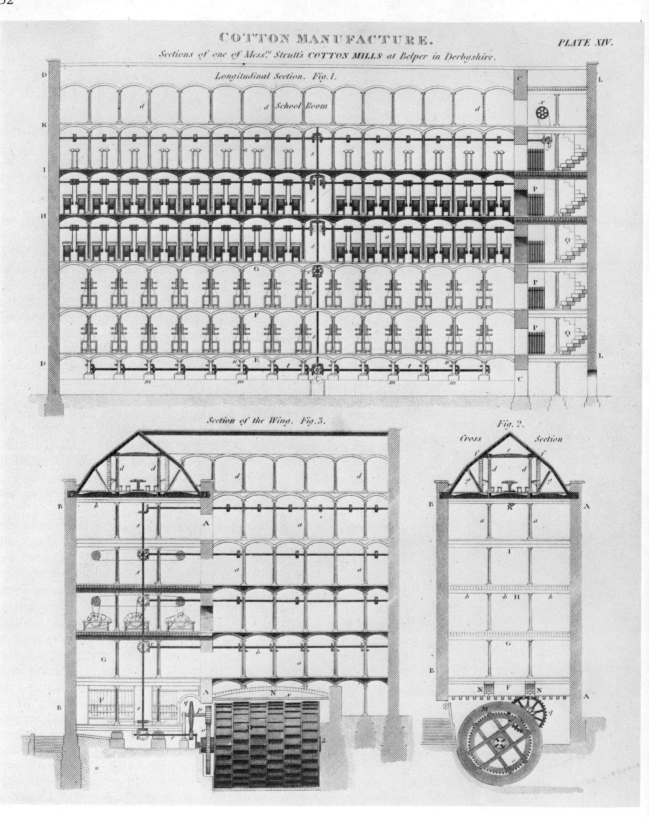

Longitudinal Section. Fig. 1.

a School Room

Section of the Wing. Fig. 3.

Fig. 2.

Cross Section

were built by Rickman using cast iron in this way — St. Michael-the-Hamlet in 1814 and St. Philip's in 1816. These three buildings formed the basis for the prefabricated cast iron churches which were 'shipped in such numbers from the ports of Bristol and Liverpool to be erected on distant shores of America and Australia' (Hughes 1964).

Although Giedion's regard for cast iron is slight since it is barely mentioned in his history of mechanization, components made of this material had reached a level of usage that would be the envy of modern system builders. Catalogues existed which listed pages and pages of structural and facade parts, all available by mail order, with some 26 foundries in America alone producing railings and other decorative features (Robertson and Robertson 1977). Factories in Europe and North America made, packaged and shipped cast iron buildings to the four corners of the world.

The potential of cast iron was set forth in 1845 by William Vose Pickett in his slim volume published in London, entitled:

> 'New System of Architecture founded on the Forms of Nature and Developing the properties of Metals; by which a higher order of Beauty, a larger amount of Utility, and various advantages in Economy over the pre-existent architecture may be practically attained.' (Pickett 1845)

Pickett emphasized the difference between construction in iron and construction using masonry, pointing out that if used properly, architecture of a different character would result. He set out four 'Primary Principles'. The first of these was a call for a cavity wall, a 'hollow iron wall with chased ornamental surface' which would be space saving and offer 'better protection from heat and cold through the admission of a stratum of air between the surfaces of the walls'. Pickett further pointed out that such a wall would take pipes for warm air, gas, smoke and water, and could be used for storage — 'closets within the hollows'.

The second of his primary principles offered 'Interstitial ornamental form' but probably most farsighted was the third:

> 'That of presenting the shelter of roofs, without the serious obstruction of space, light, air, view and sound, occasioned by piers and columns.'

Further, said Pickett, glass could be used more extensively and a greater variety and elegance would be achieved. The fourth principle emphasized that with the use of cast iron, curved forms could be introduced more easily than in masonry. Such an approach with metal was not to become common in buildings until the late 1960s.

Clearly Pickett had thought carefully about the new materials and in one section he equates utility with beauty. Even more startling, perhaps, is his section on a 'New Architecture':

2.2.1 Belper North Mill, Derbyshire, England. William Strutt, 1803-4. Sections from The Cyclopædia (Rees 1819-20)

2.2.2 Oriel Chambers, Liverpool, England. Peter Ellis, 1864. (Photo 1972)

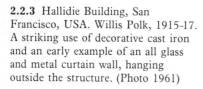

2.2.3 Hallidie Building, San Francisco, USA. Willis Polk, 1915–17. A striking use of decorative cast iron and an early example of an all glass and metal curtain wall, hanging outside the structure. (Photo 1961)

'The first and most indispensible concomitant of such beauty is a new order of forms, together with peculiar methods of expressing and combining those forms'. (Pickett 1845)

This 'new order of forms' received elegant expression in such examples as Oriel Chambers designed by Peter Ellis in Liverpool in 1864 (**2.2.2**), and the later Hallidie Building of 1915–17 in San Francisco (**2.2.3**), both of which still stand. In the former building the H-section stanchions and beams are exposed inside, and there is a very original use of glass in the top and sides of the oriel windows as well as on the courtyard elevation. A contemporary critic referred to the building as 'that large agglomeration of protruding plate-glass bubbles in Water Street' (Hughes 1964).

Daniel Badger of New York, described in directories of the time as a maker of iron shutters or just as 'an ironmonger', exemplifies the extent of the process. He appeared to operate between 1849 and 1877, producing parts for some 300 buildings in New York and many others throughout the United States and the rest of the world. His work has a number of interesting facets:

1. The standardized components were manufactured in one place and could then be shipped and assembled on the site, whether it be Rio de Janeiro, Alexandria or Havana.
2. Badger was only one of many iron foundries offering a similar service at the time and these others have largely remained anonymous. Badger's work, according to Walter Knight Sturges (1970) has become a valuable source of the period by reason of his decision to employ Sarony and Major, lithographers, to prepare his catalogue. This catalogue, by reason of its many fine illustrations, has in its own right become important.

Badger was operating a law that is of great importance to those with things to sell whether they be objects or ideas: that is that you can have as many objects or ideas as you like but, without the provision of a conceptual context, they may as well remain anonymous and/or disappear forever. As we shall see most of the reworking of ideas in the history of building systems is concerned with providing a publicly consumable piece of knowledge in which the social, technical and personal worlds are all tied together. The concept of frame and skin, one of these consumable notions which has played such a powerful role in the development of building systems, can here be seen as a direct reflection of the market and production situation. It was later idealized by Le Corbusier in the Dom-ino House project, by Gropius in his statements, by the subsequent institutionalizing of the idea with the English school system builders in the 1950s and 1960s, and latterly by Ehrenkrantz and Foster's banishment of the external wall in their own reworking of the systems idea. Such myths necessarily mediate between the 'real' and an ideal world; in the case of cast iron the realities of its functioning needed little in the way of myth making, compared to these later examples. Amongst the advantages of cast iron were that it could be easily cast and mass produced and that, compared to masonry, it had strength and lightness — although it lacked tensile strength. 'Badger Fronts', it was claimed, possessed strength, lightness of structure, facility of erection, architectural beauty, economy and cheapness, durability, incombustibility and ease of renovation.

Although occasionally floors and internal supports were used, the majority of Badger's production was confined to facades which usually, therefore, were conceived quite independently of the timber, steel or masonry structure behind. Badger's catalogue claims that whilst in England and Europe cast iron had been used for interior support, its use on the exterior of buildings is an 'American invention' (Sturges 1970) — moreover an invention of Mr Daniel Badger, President of the Architectural Iron Works. Badger erected his first cast iron structure in Boston in 1842:

> 'The columns and lintels of the first storey were of this material, but the prevailing prejudice against this bold innovation was so great that he was not permitted to engage in the work until he had given an ample guarantee that, if it should not prove a success, he would remove it at his own expense'. (Sturges 1970)

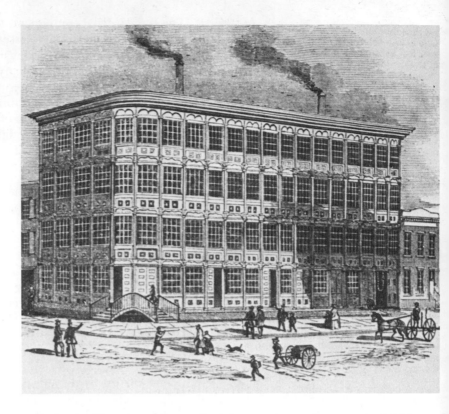

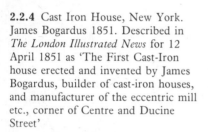

2.2.4 Cast Iron House, New York. James Bogardus 1851. Described in *The London Illustrated News* for 12 April 1851 as 'The First Cast-Iron house erected and invented by James Bogardus, builder of cast-iron houses, and manufacturer of the eccentric mill etc., corner of Centre and Ducine Street'

Badger's idea was to reproduce in cast iron whatever had hitherto been produced in stone, but at less cost: this in turn led to stone copies of cast iron originals. All the designers that Badger used have remained anonymous although it became common for architects to have castings specially made for specific projects. Arguments about whether his work is or is not architecture seem irrelevant in the light of the clear lessons that many architects were able to draw from his work. However, it is to the views of James Bogardus that we must look if we are to see the 'vernacular' of his contemporary, Badger, against a view that is as utopian as that of Gropius and Le Corbusier.

James Bogardus of New York made bold public claims for cast iron and saw in the material great possibilities, particularly with regard to fire resistance (**2.2.4**). Instead of solid masonry external walls Bogardus, like Badger, used cast iron columns and beams or arches. A five storey factory was built on these principles by him in 1853 in New York and the Harper Brothers building was completed in 1854. Bogardus was adept at publicizing his own work and in Badger's catalogue we can also see pointers to that other great interest of the more recent building systems — documentation. Indeed the publishers' forthright note in the Badger catalogue could stand for the aims of those in the business of building systems documentation:

'This volume is published at a great cost, for the twofold purpose of supplying Architects and others with plans and details for the construction of the various parts and connections of Architectural

Iron Structures, and as an advertising medium for the Architectural Iron Works; and it is designed to be presented to those
who may be profited by its study, and aid in the object of extending
the business of the publishers, and improving public taste.'
(Sturges 1970)

Such a statement might, with few modifications, have emanated from
a more recent building system — with one important qualification:
that no mention would be made of 'improving public taste'. Of all the
arguments introduced in recent years, to promote the idea of building
systems publicly, taste has not been one of them. Nevertheless it is
crucial to realize that implicit in almost all of them is an attempt to
impose very specific taste on both the system user and the public at
large. In common with much of the modern movement in
architecture, building systems designers attempted to eliminate
discussions on taste by absorbing it into their functionalist
smokescreen. In this way they have also eliminated any discussion on
choice, and this to system designers was a most important part of their
armoury. They did not claim to be 'improving public taste' but they
did lay claim, very forcibly, to acting in the public good. The question
is: how did they manage to establish this claim, whilst reducing the
opportunity for choice? For certainly the phenomena of cast iron
demonstrated that there was nothing mysterious in the notion of sets
of standardized, interrelated parts manufactured off site: however the
emerging profession of architecture saw this industry based plethora
of choice as a force degrading public taste. This puritanical reaction
assisted in the creation of that style to be known as the machine
aesthetic.

IRON
AND STEEL
FRAMES

'The frame has been the catalyst of an architecture, but one might notice that it has also *become* architecture, that contemporary architecture is almost inconceivable in its absence.'

Colin Rowe
Chicago Frame, 1956

Rowe's succinct comments on the enormous place occupied by the frame concept in modern architecture serves to illustrate also how closely the building systems idea is intertwined with that architecture. Indeed it is probably the images conjured by this relationship that lead many to consider modern architecture and building systems as synonymous.

However, it will be argued here that, although the frame as concept and fact is a vital constituent of the building systems approach, it has been supplemented by a set of additional ideas and practices which now serve to make it distinct from its original associations. The distinct division of the structural frame and the fabric of the building is one of these crucial ideas and can be seen in such diverse examples as the Japanese house and the Crystal Palace. As Rowe also points out it can subsequently be seen codified more strongly in the iconic drawing of the Dom-ino house of Le Corbusier.

The first wrought-iron joist sections were produced in France, according to Giedion, and were made by Zores in 1845 in Paris as a result of the pressures from the high price of timber, fear of fire and demand for wide spans. The development of the multi-storey building around the steel frame and the elevator during the latter part of the 19th century is well documented: the move from cast iron to the far more useful Bessemer steel was rapid and coincided with the dynamic growth of the Chicago of the eighties and nineties. Although the Bessemer method of making steel occurred in 1856, it was, according to Collins (1965) not until 1884 that the first steel sections suitable for building were rolled, with the first steel frame building not constructed in England until 1904. Whilst in the United States Louis Sullivan tells us that:

'The Chicago activity in erecting high buildings (of solid masonry) finally attracted the attention of the local sales managers of Eastern rolling mills.' (Sullivan 1895 in Mumford 1972, 1952)

These sales managers saw that the whole weight of the building could be carried on the frame. The combination of circumstance that brought together the need, the necessity to sell and the idea itself provided a new means for the architects of the time and one which they wholeheartedly grasped. However, it was used by them in such a way as to be a very direct reflection of the market situation and they quickly dispensed with much of the previous architectural rule book.

The standardization of spans, ceiling and infill units constituted a new set of rules for subsequent generations. The frame itself was used in a variety of ways but most frequently in a straightforward manner which reflected its nature. Visually it was demonstrated on the facades that the structure of the building was a frame, however it might be covered. Further, apart from a residual concern to treat the lower floor and upper floor facades in a manner more befitting the beginning and the end of a building, these facades consisted of a series of repeating identical bays.

In the Chicago frame buildings we can see what was later to become a 'holy grail' for architects. However, at this time, a period of great optimistic, expansionist industrialization, the architects and engineers were kept busy enough merely trying to keep up with the demand for building. The later attempts to apply the lessons of these developments in a more reflective mood ultimately gave credence to that style which, it was thought, was automatically dictated by the use of a frame and a reliance upon industrialized techniques.

For building systems in Britain, as for modern architecture in general, the lessons were clear. The frame, generally a steel frame, was to be the fundamental tool in rationalizing building around the industrialized concept. This predicated certain stylistic norms which, it was thought, rightly clustered around its use in this way. This belief exists today to the degree that many a system was founded by endeavouring to utilize a steel frame where it is totally inappropriate, 5M housing for example. Frequently a steel frame is used in conjunction with other elements which could do the same job, but that steel frame cannot be eliminated since it is part of 'the system'. One finds small buildings, for example, forced into utilizing a steel frame together with masonry, or timber which could equally be load bearing.

Although Chicago proved fertile ground for the development of tall framed buildings it was not alone in pursuing such a course. In addition to Bogardus' seven storey Harper building of 1854 in New York, employing cast iron there was, in Philadelphia, the ten storey Jayne building, erected between 1849 and 1852 employing a combination of iron columns internally, timber beams and a masonry external wall. According to Weisman (1970) it does not rate as a skyscraper because it had no passenger elevator, only a goods hoist. The title of the first skyscraper is accorded by him to the seven storey Equitable Life Assurance Building, 1868/70 in New York, designed by Gilman and Kendall and George B. Post, on the grounds that it rose to twice the average height of surrounding buildings and had a passenger elevator. It was in Chicago, however, that the next important step occurred. The First Leiter building of 1879, designed by engineer and

2.3.1 First Leiter Building, Chicago, USA. William le Baron Jenney, 1879. Photo J.W. Taylor. Collection of The Art Institute of Chicago

architect William le Baron Jenney, used cast iron internal columns, timber beams and cast iron external columns in brick piers (**2.3.1**). The building, of seven storeys, also had the storey height floor to ceiling 'Chicago' window, soon to become one of the trademarks of this period.

The first building with a skeleton of wrought iron was Jules Saulnier's chocolate works near Paris built 1871/72 and with the external skin simply acting as a plain curtain wall, whilst the ten storey Montauk Building in Chicago, 1881/82, designed by Burnham and Root contained a fire proofed metal frame within a masonry skin (Jordy 1976, 1972). The Home Insurance building (**2.3.2**) of 1883/5 by William le Baron Jenney was new in theory, and in practice according to Col. W.A. Starrett (Mumford 1972, 1952). Jenney took the dead load from the walls and transferred it to the frame of iron within the masonry — this frame being of cast iron columns and wrought iron I beams bolted together with angle iron brackets. By the time the building had reached the sixth floor the Carnegie-Phipps

2.3.2 Home Insurance Building, Chicago, USA. William le Baron Jenney, 1883-5. Photo J.W. Taylor. Collection of The Art Institute of Chicago

Steel Company requested and received permission to substitute, on the remaining floors, the new Bessemer steel beams they were now rolling. H.H. Richardson's Marshall Field Warehouse of 1885/87 used massive external masonry construction with cast iron columns and wrought iron beams within. 1887/89 saw the erection, by Holabird and Roche, of the Tacoma building with its twelve storey skeleton construction and facade of brick and terracotta free of the structure, and Burnham and Root's Rand-McNally building of 1889 which, says Starrett, was the 'first skeleton structure of rolled-steel beams and columns built up of standard bridge-steel shapes and riveted together'.

The second Leiter Building (**2.3.3**) of 1889/91, by William Le Baron Jenney, says Starrett

2.3.3 The Second Leiter building, Chicago, USA. William le Baron Jenney and Mundie, 1889-91. Now the Sears building (Photo 1979)

'. . . a few months later, was the first without a single supporting wall, as his Fair building in 1891 was the first to employ Z bar columns' (in Mumford 1972, 1952)

This building, which Giedion considers to be the most complete essay in skeleton construction, clearly demonstrated its skeleton frame construction on its elevations with the verticals reduced to flat columns and the horizontals to thick bands merely covering the floor zone. The owners themselves in this case, as in the Monadnock building (1884-92) by Burnham and Root, demanded large areas of glass for display purposes. In the Monadnock building this was very much against the initial desires of Root himself. As is often the case, the deed was done, according to Giedion, by a draughtsman during the absence of Root from the office.

Giedion states the three elements of 'Chicago Construction' to be:

1. a new foundation to deal with the sand mud subsoil, 'the floating foundation'
2. the introduction of skeleton construction
3. the introduction of the 'Chicago window' — elongated and storey height to completely fill the grid of the frames.

What all this meant for architecture is suggested by Giedion's claim:

'With surprising boldness, the Chicago school strove to break through to pure forms, forms which unite construction and architecture in an identical expression.' (1967, 1941)

Clearly important innovations occurred, but recalling Rowe we may well question whether it was the striving for 'pure form' that was the significant driving force. The frame offered many advantages: the potential for more square footage per floor over masonry construction — as much as one floor in a ten floor building; the fast construction times — the Manhattan building, 'a sixteen storey block with a basement' was 'designed May 1890 and completed in the summer of 1891' (Giedion 1967, 1941). In addition there was the window display potential and flexibility in planning. However, here we see Giedion enunciating some of the formal statements that were certainly to become very powerful. Architectural minds were reflecting upon what was happening and abstracting certain properties which were to become very significant indeed to the development of what was thought to be a style appropriate to the machine. Perhaps the most complete embodiment of what was to come was Louis Sullivan's Carson, Pirie, Scott department store of 1899/1904 with its emphasis on the grid of steel on the facades and vast glass areas. Many a city centre is still sprouting such facades, and its 'cage construction'.

Of some interest may be the 'design method' of the typical Chicago architect of the period as retailed by Montgomery Schuyler:

'I get from my engineer a statement of the minimum thickness of the steel post and its enclosure of terra cotta. Then I establish the minimum depth of floor beam and the minimum height of the sill from the floor to accommodate what must go between them. These are the data of my design.' (Mumford 1972, 1952)

This brief account of the steel frame has been introduced for two major, linked reasons. The first is that developments in Chicago show the influence of the growing technology itself on the approach the architects took to design problems, and second, it was from this codification of design approach that a clear architectural response to the new machine technology was forged. Rowe points out that in Chicago

'the frame was convincing as a fact not as an idea, whereas in considering the European innovators of the twenties one cannot suppress the supposition that the frame to them was more often necessary as an architectural idea before it was reasonable as a structural fact.' (Rowe 1956)

Whilst Chicago was producing this impressive array of built 'fact', other responses to the machine were developing in Europe: unlike the

Chicago frame period these quickly developed a moral tone, one strand of which elevated the frame to a central position in the modern movement. This, together with the nature of steel technology itself, meant that it acquired a dominant position in the growth of ideas about industrialization and building systems.

3

RESPONSES TO MACHINES

THE ART
OF CRAFT

'Nothing should be made by man's labour which is not worth making; or which must be made by labour degrading to the makers.'

William Morris
Art and Socialism 1884
(in Morris 1944, 1884)

The power of the emerging industrial machine to produce goods and to change society drew forth a variety of responses from the world of art, craft and design. Many of these were influenced by the critique of industrialization offered, in 1848, by Marx and Engels in *The Communist Manifesto*. In this they had stated their view of history as 'the history of class struggles' (Marx and Engels 1948, 1848) where two classes, the oppressor and the oppressed, were always in continuous opposition and from such oppositions emerged new formations:

'The feudal system of industry', says the Manifesto, 'in which industrial production was monopolized by closed guilds, now no longer sufficed for the growing wants of the new markets. The manufacturing system took its place.' (Marx and Engels 1948, 1848)

The reactions to this new manufacturing system attempted to identify a good relationship between the making of products and the people. This conjunction is most clearly expressed in the ideas of William Morris. He attempted to re-establish the concept of meaningful work for the individual in opposition to the drudgery of minding the machines which were spitting forth their succession of poorly designed products. His claims for 'Art and labour' are set out in three principles:

'First, Work worth doing
Second, Work of itself pleasant to do
Third, Work done under such conditions as would make it neither over-wearisome nor over-anxious.'   (Morris 1944, 1884)

Morris commissioned Philip Webb to design The Red House into which he moved in 1860, and it is from furnishing and decorating it himself that his firm, Morris and Company, Decorators, was started.

Morris' belief that man could only achieve dignity by making and designing things himself, and that the decorative qualities of an object must emerge from its practical intent, caused him to work with like minded people in something akin to the mediaeval workshop. Morris saw, in the extension of this notion, a release from the tyrannies imposed by industrial production techniques. Inevitably the clients that Morris found came from the more affluent, and gradually 'Arts and Crafts' classes sprang up all over the country. The work of Morris and others became another 'badge' with which a section of society sought to make itself distinct.

The central set of ideas which identify the Arts and Crafts movement subsequently, in transmuted form, became important issues for the machine aesthetic itself. The notions of high quality, fine craftsmanship, artistic intent, functional integrity, purposeful decoration and straightforwardness are those normally associated with the movement, although an examination of many of its objects shows that these ideas were interpreted very broadly. Some of the designers, C.R. Ashbee (1836-1942), W.A.S. Benson (1854-1924), Christopher Dresser (1834-1904), and architects like C.F.A. Voysey (1857-1941), M.H. Baillie Scott (1865-1945), and possibly C.R. Mackintosh (1868-1928), sit uneasily together. For although they are generally lumped together in an effort to describe 'Arts and Crafts' design, the fact is that they embodied differing attitudes with, for example both Ashbee and Benson embracing the machine, and Mackintosh employing a decoratively functional inventiveness that opened up totally new possibilities. From the ferment at the end of the century came Mackmurdo's Century Guild, founded in 1882 and the establishment of the Art Workers Guild in 1888 from the St. George's Art Society of 1884. To provide an outlet for showing the work of these designers, the Arts and Crafts Exhibition Society was established, also in 1888. This, together with the judicious use of publicity in the form of *The Hobby Horse,* founded in 1884, and the *Studio* of 1893, spread the ideas far and wide. Nevertheless the Arts and Crafts concept remains a difficult one to define with clarity since many of its practitioners slipped uneasily between machine and hand production, and many of them can be seen as gentlemen designers directing a studio or workshop in between dealing with the up and coming professional clientele. There is often in them more of the architect as he later became — designing objects that others would execute — than there is of the designer craftsman working at his bench, as envisaged by Morris.

By 1911 Ashbee had absorbed the idea that some acceptance of the machine was a necessity:

'Modern civilisation rests on machinery, and no system for the endowment, or the encouragement, or the teaching of art can be sound that does not recognise this.' (in Naylor 1971)

Moreover, he put forward the view, which would have been considered heretical a few years earlier, that 'standardisation is necessary' (in Naylor 1971). But it remained for a German, Hermann Muthesius, to enunciate the view which most decisively rejected what were seen to be the excesses of the nineteenth century machine production:

> 'This period (the second half of the nineteenth century), with its swiftly changing fashions, was the heyday of the worst ornamental aberrations and indiscriminate adoption of substitute materials. Stamped pasteboard was used for wood; stucco or even sheet-zinc to suggest stone; and tin castings simulated bronze. Even the most elementary sense of propriety was lost.' (Muthesius 1907 in Benton, Benton and Sharp 1975)

Only one year later, in 1908, came the article 'Ornament and Crime' by the Austrian architect Adolf Loos, in which he associated decoration with degeneration. 'Cultural evolution', said Loos, 'is equivalent to the removal of ornament from articles in daily use' (Münz and Künstler 1966, 1964).

Muthesius, a Prussian civil servant, was a supplementary trade attaché to the Germany Embassy in London from 1896 to 1903, and the study resulting from this, *Das Englische Haus* (three volumes completed in 1905), became an influential work for German designers. Banham points out (1960) that Muthesius 'naturally stood for order and discipline' and it is clear that he was converting to political advantage the ideas he had absorbed in Britain and, with an emphasis on standardization, harnessed this to Germany's necessity for attractive exports. The first four of his Propositions (a summary of his speech to the Deutscher Werkbund in Cologne in 1914) lay the foundations for the machine aesthetic with exceptional clarity:

1. Architecture, and with it the whole area of the Werkbund's activities, is striving towards standardization and only through standardization can it recover that universal significance which was a characteristic of architecture in times of harmonious culture.
2. Standardization, which is the result of a beneficial concentration, will alone make possible the development of universally valid, unfailing good taste.
3. As long as a universal high level of taste has not been achieved, we cannot count on German arts and crafts making their influence felt abroad.

(Muthesius 1914 in Benton, Benton and Sharp 1975)

In many ways the success of the Bauhaus embodies these visions for, in spite of protestations to the contrary, it developed a style, a machine style, which permeated the whole industrialized world. It became international to the extent that its origins in Germany's economic necessities were largely obscured. The caution sounded at the same Deutscher Werkbund Congress by Van de Velde over this combination of political expediency, standardization and style had little chance of survival particularly in the ensuing years in Germany. Van de Velde opened his Counter-Propositions with

'So long as there are still artists in the Werkbund and so long as they exercise some influence on its destiny, they will protest against every suggestion of the establishment of a canon and of standardisation.'

He was even more savage about the drive to export:

'And yet nothing, nothing good and splendid was ever created out of mere consideration for exports. Quality will not be created out of the spirit of export. Quality is always first created exclusively for a quite limited circle of connoisseurs and those who commission the work.' (Van de Velde in Benton, Benton and Sharp 1975)

But what if those exporting demands are those that commission the work? Here the battle was joined, for clearly those socialist proponents of integrity and honesty in production would not accept that good design could only develop from working for an elite, even though that was what most arts and crafts designers had done in the past.

MORAL MACHINES

'And the texture of the tissue of this great thing, this Forerunner of Democracy, the Machine, has been deposited particle by particle, in blind obedience to organic law, the law to which the great solar universe is but an obedient machine. Thus is the thing into which the forces of Art are to breathe the thrill of ideality. A SOUL!'

Frank Lloyd Wright: address to Chicago Arts and Crafts Society, 1901 in Kaufmann and Raeburn 1960

'We have, whether we like it or not, here introduced an element into human life that is mastering the drudgery of the world, widening the margin of human leisure. If dominated by human greed it is an engine of enslavement, if mastered by the artist it is an emancipator of human possibilities in creating the beautiful.'

Frank Lloyd Wright 1925
The Cause of Architecture 1975

A profound questioning of the machine, and all its implications, surrounded the arts and crafts movement, and had enlisted moral virtue as an aid in the struggle. John Ruskin's *The Seven Lamps of Architecture* first published in 1849 was all important in this both for its naivete and its force: his combined moral and technical justifications in retrospect seem very strange. Such rules as set out in the chapter 'The Lamp of Truth', 'that metals may be used as a cement, but not as a support' or Aphorism 15, 'Cast-iron ornamentation barbarous', seem curious indeed in the light of what was already well established.

Nevertheless many of his views transferred into architectural thinking in quite curious ways, and that morality which attaches to machine designed products can be seen developing out of the earlier moral rejections of the nineteenth century. It was then only a short step to accepting the machine, to insisting that there was a right way and a wrong way of using it and further, to be able to support this with a complex of argument both technical and moral. From this developed the notion that there is an appropriate machine style, and this is a view with us today. Furthermore, it is one deeply embedded in the building systems philosophy.

Many architects involved in the design of building systems have laid claim to the distinction of such systems from the mainstream of architecture, which they usually describe as being concerned merely with trivial issues pertaining to formal organization and style. They adhere implicitly or explicitly to the idea of truth to the machine, and the belief that an anonymous plenitude of buildings and components will ensue from their humbling themselves to what they see as the authority of the machine. Individual architects with a coherently ordered style of their own are often seen as imposing their personal will at all costs and are often described, by those committed to the systems philosophy, as prima donnas.

This view, of the irrelevance of the so described prima donna, applies even more to well known historical figures, particularly those of the recent past. Such a position, often strongly held, merely demonstrates architects' ignorance of their own source material. In many ways it is a blatant need to avoid the richness of architectural possibilities that history provides and is partly the result of a period of poverty in the teaching of architectural history.

To those architects brought up on this diet, no-one seems more remote from current approaches in building systems than Frank Lloyd Wright. Wilful, colourful, with very strong views, he is seen as the prima donna par excellence. However, it is Wright's contribution to the very mainstream of architectural ideas which has given rise, in part at least, to much of acceptable thought now on the use of the machine and building systems. In quite practical terms too his influence has been enormous; the use and development of horizontal and vertical grid planning, the acceptance of standardized sizes, and the refashioning of the American house in a form which encouraged the great systematization of the house building industry in North America. In addition to this Wright had, of course, proposed his American System Ready-cut method of construction as early as 1911 and, moreover, described it as a system (**3.2.2**).

First, let us examine his views on the machine. These are very much concerned with the right use of the machine and its effects on architecture, for it must be remembered that Wright developed in the ambience of Chicago liberalism and the heyday of the Arts and Crafts movement. A number of the key figures in Europe were to go to the United States during the latter nineteenth and early twentieth century. Ashbee, for example, first visited the United States in 1896 and met Wright in 1900. Wright himself stayed with Ashbee in September 1910, at Chipping Camden, Gloucestershire, where in 1902, Ashbee had established the Guild of Handicraft — liquidated in 1908. Wright asked Ashbee to write the introduction to the second Wasmuth publication of his work in Berlin in 1911.

Wright was not merely producing houses for the nouveau riche in a quiet Chicago suburb. He was clearly implicated in the social concerns of this rapidly expanding city undergoing expansion and turmoil. Lionel March gives an indication of the situation when Wright arrived in Chicago in 1887:

'He had been brought up in rural Wisconsin and had dropped out of "book learning" at university. He arrived in a period of violent social upheavals when 19th century democratic ideals came up against the incipient industrialisation and urbanism of this century. Nowhere was this confrontation more acute than in Chicago, then the most rapidly expanding city in the world.' (March 1970)

Wright met Jane Addams, who had established Hull House as a settlement for homeless immigrants, through his uncle Jenkin Lloyd Jones who, March says, was very concerned with welfare work in the city. Other family links suggest that Wright was very aware of the concerns of the reform movements of the time. Peter Kropotkin had given a lecture at Hull House on his book *Fields, Factories and Workshops* (1974, 1899), and Jane Addams was accused of promoting anarchist ideas subsequent to the assassination of President McKinley.

   Wright's involvement with the issues of the time can be seen in his address to the Chicago Arts and Crafts Society at Hull House in 1901 where he states that:

> 'The great ethics of the Machine are as yet, in the main, beyond the ken of the artist or student of sociology...' (Kaufman and Raeburn 1960)

The depth of Wright's understanding concerning the changes being wrought, or to be wrought, by the machine can perhaps best be demonstrated by his references to printing and its relation to architecture.

> 'Thus down to the time of Gutenberg architecture is the principal writing — the universal writing of humanity. In the granite books begun by the Orient, continued by Greek and Roman antiquity, the middle ages wrote the last page. So to enunciate here only summarily a process, it would require volumes to develop; down to the fifteenth century the chief register of humanity is architecture.
>
>   In the fifteenth century everything changes.
>
>   Human thought discovers a mode of perpetuating itself, not only more resisting than architecture, but still more simple and easy. Architecture is dethroned. Gutenberg's letters of lead are about to supersede Orpheus' letters of stone. The book is about to kill the edifice.
>
>   The invention of printing was the greatest event in history.
> It was the first great machine, after the great city. It is human thought stripping off one form and donning another.' (Wright 1901 in Kaufman and Raeburn 1960)

Wright's telegrammatic message here, that of the enormous significance of Gutenberg's printing press and culled from Victor Hugo's *Notre Dame de Paris* of 1832, had to await Marshall McLuhan's *The Gutenberg Galaxy* (1962) and *Understanding Media* (1964) to gain popular credibility as an acceptable model.

The concern with the correct use of the machine clearly had its connections in the misuses Wright saw around him in the Chicago of the turn of the century, and the future he saw if the artist and architect rejected it, rather than attempting to come to terms with and control it. More significantly perhaps, he could already see the positive aspects for the use of the machine, for example with wood:

'The machine has emancipated these beauties of nature in wood; made it possible to wipe out the mass of meaningless torture to which wood has been subjected since the world began, for it has been universally abused and maltreated by all peoples but the Japanese.' (Kaufman and Raeburn 1960)

Here we can see one of the foundation stones of the modern movement in architecture: the elimination of the inessential given moral force by the advent of the machine; that the direct natural use of a material is better than any elaborations or complexities that may be introduced by man. The machine will simplify and purify — indeed Wright himself says:

'Let us understand the significance to art of that word — SIMPLICITY — for it is vital to the Art of the Machine.' (Kaufman and Raeburn 1960)

Although Wright's address to Hull House was reprinted in full in *Writings and Buildings* in 1960, he had earlier referred to it in *The Architect and the Machine* (1925), where he rephrased the argument in these terms:

'The thesis I presented (to Hull House) simply stated, was — that old ideals that had well served the handicraft of old were now prostituted to the machine — which could only abuse and wreck them, that the world needed a new ideal recognising the nature and capability of the machine as a tool and one that would give it work to do that it could do well — before any integrity could characterise our Art...

But like all simple things — too common to be interesting, too hard work to be attractive, the mastery of the machine is still to come, and a harsh ugliness, in relentless cruelty, obscures the common good...

Standardisation and repetition realised and beautified as a service rendered by the Machine and not as a curse upon civilisation that is irretrievably committed to it.' (Wright 1925 in Gutheim 1975)

We see Wright setting out a morality of the machine. He is putting forth the view that there is a correct way to utilize machine processes and that this rests in the nature of the machine. He is to some extent endowing the machine with a morality. In reality there have proved to be many 'right' ways to use machine processes, as indeed there are with other processes, but it is usually the context of the problem that establishes the 'right way' in a given situation.

We also see the power of a developing myth — a myth being a device which attempts to mediate between two conflicting realities thus providing symbolic explanation. Even though the development of machine processes themselves and machine-made products have contradicted the view that there is one right or moral way to use the machine, this view is one still strongly held by architects.

The modern movement, and later the building systems designers have assimilated the idea of the moral machine but not the freedom recognized by Wright that stemmed from it when he pointed out that the machine

> 'has made possible expressive effects as plastic in the whole — as is the human figure as an idea of form.' (Wright 1925, in Gutheim 1975)

The possibilities inherent in this were ignored in favour of the post and lintel iconography and this was to become the dominant strain.

Whilst Wright's lecture of 1901 to the arts and crafts imbued Hull House audience may have come as something of a surprise in its apparent acceptance of the machine, it should be remembered that most of those clients for whom he was building 'the prairie houses' were from the new dynamic manufacturing community. In his fascinating study of Wright's clients, Eaton (1969) clearly establishes that

> 'Almost all can be classified as "businessmen"... The most striking aspect of the list, however, is the number who were manufacturers or directly involved in the industrial process.' (Eaton 1969)

The rapidly growing Chicago of the period (Oak Park grew from 4500 in 1890 to 9000 in 1900 and 18,000 by 1910) not only provided Wright with his clients but with the benefactions of industrialization. If Chicago embodied both energy and greed it also proffered a potential which Wright was quick to articulate. Wright's moral machine was on the road.

There are other aspects of Wright which are of particular interest to this account, apart from his rearticulation of the necessities of the machine in relation to man. Basically they can be reduced to three headings: first, his attitudes to construction and materials and an interest in standardization; second, his approach to three-dimensional space in planning and its relation to dimensional grids; third his relation of the building to its site, and the manner in which he controlled the environment of his buildings both by this, and by mechanical means.

The first two interests have both been seen as connected to his introduction to the Froebel toys as recounted in his *Autobiography* (Wright 1945, 1932) and by Manson (1953) and by MacCormac

(1968). As Manson recounts Mrs. Wright found the Froebel toys at the Philadelphia Centennial Exposition of 1876 and subsequently discovered that there were Froebel kindergartens in Boston and also a bookseller who sold some of the textbooks. Manson describes the basic system thus

'The toys, known in the early days of the kindergarten as "gifts", were given to the child in a series beginning with the three fundamental shapes — the cube, the cylinder, and the sphere — followed at intervals with other gifts gradually increasing in variety.'

and

'Froebel insisted that his toy structures be built carefully, with a plan marked out first upon the floor.' (Manson 1953)

These toys have clear connections with Wright's attitudes to standardization and co-ordination and also to his subtle use of square and tartan grids as shown by MacCormac (1968). Even in the early houses MacCormac shows how the intersection of elements is related

3.2.1 Charles S. Ross House, Lake Delavan, Wis, USA. Frank Lloyd Wright, 1902. Showing the relationship of Wright's design approach to the Froebel Gifts and the tartan grid. 'From the plan of the Charles S. Ross house of 1902 it is possible to abstract a perfect tartan' (MacCormac 1968)

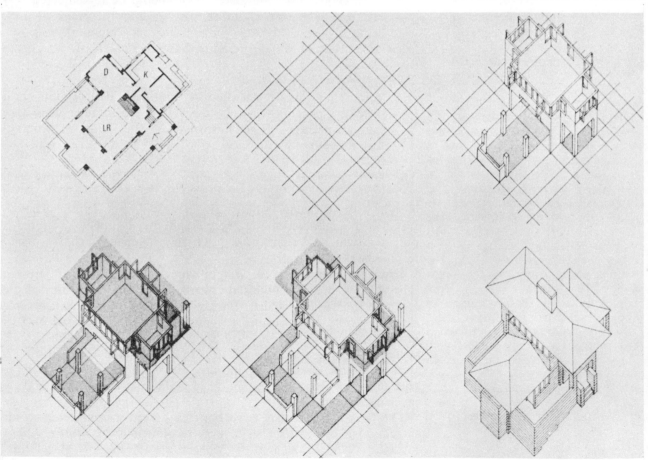

3.2.2 Richards Duplex Apartments, Milwaukee, Wis, USA. Frank Lloyd Wright, 1916. Four buildings with apartments on two floors, from the American System Ready-cut of 1911-15. Originally plaster and wood trim

to the formal constructs of Froebel toys (**3.2.1**). So, although the houses were being built in traditional materials their planning was controlled by the dimensionally related grid and thus the lengths of wall, spanning units, and so on had an inherent relationship physically and visually. Wright's Husser house of 1899 was based on a grid whilst from the Ross house of 1902:

'It is possible to abstract a perfect tartan.' (MacCormac 1968)

Walter Burley Griffin, later to win the competition for the Canberra City Plan in 1912, also used a four foot module in his Comstock Houses of 1912. However, it seems unlikely that, as Peisch (1964) claims, Griffin's solutions were arrived at independently, without precedent in Wright's work, since he joined the Oak Park studio in 1902 and had worked part-time for Wright since 1901. By 1915 Wright had designed a building system — the American System Ready-cut prefabricated flats (**3.2.2**), plans and perspectives of which appear in his books but little else. It is a timber-framed system with plaster exterior finish — some were built without Wright's supervision in Milwaukee Wisconsin in 1916 (Storrer 1974). Wright's development and use of the 'knitblock' system is also of some interest. This consisted of thin coffered and patterned concrete blocks with grooved edges along which ran ¼ inch diameter (6.44 mm) steel reinforcing rods, built in two skins with a cavity or as a single skin:

'We would take that despised outcast of the building industry — the concrete block — out from underfoot or from the gutter — find a hitherto unsuspected soul in it — make it live as a thing of beauty — textured like the trees.' (Wright 1945, 1932)

Such a system was first used by Frank Lloyd Wright in the Los Angeles Millard, Storer, Freeman and Ennis houses of 1923, but as Gebhard and Von Breton (1971) point out his son Lloyd

3.2.3 House of Carl Post, Barrington Hills, Ill, USA. Frank Lloyd Wright, 1956. One of 5 built from the first design for the Marshall Erdman Prefab

'has mentioned that it was his use of steel in the block system used in the Henry Bollman House (Hollywood 1922) which inspired his father to develop the knit block system.'

However, all these experiments may have more complex origins since, as Brooks shows:

(Walter Burley) 'Griffin developed a workable system of concrete blocks (knitlock) which was patented in 1917 long before Wright's more publicised concrete block experiments of the 1920's.' (Brooks 1972)

Bearing in mind Griffin's connections with the Prairie School of architects and his work with Wright, it seems likely that such an idea must have been discussed in the Wright studio. In his study of Wright's later, Usonian houses Sergeant (1976) shows how, with the concrete block houses the grid is extended downwards and upwards from the floor, uniting the natural forms of the site with the geometry of the house.

Much later in his life Wright made another excursion into prefabrication with the Marshall Erdman Company in 1956 (**3.2.3**). He designed four plan types, of which only two were ever used. Three were built from the number I plan in Madison Wisconsin, one in Illinois and one at Richmond, Staten Island (Storrer 1974), the houses being designed on a four foot module to an L shaped plan with a basement. However, shipping and assembly charges doubled costs and this contributed to their discontinuance by Erdman (Sergeant 1976). Both standardization and dimensional grids were pursued in other ways by European system builders. But the third of Wright's contributions was almost completely ignored by them, all the more important since it was (and is) the most valuable of the three. This was his approach to the environmental quality of the building — particularly those houses built during the prairie period. Wright's claim to the organic was no idle claim as any visitor to one of his buildings will attest. Not only do they possess a definite and subtle relationship with the place in which they find themselves but they have an ease which is something more than merely the choice of

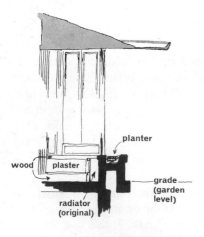

3.2.4 Baker House, Wilmette, Ill, USA. Frank Lloyd Wright, 1908. Section through living room bay window showing the integration of overhang, radiator position and planting box

materials. Indeed Wright did two things in these houses — with regard to environmental control he utilized a series of design devices which, through being part of his 'style' have had their operational meanings obscured, and he integrated with these whatever technology was appropriate (**3.2.4**).

> 'In the living room of the Baker house, plan and section, artificial heat and natural light, solid, void and overhang work together in such a way that hardly any single detail participates in only one function, nor is any single function served by only one item of equipment.' (Banham 1969)

Wright responded to climatic conditions by the provision of suitable sized overhangs as in the Robie house, by ventilation and shading using parts of the building as cool air storage tanks (Robie, and later, Falling Water) and by the integration of lighting and hot water heating (**3.2.5**) so that it all formed a part of the whole design and not merely a series of items to be fitted in as best they could.

3.2.5 Frederick G Robie House, Chicago, USA. Frank Lloyd Wright, 1906. Section and plan

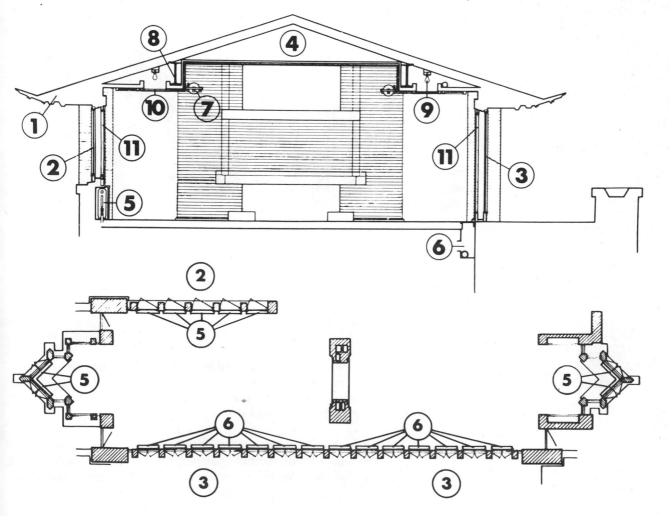

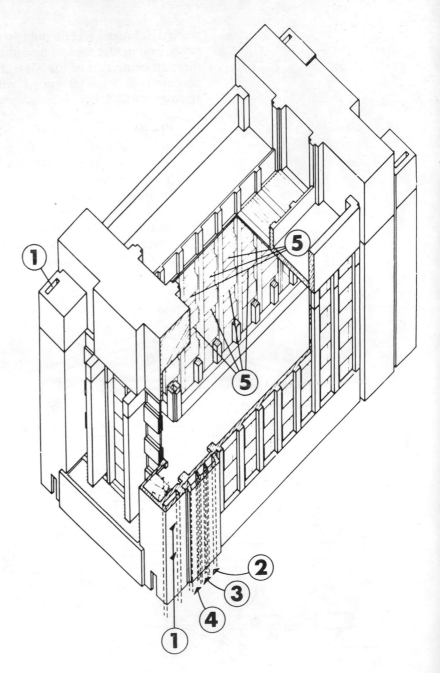

3.2.6 Larkin Building, Buffalo, NY, USA. Frank Lloyd Wright, 1903. Cutaway axonometric showing the importance of the service elements in the overall design. 1, Fresh air intake; 2, tempered air distribution; 3, foul air exhaust; 4, utilities duct; 5, tempered air outlet grilles under edge of balconies

On a larger scale his Larkin building of 1906 (**3.2.6**) demonstrates how the systems which control the internal environment of a building are at least equal to its structural system in formal importance and about which Banham observes that it

> 'must be judged a design whose final form was imposed by the method of environmental management employed, rather than one whose form derived from the exploitation of an environment method.' (Banham 1969)

If a building is a whole system, then a crucial part of that system must be the way it deals with these matters. To perfect a structural system which produces an uninhabitable building is only a partial system. Yet this is what many of the Europeans did. The latter learned many lessons from Wright, but it seems that often these were of the most superficial sort and we will find in the ensuing development of building systems that repeated attention was given merely to structure and fabric in very narrow terms indeed, usually ignoring the implications of climate, site and internal comfort.

Whilst it is therefore of some considerable importance to draw attention to what was a very systemic approach indeed, the next developments in producing building systems are of an entirely different order. It is only much more recently that renewed attempts have been made to put back into building systems that which Wright had so masterfully included.

EFFICIENT MACHINES

'Taylorism is, as we know, a system of promoting the greatest efficiency in a worker. It is one that reduces the worker to an energy-saving automaton. This system is at work throughout the New Theatre.'

Huntly Carter
on Meyerhold's theatre 1924

'I stand on one spot, about two — or three — feet area, all night. The only time a person stops is when the line stops. We do about thirty-two jobs per car, per unit. Forty-eight units an hour, eight hours a day. Thirty-two times forty-eight times eight. Figure it out. That's how many times I push that button.'

Phil Stallings, spot welder
Ford assembly plant, Chicago,
quoted in *Working* by Studs Terkel 1975

Taylorism

The 'Principles of Scientific Management', or Taylorism as it was popularly known, was developed by Frederick Winslow Taylor from his experiences on the factory shop floor and later as a foreman in the America of the late nineteenth century. He became interested in the problems between workers and management and was soon convinced that it was possible to break down the work processes; time these and establish a 'proper' output related both to the worker, and to the tools he used. Taylor was particularly concerned at the way in which management used incentives and bonus schemes whilst giving no thought to its own role in aiding the process. He presented a series of papers: 'A Piece Rate System' in 1895, which was largely ignored, and 'Shop Management' in 1903, which impressed a few manufacturers like Henry R. Towne (of Yale and Towne). The 'Principles of Scientific Management' published in 1911, was translated into Chinese, Dutch, French, German, Italian, Japanese, Russian, Spanish and Swedish. In the latter paper Taylor set out his ideas most clearly saying:

'This paper has been written:
    First, to point out, through a series of simple illustrations, the great loss which the whole country is suffering through inefficiency in almost all our daily acts.
    Second to try to convince the reader that the remedy for this inefficiency lies in systematic management, rather than in searching for some unusual or extraordinary man.

Third, to prove that the best management is a true science, resting upon clearly defined laws, rules and principles, as a foundation. And further to show that the fundamental principles of scientific management are applicable to all kinds of human activities, from our simplest individual acts to the work of our great corporations, which call for the most elaborate co-operation.

And, briefly, through a series of illustrations, to convince the reader that whenever these principles are correctly applied results follow which are truly astounding.' Taylor 1964, 1911)

The introduction to Taylor's paper carries an even more topical ring; and quotes Theodore Roosevelt, at that time President of The United States:

'The conservation of our national resources is only preliminary to the larger question of national efficiency.' (Taylor 1964, 1911)

Taylor pointed out forcibly that natural resources were diminishing fast but that it was equally important to look at the wastage of human effort in the way in which work was conducted. Taylor's experience led him to see the actual movements of men at work as often wasteful, awkward and ill-directed and, that instead of searching for competent men, one should train them saying:

'In the past, the man has been first; in the future the system must be first.' (Taylor 1964, 1911)

He saw the necessity of making it clear that the management must take at least half the responsibility for efficiency and that this would be done under four headings:

1. develop a science of man's work to replace rule of thumb.
2. scientifically, select, train and develop the man as opposed to the man picking it up as best he could.
3. an 'almost' equal division of responsibility between workmen and management.

All this was effected by minutely timing the work processes, the movements of the men and the positions of machines and tools. From this the movements and processes seen as being most economic of time and effort were set down as norms for the job with the result that production went up and the wages also went up — although not at a comparable rate, as the Special House Committee (1912) set up to look into 'Scientific Management', pointed out.

Taylorism and the American efficiency movement became powerful in the period up to the first world war, growing from a set of ideas applied to the factory and commerce to a social movement. In his study of this phenomena, Haber (1964) points out that 'basic to Taylor's program was the view that laborers did not work as hard as they should', but that this was due to backward management and wage policies.

One of the distinctive characteristics of this 'Progressive' period was what Haber calls the aim to 'uplift':

> 'In part, this means lifting up the lower classes into the middle class. Generally this elevation was to be moral, but often it acquired economic connotations.' (Haber 1964)

It is then, ironic that the revolutionaries of the Soviet Union, Meyerhold and the Constructivists, who we examine in a later section, should adopt wholesale the arguments developed to improve capitalist production. The machine, in a sense, was bigger than both of them with America and the Soviet Union each deriving a similar sort of message. In translating 'Taylorism' into 'biomechanics' in the theatre, the Russian director Meyerhold was accepting the production efficient ethic developing in the United States and so crucial to further development in the Soviet Union.

Lenin had embraced scientific management in his *The Soviets at Work,* published in New York by the Rand School of Social Science in at least two translations and five editions (Haber 1964) up to 1919, and saw Taylorism as a key part of the attempts at the centralization of authority in factories in Russia. Whilst pleased with this Taylor was not happy with the idea of 'the spirit of the super-boss and industrial autocrat' (quoted in Haber 1964) especially as many of those developing his ideas were turning them towards a sort of industrial democracy.

Whilst the views of Winslow Taylor were being accepted by the Russian revolutionaries, they were being subjected to a thorough scrutiny in their place of origin — the United States. Taylor was called to testify before a Special Committee of the House of Representatives in 1912. This was brought about by the increasing concern of the labour unions at the uses of 'Scientific Management' and particularly its unscrupulous use by many employers. Taylor's method of establishing his norms for scientific management were subject to some incisive questioning:

> *'The Chairman*: Then how does scientific management propose to take care of men who are not "first class" men (Taylor's hiring principle) in any particular line of work?
>
> *Mr. Taylor*: I give it up.
>
> *The Chairman*: Scientific management has no place for such men?
>
> *Mr. Taylor*: Scientific management has no place for a bird that can sing and won't sing.
>
> *The Chairman*: I am not speaking about birds at all.
>
> *Mr. Taylor*: No man who can work and won't work has any place under scientific management.
>
> *The Chairman*: It is not a question of a man who can work and won't work; it is a question of a man who is not a "first class" man in any one particular line, according to your own definition.

Mr. Taylor: I do not know of any such line of work. For each man some line can be found in which he is first class. There is work for each type of man, just, as for instance, there is work for the dray horse and work for the trotting horse, and each of these types is "first class" for his particular kind of work. There is no one type of work that suits all types of men.

The Chairman: We are not in the particular investigation dealing with horses nor singing birds, but we are dealing with men who are a part of society and for whose benefit society is organised; and what I wanted to get at is whether or not your scientific management had any place whatever for a man who was not able to meet your own definition of what constitutes a "first class" workman.

Mr. Taylor: There is no place for a man who can work and won't work.'

And further on:

'*The Chairman*: If society does not produce an equal balance in all the lines of production of "first class" men, must there not of necessity become men who are not "first class" in any particular line of work where they can secure employment?

Mr. Taylor: I do not think there is any man, as far as I know, who is physically fitted for work, who in this country has to go without work in ordinary times. I do not know of this case except in very dull times.

The Chairman: Is it not true and generally recognised by statisticians, that there are at all times from 1,000,000 to 4,000,000 workmen in the United States who are willing to work but unable to secure it?' (Taylor 1964, 1912)

The perspicacity of the Chairman and his concern for the whole purpose of society is clear but it appeared to have little effect on the acceptance of Taylor's ideas. It is perhaps of interest that Taylor called upon high moral purpose to introduce his ideas, just as did the Constructivists. Further, in each case the machine has been awarded a morality of its own, and to gain the 'efficient' use of the machine the human being must change his predilections. Mechanical efficiency had been merged with commercial efficiency; this developed into a movement for personal efficiency and then to a broader influence as social efficiency. The proponents of the Right Use of the Machine had developed the argument to an important level and one that was to have profound repercussions on Western culture.

Ford and the Assembly Line

Not many of the exhortatory tracts produced in the past thirty years have failed to summon up the analogy of the automobile production line when pressing the idea of prefabrication or a building system.

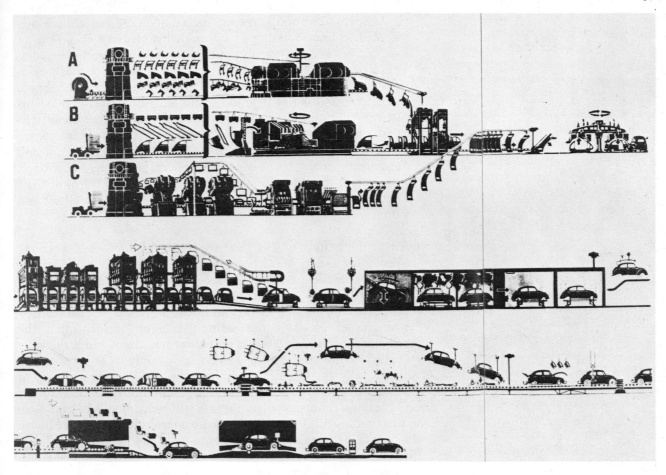

3.3.1 Volkswagen production process

Ford's production line at Dearborn, Michigan and his attitude to consumer and product have been particularly powerful. Usually the first argument used to explain or justify the building systems idea, is that houses can be produced like cars (**3.3.1**). Indeed one of the many books on mass production produced in Britain in the forties, by John Gloag and Grey Wornum, was entitled *House out of Factory* (1946) and this repeatedly referred to Ford:

> 'Henry Ford was the first industrialist to think of the motor car as a universal convenience and not as a rich man's toy.'

The call for a completely new building industry has been at the heart of most calls for prefabrication or industrialized building. Gloag and Wornum are no exception:

> 'A new building industry is needed, composed of enterprising manufacturers working in collaboration, who would approach this huge, known and promising market as Henry Ford approached the market which he believed in, though he could not measure it save by his knowledge of human nature.

A car manufacturer knowing that a market for four million cars existed, at home, could put a remarkably efficient cheap car on the market and reasonably quickly. Now a car is a rather complicated little house on wheels, and it has far more mechanism than any dwelling house carries; moreover, it is not a static structure, and it has to stand up to all the stresses and strains that mobility implies. It has to resist all the weather conditions that a house normally resists, but for a car those conditions are often intensified, because it is being driven through rain and snow and fierce gales of wind. Yet this mobile structure, with all its mechanism, and its comfortable upholstered furniture, its lighting and complete equipment, costs very little.' (Gloag and Wornum 1946)

Even in the Britain of 1946, where powerful new social currents were reshaping society and where the building industry had virtually ceased to exist for five years, this concept failed and the continuity of experience quickly re-established itself in technical terms. Further, the comparison of the housing market with that of the car, the caravan, or other consumer goods is misleading. It is no accident that the financing of consumer durables is on a short term basis and is available very easily. Whilst the housing market is certainly huge it cannot be organized easily and 'consumption' is carried out in such a variety of ways that attempts to rationalize it have so far almost all met without success in those countries where there is still some modicum of choice. The most successful is that of the North American timber-frame house and its matching finance system — but more of this later.

Finally, to equate the car with the house as a technical box of tricks is an argument that can just as easily be turned on its head. The fact that a car is NOT static, that it resists weather only for a limited period (and then with considerable maintenance) and that it is not cheap BY HOUSING standards is also true. It is an interesting facet of this type of argument that the same set of facts can be made to justify two opposite conclusions. As for being cheap, the standard small saloon works out at about 50-100% per square foot of overall floor area more than a house. If calculated on usable floor area the square foot cost of a car would be ridiculous.

The mass production philosophy probably had its most well known exponent and image builder in Henry Ford. He more than anybody embodied the most potent myths concerning what could be offered to the consumer and to society. Contrary to popular mythology, Ford, however, did not introduce mass production into the car industry although he did very successfully capitalize on what he saw going on around him. In view of the potency of the car industry as a model for the building industry it is worth examining this aspect of the mass production idea and the part that Ford played in it.

Robert Street took out the first patent describing an internal combustion engine on 7th May, 1794, but it was not until the end of the nineteenth century that the horseless carriage itself began to appear in any numbers. The fact is that in France, England and the United States considerable numbers of vehicles were by this time being built on a well organized basis. In England the Adams-Hewitt automobile was advertised in 1900 as 'a small car, cheap, with standardised parts for the common man' (Wiks 1972).

By the 1900s over 5000 cars had already been produced in France alone, whereas in the United States Ransom E. Olds had built 425 cars in 1902 and these sold at less than 400 dollars each; in 1903 Olds sold 2500 vehicles. By 1908, when the Model T was produced there were already 24 companies making cheap cars, with the Cadillac selling for less than the Ford. Ford, who was born in 1863, founded his company in 1903 and the first Model T did not appear until October, 1908 (Wiks 1972). It is thus clear that Ford was by no means original in his use of mass production, as Wiks points out. Nevertheless after a difficult start he established himself by a judicious blend of technique and salesmanship which made the Model T a household word and a myth in its own right. Further, Ford's later excursions into education, prison reform, minority groups and even the recycling of waste products endeared him to many Americans. To others, particularly the union movement struggling to establish reasonable conditions, Ford embodied all that was evil in the world of capital intensive industrial production.

The early years of the Model T demonstrate in no uncertain terms the effectiveness of the mass production argument which has subsequently become so powerful amongst system builders: this is that if you make enough of a product, and if this product is standardized it will become cheaper and cheaper. It was certainly as true of the Model T as it has become in more recent years of transistor radios and pocket calculators. However, there is an enormous difference between such objects and a building which is place dependent, although there have been many attempts (some successful) to divide buildings from their context, both historical and physical. Seeing a building in this way, rejecting both the language of the history of architecture and the influence of the locale in which it stands, has been a mainplank in the arguments of the apologists for the machine.

The impact of the earlier years of the Model T clearly spell out the quantity argument:

> 'In 1908 his Model touring car sold for $850. It went to $950 in 1909, then dropped to $690 in 1911, $550 in 1913, $490 in 1914, $360 in 1916, and to an all time low of $290 on December 2nd, 1924. At this time the roadster sold for $260.' (Wiks 1972)

Moreover, of the 5000 interchangeable parts for the Model T, a half cost less than 50 cents each. The relevance of mass production, however, is not only built on being able to produce lots of objects cheaply; it has also become potent because it enabled people to earn the money necessary to buy the products being produced in such abundance. As Wiks points out:

'For the most part rural folks liked Ford because they saw him as a business maverick who made a fortune without ruthlessly exploiting others'... 'People with little money believed Ford used his wealth judiciously. When he introduced the $5.00 working day in 1914 it seemed like Santa Claus rather than the exploitation of the labouring man. When the publisher of the New York Times heard the news, he asked "He's crazy isn't he. Don't you think he's gone crazy?" Rural people did not think Ford had gone insane, neither did they agree with the Wall Street Journal which deplored this action as unscientific and unethical; a move which would destroy incentive.' (Wiks 1972)

Ford was himself a voracious worker and his stated philosophy was concerned with eliminating the waste of time and energy so that there was more of each left over for enjoyment. In his autobiography, first published in 1922, Ford pointed out that machinery, money and goods were only a means to an end and only useful in so far as they set us free to live. He also had clear views on the marketing of his products:

'Ask a hundred people how they want a particular article made. About eighty will not know; they will leave it to you. Fifteen will think that they must say something, while five will really have preferences and reasons. The ninety-five, made up of those who do not know and admit it and the fifteen who do not know but do not admit it, constitute the real market for any product.' (Ford 1928, 1922)

However, Ford was not just interested in inventing some product, albeit the motor car, and then making millions with no further thought:

'This is not standardising. The use of the word "Standardising" is very apt to lead one into trouble, for it implies a certain freezing of design and method and usually works out so that the manufacturer selects whatever article he can the most easily make and sell at the highest profit. The public is not considered either in the design or in the price. The thought behind most standardisation is to be able to make a larger profit.' (Ford 1928, 1922)

In short, Ford was concerned both to find a market, offer a good product, and then to follow that up with first rate service, feedback and refinement. Nevertheless Ford's remark in 1909 to his salesmen who were trying, he said, to 'cater to whims', sums up many a subsequent attitude to standardization:

'Any customer can have a car painted any colour that he wants so long as it is black.' (Ford 1928, 1922)

Ford's philosophy of the elimination of waste was in accord with the business drives of the period in America, a combination of innovation, money making and service. It is important now to recall that this particular style of capitalism was service motivated as well as money motivated. Indeed this view permeates attitudes to mass production and even architecture today. In architecture it is bound up with the development of professionalism and with the professional as an independent adviser. Nevertheless it is clear that in building systems the attitudes to social purpose come as much from business approaches to mass production as from the more socialist orientated drives of serving the people. Furthermore it is also clear that the system builders can also be accused of giving people what they think they need rather than what they want although here it is clothed in more esoteric language than that which Ford was accustomed to using:

> 'If the owner of a skyscraper could increase his income 10 per cent, he would willingly pay half the increase just to know how. The reason why he owns a skyscraper is that science has proved that certain materials, used in a given way, can save space and increase rental incomes. A building thirty storeys high needs no more ground space than one five storeys high. Getting along with the old-style architecture costs the five-storey man the income of twenty-five floors. Save ten steps a day for each of twelve thousand employees and you will have saved fifty miles of wasted motion and misspent energy.' (Ford 1928, 1922)

Ford must have been well aware of the time and motion study work of Taylor, whose first paper 'A Piece Rate System' was presented in 1895. However, his approach embodies almost all Taylor's views and are applied after his early experiences:

> 'In our first assembling plant we simply started to put a car together at a spot on the floor and workmen brought to it the parts as they were needed in exactly the same way that one builds a house.' (Ford 1928, 1922)

From this Ford developed his three principles of assembly which were:

> '1. Place the tools and the men in the sequence of the operation so that each component part shall travel the least possible distance while in the process of finishing.
> 2. Use work slides or some other form of carrier so that when a workman completes his operation, he drops the part always in the same place — which place must always be the most convenient place to his hand — and if possible have gravity carry the part to the next workman for his operation.
> 3. Use sliding assembling lines by which the parts to be assembled are delivered at convenient distances.' (Ford 1928, 1922)

The application of this systematic approach to production in an endeavour to make a car available to everyone 'making a good salary' really is Ford's contribution.

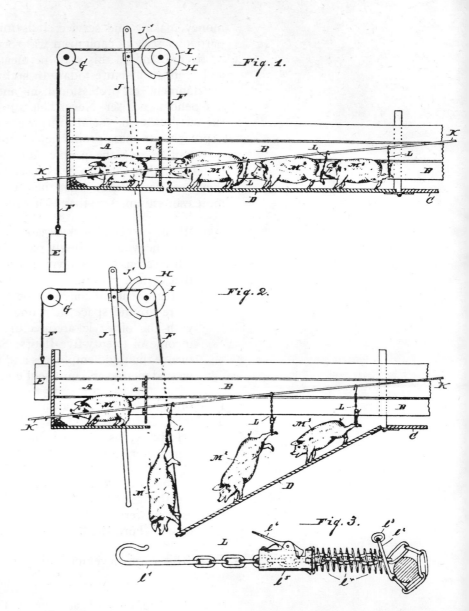

3.3.2 Apparatus for catching and suspending hogs, 1882. 'Here the living animal must be introduced into the "disassembly" line' (Geidion 1969, 1948)

Ford dates his first experiment with an assembly line as occurring about April 1st 1913. It was used for the assembly of the flywheel magneto and he claims it as the first moving line, the idea coming from the overhead trolley used by Chicago meat packers in dressing beef. He then applied the idea to the assembly of the chassis, pulling the chassis with a rope and windlass along a line 250 feet long. The six assemblers moved down the line with the chassis, picking up parts from stacks already in position at appropriate points. In this way the time for each chassis was reduced to 5 hours and 50 minutes.

Giedion (1969, 1948) traces the mechanization of slaughter and meat packing which took place in Cincinnati and Chicago in the 1860s and 1870s. The availability of animals, the vast demand for meat from the rapidly growing population all drove the industry to develop methods of dealing with animals in the fastest and most economical way. The hog was the first to undergo this process, which quickly developed and spread through the whole industry. Giedion shows a diagram of a 'disassembly' line developed in 1882 to overcome the slowness inherent in the stunning of hogs prior to dismemberment (**3.3.2**).

Ford applied these lessons to car assembly with striking results:

> 'Dividing and subdividing operations, keeping the work in motion — those are the keynotes of production.' (Ford 1928, 1922)

This division of labour into a series of fragmentary tasks is at the root of the success of mass production in carrying out work quickly and cheaply. It also raises the most fundamental questions concerning its social validity since it reduces the work of man to a series of meaningless tasks. Man thus becomes alienated from useful creative work and no longer sees labour as part of his whole self: labour has become work.

4

EUROPE AND THE IDEAL MACHINE STYLE

FORMAL MACHINES

'Nothing is ready, but everything can be done. In the next twenty years, industry will have co-ordinated its standardised materials, and technical progress will have perfected the methods of rational construction. Social and financial planning will solve the housing problem, and construction sites will no longer be haphazard breeding places of chaos and confusion, but run scientifically on a large scale.'

Marcello Piacentini 1922,
quoted in Serenyi 1975

The early years of the twentieth century saw the establishing of work study methods and of the assembly line. It also saw social upheaval across Europe with revolution and war. The application of scientific method and its child, technological progress, was matched by writers, painters, sculptors, musicians and architects in their own way. A revolution was being fermented to produce a heady brew that would change art and change the face of Europe's cities — both the change of the bomber and of the planner. Indeed these two ultimately became synonymous in people's minds, the image of the planner changing from the early years of the century, from that of the enlightened social engineer to that of a vandal (Ward 1969; Goodman 1972) who desecrates the fabric of cities at the behest of commerce and bureaucracy. Artists added their own form of impetus to developments.

After the protestations of the Arts and Crafts movement and Wright's call to come to grips with the new technology in a human way, the efforts of many artists and architects in Europe appeared to be directed in another way entirely. The European propensity for theorizing and conceptualizing developed models which for many years were to dominate architecture and to create a new architectural subculture: prefabrication, industrialized building or building systems. In retrospect, much of the work produced during the first half of the century seems to be by people who have been awestruck by the machine and its capabilities. Certainly, to use Wright's words the machine was seen as an 'engine of enslavement' but to the architects of the twenties and thirties who embraced its potentialities it also provided them with a vehicle for demonstrating their belief in the necessity for major social change. This combination caused a wholesale rejection of what had gone before in the forging of a new, more appropriate, aesthetic. In this process the memory bank which is so important to building was jettisoned in favour of the new ideals. In planning, in construction technology, and even in the way things were drawn, the past was rejected. This of course had benefits, but in the long run also caused many disasters, since it led to the destruction of craft skills in the proper sense and aided the alienation that many already felt in relation to their factory work. More importantly it created, because of its proscriptive nature, a narrower architectural language or range of possibilities than had existed before. Of course it had a rich range of new possibilities, but unfortunately many of these were unquestioningly built around some of the worst capabilities of the machine. Ironically it also placed more and more power in the hands of centralized industry — whether it was privately owned or state owned. If quantity production could provide cheap goods for all, housing included, it also had its own dictates. It was thought that merely engaging with the technology would gradually alleviate its effects and bring it under the control of its users. This ignored one of the fundamental properties possessed by each technology: that its very nature implies certain reactions. For example, mass production and the assembly line technique operates in a particular way and this in its turn sets up a net of social relations between those engaging with it. Emery and Trist (1960) point out the importance of the technological component in this sense, drawing attention to the necessity to take an open systems view of such situations — they describe such an approach as sociotechnical. In recent years we have seen, for example, Volvo and other large concerns change the nature of production itself to group assembly methods after such an analysis of the effects of the former assembly line.

With each medium that man uses there arises a different set of social relations between the human and the technological systems. Wright recognized this when he pointed out that the position of architecture as a human expressive system had changed with the introduction of the printing press by Gutenberg, as did McLuhan much later with his emphasis on the importance of the medium used. Ivan Illich has developed this further with his conviviality test for all technologies — or tools. Such tools should, he says, be limited by the 'Protection of three Values: survival, justice and self defined work' and that in a convivial society 'modern technologies serve politically interrelated individuals rather than managers' (Illich 1973). Whilst Wright continued to take a critical view of each aspect of technology and its relation to human ends, the European apostles of the new architecture fell into almost instant embrace with the machine, uncritically singing its praises in poem, prose and paint. The progeny resulting from this affair have been many and varied, but the uncritical nature of its whirlwind course has made itself felt throughout modern architecture.

In placing such a confidence in the technology to solve human problems or answer human needs many of these attempts now seem to curiously ignore or narrow human response. When appealing, through machine technology, to the greater social good they were relinquishing part of their responsibility — that responsibility was somehow to be embodied by 'the machine', whether it be a technologial system or an organizational system. In subscribing to this philosophy they were doing a great deal less than architects had hitherto done, and in eschewing responsibility in this way they were given support by the views of Marx. His view that the means of production formed the foundation of society and all other endeavour was but a super(ficial)-structure encouraged, curiously, a fatalistic view. Such an argument claimed that all developments were historically determined anyway and whatever we do can only faithfully express this underlying truth. This marxist view of deterministic social change was closely coupled to industrial inevitability. This proved a potent drug in that, in itself, it freed its proponents from the necessity of considering their specific human responsibility in each situation, and to each other as individuals. In placing the level of responsibility at the level of the collective or social body it followed that the solutions would also be on that level. Mass production technology fitted perfectly with this, as did its architectural counterpart — building systems. For the solutions here also lay at the level of the collective rather than at that of the individual. The searching for the new architectural norms was carried out through one set of spectacles, one approach. The more the propositions and the buildings conformed to this ideal view of technology, the more they were hailed as the new world. Under these circumstances it is no surprise that the ideal machine style that resulted was subsequently found wanting in human terms.

MACHINE LOVE

'We must invent and build ex novo our modern city like an immense and tumultous shipyard, active, mobile and everywhere dynamic, and the modern house like a gigantic machine. Lifts must no longer hide away like solitary worms in the stairwells, but the stairs — now useless — must be abolished, and the lifts must swarm up the facades like serpents of glass and iron.'

Sant' Elia and Nebbia 1914,
quoted in Banham 1960

In view of the profound impact that Futurist ideas have had on architectural thinking it is perhaps surprising that their sources had until recently, barely been recognized. Banham (1957) points out, in one of his opening papers concerning the Futurists, that both Pevsner and Giedion had little to say of them and yet Futurist ideas permeate the whole of architectural thinking in the West. From Marinetti's Foundation *Manifesto of Futurism* published in 1909, Futurist ideas spread across Europe very rapidly indeed. The intervention of World War I (which saw the death of Sant' Elia) clearly had an effect in clothing these ideas in the rhetoric of a new world arising from the ashes. The Futurists, and Marinetti in particular, forged a powerful set of concepts to unite what they saw as environment and man. Their environment was the new environment of the machine age: the mechanical was to be made poetry, not rejected as antihuman.

The *Messaggio* of 1914 which appeared in the catalogue to an exhibition of drawings entitled Città Nuova, under Sant' Elia's name, but with Ugo Nebbia's assistance lays the Futurist position out:

'The problem of modern architecture is not a problem of re-arranging its lines; not a question of finding new mouldings, new architraves for doors and windows; nor of replacing columns, pilasters and corbels with caryatids, hornets and frogs; not a question of leaving a facade in brick or facing it in stone or plaster; in a word it has nothing to do with settling on formalistic differences between new buildings and old ones. But to raise the new built structure on a sane plan, gleaning every benefit of science and technique, settling nobly every requirement of our habits and our spirits, rejecting all that is heavy, grotesque and unsympathetic to us (tradition, style, aesthetics, proportion), establishing new forms, new lines, new reasons for existence, solely out of the special conditions of Modern living, and its projection as aesthetic value in our sensibilities.' (in Banham 1960)

The implications of this, for architects and architecture, remained a hidden one for many years but has been profound and far reaching. By the time of the exhibition at the Royal Academy in 1973, the effects were clearly established — at least in historical circles if not explicitly for practising architects. Janet Daley exposes the two faces of Futurism:

> 'The repercussions of futurism on contemporary architecture are more than a glimmer, and they are the antithesis of liberation. Like its precursor at the turn of the century, contemporary futurism revels in a mindless infatuation with the most dehumanising forms of technological production.' (Daley 1973)

Daley draws a precise parallel between the concern of the Futurists with the use of technology and the more recent technologically based models offered by various architects, notably Archigram whose work she describes as of a 'trashier and more vulgarly affluent variety'. There are dangers in making such broad assumptions from two very different periods. Clearly in many ways the Futurists felt themselves to be concerned with the right use of technology but saw this as an artistic endeavour involving the redefinition of the boundaries of what was accepted as art. Out with the old games of style and in with the completely new vision of the machine age. They put into 'good currency' many ideas, most of which when applied in various forms, whether it be the Corbusian Ville Radieuse or Town-Planners' Everywheresville, have proved indeed to be intractable and antihuman. In enhancing the machine in this way, the Futurists also made it paramount. The area of concern was shifted more to the objects of this revolution than to its human effects. Their concern to break with the past and escape the laws of continuity has had a vicious pay-off as has the attempt at 'the perfection of technical methods, the rational and scientific use of materials' (Banham 1957), which they demanded.

To building systems designers especially, the perfection of technical method has specifically implied the exclusion of those technical methods not associated with the new machine technology and initially a rejection of any attempt to develop and build on existing practices. In thus rejecting certain techniques as morally inappropriate they also rejected the people whose lives were built around the performance of those skills. In making the rejection, the confidence of the very people needed to carry out the new building was lost. During the ensuing decades a series of blows was struck at the building industry at a technical and a human level in an endeavour to transform it to a shape more in accord with the ideal bride of Futurist vision.

It is as well to remember that Italy was not fully united until 1870. Further the rapid growth of industry in Northern Italy must be seen as a crucial backdrop to the development of Futurism. Previous attempts to revive Italian art towards the end of the century had not been particularly successful although they were to have an influence on the emergence of the Futurists. This is well covered elsewhere: suffice it to say here that the Futurists were not the first to call for the rejection of history.

Cecioni demanded that 'the divorce between modern and old must be absolute' (Martin 1968), and Segantini in 1894 stated that 'the old ideals have fallen or are about to fall... the thought of the artists must no longer turn to the past' (Martin 1968). The magazines *Leonardo* and *La Voce* published by Pressolini and Papini contributed and acted as a focus for these developing ideas until by 1912 the Futurists had achieved international attention. It is not difficult to see the impact on architecture of such a powerful case rejecting a reliance on history as a source — indeed the argument persists until today. The building systems movement is one such example, built upon the notion of largely casting off past forms and methods, in particular so-called craft methods, and searching for ways to reflect the use of machines and emerging technologies.

The energetic Marinetti forged the Futurists into a solid force with the publication of his Futurist Manifesto of 1909 — published in Paris. Marinetti's poems and performances, his embracing of dynamism, speed and space, elevated technology to almost mystic proportions. The main planks of the manifesto were:

'1. To seek inspiration in contemporary life.
2. To be emancipated from the crushing weight of tradition...
3. A contempt for the prevalent values of society and its corresponding conception of art.' (Martin 1968)

The painters Boccioni and Balla attempted to render dynamism into their work which when seen today is curiously unimpressive; incompetent almost. Severini appears a better painter although somewhat more akin to the cubists. More relevant to our argument here is the 1910 Technical Manifesto of Futurist Painting signed by Boccioni, Russolo, Balla, Carra and Severini:

'Everything moves, everything runs, everything changes rapidly. A profile is never stable in front of our eyes, but constantly appears and disappears'. (in Northern Arts and Scottish Arts Council 1972)

the argument is developed by Boccioni in 1912:

'Painting has received new blood, has become more profound and wider by bringing landscape and environment to bear simultaneously on the human figure and its object, achieving our Futurist COMPENETRATION OF PLANES.'

'And this systematisation of the vibrations of light and the interpenetrations of planes will produce Futurist sculpture the foundations of which will be architectural, not only in the construction of the masses, but in such a way that the block of the sculpture will contain within itself the architectural elements of the sculptural environment in which the subject lives.'

'Naturally we will bring forth a SCULPTURE OF ENVIRON-MENT.' (in Northern Arts and Scottish Arts Council 1972)'

Most relevant to our argument here is Sant' Elia, whose drawings of complex 'deep' city forms (see **4.2.1**) have been powerful icons in symbolizing possible futures for the city. In his manifesto of 1914, *Futurist Architecture*, he recognizably lays the ground for much that is to follow in the modern movement:

> 'We have lost our predilection for the monumental, the heavy the static, and we have enriched our sensibility with a taste for the light, practical, the ephemeral and the swift. We no longer feel ourselves to be the men of the cathedrals, the palaces and the tribunes. We are the men of the great hotels, the railway station, the immense streets, the colossal ports, covered markets, luminous arcades, straight roads and beneficial demolitions.' (Sant' Elia 1914 in Banham 1960)

Here we see a further step in the argument for the engineering content of the new emerging technologies and their building types. Sant' Elia's vision has since been brought to life with a vengeance with the long clear spans, the great freeways and not least the 'beneficial demolitions' of urban redevelopment which have all but destroyed many of the large cities of the western world.

Sant' Elia's propositions, largely contained in the drawings produced during 1913–14 and subsequently exhibited, show links with the Art Nouveau and the Secessionists (**4.2.2**) both in their formal organization and their technique — although they are stripped of the sinuousness and decorativeness associated with that work. Their main contribution is in providing a vision where the buildings are not individual objects standing freely in space in a street, but are embedded, interconnected and interrelated in a three-dimensional manner with each other and with all the means of transport. Roads, railways, lifts and aircraft zoom up and down the faces of the buildings, dive through vast structures and relate them in a huge dynamic complex. Similarly with Chiattone's drawings. In all this work hardly a person can be seen — the vast towers with the repetitive windows and structure rise from massive supports of steel and concrete. The seeds of many a later piece of architecture can be found here in Sant' Elia's drawings of the New City; in the Stepped Profile Building (1914) with its freestanding elevator shafts linked by bridges to multi-storey blocks and in 'The New City Pedestrian Crossings with Central Lifts' (1914) (**4.2.1**).

The Futurists were active in all the arts; writing, proclaiming, declaiming. Their theatrical evenings however, probably caused the most public outcry, with the police often standing by. Marinetti was the prime force in these and he was concerned with the promotion of spontaneity, technical innovation, audience participation, the breaking down of existing logic and a debunking of the sacred and solemn.

4.2.1 The New City Pedestrian Crossing with Central lifts: Sant' Elia, 1914. Pencil and watercolour

4.2.2 Electric Power Station.
Sant' Elia, 1914. Pencil, ink and
watercolour

These events, forerunners of Dada, were first improvised in 1914.
After the end of World War I, in 1921, Marinetti pushed his ideas on
theatre into another area described as

'Tactilism, a wordless art, whose only aim is harmonies of touch,
which, through the skin, will contribute indirectly to the perfecting
of spiritual communication between human beings.' (Shankland
1972)

A forerunner of the encounter group no doubt. We can also, through Marinetti's address to the Lyceum Club in 1912, see the seeds of the rejection of craft skills:

> 'When, then, will you disencumber yourselves of the lymphatic ideology of your deplorable Ruskin, whom I intend to make utterly ridiculous in your eyes. . .
>
> With his sick dream of primitive pastoral life; with his nostalgia for Homeric cheeses and legendary spinning wheels; with his hatred of the machine, of steam and electricity, this maniac for antique simplicity resembles a man who in full maturity wants to sleep in his cot again and drink at the breasts of a nurse who has grown old, in order to regain the carefree state of infancy.'
> (quoted in Banham 1960)

THE FORMALITIES OF EFFICIENCY

'Constructivism is the organisation of the given material on the principles of tectonics, structure and construction, the form becoming defined in the process of creation, by the utilitarian aim of the object.'

from the magazine *LEF* 1923
quoted in Frampton 1971

The Futurist eulogies to the machine and to industrialization from 1909 onwards, whilst exceedingly influential in pervading poetry, literature, painting, sculpture and architecture, were not matched in Italy by any strong social movement. Constructivism, on the other hand, embodied a set of ideas which, for a brief period between 1917 and the end of the twenties, saw itself in accord with the enormous social changes that were taking place in the Soviet Union subsequent to the revolution.

The seeds of Constructivism in Russia itself were developing in the years before 1917 but

> 'the precise origin of the name and its first use by these artists has not yet been established, but from 1920 onwards one finds it being used more and more in statements by this group (The Productivists), headed by Tatlin.' (Gray 1962)

Unlike the Futurists, the Constructivists brought together a series of threads in a way which was to have a powerful impact. The way in which this fusion of utilitarianism, the emerging industrialization, new ideas in art and marxist theory was made, gave it a strength which the Futurists (and indeed other movements in art) did not possess. It must have all seemed so persuasive: new society, new art, new architecture. Out with the old, replace it with the new, the innovative. In addition, a wide field of endeavour was involved: painters, sculptors, the theatre, film, architecture, literature, industrial design and graphic design. Indeed the participants were in close communication with each other and frequently moved from one field to another. For example architects designed stage sets, painters designed buildings. Their interest to us here is less because of any specific 'building system', although there were propositions, but more because of their legacy to systems designers in combining social imperatives with formal procedures. Their particular demonstration that certain social objectives were allied to specific architectural solutions centred around the doctrine of utilitarianism: that the test of any action is the greatest happiness for the greatest number.

The way in which the philosophy of utilitarianism permeated architectural thought and then became transmuted into the doctrines of 'functionalism' has an important bearing on the development of the building systems idea. The marriage of mechanization and art around the concept of buildings as functional machines is a recurring theme most clearly expressed in the work of Le Corbusier and the distinctive style emanating from the Bauhaus. However the clarification of these ideas occurred somewhat earlier, in a more vigorous form in the Soviet Union.

The changes taking place before, during and after the Russian revolution, swept through society and the world of the arts. Ideas later receiving great publicity through the work of the Bauhaus were seeded in Moscow in 1920 with the establishing of the Vkhutemas (Higher Art and Technical Studios) which combined architectural, industrial and artistic facilities under one roof with the experimental work done there widely published inside and outside the Soviet Union.

It is to the theatre that one must turn first for a clear demonstration of the links between utilitarianism, communism and the stylistic concerns of the Constructivists. Often overlooked because of its transient nature, the world of the theatre has frequently provided the climate for a clear demonstration of a new pattern of ideas. Its rituals embody narrative, polemic, physical movement and an opportunity for spatial manipulation. In this respect the importance of the stage director Meyerhold cannot be underestimated since his productions fused many of the concerns of the time into a clearly identifiable format (**4.3.1**). It is easy to see that this powerful exponent of total theatre distilled a range of ideas with such clarity that the forms

4.3.1 Meyerhold and a Synthesis of his scenery, showing the new machine and factory forms used in the new representation. A woodcut from Carter (1924)

thereby created carried great credibility and influence, far beyond the theatre. Ideas were worked out on the stage which demonstrated some of the key constructivist themes for architecture.

As director of the Moscow State Higher Theatre Workshop, created in 1921, Meyerhold was concerned to avoid conventional stage machinery and to find a type of scenery that could be put up anywhere, in the open air even. In the Constructivist exhibition of 1921, called $5 \times 5 = 25$:

> 'Meyerhold saw the possibility of a utilitarian, multipurpose scaffolding which could easily be dismantled and erected in any surroundings. Furthermore, this industrial "anti-art", which recognised practicability as its sole criterion and condemned all that was merely depictive or decorative, seemed to Meyerhold a natural aid in his repudiation of naturalism and aestheticism.' (Braun 1969)

One of the artists exhibiting in the Constructivist exhibition, Lyubov Popova joined the staff of the Theatre Workshop and designed and built the set for the production of the *Magnanimous Cuckold*. It included normal theatre flats, platforms, steps, chutes, catwalks, wheels and a rotating lettered disc. Windmill sails revolved at different speeds to the changing passions of the characters (**4.3.2**).

However, Meyerhold's interest in the Constructivists did not merely arise from a casual interest in their ability at form making. It was related to a set of ideas which related society, the effects of industry and the individual. He developed a system of practical exercises for actors which he called 'Biomechanics', demonstrated in 1922. Meyerhold saw Biomechanics 'as the theatrical equivalent of industrial time and motion study' (**4.3.3**) (Braun 1969). This view,

4.3.2 Stage set for 'The Magnanimous Cuckold'. L. Popova, 1922

which owed a great deal to the American time and motion studies of Frederick Winslow Taylor as earlier discussed, was espoused by Meyerhold as being scientific and appropriate to the demands of the machine age.

Meyerhold believed that the creative act was a conscious process based on scientific principles involving the organizing of the artist's material. He saw a new foundation for art in a society where labour was 'no longer regarded as a curse but as a joyful, vital necessity'. The philosophy of the most efficient use of body movement and work time was seen as the basis for a new art:

> 'Work should be made easy, congenial and uninterrupted whilst art should be utilised by the new class not only as a means of relaxation but as something organically vital to the labour pattern of the worker. We need to change not only the forms of our art but our methods too. An actor working for the new class needs to re-examine all the canons of the past. The very craft of the actor must be completely reorganised. . .
>
> 'However, apart from the correct utilisation of rest periods, it is equally essential to discover those movements in work which facilitate the maximum use of work time. If we observe a skilled worker in action, we notice the following movements:
>
> (1) an absence of superfluous, unproductive, movement;
> (2) rhythm;
> (3) the correct positioning of the body's centre of gravity;
> (4) stability.
>
> Movements based on these principles are distinguished by their dance-like quality; a skilled worker at work invariably reminds one of a dancer; thus work borders on art.' (quoted in Braun 1969)

a

Huntly Carter, actor, producer and playwright, in his book of 1924 gives an account of Meyerhold's work, including Biomechanics and what he called 'Construction', based upon visits to the Soviet Union subsequent to 1917:

> 'Biomechanics is really the application of the construction or mechanical theory to the actor. It assumes that the actor is a rather wonderful engine composed of many engines.' (Carter 1924)

The laws of Biomechanics, Carter continues:

> 'are founded on the study of the physiological construction of man. The system aims to produce men who understand the mechanism and laws of their structure, and can, therefore use it perfectly. It has established a principle of analysis by which each movement of the body can be differentiated and made fully expressive. Biomechanics replaces the emotional theory of acting

which assumes an ignorance of the mechanics of the human body, by a form of education in the science of technics, which trains the intellect of the actor, and develops his body by means of sport, and so produces the organised actor in full possession of the keyboard of his mind and body, and capable of adding dignity to his calling.' (Carter 1924)

Carter claims that Meyerhold was the first to apply 'Construction' and Biomechanics, and that he

'had the mass and co-operative idea of society. He regarded society as a great industrial machine, of which each individual is a functional part. Each is free in as much as he understands himself and the whole, just as a bird is free to the extent that it realises its captivity. Probably Meyerhold derived the machine idea not from Marxism but from Marinettism, with its modernolatry, the idealisation and worship of the machine, its movements and sounds and the attempt to express them in forms of art.' (Carter 1924)

From the enobling of the act of labour as an art form, to the stripping of what were considered inessentials, it is but a short step to the viewing of artefacts through the same lens. A design aim based upon the elimination of what was considered to be mere elaboration on the basic idea was to be embraced. In clothing Meyerhold stated that the actor of the future would wear an overall and wear no make-up — this was to be his everyday clothing and that used onstage. Whilst Meyerhold's claim to the invention of Biomechanics was disputed there is no doubt that his combination of Winslow's work on time and motion with the marxist view of the dignity of labour gave his version of Biomechanics a mythic quality which suffused the theatre in the same way that a similar synthesis did in other areas of the art world. The force of this can be sensed from Meyerhold's lecture of 1922:

'Since the art of the actor is the art of plastic forms in space, he must study the mechanics of his body. This is essential because any manifestation of a force (including the living organism) is subject to constant laws of mechanics (and obviously the creation by the actor of plastic forms in the space of the stage is a manifestation of the force of the human organism).' (Braun 1969)

Meyerhold's development of Biomechanics grew out of his interest in mime and movement, upon which he was writing in 1905, and Taylorism was amalgamated with his interest in the Commedia dell 'Arte and the formalities of Japanese theatre to produce Biomechanics. 'Taylorized gestures' (**4.3.3**) became the basis of acting technique, just as Taylorism gave mechanical substance to Constructivism. Again Carter is revealing on this:

b

4.3.3 (a *left*) Taylorized gesture in the Russian theatre 'Two work diagrams of angles'. Meyerhold, *c.* 1922. Actor pupils are expected to learn how to use their hands and feet on the above models. Shown are turning movements at 90 degrees. (b) 'Little Revolutionary Theatre: A machine dance applied to the "Cake Walk"'. N.M. Foregger, *c.* 1922

'today in Russia there is a distinct style known as RSFSR. It is based on a line. Just as the rococo style was based on round and elliptical lines, so the Soviet style is based on a straight line. The geometrical principle of the straight line is the shortest distance between two points, and the RSFSR style or straight line style is accordingly constructed on this principle.' (Carter 1924)

Who can now doubt the power of 'the straight line style'? Drawing its strength from a long tradition of logic, the growth of industrialization and its effects, the search for efficiency, the yearning for a machine morality — all contributed to the establishing of the Constructivist combination of morality, visual clarity, social efficiency and purpose. It is Churchman's 'minimum loop' (Churchman 1968), the search for the simple, as opposed to the 'maximum loop', the latter being the arrival at a solution after the longest journey, not the shortest, and one which is capable of embodying human experience as well as logically extrapolated simplifications (Russell 1973).

Meyerhold's use of Constructivist ideas in 1921, then, offered the first opportunity to make, in large scale physical terms, what had hitherto been largely a collection of ideas, paintings and graphic design. The stage set for the *Magnanimous Cuckold* is considered to be the first clear demonstration in these terms (**4.3.2**). Edward Braun makes the interesting observation that despite the austerity of Popova's set for this production its effect in conjuring up associations with the bedroom, the balcony and so on strongly compromised its intended ability to work merely as a functional machine:

> 'In the theatre, whose whole allure depends on the associative power of the imagination, every venture by the Constructivists led to an unavoidable compromise of their utilitarian dogma and each time demonstrated the inherent contradiction of the term "Theatrical Constructivism".' (Braun 1969)

A brief description of two more sets will indicate the debt that later architects owed to Meyerhold and his designers. That for Alexei Faiko's *Lake Lyul* in 1923 is described by Faiko thus:

> 'The back wall of the theatre was barred. Girders stuck out and wires and cables dangled uncompromisingly. The centre of the stage was occupied by a three storied construction with receding corridors, cages, ladders, platforms and lifts which moved both horizontally and vertically. There were illuminated titles and advertisements, silvered screens lit from behind. Affording something of a contrast to this background were brilliant colours of the not altogether lifelike costumes: the elegant toilettes of the ladies, the gleaming white of starched shirt fronts, aiguillettes, epaulettes, liveries trimmed with gold.' (Braun 1969)

The set for the adaptation of Ehrenberg's *Give Us Europe* appears to have embodied even more of the ideas that were later to become the currency of architects of the modern movement.

> 'The production was remarkable for its settings', says Braun, 'which were composed entirely of "moving walls". Devised by Meyerhold himself, these "walls" were a series of eight to ten red wooden screens, about twelve feet long and nine feet high, which were moved on wheels by stage hands concealed behind each one. With the addition of the simplest properties, they were deployed to represent now a set of rooms, now a Moscow Street, now a sports arena. The action never faltered and in some scenes the walls played an active part, their motion emphasized by weaving spotlights. Again Meyerhold employed projected captions, this time on three screens. As well as the title and the location of each episode, there were comments on characters, information relevant to the action, and quotations from the written works and speeches of Lenin.' (Braun 1969)

In these sets and productions one may see, sharply focused, the dilemma of the Constructivists and later that of the Bauhaus and subsequent developments, particularly in building systems. Aksyonov, a contemporary commentator summed it up in this way:

> 'So called "stage constructivism" started out with a most impressive programme for the total abolition of aesthetic methods, but once it appeared on the stage it began to show signs of being only too ready to adapt itself to its surroundings and now it has degenerated almost to a decorative device, albeit in a new style.' (Braun 1969)

In setting aside what the new wave of designers saw as a decorative style, and endeavouring to make explicit the essence of a problem in the artefact only led to a new set of stylistic conventions. Such attacks upon the order of things are part of a reordering process: when those attacks are not merely those of a few painters and designers but are paralleled by massive restructuring of the social system, as was occurring in the Soviet Union, the resulting new patterns, the emergent stylistic qualities, become strongly associated with particular sets of social goals. Whilst some of the source evidence is now available in terms of the Constructivist writers, painters and poets, the importance of Meyerhold and his theatre, and its influence on architecture has tended to be ignored.

Indeed the power of the theatre in the formulation of architectural ideas is a recurring one, as well as one that is consistently unconsidered in its histories. However:

> 'Even more directly than the other arts — or more crudely — the drama is a chronicle and brief abstract of the time revealing not merely the surface but the whole material and spiritual structure of an epoch.' (Bentley 1955)

For the period when Meyerhold was at work, one of the great influences in the theatre had been Wagner with his concept of Musikdrama and Gesamtkunstwerk — an attempt to combine high drama, music and setting in one work, an integration of all the arts, a Total Theatre. The rephrasing of this by Meyerhold, in a form more suitable to the communist revolution, created his Total Theatre. This, in its turn, contributed to the concept of Total Architecture subsequently espoused by Gropius and others in the modern movement. If the concept of Total Theatre has been a powerful influence in general terms, and in a more particular way in the possibilities offered by set design, its attitudes to movement also bear on the work of the designers. Ergonomics, the science of matching machines to men, has a close relation to movement in the way in which it was used by Meyerhold. And ergonomics in its turn has been an important part of the modern movement's love affair with the machine.

Let us now look more closely at Constructivist ideas and their application to architecture. Frampton (1971) points out that the much abused term Constructivism was a precise formulation by The Productivist Group based 'upon the subsequent definition of two compound terms'. These were 'Tektonika' and 'Faktura'.

The interaction of communist society and industrialization gives rise to 'Tektonika' whilst the synthesis of this with construction as the formulating activity, gives rise to the 'objective' realization inherent in 'Faktura'. Here we can see the aims of society and the means of construction made synonymous: the ideals of communist society amalgamated with industrialization; further than this, that new industrial techniques must be embraced and through them the remaking of society expressed. The Constructivists were against the use of traditional forms of construction, or even the partial use of them, and favoured the scientific use of industrialized techniques. The relation between these ideas and those expressed by Meyerhold is clearly seen in his observations of a skilled worker in action and his view that the resulting synthesis of movements based on these principles have a dance-like quality of their own.

Although the Constructivists were avowedly utilitarian in their claims, this must be looked at in close relation to their artefacts which might be described as far from utilitarian. In this opposition between verbal or written claims and the artefact we see one of the major tensions of the modern movement. The importance of the Constructivists to the development of architectural ideas since the twenties is clear. Their particular relevance to the growth of the building systems idea lies in the precision with which they formulated the arguments which later provided justification and authority for almost every one of the major building systems. Frampton (1971) described the constructivist 'mood as apocalyptical' and this can be seen as a recurring theme in the launching and sustaining of building systems ideas right up to the present. The term Constructivism was again defined in the Constructivist magazine *LEF* in 1923 and the crucial idea in this definition is that of the equating of the physical with the intellectual: that ideas and matter are to be treated equally in any construction.

Artefacts can be seen as summarizing social intentions and attitudes and as such are inevitably of a style. There may be elaborate arguments as to the meanings implicit in the artefact, be it building or other object, but the nature of the link between the artefact and its meanings is in no way precise. Indeed it is this summarizing quality which gives rise to the ambiguity of such objects. Philosophic, political and economic ideas can be attached to artefacts in different ways at different times and in different places. The polemics of the Constructivists were particularly effective in establishing what seemed like a direct relation between society's industrial techniques and the means and form of construction. So effective indeed that similar polemics concerning the desirability of certain industrialized techniques over other less industrialized techniques have been repeated ad nauseum by system builders particularly, and by the architectural avant garde generally over the past fifty years in making their various claims as to the necessity for reordering the problem and the design.

The marxist view of an industrial base from which springs a cultural superstructure can be seen reflected in the Constructivist view that an 'objective' realization can occur from the imperatives of society's aims and industrial techniques. Therefore, the argument goes, a faithful translation of the aims and techniques inherent in communist society will result in an honest utilitarianism. In this way the moral imperatives of the aims of the collective, industrialization and the designed object are established. Although historically this is no new link it was, in the Russia of the twenties, given a clarity that is to be of critical importance in the rise to a position of power of the modern movement, of the implementation of mechanization and the development of building systems. The right use of the machine is a philosophy that permeates the latter day building systems although its diffusion into an idea 'in good currency' has inevitably meant that antecedents have been lost. One of the recurring marks of discussions amongst proponents of recent system building is the manner in which such unsubstantiated moral arguments are employed together with an ignorance of their history and development. The latter has, of course, a very good explanation, in that any useful study of history other than the popular polemic of the modern movement versus the rest was a rare event in many schools of architecture in Britain during the post-war period.

Just as Meyerhold had insisted that the craft of the actor must be reorganized so that all superfluous and unproductive movement is eliminated, the Constructivists in design were insisting that the designed object must grow from the needs of society and its processes in a utilitarian manner with all associations and historical impedimenta removed. Such a view even extended to the use of colour with the sculptor Naum Gabo, in *The Realistic Manifesto* of 1920 stating:

'Thence in painting we renounce color as a pictorial element, color is the idealized optical surface of objects; an exterior and superficial impression of them; color is accidental and it has nothing in common with the innermost essence of a thing.' (Chipp 1968)

This rigorous view clearly had its effect on architects with, at one stage, colour almost banished altogether. More recently, a functional use of colour has become acceptable, but many architects are still uneasy with colour used as a pictorial element.

However, this reworking of the doctrine of utilitarianism, in apparently setting aside matters of symbolism and social myth

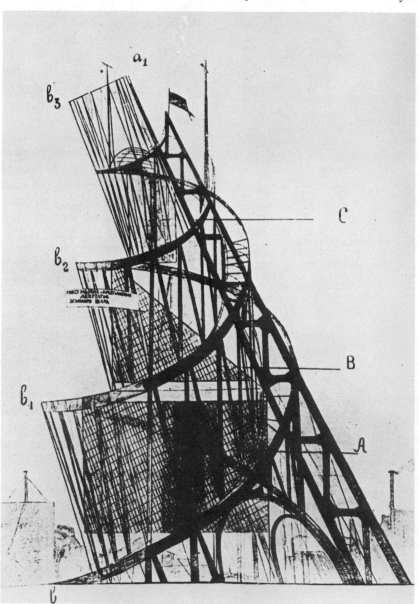

4.3.4 (a) Monument to the Third International: V. Tatlin, 1919-20 elevation

4.3.4 (b) Model of Tatlin tower made for the exhibition, Art in Revolution, Hayward Gallery, London, 1971

making, carried its own symbolic power. A symbolic aspect never adequately explained by its proponents. Indeed the compound of scientific determinism and marxism which fired them did not encourage such a view to be taken. Such things were seen to be mere decoration and elaboration on the class and industrial base of society and as such would arise naturally. Thus, what arose in this way was 'right for society' because of the original premises. Thus the circular argument is established. In his powerful critique of materialist and rationalist explanations of culture Sahlins (1978, 1976) points out that

> 'Conceiving the creation and movement of goods solely from their pecuniary properties (exchange-value), one ignores the cultural code of concrete properties governing "utility" and so remains unable to account for what is in fact produced.'

Since the basis of the Russian revolution was avowedly a radical change in the relations of the social contract it is not surprising that artists and designers took this opportunity of reappraising the formal manifestations of this change. Whilst the results of this reappraisal are of radical significance and have themselves had far reaching effects, their attempt to demonstrate a direct link between society, industrialization and form can be seen as having less to do with their doctrine of utilitarianism than with a set of powerful symbolic referents.

The most powerful Constructivist image is Tatlin's Monument to the Third International, designed between 1919 and 1920 (**4.3.4**). In this unexecuted project the tensions between the symbolic and utilitarian are at their most extreme. The tower consisted of inclined spirals within which were three glazed forms, one above the other, which were to revolve, surmounted by a smaller, hemispherical, form. The lower, cubic, form was to rotate once a year and would contain the legislative activities; the middle, pyramidal form, was to rotate once a month and would contain the activities of the executive whilst the topmost form, a cylinder, turned once a day and housed propaganda and information functions. This latter would contain cinema, radio and telecommunication equipment. The symbolism attached to the dynamic upward thrusting of the spiral coupled with the activities of government made open to the outside through the all glass walls of the three forms could be described as somewhat simplistic. These three forms translated a revolution a year, through a revolution a month, to a revolution a day for the masses by means of the outflow of information at its apex. Its potent visual quality has led to its many references permeating the architectural subculture. Amongst these are the clear divorce of the spaces containing the activities from the supporting structure; the idea that the ability to observe an activity through a glass wall is democratic; and the use of what were described as industrial materials — 'glass and iron'. All these have passed into the currency of architectural language and Tatlin's Monument succeeded in stating them in unequivocal terms.

If Tatlin's tower was a symbolic fusion of function and form around the aspirations of a communist society other projects will express this more literally. The project by the Vesnin brothers in 1923–4 for the Pravda building is a particularly interesting one to this study. The tower (**4.3.5**) had an unbelievably small plan area of 19 ft (5.8 m) × 19 ft (5.8 m) and rose to seven storeys at its highest point. It is designed with an unashamed steel frame exposed on the outside with a series of standard bays and exposed diagonal windbracing. A glazed double lift tower is exposed on one side and the tower carried a rotating display panel for propaganda statements, a back projection screen for the latest news, a clock, a loud speaker system, and surmounting all a searchlight and red flag. In this small building one can see a model for integrating a wide range of elements into a synthetic (or Constructivist) whole. The first multi-media building perhaps, or a foretaste of Orwell's 1984. Frampton makes the claim

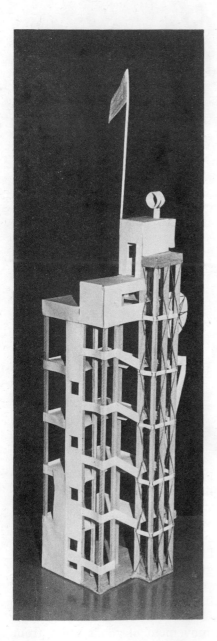

4.3.5 Project for the Pravda Building. A.A. and V.A. Vesnin, 1924 Model *c.* 1974

that, of the other works of the period in Europe that laid claim to importance, the Vesnins' tower is the only one that clearly makes this sort of total statement.

'The direct structural articulation of the Pravda building, the transparency of its facade, the expressive mobility of its components and the empirical determination of its arrangement are each in turn characteristic of the Constructivist aesthetic as it was to evolve during the early twenties.' (Frampton 1971)

One can see its implications for Miesian rationalism and for a whole generation of building systems.

THE
ICONOGRAPHY OF
THE MACHINE

'Given the trend of our age to eliminate the craftsman more and more, yet greater savings by means of industrialisation, can be foretold, though in our country they may for the time being still appear Utopian.'

Gropius to AEG 1910

'Pessac was conceived to be built of reinforced concrete. The aim: low cost. The means: reinforced concrete. The method: standardization, industrialization, taylorized mass production.'

Le Corbusier in Boesiger 1960

'A Completely New Method of Construction'

If the Constructivists made out a case for moral utilitarianism in architecture, it was Le Corbusier who developed the argument, with the writings and objects that forged a machine style. Le Corbusier managed to create a unified aesthetic and a philosophy which, in Europe at least, became synonymous with modern architecture. We have seen these ideas run their course, from messianic inception, through a minority following, gradual acceptance as the preferred solution by City Fathers everywhere, to a reaction and revulsion of considerable proportions. Le Ville Radieuse became less than radiant in the hands of politicians, housing managers, engineers and architects. Similarly 'une machine d'habiter' became a slogan for buildings that looked as if they had been made by machine processes, and lent authority to architects searching for an appropriate modern aesthetic.

The Dom-ino House (**4.4.1**) conceived by Le Corbusier in 1914 is probably one of the most powerful icons in the growth of the modern movement and the building systems idea, albeit an icon whose roots are usually overlooked when the case is being made again for those same ideas. Described by Le Corbusier as a 'system of construction which envisaged the problems of post war re-construction' (Boesiger 1960) it contained much of the essence of an argument later to become familiar on the lips of systems builders. The Dom-ino House concept, says Le Corbusier:

'would result in a completely new method of construction: the windows would be attached to the structural frame, the doors

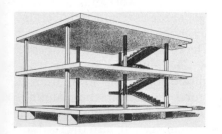

4.4.1 Dom-ino House, standardized framework. Le Corbusier, 1914

125

would be fixed with their frames and lined up with wall panels to form partitions. Then the construction of the exterior walls could begin.' (Boesiger 1960)

The crystallizing of his ideas into this drawing, showing the standardized framework, has been a key force in the promulgation of the concept. Simple and direct, the drawing shows simple slab floors supported on a slender frame, and the point is made that the floor layouts may thus be independent of the structure. It also follows from this that the foundations can be reduced to simple cubes beneath each column. Although the concept is described as being carried out in reinforced concrete the nature of the drawing is not so specific. Many designers will have carried this powerful image with them long after they have forgotten the supporting argument — if indeed many of them were ever conversant with it in the first place.

The framework is further described as being made from a series of standardized elements fixed together in differing combinations to give variety:

> 'The contractor would deliver the frames marked and grouped upon the order of the architect-planner or, more simply, upon the order of the client. Another contractor would furnish all the additional elements, which could be mass produced: the windows, doors, etc.' (Boesiger 1960)

Not only is there here the philosophy that subsequently flowered in the Hertfordshire Schools, CLASP, and other building systems but an element which all these architect/client controlled approaches did not incorporate and which has, since the early 1960s become again a powerful force: that of participation. Le Corbusier here suggested that standardization in this way would allow the client as well as the architect to select components from a series of standard ranges. Later attempts to focus on user oriented housing led Habraken in 1961 to symbolically reject the power of the Dom-ino House (**4.4.2**) image (Pawley 1970) although almost none of the architect controlled systems, whatever their intentions, have managed to give freedom in this sense. Ironically for the socially conscious architect, where this has happened, it has been by means of a commercial system where the buyer selects from a series of components or plan types. At a smaller scale, some individual, but equally significant work, takes a very different view of standardization, the role of the user and architect. Of these approaches, that embodied in the timber houses of Walter Segal is one of the more significant (**4.4.3**). The compression of a number of ideas into the single concept of the Dom-ino House is further illustrated by the references to the ability of the frame structure and slab floors to give freedom from structural restraint in the planning. Thus, not only are standardization, component building and user participation encapsulated in this powerful image, but also the seeds of the flexibility argument that has subsequently so often been associated with them. With 'slab' floors and a frame, plan elements are seen to be

4.4.2 The Dom-ino framework rejected N. Habraken, 1961

4.4.3 House, Chailey, Sussex, England. Walter Segal, 1972 (Photo 1971 during construction)

freed from the tyranny of structural support and can thus be more responsive to requirements. The argument is then extended, as will be seen, to lightweight internal walls, moveable partitions and pneumatic space dividers. Similarly, it is argued, the external skin can then be freed of its structural connotations and so permit a greater freedom: the freedom to tap the stylistic concerns of the time. These were, of course, the skin of glass or wafer-thin concrete with its metal framed horizontal windows. The case for the latter was also made in an eloquent series of drawings (**4.4.4**) showing how the 'new' construction and the advantages of the horizontal view strip went together (Le Corbusier, 1960, 1929). Curiously the diagrams also claim that the uniform, even light of the horizontal window is better than the variety of lighting provided by the more traditional vertically proportioned openings. Yorke's influential book *The Modern House* (1943, 1934) reprints these diagrams from Le Corbusier. The problems of environmental comfort and technical performance that accompanied these ideas when implemented in the twenties and thirties became no less apparent when used by the system builders of later years. It is interesting that the analogue of the Dom-ino House, in the form of this bare structure, has become such a strong source for later action when the many other related concepts also developed by Le Corbusier have been passionately rejected by those involved in harnessing architecture and the machine more closely. The building system designers, in rejecting those parts of the Corbusian aesthetic which they associated with him as a prima donna, unwittingly became prey to the most persuasive part of his philosophy.

Le Corbusier's links with the traditions of residential architecture are revealingly explored by Colin Rowe in his comparison of the villa at Garches and Palladio's Villa Foscari, the Malcontenta. Rowe demonstrates how Le Corbusier has laid the Palladian villa on its side (**4.4.5**):

'In the frame building it is not, as in the solid wall structure, the enclosing walls that are dominant, but the horizontal planes of floor and roof. The quality of partial paralysis, which Le Corbusier has noticed in the plan of the solid wall structure, in the frame building is transferred to the section. Perforation of the floors giving a certain vertical movement of space is possible; but

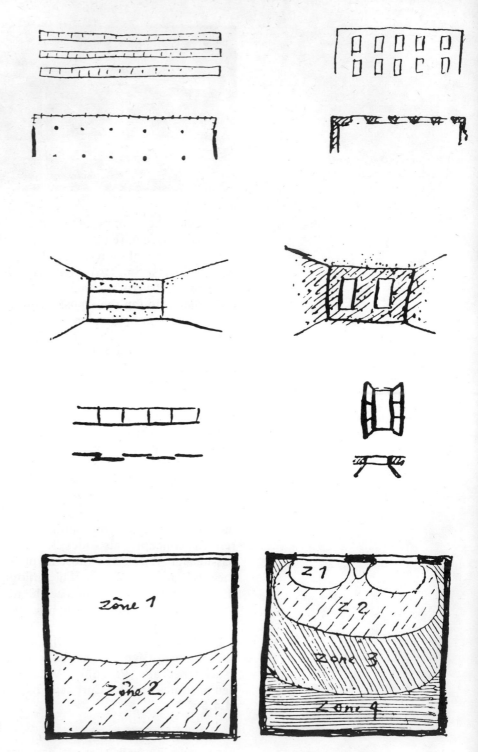

4.4.4 Comparison showing the claimed advantage of the fenêtre en longueur over the traditional hole in the wall window by Le Corbusier

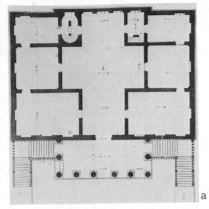

4.4.5 Villa Foscari, Malcontenta. Andrea Palladio, 1559-60. (a) Plan. (b) Elevation on to the Brenta Canal (Photo 1978)

b

4.4.6 'Citrohan' House. Le Corbusier, 1920 (a) Plans. (b *right*) Perspective

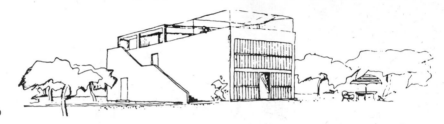

b

a

Coupe

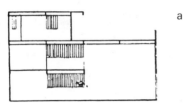

Terrasse

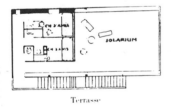

Entresol

Rez-de-chaussée

the sculptural quality of the building as carving has disappeared, and there can be none of Palladio's firm sectional transmutation and modelling of volume. Extension must be horizontal, following the established planes; free section is replaced by free plan, paralysed plan by paralysed section; and the limitations in both cases are equally severe; as though the solid wall structure had been turned on its side, the former complexities of section and subtleties of elevation are now transferred to plan.' (Rowe 1947)

By 1920 Le Corbusier, in the Maison Citrohan (**4.4.6**) had brought together his major arguments for the mass produced house. Banham (1960) has shown how, much earlier, Le Corbusier had erected the first concrete framed villa in Europe, the House at Chaux de Fonds, Switzerland (1916), in which with the frame and roof up, 'it was possible to proceed with building right through the winter' (Banham 1960). The Citrohan house project (a pun on the word Citroen) was a proposal for houses produced on the same basis as cars — a standardized, mass produced studio type house and readily available.

'houses must go up all of a piece, made by machine tools in a factory, assembled as Ford assembles cars, on moving conveyor belts.' (Le Corbusier quoted in Banham 1960)

4.4.7 Drawing: Le Corbusier, *c.* 1920

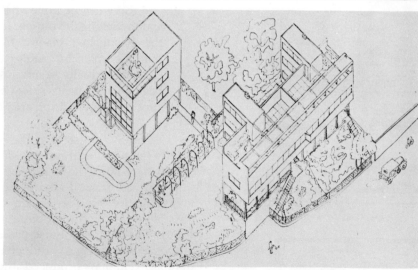

4.4.8 'Two houses for the city of Stüttgart in the Weissenhof Colony'. Le Corbusier, 1927

That Le Corbusier placed a high value on the relation between his painting and his architecture is not in doubt, even to the extent that he divided his day so that he could paint in the morning and design in the afternoon. That he saw the frame structure as a device for allowing

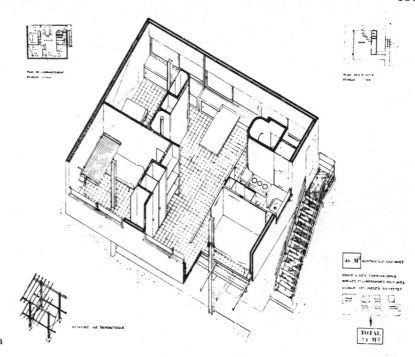

4.4.9 (a) Loucheur House Project. Le Corbusier, 1929. Axonometric of one of the alternatives. (b) Lagny house project. Le Corbusier, *c.* 1959. A later development of the Loucheur House

a

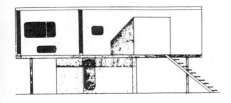

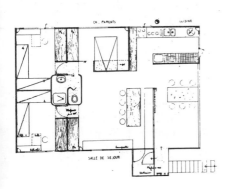

b

him to plan using the formal elements of his painting is more difficult to demonstrate conclusively. It is clear that there is an undeniable relationship between the two (**4.4.7**), the rectangular plan becoming the picture frame for the free disposition of elements. The many threads of the argument can be seen developing through Le Corbusier's later work: Maison Cook (1926), the Weissenhof houses (1927) (**4.4.8**), the culmination of his Citrohan House ideas in the metal Loucheur (1929) and Lagny (1950s) projects (**4.4.9**), and the Villas at Garches and at Poissy (1929 and 1929/31) respectively. The project for Immeubles-Villas (1922) demonstrates an extension of the Citrohan and Dom-ino concepts in a multi-storey apartment building, each double height unit being a self-contained home with a large terrace. This later saw development in the Unités and acted as a model for countless versions of the ideal community slab block.

However, Le Corbusier's concern with industrialization was to be seen at its most strident at Pessac (**4.4.10**). Here, in 1925, he was invited to build a group of houses for the Bordeaux industrialist M. Frugès who, to quote Le Corbusier, put the problems thus:

> 'I am going to enable you to realise your theories in practice — right up to the most extreme consequences — Pessac should be a laboratory'. (Boesiger 1960)

As Boudon (1969) records in his study these houses were modified with all manner of devices from shutters, through pitched roofs to paint. Le Corbusier's symbolization of the relation of industrialization and architecture has thus, over a period of years, been modified by actual industrialized components from the market place. This casts an

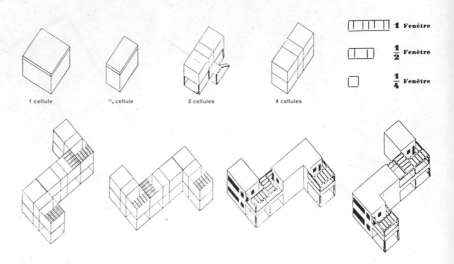

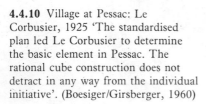

4.4.10 Village at Pessac: Le Corbusier, 1925 'The standardised plan led Le Corbusier to determine the basic element in Pessac. The rational cube construction does not detract in any way from the individual initiative'. (Boesiger/Girsberger, 1960)

ironic aside on the manner in which many later proponents of the machine and architecture marriage have disclaimed association with the harsh determinism of Le Corbusier's buildings and their constructional weaknesses, whilst at the same time drawing on the very symbolic conjunctions presented by them.

Le Corbusier had made a close study of a survey of dwelling types in France published in 1894, and introduced by the founder of the French Institute of Statistics, Alfred de Foville. This study, says Brian Brace Taylor,

> 'betrays a deterministic philosophy of reform which subsequently appears in Le Corbusier's concept of planning: reform the conditions of habitation and you can eventually improve man's moral behaviour.' (Taylor 1972 and 1975)

Taylor also points out that Frederick Winslow Taylor's book, *Principles of Scientific Organisation of Factories,* was published in French in 1912 and that Le Corbusier referred to the concept of Taylorism over the years as being central to his ideas on mass produced housing. Clearly Le Corbusier had been impressed by the moral implications of the views of F.W. Taylor and his attitude to men and machines. Even the fee structure ultimately agreed between Frugès and Le Corbusier for Pessac was brought within the overall framework of ideas. After much discussion a 'modular' system of payment was agreed under which, for every 5 × 5 × 3 metre module built, the architect received 100 francs (Taylor 1977).

One building often overlooked is the Clarté apartment building in Geneva (1930/32) (**4.4.11**), perhaps because it contained little of the spatial gymnastics of his other buildings or projects. This nine storey block nevertheless has a number of interesting features, not least its use of a steel frame:

> 'constructed entirely of standard elements, upon a frame of

4.4.11 'Clarté' apartment house, Geneva, Switzerland. Le Corbusier, 1930-2. 45 apartments with double floor heights

standard steel sections electrically welded and conforming to a strict module of columns, beams and windows.' (Boesiger 1960)

Le Corbusier's writings have been as or more influential than his buildings and projects, and his book *Vers une Architecture,* first published in Paris in 1923, sets down his views on industrialization in a cross between a piece of concrete poetry and a mathematical proof:

MASS-PRODUCTION HOUSES

'A great epoch has begun
There exists a new spirit

Industry, overwhelming us like a flood which rolls on towards its destined ends, has furnished us with new tools adapted to this new epoch, animated by the new spirit.

Economic law inevitably governs our acts and our thoughts.

The problem of the house is a problem of the epoch. The equilibrium of society today depends upon it. Architecture has for its first duty, in this period of renewal, that of bringing about a revision of values, a revision of the constituent elements of the house.

Mass-production is based on analysis and experiment.

Industry on the grand scale must occupy itself with building and establish the elements of the house on a mass-production basis.

We must create the mass-production spirit.
The spirit of constructing mass-production houses.
The spirit of living in mass-production houses.
The spirit of conceiving mass-production houses.

If we eliminate from our hearts and minds all dead concepts in regard to the house, and look at the question from a critical and objective point of view, we shall arrive at the "House-Machine", the mass-production house, healthy (and morally so too) and beautiful in the same way that the working tools and instruments which accompany our existence are beautiful.

Beautiful also with all the animation that the artist's sensibility can add to severe and pure functioning elements.' (Le Corbusier 1927, 1923)

There the argument is set down, a set of mental associations made, the effects of which are still being worked out in all parts of the world where industrialization has appeared. Le Corbusier, of course moved on, leaving behind such simplistic notions and producing buildings such as the weekend house near Paris of 1935 and the Maison Jaoul at Neuilly in 1954/56, which showed an organic and subtle use of materials which startled his followers. Nevertheless his mass production theorem has remained a powerful influence: the new epoch; a new spirit; industry and the inevitability of economic law; housing; mass production; the rejection of tradition; the 'House Machine'; physical and moral health. All this would, apparently, give rise to a beauty comparable to that of a 'working tool' although, as the playwright Bertolt Brecht was fond of pointing out, such beauty resided in the interaction between that tool and the user over a long period.

The iconography created by Le Corbusier carried a written rationale and a visual coherence which brought together many developing threads. From the seeming chaos and turmoil of the industrial revolution a set of rules had been created which architects seized upon. As Yorke was later to say in his influential *The Modern House*:

'There came a period of purification and, largely under the influence of Walter Gropius and Le Corbusier, the unnecessary was eliminated.' (Yorke 1943, 1934)

This simplification became the central cry of the modern movement and of its system builders. They may have turned the icon to face the wall but its power nevertheless became permanent.

'Crystal Symbol of a New Faith'

If Le Corbusier was one of the priests of the new iconography of the machine, the Bauhaus was its institutional model. With its general history well documented it is here intended to examine the approach to standardization in building developed during its existence, since this has been a powerful force in shaping subsequent approaches to the development of building systems.

The fecundity of the Bauhaus coupled with the effect of its closure in 1933 led to its teachers, pupils and ideas spreading throughout the western world, its approach to design and design teaching thus becoming a powerful force in schools of art and architecture. We have now arrived at a position where these ideas can be seen embodied in buildings in most towns and cities of western Europe, eastern Europe and the Americas. The potency of the myth has been such that the particular rationale developed in support of the approach and its artefacts has been so firmly embedded in architectural thinking that to question its basis has often been seen to be heretical. Since much of the received Bauhaus argument has rested upon the right use of the machine, those involved in designing, developing and using building systems have been heavily reliant on its particular mythologies. This section will look at the basis for some of these in written, drawn and built forms.

The Bauhaus was first established in Weimar in 1919, moving to Dessau in 1925 in a new building designed by Gropius and Adolf Meyer, and, after closure in Dessau moved to Berlin where it did not survive beyond 1933. Although Gropius was an architect, and architects taught at the Bauhaus, there were only limited foundation courses in architecture until Hannes Meyer took over in 1928. This in spite of the original programme of 1919 which claimed that 'the final goal is building'.

The cover of that programme carried a woodcut by Lyonel Feininger (**4.4.12**) of a cathedral and claimed that

> 'architects, painters and sculptors must once again learn to know and understand the multiform shape of buildings in their totality and in their parts.' (Pevsner 1968)

and it called upon architects and artists to unite with the craftsmen.

The mediaeval cathedral buildings had called their headquarters 'Bauhutte' and it was from this that the name Bauhaus was adapted. Gropius' manifesto circulated in 1919 set the scene:

> 'Let us create a new guild of craftsmen, without the class distinctions which raise an arrogant barrier between craftsmen and artist. Together let us conceive and create the new building of the future, which will embrace architecture and sculpture and painting in one unity, and which will rise one day toward heaven from the hands of a million workers like the crystal symbol of a new faith.' (Royal Academy of Arts and German Federal Republic 1968)

Here was another call for total architecture — Gropius's book published much later in 1955 was entitled *The Scope of Total Architecture*. The use of the Cathedral was indeed symbolic of the whole Bauhaus approach. The concern with unity, wholeness and totality was at that time a powerful force in the Weimer Republic (Gay 1968) and the tendency inherent in the idea of the relation of the parts and the whole in a unity is, in the strict sense of the word, a

4.4.12 Bauhaus programme, cover woodcut Lyonel Feininger, 1919

religious one. Certainly, if as the dictionary tells us, a religion means to enter into a monastic condition and the practice of sacred rights, then the Bauhaus contained both. Further the concept of a religion involves the recognition of a superhuman controlling power and involves actions that one is bound to make in accordance with this recognition. Logic, science, technology and the necessities of mechanization were thus called upon in the cause of total architecture. The most precise statement of this occurs in Gropius' book *The New Architecture and the Bauhaus* first published in 1935 in which it is stated:

> 'It is now becoming widely recognised that although the outward forms of the new architecture differ fundamentally in an organic sense from those of the old, they are not the personal whims of a handful of architects avid for innovation at all cost, but simply the inevitable, logical product of the intellectual, social and technical condition of our age.' (Gropius 1965, 1935)

Describing the processes and products as the rational and only outcome of the historical process was in the great tradition of marxist determinism: the inexorable march of history giving rise to those contradictions which could only be resolved by the acceptance of the inevitable logic of the situation. 'The 'inevitable logic' on this occasion demanded that the dictates of industrialization be understood and used by designers rather than ignored. However, this use was of a particular nature, with its formal roots embedded deeply in the art and handicraft traditions of the time. At the Bauhaus precisely handcrafted artefacts were produced which often paid lip service to mass production techniques and the popular imagery that such methods employed to gain a mass sale. That both Le Corbusier and the de Stijl movement had a considerable influence on the Bauhaus is not in doubt and Pevsner suggests that the move from the early expressionist forms to those of a more rectilinear and primary nature was due in part to Theo van Doesburg's presence at Weimar in 1921 and 1922 (Royal Academy of Arts and German Federal Republic 1968).

Van Doesburg visited the Bauhaus as early as January, 1921, but although there was talk of a Professorship he was never appointed. However, the manner of his influence may be seen expressed in his letter to Anthony Kok.

> 'Every evening I have talked to the students and spread the vermin of the new spirit. Within a short time De Stijl will reappear in a more radical way.' (Baljeu 1974)

Certainly the series of models produced by students on the 'De Stijl course' at Weimar in 1922 have a relationship with Gropius' work on standardization and Baljeu claims them to be more plastic and organic (**4.4.13**). Van Doesburg had edited the first issue of the magazine *De Stijl* in 1917, and his early paintings make use of rectangles, straight lines and primary colours. His 'Composition in discords' of 1918

4.4.13 'A study of a purely architectonic sculpture, based on its ground plan. Model at Weimar by Bernard Sturzkopf, student of architecture in Van Doesburg's De Stijl course at Weimar, 1922' (Baljeu 1974)

consists of an irregular grid of black lines laid over the rectangular shapes in red, pink, blue and green, a theme developed by Mondrian and one that was to be recurrent in the curtain walling of later years

Kandinsky's arrival in 1922 gave a further impetus to the Constructivists and their interest in primary forms, around which had been developed an argument relating this to the new social order together with the beauty and dignity of labour and its processes. In view of the striking similarity between the Bauhaus programme and those of the Moscow Vkhutemas it is somewhat surprising that the former has gained the recognition that should, perhaps, have gone to the latter. Certainly there is very little mention of the Constructivists in all the Bauhaus literature. Gropius and others at the Bauhaus have since been attacked for a type of architectural reductionism in that they employed the rhetoric of the apparent necessities of industrialization to support a set of limited stylistic objelctives. Certainly this reduction is the reading made by building systems designers although, of course, many of them have been ignorant of the source of their authority. However, Gropius himself makes it quite clear in *The New Architecture and the Bauhaus* that he holds a more fundamentalist belief:

> 'That is why the movement must be purged from within if its original aims are to be saved from the strait jacket of materialism and fake slogans inspired by plagiarism or misconception. Catch phrases like "functionalism" and "fitness for purpose = beauty" have had the effect of deflecting appreciation of the New Architecture into external channels or making it purely one sided. This is reflected in a very general ignorance that impels superficial minds, who do not perceive that the new architecture is a bridge uniting opposite poles of thought, to relegate it to a single circumscribed province of design.' (Gropius 1965, 1935)

Again we see the concern to create a unity from two opposites: that synthesis of the proposition and its contradiction put forward by Hegel and developed by Marx is here being reworked in architectural terms. The general systems theorists of later date appear to be restating a similar set of ideas, albeit not confined to building. Gropius saw the building itself as 'the final goal' and thus his synthesis, the whole, was to be contained in the form of the building. His avowed concern with the processes of its design and construction support this. The systems theorists on the other hand see the building as part of a larger synthesis in turn. The building cannot, or should not, be a whole in itself since this is to define it in such a way as to downgrade other parts of the synthesis. If the totality of the building itself is the aim, then the people who use it, the specific locale and environment, become relegated to a secondary status. Thus Gropius, in placing the emphasis on the building as an artefact, and describing the 'inevitable logic' of how it should be achieved, opened the way for 'the new architecture'. Accurately named the 'International Style' it had set aside the normal concern of the architect for people and their

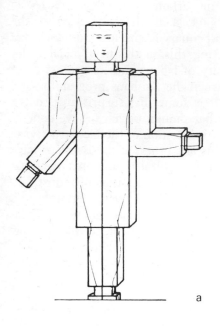

a

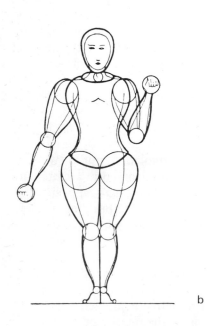

b

4.4.14 Drawings for costumes. Oskar
Schlemmer, 1924. (a) Walking
architecture; (b) the jointed doll; (c) a
technical organism; (d) demateriali-
zation

differences, and for places and their differences and substituted the
idealizations of machine technology.

In this way Gropius had set about reworking the architect's role in
terms of a new set of imperatives. To do this it was necessary to
discredit what had occurred previously and to reorder or repattern the
understanding that architects and others had of architecture. It was
then necessary to reformulate the argument around objectives seen as
relevant by the new participants. The physical establishment of a
school of design, the Bauhaus, identified one potential; the bringing
together of a seemingly disparate group of people to work together
identified another. Gropius, with these potentials, identified a pattern
or structure of ideas, processes and products that created a sense of
order to the observer. Since he was the first director and selected the
first staff members he was in a prime position to reduce apparent
randomness. Couple all this with skilful administration and good
publicity and one has the makings of a myth.

One example of the power generated by such a myth can be seen in
the way in which the work of Oskar Schlemmer has fared (**4.4.14**). By
several accounts his theatre was most powerful and influential but due
both to its tenuous relation to the apparent main concerns of the
Bauhaus (rationalization for example), and its inherent ephemerality,
Schlemmer has never featured largely in accounts of key Bauhaus
work and influence. However, it was Schlemmer who, by 1922, had
already discarded the craft approach:

> 'Away with medievalism, then, and the medieval concept of
> handicrafts and ultimately with handicrafts themselves, as mere
> training and means for the purposes of form.' (in Pehnt 1973)

Itten's famous basic design course may be cited as another example,
although in this case the problem is somewhat different. Whilst basic
design courses have subsequently proliferated around the world, the
fact that for Itten and his method of teaching it was closely bound up
with his mysticism has largely been set aside as irrelevant. The search
for wholeness, the unity, the totality, Itten's mysticism, Feininger's
Cathedral woodcut must all be seen as key parts of the creation of the
Bauhaus myth. The writing of the time says little of such things and
almost always emphasizes the drive to rationalization. This in itself is
worth examination.

The meaning of the word rationalize is to explain, explain away
or to bring into conformity with reason. Gropius' view on rationalization is in this present context an intriguing one. After the above
quoted passage, where he saw the New Architecture as 'a bridge
uniting opposite poles of thought' he continued:

> 'For instance rationalisation, which many people imagine to be
> its cardinal principle is really only its purifying agency.' (Gropius
> 1965, 1935)

What aspects of architecture were thus being ceremoniously cleansed by Gropius?

> 'The liberation of architecture from a welter of ornament, the emphasis on its structural functions, and the concentration of concise and economical solutions, represent the purely material side of that formalising process on which the practical value of the New Architecture depends. The other, aesthetic satisfaction of the human soul is just as important as the material. Both find their counterpart in that unity which is life itself.'

In the last sentence of the passage he makes an important distinction:

> 'For whereas building is merely a matter of methods and materials, architecture implies the mastery of space.' (Gropius 1965, 1935)

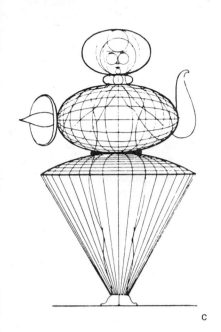

Thus all is made subservient to the architectural organization of space: there is here a clear implication of that division between building and architecture so eloquently expressed by Pevsner some years later:

> 'A bicycle shed is a building: Lincoln cathedral is a piece of architecture.' (Pevsner 1951, 1943)

In retrospect it is a very curious reworking of the traditional image of the architect: on the one hand a wish to see 'good design' permeate industrialized production and reach the mass of the people; on the other retention of the traditional role of the architect in the making of decisions about such design and about the organization of space. The 'good design' turned out to be a rather particular set of rationalizations which, as has been seen, were to eliminate ornament and concentrate on 'concise and economical solutions'. It is worth looking at what some of these solutions were, since they have considerable bearing on what was to happen later.

   The new materials of steel, concrete and glass were seen to be superseding traditional methods of construction and giving rise to new methods: lighter and longer spanning structures of great transparency were now possible. The technological imperative in Gropius' argument called for new solutions just as it had for the Constructivists and just as it does for the system builders and others. An appeal is made to an integrity where the means are to be seen to precisely match the ends. A moral imperative is thus introduced which insists that the building match the polemic. This fusion of ideas is one of considerable significance, and through it the Bauhaus philosophy became all pervasive. Gropius describes the new approach to be used in these terms:

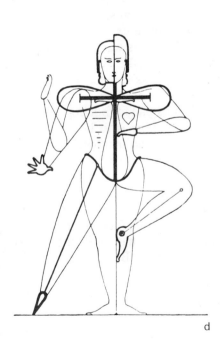

> 'One of the outstanding achievements of the new constructional technique has been the abolition of the separating function of the wall. Instead of making the walls the element of support, as in a

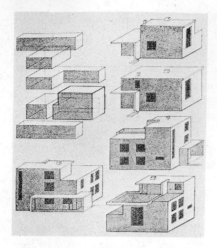

4.4.15 Serial houses. Walter Gropius, 1921

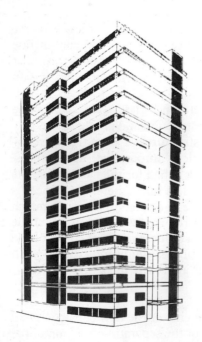

4.4.16 Apartment building. Georg Muche, 1924. Project at Bauhaus

brick built house, our new space-saving construction transfers the whole load of the structure to a steel or concrete framework. Thus the role of the walls becomes restricted to that of mere screens stretched between the upright columns of this framework to keep our the rain, cold and noise.' (Gropius 1965, 1935)

Having described how new structural methods will reduce the area occupied by the structure he goes on:

'It is, therefore, only logical that the old type of window — a hole that had to be hollowed out of the full thickness of a supporting wall — should be giving place more and more to the continuous horizontal casement, subdivided by thin steel mullions, characteristic of the New Architecture.' (Gropius 1965, 1935)

Neither does the roof escape scrutiny:

'In the same way the flat roof is superseding the old penthouse roof with its tiled or slated gables.' (Gropius 1965, 1935)

The advantages of the flat roof are then listed which, in view of the difficulties still being experienced by proponents of flat roofs everywhere, suggest that Gropius should have spelt out their disadvantages as well. The development of air transport is then cited as a good reason for cleaning up the design of roofs, another aspect of visual hygiene that was not to be ignored. In this way the delineation of a style can be seen quite clearly as can the relation with Le Corbusier's concept of the new architecture, as seen in the Dom-ino House design of 1914.

The idea of the light structural frame on its pad foundations, with its platform floors and roof and light dry envelope is to be a recurring theme both at a conceptual level and at an operational level. For many system builders its imperative has been a powerful one, defeating almost all others in most building types other than housing. Even there, however, as we shall see, it was repeatedly tried and frequently failed. Such a method of construction has, of course, many of the advantages claimed for it: it also has many disadvantages. It is for the designer to select from available technologies that most appropriate for the job in hand. The pervasive effects of the Bauhaus arguments, however, made this increasingly difficult. Architects, and ultimately many others connected with building, began to apply the 'inevitable logic' in all situations. The system builders, particularly, made the connection between society, the machine and style in a direct manner: their idealization of the machine encouraged an acceptance of the Bauhaus style as the only valid approach.

A series of projects at the Bauhaus examined the implications of rationalization and standardization. Gropius' serial houses of 1921 claimed the 'combining of maximum standardization with maximum variability' (Royal Academy of Arts 1968). The often illustrated diagram (**4.4.15**) shows the six 'box' elements and some of the

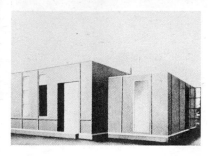

4.4.17 Home with flexible plan and expansion potential for Carl Castner Corp, Leipzig. Erected for inauguration of Bauhaus building. Georg Muche, 1926. Concrete foundation, steel structure, 3 mm siemens steel plates 'wedged in', insulation, air space sealed by torfoleum plates and plaster slag boarding

configurations which apparently result, although it is somewhat difficult to establish to what level the standardization has been pursued since windows and canopies appear to contravene the box perimeter. The project for an apartment building by Georg Muche (1924) consisted of a steel frame with prefabricated concrete slabs (**4.4.16**) and embodies the qualities of the multitude of high rise blocks which subsequently sprang up in cities.

The prototype steel house (**4.4.17**), erected at Torten near Dessau in 1926 for the inauguration of the Bauhaus building, had 'a flexible ground plan with expansion potential. Steel structure on concrete foundation, 3 mm siemens steel plates wedged in, insulation: air space sealed off by torfoleum plates and plaster slag boarding' (Royal Academy of Arts 1968). This project followed the metal house of 1925 and preceded Breuer's Bamboo house project of 1927 (**4.4.18**) which contained moveable internal walls and was constructed of standardized asbestos concrete slabs for external and internal walls on a steel frame. Breuer's drawing of 1924 (**4.4.19**) for a modular furniture system on a module of 33 cm is also of interest here.

Gropius' development at Torten was based on the serial house idea (**4.4.20**) and is a rationalized version of a cross wall construction system, whereas the 'Kleinhaus' of 1927 (**4.4.21**) shows variations on a square ground plan. These, like Gropius' masters houses (**4.4.22**) at Dessau (1925/26), seem to owe more to the machine aesthetic of Le Corbusier than to the realities of mass production. On the other hand Hilberseimer's study of growth and standardization in his project of 1930/31 (**4.4.23**) appears to be closer to the realities of the housing

4.4.18 Bamboo Houses, type 1. Marcel Breuer, 1927. Open plan with moveable walls. Steel frame and standardized asbestos concrete slabs as cladding and partitions

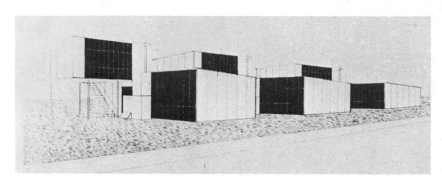

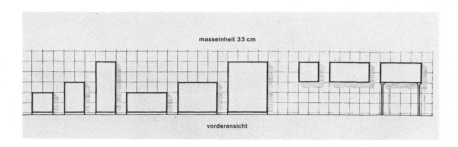

4.4.19 Modular box system for furniture. Marcel Breuer, 1924

4.4.20 Serial houses for Törten, Dessau, 1926. Model showing pre-fabricated components

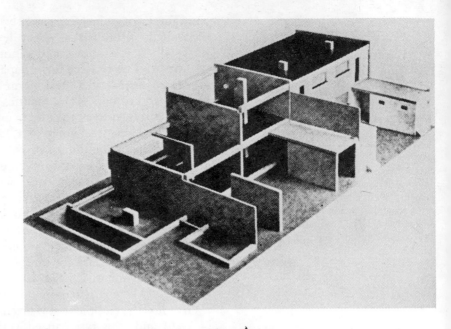

4.4.21 (a) Kleinhaus. Walter Gropius, 1927. 6 variations of a square ground plan

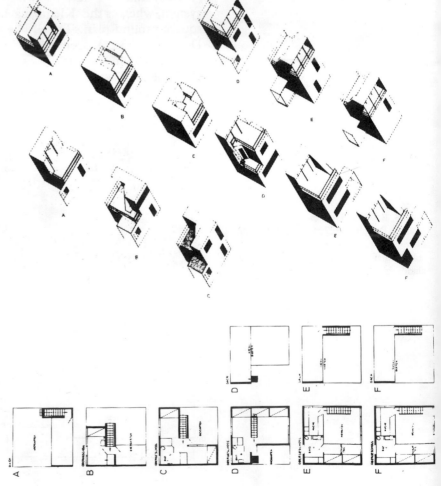

a

4.4.21 (b) Experimental house, Weissenhofsiedlung, Stüttgart. Walter Gropius, 1927. (c) Experimental house, Weissenhofsiedlung. Vertical plan and section.

b

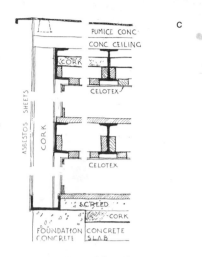

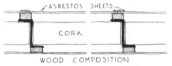

c

4.4.22 Masters' houses, Dessau. Walter Gropius, 1925-6

market, and indeed is similar to the approach adopted in more recent years by the Levitt Brothers in the USA, for their 'tract' or speculative housing.

Gropius' ideas were put into practice with the Weissenhof prefabricated housing at the Stuttgart Werkbund Exhibition in 1927, and in a different form with his copperplate houses of 1931 at Finow. These ideas can also be seen in the house developed with Wachsmann in the United States, in 1942, for the General Panel Corporation: a timber system with an ingenious three-dimensional joint which could be used as panel/floor or panel/ceiling (**4.4.24**), and four basic panels (Herrey 1943). The argument underlying all this was set out quite clearly at the time by Gropius in Bauhaus book No. 3:

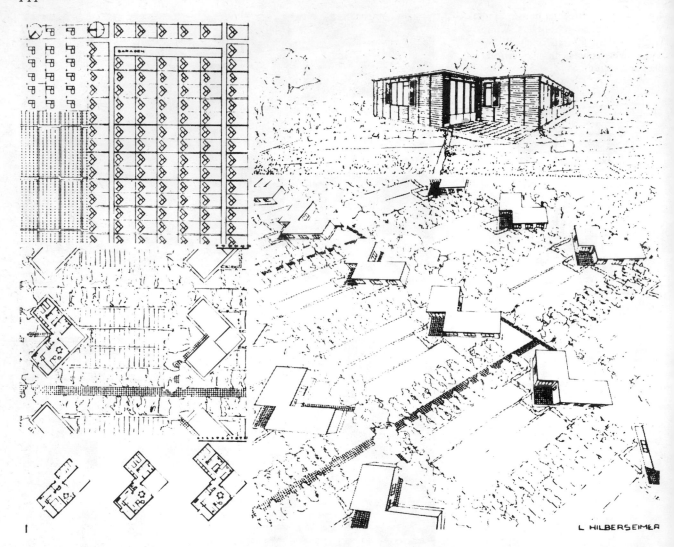

4.4.23 Plan for Dessau. Ludwig
Hilberseimer, 1932

'From the artistic point of view the new method of building has
to be accepted. Standardisation of the building elements will
result in new housing units and sections of cities having a uniform
character. There is no danger of monotony, for if the basic
requirement is fulfilled that only the building units are standardised the structures built thereof will vary. Their "beauty" will
be assured by properly used material and clear simple construction. It will largely depend on the creative ability of the
architect to what extent the arrangement of the "giant building
blocks" will form well designed space in these structures. Certainly
the standardisation of parts will not limit individual design
(gestaltung) which we are all striving to achieve... there is
enough room for the characteristics of the individual and the
nation to express themselves and yet everything bears the mark
of our time.' (Royal Academy of Arts 1968)

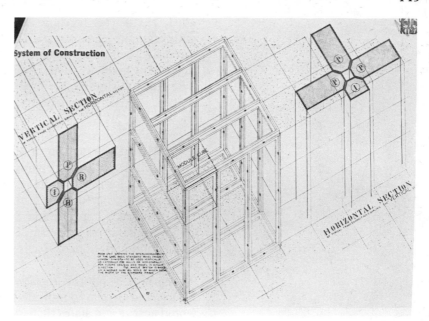

4.4.24 House for General Panel Corporation. Gropius and Wachsmann, 1942

In spite of this both Le Corbusier, and Gropius during the Bauhaus period, virtually excluded environmental comfort and services from their call for new attitudes to technology as has been well documented by Banham (1969):

> 'For Gropius and the Bauhaus connection, lamps and heaters alike seem to have been simply sculptural objects, to be composed according to their aesthetic rules, along with the solids and voids of the structure, into abstract compositions.' (Banham 1969)

Amongst other examples Banham draws attention to Gropius' introduction of a naturally lit glazed ceiling in the Dessau Employment Exchange (1927) which contrasts with the curious pendant artificial light fittings (**4.4.25**) almost appearing as an afterthought. Aptly summarized by Banham

> 'It all suggests that the "New Synthesis of Art and Technology" which Gropius was preaching at the time could contain oversights that left it less than the total discipline of design with which his followers have credited him.' (Banham 1969)

The 'plan' for the Bauhaus was, however, set out quite clearly by Gropius long before, in 1910, in his programme for AEG, first published by the Architectural Review in 1961. In this Gropius, aged twenty six, had stated what he meant by a purifying rationalism. In March, 1910 whilst he was still with Peter Behrens he wrote a memorandum to AEG, the electrical combine, outlining his ideas and entitled 'Programme for the Establishment of a Company for the Provision of Housing on Aesthetically Consistent Principles' (Gropius

4.4.25 Employment exchange, Dessau, Germany. Walter Gropius, 1927

1961, 1910). Having drawn attention to the opposed activities of the entrepreneur in housing and the architect who, says Gropius, 'is interested in raising the cost of a job' he cites the attention to detail that can be given by the architect if he were to utilize mass production. The cultural significance of this, he goes on, lays in 'the concept of a Zeitstil' — the manner or style of the time. Such a style for the age shows:

> 'Methods based on craftsmanship are antiquated and must be replaced by the acceptance of a modern concept of industry. The search for the odd, the wish to be different from one's neighbour, makes unity of style impossible. . . . Our age, after a sad inter-regnum is approaching Zeitstil which will honour traditions but fight false romanticism. Objectivity and reliability are once more gaining ground.' (Gropius 1961, 1910)

Gropius calls on the traditions of repetitive housing in Holland, France and Britain to indicate the economic and aesthetic benefits that would accrue and states that:

> 'The new Company intends to offer its clients not only inexpensive, well built and practical houses and in addition a guarantee of good taste, but also takes into consideration individual wishes without sacrificing to them the principle of industrial consistency. (Gropius 1961, 1910)

Gropius then outlines what has since become the 'ideal' model for the building systems designers. He described how the client will select his house from a range of materials and parts which will all be dimensionally co-ordinated. For each item there will be a range to choose from but to the same sizes and everything necessary for the house will be 'put down and catalogued as variants'. It is emphasized that 'all parts fit exactly, as they are made by machine to the same standard dimensions. For the same reasons they are interchangeable'. Furthermore a case is emphatically made for the standardization of dimensions as a prerequisite and the implications of this on advance ordering and stockpiling are not ignored:

> 'Contracts with suitable specialist manufacturers secure that all objects and parts satisfy the standards laid down by the company and are, if possible, permanently in stock.' (Gropius 1961, 1910)

Gropius completes his ambitious programme for AEG by outlining the way in which whole sites will be designed, and the organization needed to design and sell the idea. There would, he says, be an art department which would be quite separate from the commercial department and the company should have a wide range of publicity and information including travelling exhibitions and 'public lectures with lantern slides'. The programme concludes with comments on publicity, which, says Gropius, is 'of the greatest importance to the enterprise. For in contrast to other building enterprises overheads hardly change when turnover increases, and so every order taken after overheads have been paid for can be regarded almost entirely as profit.' (Gropius 1961, 1910)

It is clear that Gropius sustained the ideas contained in the AEG document over a long period: the Bauhaus itself can almost be seen as the 'Company' which is here proposed, since it projected the coherent image or 'house style' by means of its very well managed publicity. This has firmly equated several keywords into one interlinked concept: mechanization, standardization, dimensional co-ordination, mass production, efficiency, low cost, working class housing. In this way the 'Bauhaus Company' has been a profound model for institutions faced with building problems: its strength has been all pervasive until recent years to the point where no self-respecting housing agency has not espoused much of its doctrine. That it has seldom been as effective in practice has not daunted those involved, since repeated attempts have, and are being made to insist that it works. Having established the validity and working of the mass production idea for housing, demonstrated the importance of co-ordinated dimensions and catalogue availability, Gropius then outlined another factor that was to become crucial — this concerns the independence of the house from its site. He mentioned in passing the importance of climatic and historical considerations but manages to convert these into a proposal for universal solutions:

> 'The houses as designed are independent, coherent organisms not tied to any site, devised to fit the needs of modern civilised man in any country, not even only Germany.' (Gropius 1961, 1910)

The support for this argument is the increasing ease of transportation which, in some obscure way, gives authority for ignoring the constraints of the specific locale. It is clear that the creation of norms in the Gropius' sense is in opposition to the response of a building to the place that it occupies on the surface of the earth. The seed of another part of the myth of industrialization is also laid in the references to working class housing:

> 'The advantages of the principle of industrial production are particularly obvious in working class housing.' (Gropius 1961, 1910)

It is no accident then that many public authorities the world over associate mass housing and industrialization as one concept. The ground has been well prepared by Gropius and others. The late 1920s for example, saw one of the earliest experiments with a precast concrete prefabricated system on a large scale, the Massivblock system. This occurred in the new suburbs of Frankfurt with

Praunheim, Romerstadt and Westhausen completed in 1930 under the direction of Ernst May who had been made Director of Municipal Construction. Although May had worked for Raymond Unwin in Britain his two and three storey terrace houses and apartments could not have been further from the individual homes of the garden city. Of small floor area they introduced open planning, central heating and roof drying areas. Pawley (1971a) points out that this work had an enormous influence on architects working on mass housing, and that versions of it were subsequently used in the Soviet Union, where May went when the Nazis took power in Germany in 1933. Whether it be in France, Peru or in Operation Breakthrough in the USA, when the question of gaining political support for housing those on lower incomes becomes an issue, building systems are invariably proposed as the solution.

Hannes Meyer, who joined the Bauhaus in 1927, becoming its head in 1928 upon the departure of Gropius, also had some experience of mass housing. He also departed for the Soviet Union, in 1930, having been dismissed from the Dessau Bauhaus without notice, it being claimed that he had introduced marxism. In 1919 Meyer had been asked to design a co-operative community in Switzerland at Freidorf, near Basle. He describes how, influenced by his spare time redrawing of all Palladio's villas onto A sized paper, he had designed the estate

> 'on the modular system of the architectural order. By means of this system all the external spaces (squares, streets, gardens) and all the public internal spaces (schools, restaurants, shops, meeting rooms) were laid out in an artistic pattern which would be perceived by those living there as the spatial harmony of proportion'. (Schnaidt 1965)

However, by 1926 Meyer had developed a very specific attitude to the influence of art, but one that was in line with the rigours of planning at Freidorf. He saw the straight lines emanating from the social and economic 'fields of force' transforming the values of the day and with it the external world. This revolution called for

> 'a change in our media of expression. Today is ousting yesterday in material, form and tools', and
> 'The art of felt imitation is in the process of being dismantled. Art is becoming invention and controlled reality.' (Schnaidt 1965)

At the Bauhaus Meyer set out a very clear set of aims and a programme which explicitly drew in scientific study of materials as well as the social sciences. His strongly utilitarian attitude is clearly seen in his manifesto of 1928 *Building* in which he develops ideas laid down in 1926.

> 'All things in the world are a product of the formula: function times economics...

Architecture as an embodiment of the artists' emotion has no justification. . .

Architecture as continuing the building tradition means being carried on the tide of building history. . .

Building is only organisation: social, technical, economic, psychological organisation.' (Schnaidt 1965)

This rigorous utilitarianism is probably the most thorough of all those espousing the functionalist cause and perhaps because of this it embodies the contradictions of the position most clearly. On the one hand there is the view that architecture will arise naturally from the social and technical conditions of the age and on the other the fact that all the expressions of this were the creations of individuals who actively sought to express a particular ideology — in fact to create artefacts that would symbolize this new world order.

Writing in *My Dismissal from the Bauhaus, 1930,* Meyer drew attention to the position at the Bauhaus upon his appointment:

'A university of design which made the shape of every tea-glass a problem in constructivist aesthetics. A cathedral of socialism in which a mediaeval cult was practised.' (Schnaidt 1965)

The ideal machine aesthetic which ultimately became associated with the Bauhaus is, as Pehnt has eloquently shown, deeply rooted in German Expressionism. Citing the mystic concern with number ratios and the arts and crafts nature of much of the later Bauhaus work, Pehnt suggests that this 'may be what prompted Hannes Meyer, one of the few consistent radical functionalists to call Dessau a second Dornach' (Pehnt 1973). Dornach was (and is) the centre of Rudolf Steiner's Anthroposophical Society, and where those forceful examples of German Expressionism are to be found — the Goetheanum.

The distance between the organic convolutions of the latter and the rigours of the Bauhaus machine aesthetic is very short indeed. Both have their sources in the same ideological and religious resurgence of the Germany of the late nineteenth century.

The Mechanic in the Garden

Both the propositions of Le Corbusier and those of the Bauhaus were much concerned with the imposition of prevailing avant garde art values on both the process and the product. However Jean Prouvé, the French designer, came from a background firmly based in the process: that of the process of metal fabrication. This partly accounts for the ambivalent place occupied by him in the development of industrialized approaches to building. From a business concerned with architectural metalwork Prouvé developed a practice as a designer and fabricator of whole buildings or parts of buildings. He saw himself as very much a part of the movement to bring the techniques of the factory into the building industry, emphasizing that his approach was a pragmatic one based largely on working directly with the materials themselves and with the workers involved in the manufacturing process. The three 'principles' which, he claims, 'led [him] to become unconditionally part of my own era' are:

1. 'Following the evolution of science which governs the development of techniques, and partly by accumulating information and studying materials and their treatment.'
2. 'Watching work in operation. Further, by seeking inspiration, and discovering the options available, through the practice of advanced techniques.'
3. 'By never postponing decisions, so as neither to lose impetus nor to indulge in unrealistic forecasts.' (Huber and Steinegger 1971)

A further reflection of the practical orientation of his early experiences can be seen in his views on 'utopian projects' which, he stated, one should not indulge in since evolution can only result from practical experience. He even went so far as to state that studies independent of practice should be forbidden. He reinforced this by the view that 'all that is extraneous seldom conforms to requirements and leads to loss of time' (Huber and Steinegger 1971).

While the reverence for the practical can be admired, the manner in which it is put suggests too strongly the acceptance of the ethic of the production line: time is money, theory is an irrelevant luxury, work or thought which is not seen to be immediately relevant is valueless. In spite of all this, Prouvé, in his arid way, embodies a utopia more firmly than such seemingly down to earth language might suggest. The opening statement 'I was led by them (his early experiences) to become unconditionally part of my own era' clearly suggests that he was not the merely practical man suggested by his rejection of 'utopias'. Like many practical men he seems, somehow, by not recognizing the forces on him, to have been a victim of a very particular utopia. This, I believe, can be seen in his work and career, where he saw the building process and indeed architecture through the window of his training as a metalworker. He could produce technically interesting solutions to curtain walls, windows and entrance porches which carried both his expertise as a designer/fabricator and a certain view about the right use of these processes applied to building. On occasions this came off but frequently, as Banham points out, the solutions remain as technical essays:

'The value of the fresh mind and the high technical standards that a technician like Prouvé can bring to architecture is evident, but it rests with architects to so phrase their demands upon him that a thin bent detail will become a slender curved one.' (Banham 1962a)

Banham's choice of words here demonstrates the difference, like it or not, between architecture and mere technical virtuosity. Further evidence of Prouvé's claims can be seen in the slogans at the opening of each chapter in the book and place him securely in the utopian machine tradition:

'Note that the most highly industrialised objects — on wheels, in flight or fixed on the ground — are most subject to renewal and constantly improve in quality, even in terms of process. Building is the only industry that does not advance'

and

'Every object except a building is made by a single organic entity, a single industry equivalent to one firm'

and

'Are our towns, our schools, our public buildings and our houses worthy of our mechanical and atomic age?'

also

'The individual dwelling must be light and dynamic, which is an expression of large-scale production and therefore characteristic of industries.' (Huber and Steinegger 1971)

Such familiar arguments have been wielded again and again by the machine utopians in their efforts to refashion architectural form and have become some of the basic arguments for the building system inventors. These arguments can be seen operating in two ways: first in the process of industrialization which, in building, has seen a gradual increase in the use of preformed components, subsystems or whole buildings. Thus timber windows are first made on site, then off site, then to standard sizes on batch process; similarly with staircases and many other components. Even Prouvé's ideal, that of the whole building manufactured by one firm is standard practice in the US mobile home or module home market, and was already well established in the inter-war years. Second the argument operates at what might be called the level of architectural hypotheses. Here a building, a project, or an idea is built or put forward so that the way it is organized presents a hypothesis for the possible future course of architecture. It is, in Boulding's terms (Boulding 1956) an image of the world. It posits a set of values or acts as a model suitable for debate or emulation.

There is a constant dialogue between these two approaches — between what is actually happening and what someone thinks is happening, or would like to happen. An abundance of such models enriches architecture: a reduction of models in currency at any one time is an impoverishment. The dominance of the machine model over the past fifty years, whilst it has offered some exciting results has also carried with it a limiting rhetoric. This seems to have encouraged

4.4.26 La maison du Peuple, Clichy, Paris. Consultant Jean Prouvé, architects Beaudouin, Lods and Bodiansky, 1938-9. (Photo 1980)

many architects to accept many of the imperatives of the machine that were only transient: for example that the use of the machine must mean mass production, must mean standardization. Such notions are already part of the first machine age.

A few examples of Prouvé's work will indicate the sophistication of his approach to metal fabrication. The Maison du Peuple built at Clichy in 1938/39 (**4.4.26**) is of interest for the manner in which it uses sheet steel and for its use of moveable partitions. Since the building was designed by Beaudouin, Lods and Bodiansky with Prouvé acting as a consultant on much of the metalwork, precise contributions are unclear. Since the work of these three architects has a quality consistent through a number of buildings and, bearing in mind other developments at the time in Europe, it is probably fair to accept Banham's view that Prouvé works best in response to strong architectural ideas. Prouvé's hand is certainly in evidence in the external panels which had their thin external and internal skins equalized by interior springs, thus giving them a double convex shape and eliminating the necessity for internal framing (**4.4.27**). The large moveable internal partitions are of particular interest in view of subsequent work in this field. The partitions were secured by spring fixings which, says Prouvé, were being produced in 1930 (**4.4.28**). The way in which Prouvé sees his work in this building and others, is given clear expression in the selection of material for this one book on

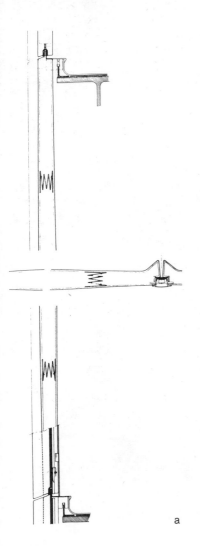

a

b

4.4.27 (a) La Maison du Peuple. Use of sheet metal and springs to equalize the planes giving a double convex form to the panels. (b) The sprung panels (Photo 1980). (c) Detail of fins (Photo 1980)

c

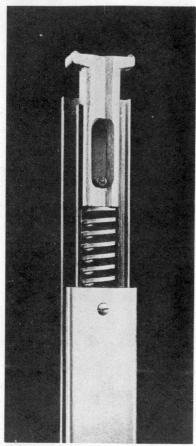

4.4.28 La Maison du Peuple; Moveable partitions, spring fixings. First produced *c.* 1930

4.4.29 Free University, Berlin. Consultant engineer, Jean Prouvé. Architects Candilis, Josic, Woods, Schiedhelm, 1964 on. Built of industrialized components with the intention that units can be dismantled and rearranged as required. Horizontal and vertical dimensions based upon Le Corbusier's Modulor with two modules of 700 and 1130 mm for the wall panels, permitting those dimensions that can be made from their combination in various ways. (Photo 1977)

him. Particularly significant is that details of the partitions and the exterior of the building are shown without any plan or indication of internal organization.

Prouvé's approach is clearly rooted in the inter-war years and the search for a machine aesthetic, and his later career develops these same ideas. In 1950 Prouvé discontinued his association with the factory he had set up at Maxeville in 1944 and turned to being a consultant designer. One of his most interesting pieces of work after this time was done in association with Candilis, Josic, Woods, Schiedhelm on the

4.4.30 Holiday house at Beauvallon. Jean Prouvé, 1967

Free University of Berlin. This building is virtually a building system in itself and with a grid planning concept which uses all the established rules of the industrialization game. The only fixed facade elements are the deep horizontal bands (**4.4.29**) of the floor and roof zones. All the infill elements are, theoretically, adjustable: glass, solid panels and frames, opening frames, projecting storage elements in a component range using Cor-ten steel sheet on the exterior. In this way those concepts of growth, change and indeterminacy so beloved of architects in recent years, are prominently displayed. However, no intellectual argument can atone for the resulting banality of external appearance. The houses at Beauvallon and St. Die (1967) also embody many of Prouvé's central ideas. Unlike some of his earlier houses, at Meudon for example, they show an attempt to bring together some of the realities of the machine with some of the myths of the machine age. His association and friendship with Le Corbusier is clearly seen in these plan forms, with their free external walls and freestanding service cores. The Beauvallon house (**4.4.30**) and the St. Die house are of steel frame construction with an external skin of lightweight panels clearly differentiated from the ground by the line of the floor slab which hovers over the untamed site. Once again the machine finds its way into the garden. To achieve the image required, that of the free lightweight skin, Prouvé produces some amazing details: ingenious, complex and, one suspects, problematic (**4.4.31**). The ability of metal to transfer heat rapidly seems to be totally ignored, with little attempt to deal with insulation, condensation and weathering. Screws appear in the most exposed positions.

The plans of the houses also exhibit some interesting characteristics. Like many frame houses, columns appear in the oddest places, there are doors opening into dining spaces (**4.4.32**), strange race-track corridors, and, one suspects, considerable noise problems. Interestingly, a similar solution, in principal, can be seen in the

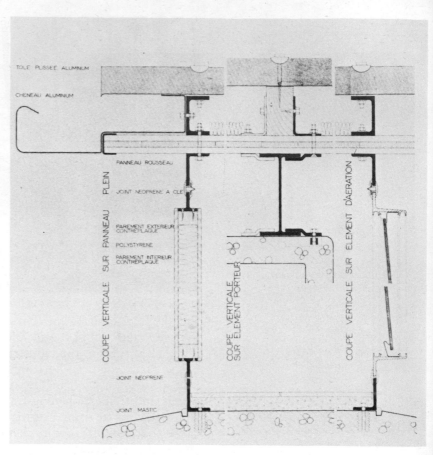

4.4.31 Holiday house at Beauvallon, sections

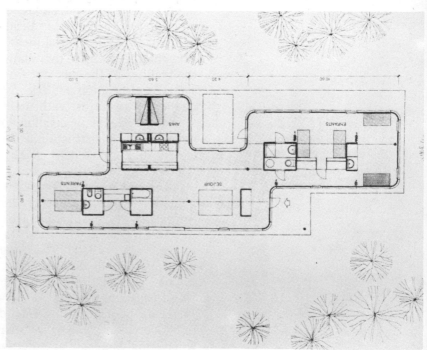

4.4.32 Holiday house at Beauvallon, plan

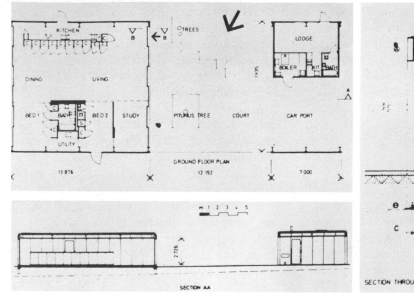

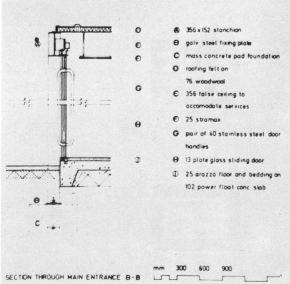

4.4.33 House at Wimbledon, London, England. Richard and Su Rogers, 1970. (a) Plan, elevations, wall section. (b) View of elevation. (Photo 1971)

Rogers' house of 1970 at Wimbledon (**4.4.33**). In both houses there is a steel frame and lightweight external panels with a free internal space around fixed service cores. In both cases very similar supporting theories are put forward: light, dry, off-site construction for a (mythical) speed of erection, with an interior of free-flowing space. The curved corner windows and panels complete the machine analogy in both cases.

In many ways these exemplars are a disservice to architecture and to the real and valid use of light, dry, fast construction. They are the designer's subscriptions to the architecture club. They are models which can say to the perceptive 'here are some ideas here you might like to consider', but unfortunately what they usually encourage is the total rejection of one set of values for the total acceptance of another. In his houses we see Prouvé making his utopian bid, reintroducing some valuable ideas to building, but equally rejecting many valuable

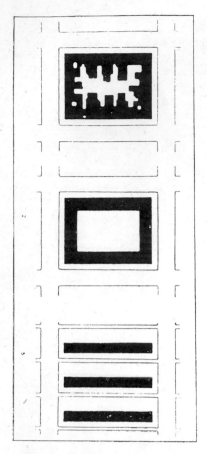

4.4.34 'From uncontrolled development-building site split up for exploitation by many individuals to controlled and development-rational orientation and economical lay-out and structure'. Illustration from The Modern House (1943, 1934) by F.R.S. Yorke

lessons which his references to evolution and background in metalwork should have warned him against. Prouvé's position is an ambivalent one. A 'self-styled constructeur' (Kidder Smith 1962, 1961) he embodies many of the most cherished dreams of the apologists for a machine aesthetic: he is technologically biased and demands a 'new' approach to building problems, one more akin to that used for automobiles and consumer goods. Rarely have his works been included in the standard architectural histories and this is in part due to his own desire not to act as an architect. Kidder Smith, in introducing Prouvé's Spa Pavilion on the shore of Lake Geneva, describes him as 'one of the most inquiring architectural-constructional minds in France today' (1962, 1961) and such a description seems fair for someone who was less interested in those concerns which usually mark architects from 'constructors' — that is an interest in place and people.

The development of a coherent set of attitudes forming an idealized machine style were slow in gaining ground in the Britain of the inter-war years. Probably the most influential were Yorke's *The Modern House* (1943, 1934) and *The Modern Flat* (Yorke and Gibberd 1937) and it is in these that one can see the argument set out, albeit in the form of a manifesto. Photographs of Chicago traffic, flying boats, ships and buildings are offered frequently with little explanation of the purpose for their inclusion. Like the writing itself they are in the architectural tradition of a collage of images whose very simultaneous presentation invites an assumed relationship.

In the introduction to *The Modern House* Yorke points out that:

'There have been many projects for the layout of groups in flat blocks, and for the design of economic and convenient living quarters as standardised units in the scheme, prepared by such men as Gropius and Le Corbusier, but there has been little actual development.' (Yorke 1943, 1934)

At this point in the text the diagram captioned 'From Uncontrolled Development' is inserted to show how rationalism and visual order will displace the organic growth patterns of the traditional city block (**4.4.34**). Another, captioned 'The Modern Scene' is especially interesting (**4.4.35**). That it could seriously be put forward as an aim, or a model is now difficult to comprehend unless one understands the context as discussed in this chapter. Uncompromisingly pristine and new it has been 'purified' of tradition and demonstrates its standardized units. Most importantly, however, the electricity pylon stands proudly, near the centre of the picture. 'Communism', said Lenin, 'is socialism plus Electricity'. Yorke espoused the equation: mass production plus standardization plus prefabrication equals modern architecture, and the chapter entitled 'Twentieth Century Architecture' ends with the house made like an automobile and the words:

'Twentieth-century architecture is dictated by new methods of construction and new materials, and by unprecedented practical

4.4.35 'The Modern Scene: housing at Dessau. Architect Walter Gropius'. As in *The Modern House.* This is the same photograph that appears in *The New Architecture* and *the Bauhaus* by Gropius published in 1935. It was also included in a paper read by Gropius to the Design and Industries Assocation in May 1934, reproduced in the RIBA Journal for May 1934

requirements, a new outlook on life, a new sense of space and time. If we are to act in accordance with tradition we must allow these factors to determine the twentieth-century aesthetic.' (Yorke 1943, 1934)

This is a concise and well put statement, but redolent of determinism. Should an architect, or indeed any individual allow such factors to 'determine the twentieth-century aesthetic' and set aside the many others Yorke has omitted? That many did so is now a matter of history. Many went further than even Yorke anticipated perhaps, and in setting aside many of the traditional humanistic qualities of architecture in favour of those of a more mechanistic flavour, brought about a near fatal union between building systems and centralized, institutionalized bureaucracies.

5

TRANSATLANTIC TONIC

ACHIEVEMENTS AND DREAMS

'I suppose that Europeans, accustomed to a world that changes more calmly and slowly, are not much interested any more in imitating its surface. It becomes more exciting to see appearances as a mask, a disguise or illusion that conceals an unexpected meaning. The theme of illusion and reality is very common in Europe. In America, illusion and reality are still often the same thing. The dream is the achievement, the achievement is the dream.'

Gavin Lambert
The Slide Area, 1963, 1959

'The "Metal Age" or the "Machine Age" — either name defines our epoch.'

Albert Farwell Bemis in
The Evolving House, Vol. III: *Rational Design,* MIT 1936

'In recent years there has been a rapid increase in the factory-fabrication of house components, particularly in America. Wall sections, complete with insulation, and internal facing, window-frames and radiators, are now manufactured as units ready for installation. Prefabricated bathroom and kitchen assemblies will soon be on the market. These instances indicate the trend of building in America at least, and we know from past experience that American innovations soon spread through Europe.' (Yorke 1943, 1934)

Thus opens the fourth edition of Yorke's influential book *The Modern House,* pubished in 1943 with the inscription 'with additional American examples', many of which occur in a section entitled 'Experimental and Pre-fabricated Houses'. For the most part the examples in this section seem to be little different from those in the main body of the book which is largely composed of experimental houses of one sort or another. This in itself is significant for the distinction between 'modern architecture' and experimental hardly existed. Indeed these terms were almost synonymous. To be modern it was necessary to reject almost all traditional forms and, if possible, constructional methods. According to Yorke again:

'Twentieth century architecture is dictated by new methods of construction and new materials, and by unprecedented practical

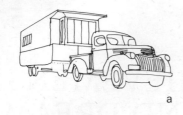

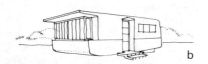

5.1.1 Trailer House. Tennessee
Valley Authority (TVA), 1942.
Designed for transport in two halves.
(a) The half house on its trailer;
(b) The complete house

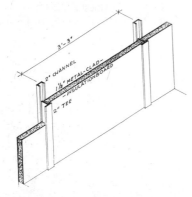

5.1.2 Illustration of lightweight
construction from *The Modern House*
Exterior walls and partitions of metal
sheathed insulation board

5.1.3 Factory fabricated bathroom.
William Wilson Wurster. From *The
Modern House* Yorke (1943, 1934)

requirements, a new outlook on life, a new sense of space and
time. If we are to act in accordance with tradition we must allow
these factors to determine the twentieth century aesthetic.'
(Yorke 1943, 1934)

The section on prefabrication included examples ranging from the
Tennessee Valley Authority's Trailer House of 1942 (**5.1.1**) to
packaged bathrooms and kitchens. More significantly many of the
American examples were timber-framed houses. There was, of course,
great emphasis on lightweight construction and a number of examples
of construction for 'Lightweight Roofs and Floors and Walling
Systems'. These set down methods of construction long since
assimilated into building technology and which formed a basis for
much of the lightweight systems work (**5.1.2**).

However, as early as 1921/22 Rudolph Schindler had experimented
with prefabrication in concrete on his own house. The 1 metre wide
slabs are one of the earliest examples of the storey height precast
concrete unit. Subsequently Schindler experimented with grids and
extensively with timber-frame construction, proposing in 1947 'The
Schindler Frame'.

Several architects who subsequently became well known are included
in Yorke's book; William Wurster and Ralph Rapson (**5.1.3**) produced
prefabbed kitchen and/or bathroom units and Fuller's Dymaxion house
of 1929 rubs shoulders with Gropius' experimental house at the
Stuttgart Weissenhofsiedlung of 1927 and Neutra's Lovell House
in Los Angeles, also of 1927. The latter's claim to admission was
that 'window sizes determine the units of measurement for the
standardised structural steel members' (Yorke 1943, 1934), which
were then cased in poured concrete. More interesting perhaps is the

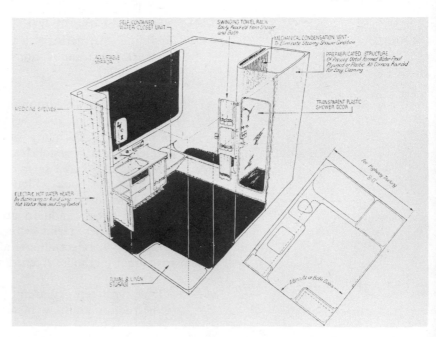

5.1.4 Aluminaire House. Kocher and Frey 1931. Erected at Architectural League show, New York. Re-erected at Syosset, Long Island, USA

work of Kocher and Frey who produced a number of houses, most notably the Aluminaire House of 1931 (**5.1.4**) and the Weekend House of 1932 (**5.1.5**), the latter making ingenious use of canvas as an external cladding material.

The importance of all this is in illustrating a tendency that was becoming increasingly relevant: to draw attention to the emerging industrial componentized industry that existed in the United States and to join this to the formal solutions of the 'new' architecture emerging in Europe. In Yorke's book we see at work one of the major

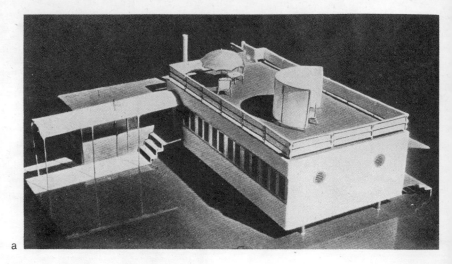

a

b

Typical wall section labels:
FIREPROOF
½" INSULATION BOARD
ALUMINUM FOIL
2" x 4" WOODSTUDS
PAINTED CANVAS
AIR-SPACE
½" INSULATIONBOARD
DYED CANVAS
4'-0"

TYPICAL WALL SECTION

5.1.5 Experimental weekend house. Kocher and Frey, 1932. (a) Photograph of model; (b) typical wall section showing a finish of 'painted fireproof canvas'

pressures emanating from Europe — that of endeavouring to draw all innovation into a stylistic norm. Just as Ford's production line became sloganized in Europe, and Taylor's work earlier had been idealized by designers, we see in this conjunction between production methods and form a force that influenced the development of prefabrication in Britain right through to the creation of the institutional systems. More important, however, than these claims of the 'new men' were a number of developments in America that did contribute to an integration of industralization and user demand. In their way Fuller's proposals stand aside from both lines of development: both the innovative and that more tied to the technology of mass production. In contrast Bemis' study of American Housing in his three volume *Evolving House* (Bemis and Burchard 1933; Bemis 1934; 1936) is more rooted in real development without being bound by it — his proposals for a cubic four inch module were taken up and applied. But, more importantly, whilst all this was going on the timber-frame house was itself developing and, dramatic gestures aside, was responding to both user demand and the new technology whenever appropriate: this was to prove a more powerful and durable example of a systems approach.

RUDOLPH SCHINDLER

'The factory must remain our servant. And if a 'Machine-made House' shall ever emerge from it, it will have to meet the requirements of our imagination and not be merely the result of present production methods.'

Rudolph Schindler
Dune Forum, 1934; quoted in Gebhard 1971

It has frequently been the case that architects use a single building to demonstrate a total philosophy or approach to building opportunities. The way in which concepts originated in this way have transferred themselves into 'good currency' is a frequent occurrence in the development of building systems. Both Wright and Le Corbusier have already been cited in this respect — they produced powerful icons in written, drawn and built form, concerning the rationalization of construction techniques. Invariably for such men this has gone hand in hand with pleas for the humanizing of technology — an aspect that has often been missed where these ideas have been reintroduced in a more general way.

Rudolph Schindler was also such a man, and a contributor to the flow of ideas concerning industrialization whose work has received little attention in the history books. Banham refers to the 'total silence' of the historians and claims that what Schindler means 'historically' is this — that modern architecture would have happened in California even if De Stijl, Corb, Mies, Gropius and the Museum of Modern Art had never existed.

It is important on two counts to include Schindler in any history of the development of the systems idea: first, being a European settled in America (he worked for Wright and assisted Neutra to come to the United States) he was very conversant with the developing discussions surrounding the modern movement of the early years of the century in Europe. Second, having settled in California he entered a tradition of a very different sort which, as we have seen elsewhere, has given rise to such work as Maybeck's and the Eameses, and can encourage and accommodate such propositions about the possible uses of available technology. Schindler is therefore relevant both because of, and despite, his European background. He brought a series of spatial and technological ideas to the building of houses and equally was absorbed into an ongoing stylistic and constructional approach that flourishes in California. Both of these can be seen at work in the Schindler-Chase house in Kings Road, Hollywood, built in 1921-2 (**5.2.1**).

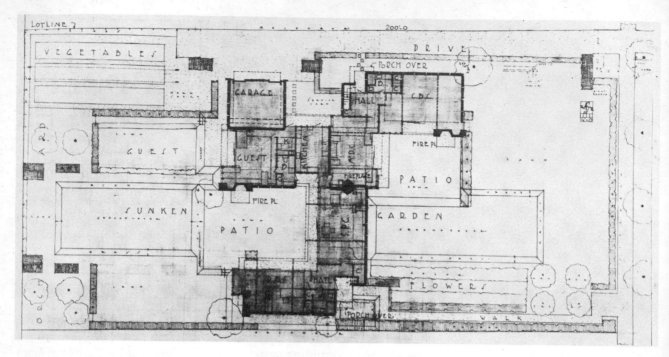

5.2.1 House for Schindler and Clyde Chase, Hollywood, USA. Rudolph Schindler 1921-2 Plan. Note the zoned areas

Schindler, a Viennese, whose father was a furniture maker, was born in 1887. As a student he was clearly impressed with the secessionist painters Klimt and Schiele (as can be seen from his nude studies) and the architects Adolf Loos ('ornament is crime') and Otto Wagner. He subsequently went to the USA and worked for Wright, leaving after a fierce argument. He set up practice in Hollywood and built some two hundred buildings until his death in 1953. In the twenties he and his wife formed a close friendship with Dr. Philip Lovell who published a series of articles in the Los Angeles Times under the title, 'The Care of the Body' (Gebhard 1971). Lovell was prominent in the physical hygiene movement that has seemed such a powerful force in the architectural philosophy of the twenties. Physical exercise, natural foods, progressive education were all causes which Lovell actively, and apparently lucratively, pursued. The links between Lovell's interests and the modern movement's concern with sunlight, ozone and hygiene clearly struck chords in Schindler. Schindler's beach house is an important embodiment of this fusion of notions and a significant building. This said, however, it is not of such direct relevance to our concerns here as his own house in Kings Road, Hollywood. Clearly Schindler's concern with sunlight and air, opening up the house to the outside, can be seen running through the whole modern movement and later codified into run of the mill buildings. School buildings embody many of these concerns with the obsession for large windows that slide open, cross-ventilation and ultimately the 2% daylight factor which was built into the school building regulations in Britain. From the planning point of view alone the Schindler house is of interest (**5.2.1**): each of the four individuals has a private retreat, each pair has their own entrance, and there is a communal kitchen for both couples.

5.2.2 Schindler/Chase House. Schindler's studio showing storey height precast concrete slabs and narrow windows at joint positions. (Photo, 1974)

5.2.3 Schindler/Chase House: Storey height precast concrete slabs being lifted into position

No bedrooms were provided but open sleeping porches in the Californian tradition were positioned over each entrance. This is a much more subtle piece of planning than the mere spatial gymnastics employed by many of his contemporaries, with its use of linked private spaces rather than anonymous flowing space. Keen observers of the place of women in such a situation may be amused by the fact that the two private spaces for Mrs. Schindler and Mrs. Chase were adjacent to each other and to the kitchen. Further, their rooms were also the circulation space for their two men to gain access to the kitchen.

Technologically the houses are more relevant to our argument in that the construction of each of the four spaces is identical and composed of a solid back wall of precast concrete slabs (**5.2.2**) cast on the ground and tilted into place (**5.2.3**); the side facing the private spaces is constructed of lightweight timber, glass and panel. Gebhard (1971) says:

> 'The repetitive slab walls suggested modern technology, and their rhythmic appearance throughout the house expressed the repetitive process of machine production',

whilst Banham (1971b) points out that Schindler's use of concrete at this time in California was in itself unusual since the timber tradition was so well established:

> 'It is used as something between an adobe wall and a plank fence.'

The 2m × 1m slabs were cast on the ground and were tapered from top to bottom, they were then tilted up but not edge butted. Schindler solved the joint problem that has bedevilled component construction simply by leaving a gap about 20 cm in size between each slab and filling this with fluted glass. In this way he turns a technological problem to positive architectural advantage — which is more than one can say for most component joint solutions. His use of timber in the rest of the building is straightforward and presages his later dexterity in this west coast vernacular. He turned more and more, either for expedience or architectural reasons, to the use of the ubiquitous timber frame and stucco of the area but modified to his particular purposes. His contribution then, to this tradition is important for he not only developed, in his own right, the use of rationalized timber-frame construction as an open system of construction with many stylistic possibilities, but he also contributed technically in introducing the structural plank ceiling over long spans in place of beams and joists. A version of the approach is set out in his article 'The Schindler Frame' (1947) in which he says:

> 'In building a contemporary house, the "Schindler Frame" utilises ordinary framing members and established framing techniques.
> 'Although the "Schindler Frame" unavoidably repeats certain characteristic details, it allows such freedom in the use of the more important features of space architecture that it should prove a boon in developing it, and might well help to give contemporary houses what the past called "style".' (Schindler 1947)

Although Schindler had a good control of the fabric and structure in creating environment in his early houses, including his own and the notoriously bad Pueblo Ribera Court at La Jolla, he stood out against the already well established American tradition of mechanical heating and ventilation. For even though the Southern Californian climate was not harsh it suffers extremes of heat, cold, mists and rain — Gebhard points out that 8-12 inches (203-305 mm) of rain can fall in 24-30 hours:

> 'Schindler aimed for the norms, ignored the extremes, and in the process compromised the full livability of his environment.' (Gebhard 1971)

By the late twenties Schindler was forced to integrate mechanical heating and cooling into his designs. This was not soon enough to avoid the problems of Ribera Court, a group of rentable houses where Schindler insisted on using single skin concrete walls instead of the favoured timber frame. The junction details between timber and concrete were weak and the roof leaked and the client was obliged to sell off the development, thus proving correct the forecasts of the lending agencies, who initially would not provide the money for such a venture. The history of the modern movement is strewn with such examples of the wilful ignoring of environmental and constructional design matters under the pressures of a dogmatic architectural view. Unfortunately there is ample historical support for those system builders in modern times who, with leaking walls and roofs and inadequate thermal control, point out that all 'great buildings' leak. Such innovations may be appropriate in the single building (especially an architect's own house), excusable even when the client (as at La Jolla) is ideologically with you and understands the problems, but totally inexcusable when practised on the vast scale of, say, the English school building systems.

Schindler gave considerable attention to the dimensional problems of building and the use of modules in design. In view of his background in Europe and later with Wright this is perhaps not surprising. What is of interest is that this contributed to the pragmatic use of such ideas in timber-frame construction in the United States and to the establishing of a tradition which ultimately permeated, in other forms of course, the whole house builders' market. Bemis codified much of this in *The Evolving House* and these ideas subsequently became built into US standards. Schindler's grid was generally a 4 foot (1.22 m) one which he claimed related to the human dimension. It also conveniently related to material sizes and this is very relevant. As we shall see later, developments in Europe were more concerned with bending the machine to a set of ideal sizes and this approach failed repeatedly right up to Britain's recent attempts to encourage rationalized metric component ranges. Schindler used two modules for ceiling height (8 ft 0 in — or 2.4 m) and a door height of 1¾ units. He apparently also employed an intriguing method of drawing plans at a number of levels in the building, as did Wright (Sergeant 1976), to explain the complexities of the construction — something that perhaps other architects could well emulate.

Schindler clearly set out where he stood in relation to the machine, most notably in an article in *Dune Forum* in 1934, going on to make some very perspicacious remarks on Fuller:

'The work of Buckminster-Fuller in propagating the tremendous possibilities which the use of our technique of production may have for building construction is, invaluable. If he creates his Dymaxion house, however, entirely from the viewpoint of facile manufacture, letting all considerations of "what" take care of themselves, he is putting the cart before the horse. The space architect has primarily a vision of a future life in a future house...

'...Although Mr. Buckminster-Fuller realises the coming importance of space considerations in architecture, his Dymaxion House is not a "space creation". However "ephemeral", to use his own term, it may be, it is born of a sculptural conception. Its structural scheme is akin to the one of the tree, and although its branches and members may try to wed it to space by the tenderest interlockings, the "room" they enclose is not an aimful space conception but a by-product without architectural meaning.' (Gebhard 1971)

Schindler's significance is in understanding and uniting the European and the North American experiences. Clearly his experiments with concrete, although limited, were far sighted, particularly his use of storey height precast panels in 1921/22. However, it is in his appreciation of the potential of the timber frame as an organic system that he has most to offer in the present context. Schindler, through his experience of the west coast of America could see that which many of his contemporaries were unable to see because of their love affair with the technology of mass production. Homes built of timber frame embodied rationalizations and 'rules of thumb' and yet still offered opportunity for the designer, builder and owner to personalize the result. That Schindler did this in a particularly idiosyncratic manner should not blind us to the many advantages of the timber-frame system, a topic to which we return in a later chapter.

RICHARD BUCKMINSTER FULLER

'Bucky will speak for as long as he likes and there will be no questions.'

Cedric Price introducing Fuller to an audience at the Institute of Contemporary Arts, London

The contribution to the debate about the impact of science and technology on design and architecture made by the ideas of Fuller is enormous. Many of our concepts concerning the uses of machine technology in the building industry have been framed by his propositions, which first began to appear in 1927 when, as Fuller says, 'I resolved to do my own thinking' (Fuller 1972). Even before this, however, he had been involved in a commercial building system, the Stockade System, with his father-in-law. Indeed it was this experience that partially led him to decide to rethink his whole attitude on the basis that 'The individual can take initiatives without anybody's permission' (Fuller 1972). The flow of ideas that came from this decision embraced housing, transport, synergetic-energetic geometry, cartography and geodesic structures. If Fuller has become most well known for the latter, his domes, this merely supports his own views that although the underlying forces are invisible, most people and the media in particular, are mostly impressed by the visible.

The current concern with ecology, and the whole earth's resources, is a theme that has been central to Fuller's approach and, as he is fond of pointing out, one that he drew attention to in 1930/32:

'If you look at Shelter magazine, which I published between 1930 and 1932, you will find that the title of the opening chapter is "ecology". At that time, I also felt that the word "architecture" tended to make humans think only of classical orders rather than solving total humanity's evolutionary shelter problems by competent and comprehensively adequate anticipatory design of tools with which humanity could cope intelligently in solving their problems to everyone's progressively highest advantage and satisfaction. I then decided to rename "architecture" and called it "Environmental Designing".' (Fuller 1972)

Since then environmental designing has become a well understood activity involving a whole range of disciplines other than architects.

175

There are courses in environmental design and Faculties of Environmental Design in institutes of higher education across the world. The above quotation illustrates another of Fuller's characteristics, that of putting words together in a rather unusual way. He is also prone to create his own new words or phrases if he feels that existing terminology is inadequate. The worst excesses of this have often confused those who have not taken the trouble to examine his meanings and on occasion it has been described as gibberish. Words, however, are often an interesting indication of the change inherent in a society and a language which changes little is a counterpart to a culture which is also unchanging.

Fuller's concern to find precise descriptions of his thoughts and ideas inevitably led to new word formulations and this in turn paradoxically made his message more difficult to understand. However, only those without poetry will fail to gain something from his writings or long 'thinking aloud' sessions, however little they understand of the explicit message. Much of his approach must come from his family background. His great-aunt, Margaret Fuller was, says Marks (1960), 'a high priestess of Transcendentalism'. She edited *The Dial,* a literary journal, and first published Emerson and Thoreau.

Fuller's first exploration into building was with his father-in-law James Monroe Hewlett. They invented the Stockade System of building and between 1922 and 1927, the firm built 240 buildings using this system (**5.3.1**). The blocks were made of:

'fibrous material such as excelsior or straw, bonded with magnesium-oxy-chloride cement. The fibres, impregnated by the cement, were blown into moulds which felted them together.

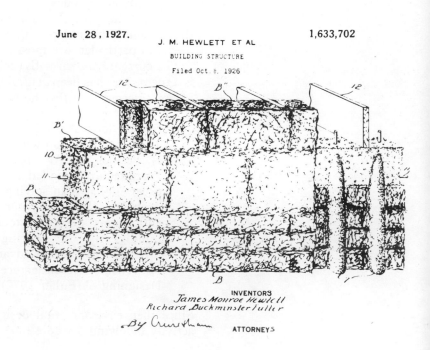

June 28, 1927. 1,633,702

J. M. HEWLETT ET AL

BUILDING STRUCTURE

Filed Oct. 8, 1926

INVENTORS
James Monroe Hewlett
Richard Buckminster Fuller

By Crutham ATTORNEYS

5.3.1 Stockade Building System: J.M. Hewlett and R.B. Fuller, 1927. Patent drawing

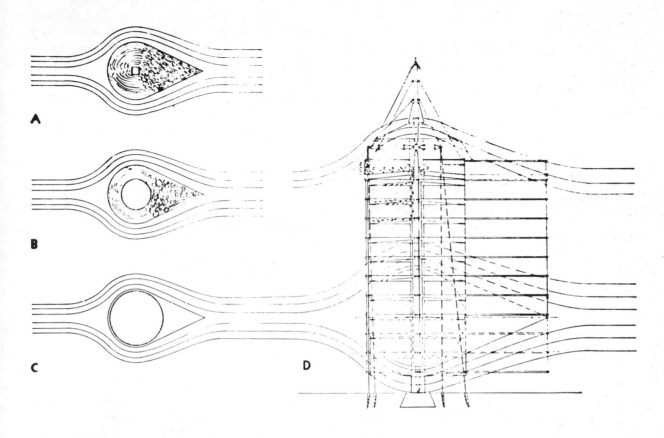

SHELTER, NOVEMBER, 1932

5.3.2 10-deck house design. R. Buckminster Fuller, 1927. Diagrams illustrating the effect of a streamlining shield: a, a cube; b, a cylinder; c, an efficiently streamlined unit; d, 10-decked 4D structure with streamlined windshield

The 16″ × 8″ × 4″ (approx. 400 × 200 × 100 mm) blocks weighed in the neighourhood of 2 pounds (0.91 kg) each; they were so light that they could be thrown to the second floor scaffolding, so tough that they would not break if they fell.' (Marks 1960)

The acoustic wall and ceiling materials now used in the building industry are derived from this, the Celotex corporation ultimately purchasing the idea.

From these experiences Fuller formulated his attitude to building:

'That was when I really learned the building business, and the experience made me realise that craft building — in which each house is a pilot model for a design which never has any runs — is an art which belongs to the middle ages. The decisions in craft-built undertakings are for the most part emotional — and are based upon methodical ignorance.' (Marks 1960)

5.3.3 (*opposite*) 4D Dymaxion house.
R. Buckminster Fuller, 1927

In this one can see Fuller firmly attaching himself to the dictates of the machine, seeing building as synonymous with the car and other mass produced goods, and implicitly rejecting understandings which rely on continuity and emotion. The first examples of this thinking can be seen in his 10-deck house design of 1927 (**5.3.2**) in which environmental conditions are taken into account: the streamlined shield brings the building's heat losses proportional to the air drag and, claims Fuller, could reduce the heat losses to very little. From this work emerged the Dymaxion house (**5.3.3**), probably with the Dymaxion car Fuller's most publicized work. The simplest analogy for the Dymaxion house is that of the wire wheel, turned on its side and with the hub extended to become the mast, which is in compression and which is acting as a support and a service core. The Dymaxion house demanded materials and standards which at the time of its design could not be met:

'I could see that it would be a minimum of 25 years before the gamut of industrial capabilities and evolutionary education of man — as well as political and economic emergency necessities — would permit the emergence of the necessary physical paraphernalia of this comprehensive anticipatory design science undertaking.' (Marks 1960)

The Duralamin mast is set in a foundation and the two floor perimeter tubes supported by tension cables. The floor is laid on a pneumatic bladder and diestamped bathroom units are placed against the service mast. The ceiling units distribute light and air and the doors are operated by photoelectric cells — which only became available subsequent to this design. The external walls were of transparent plastic with aluminium sheet, camera shutter type roll curtains. The exhibited model showed a nude woman lying on the pneumatic bed in order to 'dramatize the fact that within the house, temperature, humidity, and air flow are maintained at optimum levels, making clothing and bed covering unnecessary' (Marks 1960). The house is completed with a Duralumin hood and the whole design weighed 3 tons (3.048 tonnes) with a floor area of 1600 sq ft (148.64 sq. m).

In the Dymaxion house Fuller tried to show that, when attacked fundamentally, the problems of 'livingry' (as opposed to weaponry) could be solved by mass production technology. It clearly showed the work of the European machine apologists for what it was — a preferred machine aesthetic — not a radical programme for harnessing industry to solve the shelter problem.

It was Fuller who drew attention to the way in which the world was 'shrinking' yearly, as faster and faster modes of transport developed. He propagated the view that the new ability to transport materials long distances made the use of local materials obsolete. In this way the new mass production technology would become truly international, and would transcend national differences. Such a view motivated many architects to actively encourage technologies which ignored

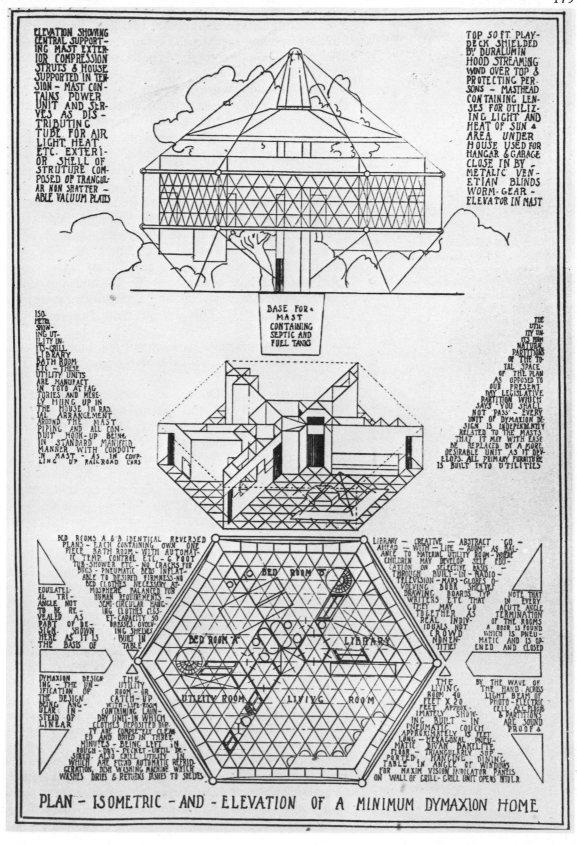

ELEVATION SHOWING CENTRAL SUPPORTING MAST EXTERIOR COMPRESSION STRUTS & HOUSE SUPPORTED IN TENSION - MAST CONTAINS POWER UNIT AND SERVES AS DISTRIBUTING TUBE FOR AIR LIGHT HEAT ETC. EXTERIOR SHELL OF STRUTURE COMPOSED OF TRIANGULAR NON SHATTERABLE VACUUM PLATES

TOP 50 FT. PLAYDECK SHIELDED BY DURALUMIN HOOD STREAMING WIND OVER TOP & PROTECTING PERSONS - MASTHEAD CONTAINING LENSES FOR UTILIZING LIGHT AND HEAT OF SUN - AREA UNDER HOUSE USED FOR HANGAR & GARAGE CLOSE IN BY METALIC VENETIAN BLINDS WORM-GEAR ELEVATOR IN MAST

BASE FOR MAST CONTAINING SEPTIC AND FUEL TANKS

ISO-METRIC SHOWING UTILITY INITS-GRILL LIBRARY BATH ROOM ETC. THESE UTILITY UNITS ARE MANUFACTORIES AND MERELY HUNG UP IN THE HOUSE IN RADIAL ARRANGEMENT AROUND THE MAST PIPING AND ALL CONDUIT HOOK-UP BEING IN STANDARD MANIFOLD MANNER WITH CONDUIT IN MAST - AS IN COUPLING UP RAILROAD CARS

THE UTILITY UNITS FORM NATURAL PARTITIONS OF THE TOTAL SPACE OF THE PLAN AS OPPOSED TO OUR PRESENT DAY LEGISLATIVE PARTITION WHICH SAYS YOU SHALL NOT PASS - EVERY UNIT OF DYMAXION DESIGN IS INDEPENDENTLY RELATED TO THE MASTS THAT IT MAY WITH EASE BE REPLACED BY A MORE DESIRABLE UNIT AS IT DEVELOPS. ALL PRIMARY FURNITURE IS BUILT INTO UTILITIES

BED ROOMS A & B IDENTICAL REVERSED PLANS - EACH CONTAINING OWN ONE PIECE BATH ROOM - WITH AUTOMATIC TEMP CONTROL ETC - 6 FOOT TUB-SHOWER ETC - NO CRACKS FOR BUGS - PNEUMATIC BEDS INFLATABLE TO DESIRED FIRMNESS-NO BED CLOTHES NECESSARY ATMOSPHERE BALANCED FOR HUMAN REQUIREMENTS SEMI-CIRCULAR HANGING CLOTHES CLOSET-CAPACITY 50 DRESSES, OVOLVING SHELVES BUILT IN TABLE

LIBRARY - CREATIVE - ABSTRACT GO-AHEAD - WITH - LIFE - ROOM - AS BALANCE TO MATERIAL UTILITY ROOM - WHERE CHILDREN MAY DEVELOP SELF EDUCATION ON SELECTIVE BASIS THROUGH BUILT-IN RADIO TELEVISION - MAPS - GLOBES OVOLVING BOOK SHELVES DRAWING BOARDS TYPEWRITERS ETC THAT THEY MAY GO TOGETHER AS REAL INDIVIDUALS NOT CROWD NONENTITIES

EQUILATELAL TRIANGLE NOT TO BE REVEALED AS PART OF DESIGN SHOWN HERE AS IT IS THE BASIS OF

NOTE THAT IN EVERY ACUTE ANGLE TERMINATION OF THE ROOMS A DOOR IS FOUND WHICH IS PNEUMATIC AND IS OPENED AND CLOSED

DYMAXION DESIGNING THE UNIFICATION OF THE DESIGN BEING ANGULAR INSTEAD OF LINEAR

THE UTILITY ROOM-OR CATCH-UP WITH-LIFE-ROOM CONTAINING LAUNDRY UNIT IN WHICH CLOTHES DEPOSITED DIRTY ARE COMPLETELY CLEANED AND DRIED IN THREE MINUTES - BEING LEFT IN ROUGH-DRY-POCKET-UNTIL DESIRED ALSO GRILL UTILITY IN WHICH ARE FOUND AUTOMATIC REFRIGERATION, DISH WASHING MACHINE WHICH WASHES DRIES & RETURNS DISHES TO SHELVES

THE LIVING ROOM 40 FEET X 20 FEET APPROXIMATELY SHOWING BUILT - IN PNEUMATIC COUCH APPROXIMATELY 15 FEET LONG - HEXAGONAL PNEUMATIC PIVANT BAKELITE FLOOR - TRIANGULARLY SUPPORTED HANGING DINING TABLE IN ANGLE OF WINDOWS FOR MAXIM VISION INDICATOR PANELS ON WALL OF GRILL-GRILL UNIT OPENS INTO L.R.

BY THE WAVE OF THE HAND ACROSS LIGHT BEAM OF PHOTO-ELECTRIC CELL ALL FLOORS & PARTITIONS ARE SOUND PROOF

BED ROOM B
BED ROOM A
LIBRARY
UTILITY ROOM
LIVING ROOM

PLAN - ISOMETRIC - AND - ELEVATION OF A MINIMUM DYMAXION HOME

180

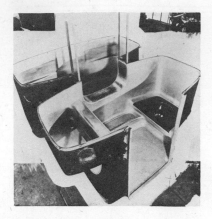

5.3.4 Dymaxion bathroom, the Phelps-Dodge version. R. Buckminster Fuller, 1937. 'Fuller improved this model by making the front and back "room" of the bathroom in identical oval patterns, thus enormously reducing tooling costs'. (Marks 1960)

local conditions and methods of building. Indeed many involved in building systems excused the often considerable limitations of such systems by pointing to the economies of scale that accrued from universality.

The Dymaxion house is in the tradition of the American single family home in that it is a self-contained unit standing on its own, and Fuller produced no proposal for grouping or integrating them into the existing fabric. The proposal was concerned only with its technology, and a new technology at that. In its way it also ignores some truths surrounding the idea of a home in the minds of most people, and this theme runs through much of Fuller's work. It is as difficult to imagine untidy family life in the Dymaxion house as it is in the Citrohan or the Bamboo house. Although it aroused a great deal of interest the Dymaxion house remained only a proposal, and Fuller moved on to examine other areas. The Dymaxion bathroom (**5.3.4**) was one of these, the Dymaxion car another. The mass produced self-contained bathroom was part of the 4D house of 1927 but the first prototype (for American Radiator Company's Pierce Foundation) was not made until 1930. This was not publicly shown because:

'The manufacturer was convinced that the plumbers' union would refuse to install the bathrooms.' (Marks 1960)

Ultimately Phelps-Dodge, in 1936, produced twelve prototypes which contained all the normal facilities together with air-conditioning. The unit was 'plugged in' by the connection of the services manifold. The prototypes were in metal but Fuller had designed the unit, says Marks, to be produced in plastics when that technology had developed sufficiently. In spite of its apparent advantages the Dymaxion bathroom was never produced in quantity, Fuller blaming this on the 'general inertia of the building world'. (Marks 1960)

However, the basis of Fuller's approach is prediction, and he had already worked out the innovation lag in various industries — that is the time it takes for new ideas to become acceptable. In the case of the Dymaxion house it was 25 years, and so when he was consulted in 1944 concerning the manufacture of such a house using the labour and facilities of the wartime aircraft industry he pointed out that:

'The time is premature for a consideration of this new industry as a commercially exploitable undertaking.' (Marks 1960)

Nevertheless Fuller moved to Wichita, Kansas, the home of the Beechcraft aeroplane company to endeavour to apply his ideas and to stop the gradual loss of labour in the industry. The Wichita house (**5.3.5**) which resulted from this was a development of Fuller's Dymaxion Deployment Unit (1941) which he designed for The Butler Manufacturing Co: ultimately used as wartime dormitory accommodation and radar shelter it embodied many of Fuller's earlier thoughts — particularly those on the use of shape to control energy

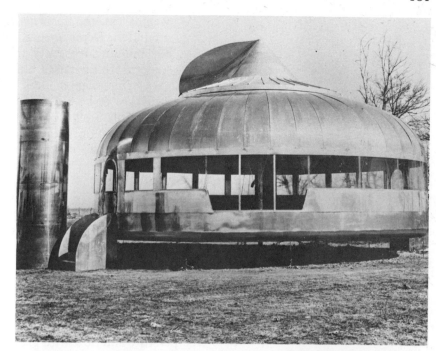

5.3.5 Dymaxion Dwelling Machine or Wichita House. R. Buckminster Fuller 1945. 'The total structure weighed approximately 6,000 pounds. This figure agreed with Fuller's 1927 estimate of Dymaxion house of the same dimensions...' (Marks 1960)

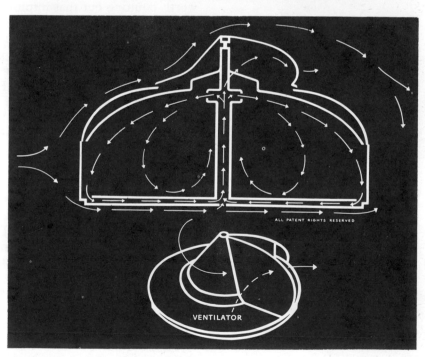

5.3.6 Dymaxion Dwelling Machine or Wichita House. 'Diagram showing how the external air flow, travelling its greatest distance over the top of the structure, created a vacuum drag at the ventilator, which in turn dragged the internal air flow pattern.' (Marks, 1960)

relationship between inside and outside (**5.3.6**). Marks records how, in a temperature of 100°F (38°C) the interior of the DDU was very cool indeed, with a very particular pattern of air movement.

'...the sun radiation (diffused into the atmosphere around the DDU) is heating the atmosphere, causing it to expand. Hence it weighs less, and floats upward, initiating a rising thermal column around the Deployment Unit. This thermal drafts air from around the building to satisfy the upward flow, pulling the air from underneath the raised building, as well as from the surrounding area. This phenomenon, in turn, causes air from the top of the structure to be pulled downward inside the structure (through the ventilator at the top) to satisfy the partial vacuum created by the exhausting of air from under the structure. Also a cold draught spirals downward through the central core of the rising thermal and is drawn in through the top ventilator.' (Marks 1960)

Such an approach to design is clearly of a totally different order to that of the Europeans discussed in the last chapter. In spite of their invocation of the importance of science and machine technology none could match such an approach, although it would be wrong to assume that because the means of environmental control on a dwelling are not made absolutely explicit they have not been considered. Frank Lloyd Wright is certainly an example of an architect who, partly consciously and partly with great intuitive skill, was able to control air temperature and flow with great subtlety. In this context it is perhaps worth pointing out that intuition means what it says — in-tuition or tuition inside, that is, something that the human has inwardly learned. In Wright's case his background was important but equally important was the fact that he just designed house after house very rapidly, intuiting (as it were) more from each one. It is interesting to set this against the approach of the Europeans who, in quantity alone, designed far fewer buildings and further, to set this against Fuller who had actually designed very few houses indeed.

The Wichita house itself (1944) weighed approximately 6000 pounds (2721.6 kg) and consisted of a 22 ft (6.7 m) mast from which the floor ring and radical beams were hung. When assembled the roof dome becomes rigid and is tied down by stainless steel tension rods. The cladding was an aluminium alloy and a continuous Plexi-glass window ran around the outside. The ventilation system followed a similar pattern to that of the DDU.

Beech Aircraft made 'firm proposals to soft-tool manufacture the Dwelling Machines at a cost of only $1800 each' (Fuller in Marks 1960). This was made possible by the soft-tool approach developed in the aircraft industry where 'change is normal' (Marks 1960). These tools are those especially developed for limited run production and thus do not in themselves become such expensive investments that components and design are inhibited from change. Fuller here recognized a crucial development of machine technology missed by those European proponents who were obsessed with the standardization of the product. For them this in turn led to the virtual standardization of appearance. However, the Wichita house did demand a vast tooling up resource to make production economically justifiable and this in turn relied on the market. Marks summarizes its fate:

'Lacking the 10 million dollars necessary for tooling the Dymaxion Dwelling Machine for a 20,000 units per year production, the development ended the Wichita experiment. Two prototypes were ordered by the Air Force and later resold to Fuller's project. They were ultimately acquired by a Kansas oil man, who combined them to form the house shown above, sans rotating ventilator. "His architectural additions and modifications in effect," said Fuller, "forever grounded this aeroplane".' (Marks 1960) (**5.3.7**)

This sad postscript summarizes many of the problems already encountered by those intent on applying industrial mass production techniques to housing:

1. the massive tooling up costs necessary which in turn demand
2. a large continuous guaranteed market and
3. that apparent functional efficiency, however low the cost, is not in itself sufficient to most people. To them a home has meaning and that meaning is involved both in the initial choice and in its subsequent personalization. 'The Kansas oil man' referred to above managed to incorporate the beautiful but ferocious efficiency apparent in the Wichita house in his own idea of what a home was — not so easy perhaps for less well heeled purchasers. This fundamental factor is one never discussed by Fuller and which he may well consign to the inertia of society.

It also points up the two parts of Fuller's work which are usually fused in people's minds — first a study of the underlying forces that shape a situation. His 'Universal Requirements of Human Dwellings' of 1930 based upon general systems theory is an example of this approach to the housing problem. Second, there are the products, the inventions or

5.3.7 Dymaxion Dwelling Machine or Wichita House. The only two prototypes 'acquired by a Kansas oil man, who combined them to form the house shown above, sans rotating ventilator. "His architectural additions and modifications in effect," said Fuller, "forever grounded this aeroplane"' (Marks, 1960)

artefacts that he has produced. These are so total, so well argued, so well supported by data, that they are in themselves mistaken as solutions. Indeed, it may be that Fuller, in endeavouring to implement them himself also mistakes them for solutions. As Marks says, after the Wichita house:

> 'He (Fuller) was also left with a determination never again to put his projects in the lap of a promotion whose life or death was determined by the speculative instincts of others.' (Marks 1960)

If his Universal Requirements were so all embracing why is it that they were unable to embrace such fundamental and well known difficulties as those of marketing? The answer lies in part in the fact that the objects were not all embracing and further they carried (as they must) the imprint of total design as surely as any of Le Corbusier's proposals.

Fuller's importance is in his fundamental studies of a problem and in his reformulating of possible directions. As he himself is fond of pointing out, many of his predictions have materialized and his inventions have sparked off developments which have made an impact on housing. In contradiction to his own gospel, his many disciples have often seemed to be more interested in the objects than in the ideas. In fact Fuller would probably have served his own philosophy better had he not produced any objects but confined himself to using his 'total' philosophy to modify the system by the insinuation of low profile change. But his obsession with prediction is probably the clue; in common with all witch doctors his credibility rests in his ability to foretell events. His constant reminders of how right he was in such and such a year show his concern with this.

More importantly for us here, however, is that by this means he has rightly or wrongly given credibility to a number of notions which support those who would see buildings as machines of one sort or another. Prediction gives control. Accurate prediction can give total control. But prediction is often no more than a prescription for future events, and one which others try to fill. Many have tried to fill Fuller's prescription for an efficient, mass produced rational solution to the provision of housing. The reason it has not been successful suggests some fundamental omission — that omission is people and how they act. Indeed, it would be a cause for alarm if the prediction were to prove correct because total control would likely then have been achieved, life would accord totally to rational principles, and real choice and unpredictability would have disappeared. It is certainly true, as Fuller says that 'the individual can take initiatives without anybody's permission' but to be effective he certainly must have permission or agreement. Further than this, he has to have their active participation, so that what is designed, what is produced, contains some measure of meaning for them. That meaning must inevitably embody the familiar to some degree: in setting aside the familiar Fuller was able to satisfy his own predictions concerning the length of time necessary for acceptance.

BEMIS AND MODULAR DESIGN

'Mass productive methods have come to stay, because they are simply the further development of the division of labour. It seems to be a law of life that function or labour is divided and subdivided, specialised and further specialised, infinitely and forever. Further extensions of mass-production into both old and new fields may be confidently predicted.'

Albert Farwell Bemis
Foreword to Volume I
of *The Evolving House*, 1933

The work of Albert Farwell Bemis in the United States during the late twenties and early thirties forms a foundation for much of the modern polemic that has surrounded modular co-ordination. For example the vociferous campaign conducted by many architects during the 1950s and 1960s produced a considerable literature, much of which opened with references to Bemis' pioneering work. The Royal Institute of British Architects' own publication, *The Co-ordination of Dimensions for Building* (Martin 1965) contained a section entitled 'Brief history leading to the present stage of development' commencing with an account of the Bemis work. A history of dimensional co-ordination that commences in 1936 with Bemis is, one might think, brief indeed: however, the purpose of the publication was to convince and to convey basic current thinking, and Bemis' concern for rationalization in house building was very appropriate to the attitudes of the moment.

'In Rational Design Bemis foresaw that the parts of houses could be mass produced. He proposed for the first time the use of a cubical module and he conceived the idea of a matrix made up of these modules covering the whole of the building in three dimensions.' (Martin 1965)

Rational Design was the third and last of Bemis' three volume work *The Evolving House*. Volume I, subtitled *A History of Home*, and largely the work of John Burchard 2nd, attempted 'a review of the evolution of the home and the social and economic forces which have influenced its development' whilst volume II contains 'an analysis of current housing conditions and trends and comparisons with other industries' (Bemis 1934).

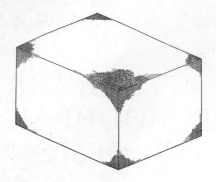

Most commentators have concerned themselves with Bemis' third volume, *Rational Design,* since this is the one which proposes the cubical module (**5.4.1**) and which describes the benefits which will accrue from its adoption. For our purposes, however, it is necessary to look at the context of Bemis' propositions so that they may be related to the concerns of those in both Europe and America. The Foreword in volume I describes Bemis' intent for the third volume thus:

> 'Finally, a solution of such problems will be offered in the third volume in the form of a rationalization of the housing industry, thus harmonizing the means by which our homes are provided with those mostly used in supplying the other major needs.' (Bemis 1933)

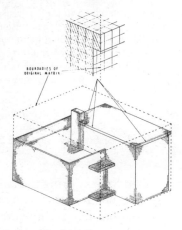

Here Bemis is drawing the time honoured parallel between the methods and capabilities of other consumer industries and the building industry. As with Fuller he was calling for the rationalizations used in mass production and marketing to be introduced into the provision of housing. Bemis goes on to nail his colours to the mast in no uncertain fashion:

> 'This rather large task, I am frank to say, I have approached with the distinct preconceived idea that the chief factor of the modern housing problem is physical structure. A new conception of the structure of our modern houses is needed, better adapted not only to the social conditions of our day but also to the modern means of production: factories, machinery, technology, and research. Other industries have made use of such forces, to a far greater extent than the building industry has done.' (Bemis 1933)

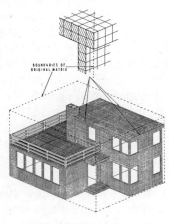

At this distance in time it is difficult to imagine how, after such a thorough historical and contemporary analysis of housing and its provision, Bemis could deduce that the 'chief factor... is physical structure'. Much evidence has since come to light that the physical fabric of a house (or housing) is far less significant than, for example, the method of financing adopted. Any monetary savings that may be made by rationalization of production and prefabrication are, in housing, able to contribute little to the solution of 'the housing problem'. A report by the Consultants Arthur D. Little Inc., *Housing: Expectations and Realities* (Wilbur 1971) points out that the savings to be made in technologically based solutions are marginal:

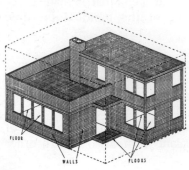

> 'Thus, the ultimate construction cost of system-built structures in this case would be 20% less than that for conventionally built structures. Since construction costs for the latter represent 70% of the sale price, systems building might theoretically save 14% on traditional construction. If even half these theoretical savings are possible, maximum savings would be no higher than 10%. If a dwelling unit sells for $20,000, an owner would pay $154 per month for 25 years on an 8% mortgage with no down payment.

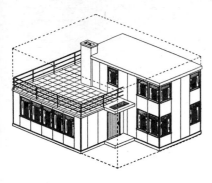

5.4.1 The 4 inch cubical module matrix: Albert Farwell Bemis: 1936. Bemis's most potent idea seen here from abstraction to building

If technology could reduce this cost by 10%, the monthly payments would be reduced to $139. However, other ownership costs, such as taxes or utilities, would not change, so this 10% decrease in mortgages payment would probably result in 5% ownership cost.' (Wilbur 1971)

This is supported by Donnison (1967) in his fascinating study *The Government of Housing:*

> 'There has been much talk in recent years about the need for an industrial revolution in the construction industry, leading in some quarters to hopes of a spectacular 'break-through' producing major cost reductions, particularly on the housing front. Such hopes are likely to be disappointed, at least in their more extravagant form. Many countries have devoted years of effort and large sums of public money to this task. But in Western Europe, where conditions are most nearly comparable to our own, no country has built houses at appreciably lower cost with the aid of radically new systems — with the possible exception of France where the evidence is not entirely convincing.' (Donnison 1967)

Bemis' work grew firmly out of the times, and he can hardly be criticized for backing the same technological horse as the European idealists of the Bauhaus, or the high technocracy of Buckminster Fuller. Nevertheless it is important to place his work in its tradition, since he was clearly a latter day apologist for the mass production and efficiency viewpoint which had its roots firmly in the nineteenth century. Furthermore, he was clearly influenced by the political events of his time and in his Foreword threw them all in as support for his plea for more rationalization:

> 'The great communistic experiment of the Soviets, the autocracy of Mussolini, the spiritual democracy of Gandhi, the flounderings of all entrenched political and economic forces, including those of the US, the philosophy and suggestions of the scientists, including "technocracy", are all valuable contributions to this end. Rationalisation between world production and distribution through which we shall make better use of our great advances in productive technique for the general public good is clearly on the horizon.' (Bemis 1933)

There is a simple faith that rationalization is a 'good' end in itself and that in some ill defined way everything was contributing towards this end. Probably more significant are those words of Bemis quoted at the beginning of this section, where it is clear that, in common with other thinkers of his time he accepted the marxist concept of the division of labour, and increasing specialization, with the additive of a faith in the success of mass production methods.

In volume II Bemis analyses the social and economic aspects of housing in some detail under the following headings:

I The Economic Importance of Housing
II The Present Day House
III Annual Cost of Shelter
IV Disabilities in the Housing Industry
V The Organization of the Building Industry
VI Relative Efficiency of the Building Industry
VII Analysis of the Cost of the Home
VIII Cost Restrictions on Building
IX The Architecture and Housing
X Financing of the Home
XI Government Intervention in Housing
XII Rationalization

A particularly interesting aspect of the second volume is Bemis' analysis, 'Disabilities in the Housing Industry', which looks at the industry as a biologist or doctor might look at it. In this Bemis had clearly been influenced by Dr. Walter B. Cannon who, he says, 'recently drew an enlightening comparison between the functioning of the human body and the economic structure' (Bemis 1934). Cannon is important as the originator of the concept of homoeostasis through which the body is seen as a whole system held in a constant state although there continues to be a constant flow of inputs and outputs. This has become a concept fundamental to General Systems Theory (Bertalanffy 1968; Katz and Kahn 1969, 1966) and a particularly useful one in endeavouring to define the responsiveness of a system.

Bemis classified the ills of the housing industry under seven 'disabilities';

1. General Disabilities
   (a) Local nature
   (b) Lack of organization
   (c) Excessive plant capacity
   (d) Seasonal production
2. Architectural Disabilities
   (a) Lack of professional advice
   (b) Tradition and lack of standard specifications
   (c) Unnecessary estimates
3. Constructional Disabilities
   (a) Lack of integration
   (b) Work on site
   (c) Antiquated assembling methods
   (d) Custom work
4. Managerial Disabilities
   (a) Small operators
   (b) Failure to use labour-saving devices and modern methods
   (c) Lack of ability
   (d) Bad practices

5. Labour Disabilities
   (a) Excessive numbers of crafts and jurisdiction
   (b) Strikes
   (c) Restriction of output
   (d) High wage scales
   (e) Labour versus management
   (f) Seasonal unemployment
6. Legislative Disabilities
   (a) Building codes
   (b) Usury laws
   (c) Tax legislation
7. Consumer Disabilities
   (a) Lack of knowledge
   (b) Insistence on individuality
   (c) Insistence on speed
   (d) Concentration of leasing dates

It is clear that, although the 'seven disabilities' draw attention to important problems still with the building industry, they are couched in such terms that many are questionable. In fact the list is compiled with the over-riding consideration of the benefits of mass production methods: over-riding to the point that Bemis cites the 'insistence on individuality' and speed as 'Consumer disabilities'. Earlier, he cites 'Custom work' as a disability and high wage scales. Indeed the list could well have been compiled by the British National Building Agency during the 1960s or by the US HUD agency for Operation Breakthrough.

Whilst it is possible to understand Bemis' euphoria for the ultimate success of mass production methods, the relegation to a minor role of those parts of the industry which give the home-owner a choice do not bode well. The great success of Ford and others in providing standardized basic products encouraged Bemis that such methods could be applied to housing. His 'Trades House' versus automobile manufacture illustrate this (**5.4.2**). Somewhat surprisingly this led him to the cubical module with additive dimensional consistency.

> 'The principle of the cubical modular method is so simple that the terms necessary to explain it, such as RECTANGULARITY, MODULARLY CUBICAL MODULE, and MATRIX, will require usage before the conception can be accepted as the merest common sense. It is in the weaving process, in tapestry, brickwork, and tiles as well as in the processes of mass-production, the nature of building materials, the rectangularity of building structure' (sic). (Bemis 1936)

Bemis' proposals emerged from a 'Survey of Efforts to Modernise Housing Structure' and the results of this are set out in a supplement to Volume III of his book and was compiled by John Burchard 2nd. The survey includes all the examples of the period one might expect in a work extolling the virtues of the machine age and of mass production

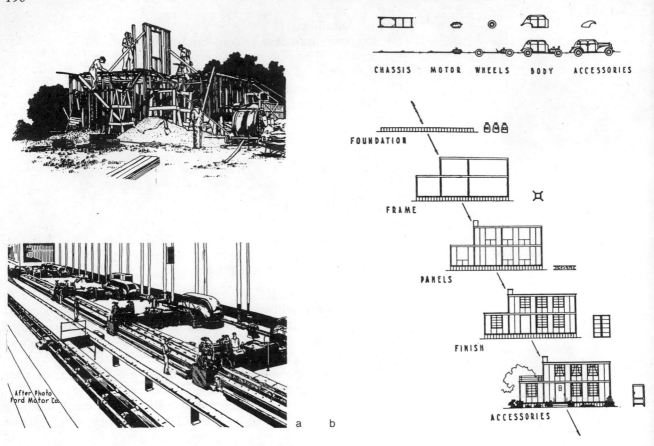

CHASSIS   MOTOR   WHEELS   BODY   ACCESSORIES

FOUNDATION

FRAME

PANELS

FINISH

ACCESSORIES

a   b

5.4.2 (a) 'Contrast between house building and motorcar making'. From 'The Evolving House' by Bemis (1933-36) with the caption 'The disordiliness of a typical house assembly and the order of the production line of a motor-car assembly plant.' (b) 'Rationalized house assembly' (Bemis 1933-36)

methods — Fuller's Dymaxion House, steel framed houses by Howe and Lescaze in New Hartford (1933), Le Corbusier and Jeanneret's Loucheur house, houses by Gropius and Wright and prefabricated houses in Britain of the early twenties. Also reference is made to Hewlett's Stockade House, conceived 'some time prior to May 1922' but no mention is made of Fuller's involvement with this building system.

Bemis' commitment to mass production was a typical one for the time and although, unlike its European counterparts it did not take an overt ideological stance, it is clear in reading the three volumes that in a pragmatic way he was similarly aligned. As with many commentators on building industry problems he makes frequent reference to the successes of automobile manufacture as a model to emulate, with the now familiar diagrams of the car assembly line. Indeed, Bemis makes that direct analogy with the car industry which subsequently became so powerful — and admiringly quotes Ford:

> 'Mass production is not merely quantity production... nor is it merely machine production... Mass production is the focussing upon a manufacturing project of the principles of power, accuracy, economy, system, continuity and speed.' (Bemis 1936, quoting Ford from *Encyc. Brit.*, 14th Edn, 1929)

Bemis saw his cubical module as the 'focus for standardization' and points out how all the parts of a house, whether factory made, or made on site, could relate to it. Bemis proposed that the size of this basic unifying module should be 4 inches, that is the structural wall thickness using 4 inch × 2 inch (approx. 100 × 50 mm) timber studs, which was (and is) the traditional structural system for housing in North America. It was Bemis who designated this linear module, M (**5.4.3**). It is curious that, with Bemis' commitment to mass production

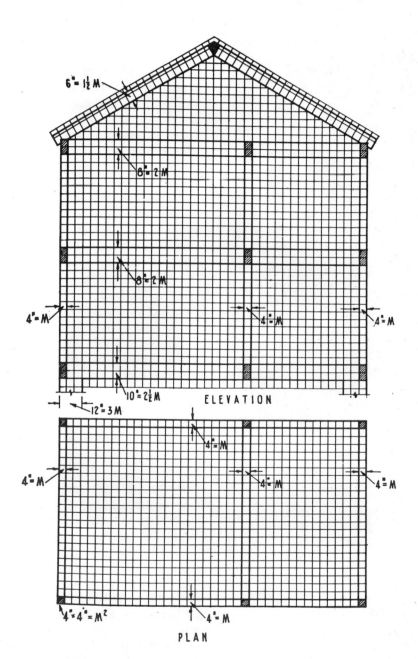

5.4.3 'Plan and section of a typical wood frame structure (platform type)' (Bemis 1936) showing the use of the 4 inch module, M

and the uses of metals, he should opt for a dimension arising from timber-frame construction. Nevertheless, his proposal has proved remarkably durable, perhaps because he ignored this aspect. The 4 inch module (or 100 mm in Europe) was subsequently adopted by all the proponents of component building in Europe. In Britain it was adopted by the Modular Society, and after a long campaign, became incorporated into British Standards — initially BS 4011:1966.

On the surface, Bemis' three thorough, rather dull volumes stand at the other end of the spectrum to Le Corbusier's *Vers Une Architecture* of 1923, but essentially it is imbued with the same philosophy. As I have attempted to show in previous chapters the European projects and proposals were almost always couched in terms of ideological calls to action. Bemis, in common with most approaches North American, tended to emphasize the observable inevitable progress of the machine age and its implications for building. Examples were given, the industry analysed and proposals made. Americans had grown up with the machine, there was no need to elevate it to a metaphysical level. However, it is clear that Bemis held very similar views to those who had gone before in Europe, even if those views were couched in a different style. He was committed to what he called 'truthful expression in present day buildings' and in an unusually colourful passage sets out the philosophy associated with functionalism. 'The progressive architect... recognises that period architecture and false imitation of the antique are simply signs of artistic decadence' (Bemis 1936). Bypassing the question concerning what his reaction to honest 'imitation' might be, we are left with a message no less committed than was Le Corbusier's, to an architecture which directly reflected its means of production — or more accurately an idealized picture of what that might constitute.

Bemis obviously had a view every bit as specific as that of the Europeans but, in typically American pragmatic fashion, he set out his argument with a weight of analysis and a minimum of polemic. That he saw the whole building as a total system is evident in the following:

> 'Structure has been used, not as a synonym for edifice or for the work of erection, but, more exactly, for the system and for the combined essential parts that produce a building.' (Bemis 1936)

Furthermore he also felt that through such rationalizations as those he was proposing the house, and housing, could be more appropriate to his list of social changes which were earlier quoted and that this would be reflected in its very appearance:

> 'Finally from the aesthetic viewpoint it (the house of the time) does not adequately express the spirit of the present era or utilise the wealth of adornment available through new materials, colours, and textures.' (Bemis 1933)

Bemis' proposals for the use of a co-ordinating module were quickly taken up in the United States, with manufacturers and builders gradually incorporating the ideas into their everyday vocabulary. By the early 1950s many catalogues could be found with the product details shown related to the four inch grid lines and with all sorts of other aids, like pull-out trace sheets, which show that some of Bemis' ideas became a reality for designers. In Britain the acceptance of the idea followed a long and drawn out pattern. The course this was to take is set out in subsequent chapters, suffice it to say here that in returning across the Atlantic the idea once again took on its full ideological colours with the proponents of industrialized building using it as a weapon in their attempts to rationalize the building industry out of what they saw as its traditional bad habits. The industry first saw this as a threat and then, as the idea was given centralized authority and legislative teeth they quickly saw the profits to be made in the innovations that became mandatory.

As I shall argue, Britain's path in all this has been significantly different to that in North America or in Scandinavia. Indeed the only way in which it may even be honoured with the analogy of a path is one that crosses a particularly treacherous and uncharted mountain range.

6

BRITAIN FLIRTS WITH MECHANIZATION

THEY CALLED IT PREFABRICATION

'The well-integrated, standardised prefabricated, not built but assembled, house is in conflict with mass prejudice, which, as we know, first would have to be dissolved.'

Richard Neutra
Routes to Housing Advance in *Circle*, 1937

Britain, always subjected to influences from both the United States and from Europe, frequently found the pragmatics of the former in conflict with the idealized concepts emanating from the latter. The growing prefabrication of building components in American timber-frame construction, standard windows and doors, the TVA projects, and other experiments made an unhappy match with the stylistic overtones which Europe's machine apologists had sought to use in demonstrating how relevant they were being to social demands.

Had Britain's architects been able to use the pragmatics of American industry they would have sought to understand and then to utilize the existing and developing abilities of the building industry. During the thirties for example, speculative house builders in Britain were constructing houses at a considerable rate (**6.1.2**). To be sure, they were criticized by architects, as being poorly planned, jerry built and derivative in style. Nevertheless they were popular and today, in their arcadian settings constitute a much desired purchase.

In the construction of such houses there was enormous rationalization, albeit seemingly casual. Vast estates were built cheaply because bricks and other materials were bought on a 'bulk purchase' basis by the builders: doors, windows, bay fronts, and staircases were usually all to standardized designs, whether made on site or off. However, little of this matched what the modern movement in architecture thought industrialization was all about. Architects

6.1.1

Year	Local authority	Private builders	Total
1930	55,000	110,000	165,000
1931	65,000	130,000	195,000
1932	68,000	130,000	200,000
1933	45,000	170,000	220,000
1934	52,000	255,000	310,000
1935	40,000	275,000	315,000

(*AJ* 19/12/1935)

6.1.2 Speculative housing. The Highbury Estate, Portsmouth, 1931. Houses sold for £550 (Photo 1980)

committed to this view wanted their buildings to LOOK as if they had been made from prefabricated parts, not merely constructed on rationalized lines. Since designers are visually trained, they see processes and results largely in this way. Furthermore, although many building industry procedures and methods are very orderly and rationalized, they are often not seen as such because they do not conform to the current visual frame of reference of the architect. Therefore, for many architects brought up in the traditions of the modern movement such rationalizations do not exist.

For these architects, the idealizations surrounding mechanization were, in large part, to be seen in the work of Le Corbusier and that of the Bauhaus. However, before the impact of these influences on Britain the period following the 1914–18 war saw attention turned to developing non-traditional house building techniques and this gave rise to a series of ill-fated attempts to produce homes by prefabricated means. Frame construction, with its lightweight skin, or load-bearing components used like a box of bricks began to appear. The importance of dimensional co-ordination was apparent and its relation to standardization. More significantly a philosophy had been created which carried a conceptual model of what constituted the preferred solutions.

From being a series of premises closely bound up with the development of modern architecture itself, a single set of ideas now became isolated as the possible next step. This set in motion a particular strain of development in Britain which was, in under thirty years, to become a set of institutionalized procedures for producing buildings which well matches the rigid hierarchical organization of central and local government in Britain. At the time few could have foreseen the implications of the experiments being made — least of all those so enthusiastically engaged in it. Quarry Hill, for example, at Leeds was not only a social experiment in that it was a large local authority owned housing estate but it was to be constructed in a totally new method — that of the Mopin system from France — and it was to have the latest in rubbish disposal, the Garchey system.

In school building the signs were on a more modest scale and can best be seen in the work of Stillman in West Sussex. Here the lightweight construction and industrialized components were put together to give visual expression to the underlying philosophy. These and the experiments by Gibson and Neel at Coventry provided a foretaste of what was to come after hostilities had ceased in 1945. With the war period to germinate such ideas, and the new social impetus offered by the subsequent period of reconstruction, Britain had reached a point where they could be put to test on a larger scale.

The post-war prefabricated houses, and the prefabricated approach to schools developed at Hertfordshire County Council demonstrated what could be done by a humane, enthusiastic and dedicated group of architects when given the chance. The recently reorganized London County Council itself conducted its own experiments into prefabrication at Picton Street (Elmington Estate), Camberwell and at Roehampton and elsewhere. With encouragement from government, and from architects', manufacturers developed many proprietary systems of their own, many of which became very successful. This success had profound implications for the architectural profession who, having nurtured these children of industrialization, had ultimately to find a way to kill them off: the client-sponsored or institutional system was that way.

LIGHT AND DRY

'Prefabrication means making as much of the building as
possible under cover and out of the rain.'

Hugh Anthony
Houses: Permanence and Prefabrication, 1945

Homes Fit for Heroes

The period after World War I in Britain gave rise to a wide range of
experiments in prefabricated housing. White (1965) points out that it
is not clear that the claimed shortage of houses was of serious
proportions, although the drop in available building craftsmen is
significant: from 115,955 bricklayers in 1901 to 53,060 in 1920. The
structure of the industry was that of small groups of skilled craftsmen
led by a master craftsman and, says White:

> 'since the work they turned out was mainly of good quality, and
> the prices were low compared with more highly finished
> engineering products, there was no urge to alter this structure.'
> (White 1965)

In view of the success of this type of small industry structure in
producing the enormous numbers of houses built between 1870 and
1914 it is curious to find the government sponsored committees
pressing for a radical change in constructional techniques and
organizational method. Nevertheless, houses of *in situ* concrete, precast concrete, steel and timber were developed with government
backing and, as White points out, this attempt was symbolic of a new
approach to such problems. It was prompted by the social concern
following upon the 1914-18 war and the need to be seen to be doing
something for those soldiers who had been involved in the carnage of
that conflict. It was necessary not only to be building homes, which
the industry was well able to do, but to be seen to be building homes.
This conjunction of events, which was to be repeated after 1945,
brought together the needs — Homes for Heroes — and the idea of
radical change in building technique. A question that must be posed is
why, with an industry that had shown itself capable of quality,
quantity and low cost, the major effort was directed to methods which
almost totally ignored this? Clearly both the idea and the reality of
factory production created a cultural climate for this to occur.
However, White is correct in pointing out the symbolic force of what

happened: the government had to be seen to be supporting a change from the old ways and what better way to do this than by producing houses that were made in a totally new manner?

In this way, factory produced houses entered the political arena in Britain, to reappear with monotonous regularity time and time again. For politicians concerned to demonstrate their belief in radical change and concern for the mass of the population, what better way than through a mass produced house? That many of the new methods of the post-1914 period were technical failures is clear, that almost none of them survived as viable building methods after the crisis period is of particular importance, if both designers and politicians are to learn anything from these efforts. Subsequent efforts in 1945 and 1961 have all followed a similar pattern, with much of the system built housing of the 1960s already showing technical deficiencies. Furthermore, whatever the statements about disappearing craftsmen in 1920, they still survive, along with a predominantly brick-based industry which in its own way is very good at building certain sorts of houses. Perhaps if politicians, administrators and (sorry to say) many architects had been more interested in understanding the real nature of an industry that could produce quality and quantity at a low price, they could have shown how the effects of industrialization could be introduced in a positive way so that everyone would benefit and the industry grow into a new situation.

To its credit the Tudor Walters Report of 1918 (largely drafted by Raymond Unwin) drew attention to the dangers of ignoring tradition where it worked, and suggested exactly the approach outlined above. This can best be seen in its encouragement of a rustproofed, standardized, metal window range, which Crittall responded to with their cottage windows of the early twenties. Following 1918, and the Tudor Walters Report, a series of committees was set up to look into new ways of building. The Buildings Materials Committee was set up in July 1918 and disbanded in February 1919, after achieving little. In 1919 the Committee for Standardization and New Methods of Construction was asked to look into standardization for state-aided housing schemes and to consider proposals for new materials and methods of construction. This committee advertised in the press for proposals for new building methods and this brought many suggestions, these being published after examination. This committee was disbanded in 1920, and, as White points out:

> 'it is rather surprising to find an entirely different committee appointed by the Minister of Health in 1924 to look into the same question of innovations for working class housing.' (White 1965)

In 1924 this committee, the National House Building Committee, was set up to endeavour to increase the numbers of houses built annually by local authorities, but although it recommended component standardization it did not stress prefabrication as previous committees had done.

The house building systems of the twenties, and they are referred to as systems, surveyed by the Burt Committee in their 1944 report are categorized by main construction material: of concrete, timber, steel framed and metal clad — almost all had serious drawbacks. Of the concrete houses the no-fines method, which still survives, was the most straightforward and flexible. The walls, and sometimes the floors, were of 8 inch or 9 inch no-fines concrete with a clinker aggregate using a proprietary steel plate shutter system in two lifts. Burt's comment on this was that assuming that supervision was good and there was a high degree of organization such houses could be constructed rapidly using largely inexperienced labour. Government support of 'deskilling' can probably be said to date from this post-1918 period. Rather than create a vast need for less skill it might be thought that the proper political concern should have been with creating work for those with skills or for training those skills needed. After all, the period before 1914 had seen a great burgeoning of house building skills (as did the period before 1939) which resulted in very good rates of house building. The no-fines house had two other significant disadvantages: additional insulation was needed to achieve the standard of an 11inch cavity wall and there was no saving in cost over the use of brickwork. However, no-fines houses were still being built at the time of the Burt Committee report, unlike most other examples.

There were several ingenious concrete houses using precast wall and frame sections: the Duo-slab (**6.2.1**), Boot (**6.2.2**), Winget (**6.2.3**) and Underdown (**6.2.4**) houses. Although the summary of the report states that there are few serious problems, a close reading will reveal several. All the houses, which were rendered externally, suffered from rendering failure. The erection of the houses turned out to require more skilled labour than was thought — not surprising when one looks at the demands of the construction. One clear piece of advice that emerged was that those that had a frame which carried the total loads were satisfactory, whereas where panels were used to carry loads the differential stresses set up caused problems. Such findings no doubt encouraged later attempts to use a frame in housing systems, even though economically disadvantageous.

Of the several types of timber houses used (**6.2.5**) there were almost no serious problems. Indeed the report stated that:

> 'On all sites the tenants stated that they found the houses comfortable and equal in all respects to normal brick houses. No major defects have developed in any of the houses.' (Burt 1944)

It may be thought that such positive evidence of success would offer a clear lead to the Burt Committee and to subsequent post-1945 housing development, to do everything to encourage timber housing. Although timber was short, it would have been possible to have laid the foundations for a viable alternative mode of building to brick, and thus reap some of the benefits that accrue from this as in Scandinavia and North America.

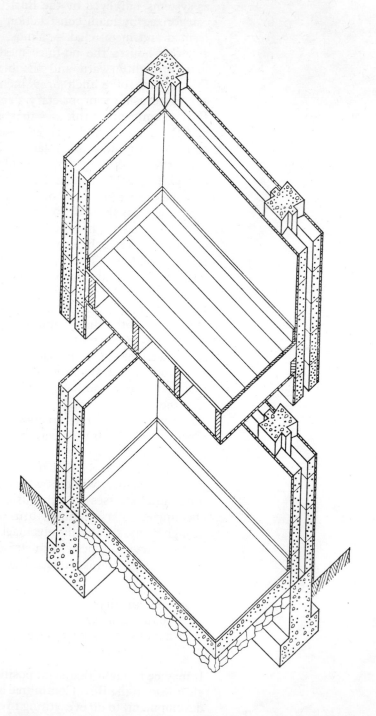

6.2.1 The Duo-Slab house. Sir Edwin Airey. First houses built in Liverpool, England, 1922

CAVITY WALLS OF PRECAST CLINKER CONCRETE SLABS. CONCRETE PIERS POURED IN SITU. TIMBER JOIST FLOOR ON WALLPLATES BOLTED TO INNER SKIN

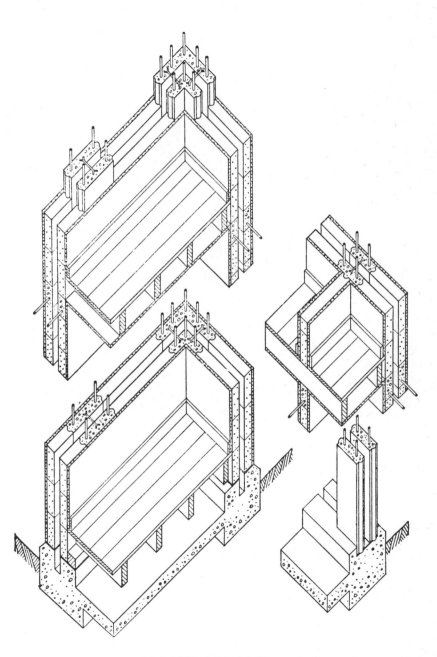

6.2.2 The Boot Pier and Panel
Continuous Cavity House: Henry
Boot and Sons Ltd. First houses built
in Liverpool, England, 1928

PRECAST REINFORCED CONCRETE PIERS WITH A CAVITY
WALL OF 3″ CLINKER CONCRETE SLABS. INTERNAL AND
EXTERNAL RENDERING. TIMBER JOISTS AND FLOORING

206

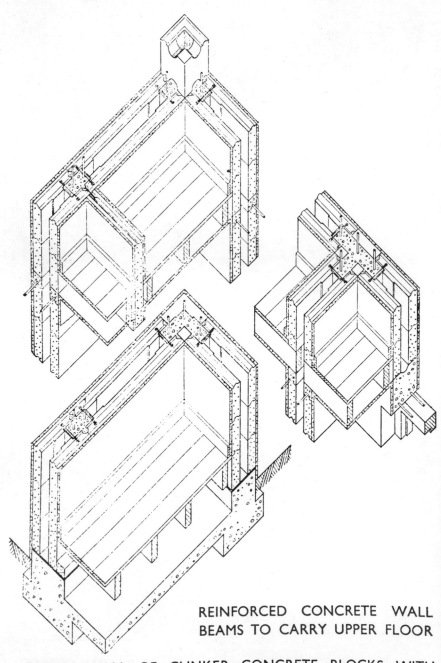

6.2.3 The Winget Pier and Panel House. Winget Ltd and M F Hill. First houses built in Hull, England, 1928

REINFORCED CONCRETE WALL BEAMS TO CARRY UPPER FLOOR

CAVITY WALL OF CLINKER CONCRETE BLOCKS WITH REINFORCED CONCRETE PIERS POURED IN SITU. WALLS RENDERED EXTERNALLY AND PLASTERED INTERNALLY

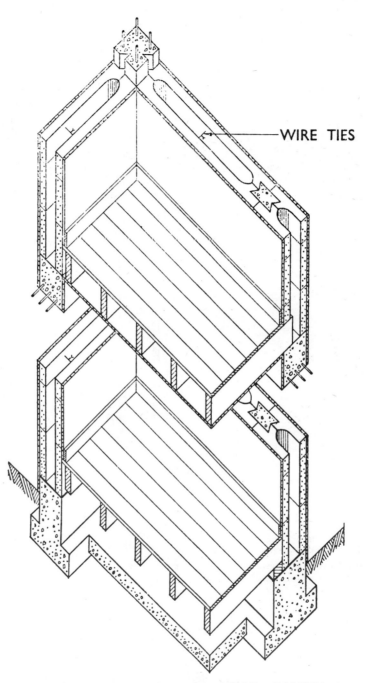

WIRE TIES

CONCRETE BLOCKS FORMING CAVITY WALL, WITH REINFORCED CONCRETE PIERS. REINFORCED CONCRETE BEAMS CARRYING UPPER FLOOR

6.2.4 The Underdown house. First houses built in Cambridge, England, 1926

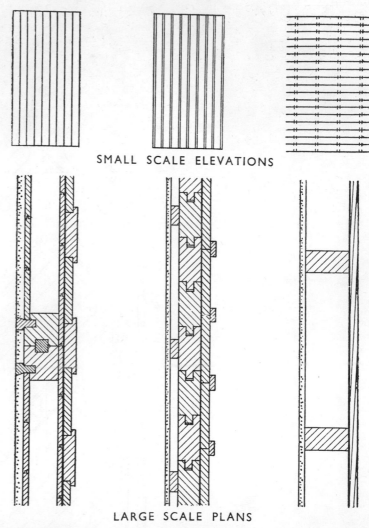

SMALL SCALE ELEVATIONS

LARGE SCALE PLANS

NORWEGIAN

PREFABRICATED PANEL:
3″×3″ TONGUED STUDS;
COVERED EXTERNALLY
WITH REBATED PANELS
AND BUILDING PAPER
ON ½″ BOARDING; IN-
TERNALLY WITH ¾″
FIBREBOARD ON ¼″
BOARDING

SWEDISH

PREFABRICATED
PANEL: 3″ T. & G.
PLANKS LAID HORI-
ZONTALLY; COVERED
EXTERNALLY WITH ¾″
BOARDS, COVER STRIPS
AND BUILDING PAPER;
INTERNALLY WITH ½″
PLASTERBOARD ON
BATTENS

ENGLISH

CONSTRUCTED ON
SITE: 4″×2″ STUDS;
COVERED EXTERNAL-
LY WITH CREOSOTED
WEATHERBOARDING
AND BITUMEN FELT;
INTERNALLY WITH ¾″
FIBREBOARD

6.2.5 Timber-framed houses built at various sites throughout Britain from 1926. Diagrams show typical external wall constructions

Steel framed houses like the Denis Poulton (**6.2.6**) or the Dorlonco, used standard or light rolled steel members as the frame and this created impact noise problems. The report points out that these systems were not fully exploited since they utilized load-bearing wall elements which carried no loads and yet had a steel frame.

However, this was another lesson not learned since there were subsequent attempts to contradict this rather obvious truth — notably the Ministry of Housing's 5M System, discontinued in 1968. Metal

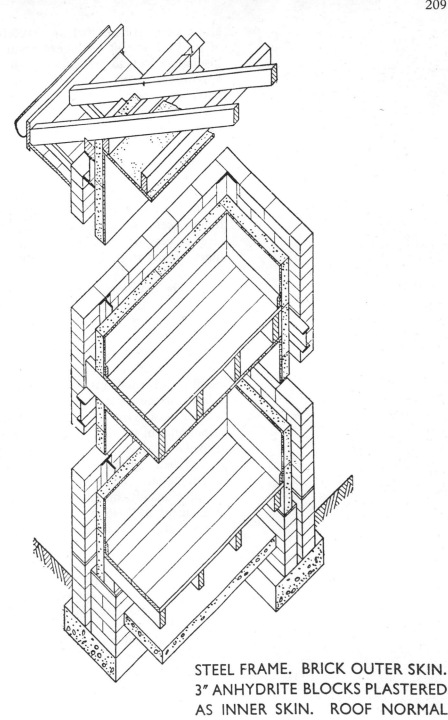

**STEEL FRAME. BRICK OUTER SKIN.
3″ ANHYDRITE BLOCKS PLASTERED
AS INNER SKIN. ROOF NORMAL**

6.2.6 The Denis Poulton House.
Architect, Denis Poulton. First houses
built at Haversham, England in 1937

clad houses like the Wier, Atholl and Telford, were expected to create
rust problems, but Burt points out that when efficiently jointed, and
with the surface suitably treated (in Scotland in 1933 a method of
painting called paint harling was successfully introduced), 'serious
corrosion of the sheets was prevented'. However the key point was

also drawn out here; that 'steel sheets contribute little to the function of the wall, except to the exclusion of rain'. Therefore strength and insulation had to be provided in some other way. Further the appearance clearly presented a problem. The steel used for roofs and party walls, was, it seems, a total failure. One curiosity in this group was the Thorncliffe house (6.2.6) which consisted of 3 ft × 3 ft × ⅜ inch (76 × 76 × 9.5 mm) flanged cast iron plates keyed to take a ¾ inch (19.1 mm) layer of 1 to 3 cement sand. The finished plates weighed 2 cwt (101.6 kg) and the weight of cast iron in one house amounted to 7 tons (7.112 tonnes). The last ones were built in Derby in 1927 at a cost of some £15 more than brick houses.

The summary in the Burt report on these inter-war prefabricated houses seems somewhat confused: it says nothing positive about the clear success of the timber housing and yet extols the few virtues of the houses using steel. The reason for this may be seen in the following:

> 'The development of both steel-frames and metal cladding may lead to the employment in housebuilding of labour and factory space now absorbed in war industries.' (Burt 1944)

However, the most important suggestions concerned standards. Part I set out, for the first time, performance standards for strength, stability, thermal insulation and acoustic insulation. Further it referred to these standards as a 'performance basis' — an idea that only became used in the late 1960s. Its final sentences drew attention to the importance of standardization and pointed out that attention should be turned to achieving a high degree of organization in workshop and on site, as well as to developing new forms of construction and uses for new materials.

Unfortunately this advice was barely heeded and almost all the effort subsequent to 1945 went into the latter, with many new systems created and new methods emphasized for their own sake, until ultimately the inevitable feedback brought this approach once again into question. By the end of the twenties the post-1918 prefabrication drive had all but disappeared since such methods could not compete with traditional ones, although some houses continued to be built into the 1940s. More importantly to us here:

> 'It had contributed nothing to any other class of buildings, nor had any move been made towards the industrialisation of building, which remained essentially, traditional both in design and execution.' (White 1965)

No attempt was made during this period to assess the many experiments which were encouraged, and it remained for the Burt Committee on House Construction (Burt 1944) to assess what had been achieved. In this 'Alice in Wonderland' world of government committees and building systems, it will be no surprise to find the Burt Committee, charged with considering 'materials and methods of construction suitable for the building of houses and flats' (Burt 1944), devoting the majority of its report to detailing 18 house building systems developed in the twenties, most of which were technically inadequate, and all of which had long since become defunct, with the notable exceptions of timber-frame construction and the no-fines technique. The implicit effect of this was to encourage interest in such

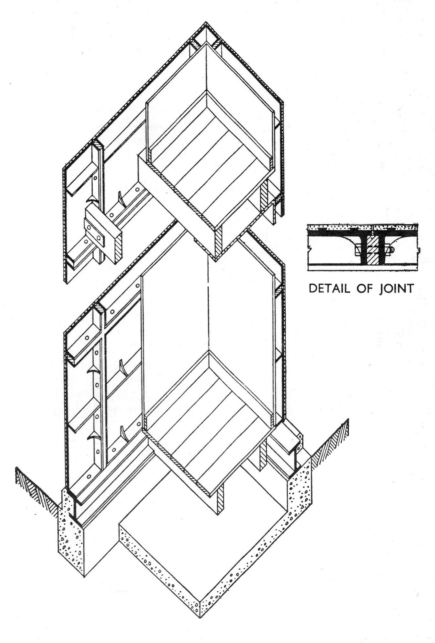

DETAIL OF JOINT

CAST-IRON STRUCTURAL PANELS WITH VIBRATED CONCRETE. FACING EXTERNALLY. INNER SKIN OF PLASTERBOARD, ETC., ON WOOD FILLETS

6.2.7 The Thorncliffe House. Newton Chambers and Co Ltd. First houses at Mortomley, England, 1927

methods in spite of their proven inadequacy, and that is, of course, exactly what happened after 1945. A typical part of the attitude of this report, on precast concrete houses, will illustrate this somewhat Orwellian approach.

'The serious defects which occurred are not inherent to (sic) the system, but resulted from inferior design or in the use of unsound material.'

However

'with good organization, houses of this type should be built more cheaply and quickly than brick houses, and there is scope for the employment of labour, inexperienced in the building industry.' (Burt 1944)

In this way the scene was set, the argument advanced, for yet another political drive to force the industry to change its ways.

Quarry Hill

'The peculiar building system of Quarry Hill flats was, like the scale, style and layout, part of the design utopianism. It did not in the end prove an economy, because of a failure to appreciate the difficulties of importing an unknown system and the special technical and social hazards of prefabrication.' (Ravetz 1974)

Inspired by the Viennese estates of the twenties, and employing a French building system, Quarry Hill Flats in Leeds embodied a range of notions which have surrounded both building systems and housing ever since (**6.2.8**). Commencing on site in 1935, the first block, Lupton House, opened in March 1938 but by 1940 the estate was still unfinished and remained so. As an example of a 'model' Local Authority estate, Quarry Hill offers a grim warning, although Alison Ravetz (1974) in her thorough study points out that it is 'uniquely typical' in its overall characteristics. Like many large new estates, bold claims were made at its inception, and it was a source of continuous press comment throughout its life. Much of this derived from the original decision to use the Mopin building system and highlighted the estate's problems in a way that many of its other problems could not, since they were (however important) common to many local authority schemes. Indeed it was these, rather than other shortcomings, that caused the estate to be thoroughly renovated after 25 years and completely demolished after 40.

The Mopin system, already used by Beaudouin and Lods at Drancy in the early thirties, consisted of a very light, hot rolled steel framework of double channels welded together, which were connected by rolled steel joists around the perimeter (**6.2.9**). White (1975) claims Quarry Hill to be one of the first uses of welded structural steelwork in the country. The frames were at twelve feet (3.66 m) intervals and

6.2.8 Quarry Hill Estate, Leeds, England. R.A.H. Livett, City Architect, Leeds and the Mopin system, 1935-40. Photograph of model of total scheme as envisaged

each bay was divided into three by means of precast vibrated reinforced concrete posts bolted to the perimeter beam. Precast concrete cladding slabs with a white spar finish were then slotted onto projecting nibs on the posts, *in situ* concrete then casing parts of the frame. Steelwork carried precast concrete floor slabs (or biscuits as they later became known in other systems) which were 1¼ in (31.8 mm) thick with a 2 in (50.8 mm) edge thickening. Inner lining, party walls and internal stanchion casings were all of anhydrite plaster blocks, usually 2 ft 0 in (0.51 m) × 12 in (0.305 m) × 2½ in (63.5 mm) thick, finished with a skim coat of plaster.

One of the key innovations, and one which contravened regulations in Britain at the time, was the way in which the steel framework and concrete was calculated to work together. This was acceptable in France but the BRS in a report pointed out that loads other than floor loads could not be effectively computed. The same report drew attention to insufficient fire protection, unprotected joints between upright members at floor levels and the unusual organization needed to achieve quality control. Finally they advised that the system should be tried, but on a large contract of two storey houses. Ravetz points out that the Building Research Station predicted 'all the main weaknesses of the system and many difficulties of the contract' which

214

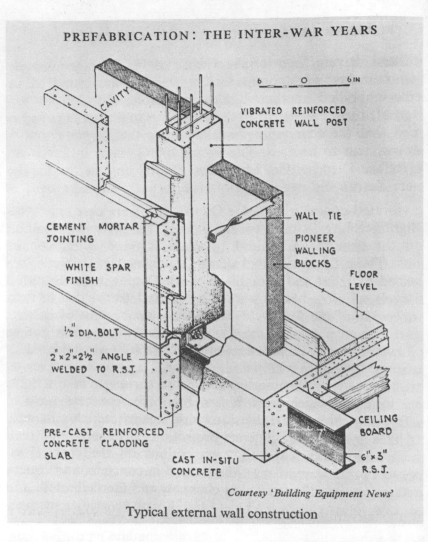

PREFABRICATION: THE INTER-WAR YEARS

VIBRATED REINFORCED
CONCRETE WALL POST

CAVITY

CEMENT MORTAR
JOINTING

WHITE SPAR
FINISH

WALL TIE

PIONEER
WALLING
BLOCKS

FLOOR
LEVEL

½" DIA. BOLT

2"×2"×2½" ANGLE
WELDED TO R.S.J.

PRE-CAST REINFORCED
CONCRETE CLADDING
SLAB.

CAST IN-SITU
CONCRETE

CEILING
BOARD

6"×3"
R.S.J.

Courtesy 'Building Equipment News'

Typical external wall construction

6.2.9 Quarry Hill Estate. External wall construction showing precast concrete slabs

Livett, the Director of Housing and City Architect, might well have considered.

The total number of flats at Quarry Hill was 938 (930 only used for residence) with some blocks rising to eight storeys. Both the passenger lifts and the Garchey water-borne refuse system were claimed to be possible due to the savings to be made from the system. The style was a clumsy confusion of 'The International Style' and the monumental modernism of the period, with alternate bands of white and brown slabs (**6.2.10**). The parabolic entrance ways and general massing were an attempt, in the continental fashion, to give the housing of the workers the status it was thought to deserve.

The use of the Mopin building system was, says Ravetz, 'almost totally negative and obstructive', with delays, threatened lawsuits, and, ironically, shortage of materials. Ravetz also perceptively observes that a 'prime cause' of the many problems on the contract was the use of an imported building system. In addition to the differing contexts which give rise to such solutions there are the considerable problems of implementation in an alien environment. This salutary experience was ignored in the sixties when a number of imported housing systems were tried with, ultimately, very similar results.

Some aspects of the Quarry Hill design, such as the curved blocks, were new to the system and inevitably the British contractor, Tarran, failed 'to appreciate the need for organization' and thus the flow of materials, factory production and a misuse of concrete caused many arguments and delays. All this gave rise to the expected problems. By 1956 the 'mean unit cost of "normal general repairs" regularly exceeded that for prewar houses, many of which were nearly twenty years older' (Ravetz 1974)

The trouble and costs caused by the wall slabs and the Garchey system continued until 1955 when the Council began systematic replacement of slabs by cement rendered copies, and by 1959 a full time staff worked on this replacement programme. Further, stripping the slabs had revealed an alarming amount of corrosion, with the result that in 1960 the City Council appointed consulting engineers to

6.2.10 Quarry Hill Estate, Leeds, England. Photograph of elevation

report on the structure, who ultimately advised a test stripping of one complete block. This revealed design faults at parapet, slab joints and projecting canopies. Worse, it was found that water had penetrated to the steelwork used underground and that the Garchey system was prone to frequent fracture. Calculations showed that the main claim of the design of the system — that all parts worked together to form a whole — was fallacious since if any one part of the structure became inadequate — such as was occurring with steelwork corrosion — the whole structure was in doubt. The Council decided to carry on extensive remedial works and, after two years, these were completed in 1964. However, by 1975 demolition of the entire scheme had commenced.

Quarry Hill is a classic case of placing too heavy a reliance on technology to deal with social problems. As Ravetz indicates, although it has interest, it is little different in terms of social issues to most other local authority estates and, after all the initial utopianism has faded, as a 'habitat' it is little worse than others. Indeed the use of a system had the benefit of attracting continuously a stream of foreign visitors which gave the estate an identity. There is also the question of feedback, for although the test block was not stripped until 1960, the many problems during construction and after were known, but in no way seemed to inform the subsequent drives to industrialize. White (1965) drew attention to some of the problems but did not draw the general conclusion, describing it as a successful 'venture in unconventional building'. It is a further, if dramatic example of the way in which architects and politicians seem unable to take note of quite simple feedback. Here was a set of practical lessons which received little of the sort of publicity which would have helped future work.

For the purposes of this study we have dwelt overly perhaps on the technical aspects of Quarry Hill: however, it should not be forgotten that 'building innovation, particularly when the importation of foreign systems is involved, (and) had lasting effects, material and non-material, on the estate and life within it' (Ravetz 1974).

Sussex Schools

The ideas crystallized by the Bauhaus, and particularly by Walter Gropius, have had a profound and significant influence on the development of the theory and reality of building systems. In Britain the development of school building systems owes a particular debt to this source, although it is one perhaps that many wish had been paid off more quickly. With school building systems embodying these views accepted as a part of government policy, and used extensively in the school building programme, it is not surprising perhaps that many aspects of this particular design approach have received scant attention. In this regard that part of the model or image so eloquently delineated by Gropius and which referred to the merits of the norm, of standardization, rationalization and of steel frame construction with lightweight cladding and flat roof are the stuff of CLASP, SCOLA and most of the other school building systems.

An early translation of the image into built form can be seen in the work of C. G. Stillman whilst he was County Architect at West Sussex during the 1930s. In the latter part of this decade several schools were built which embodied the Gropius image clearly and concisely. Sidlesham, completed in 1936, and the first school in England in 1938 to provide hot mid-day meals for its children (**6.2.11**), Selsey (**6.2.12**) completed in 1937 and North Lancing completed in 1938. According to the *Architects' Journal* for 26th November, 1942:

> 'It was in 1936 that educationalists began to take an interest in WSCC (West Sussex County Council). In that year they heard that C. G. Stillman, the County Architect, had developed and built some experimental classrooms at Sidlesham. These classrooms were unusual. They were of light steel standard construction and were the first of their kind to be built in the country.' (White 1965)

Indeed a close examination of both the Sidlesham and Selsey schools is revealing. Sidlesham for example consists of a classroom wing with

6.2.11 School at Sidlesham, Sussex, England. C.G. Stillman, County Architect to West Sussex, 1936. South elevation (Photo 1969)

6.2.12 School at Selsey, Sussex, England. C.G. Stillman, County Architect, 1937. South elevation (Photo 1969)

its long side facing due south and its corridor running along the north side. This corridor has a low ceiling height to give clerestory light to the classrooms. The structure is a light steel frame with exposed lattice beams spanning the classrooms (**6.2.13**). The dropped corridor roof allows the cross-ventilation and additional natural light considered of such importance.

The south side of the classrooms consists of a lightweight skin of horizontally sliding steel windows with a continuous spandrel wall beneath. On a visit to the school in 1969 on a soft spring day, with the windows slid open, it was possible to regain the design intentions with some delight. The internal partitions were all of lightweight construction reinforcing the idea of the interior being flexible. The doors like the windows were steel, glazed, and the passage of time had endowed the thin members with an Art Nouveau quality not perhaps intended by their rigorous designers.

The south elevation of the building demonstrated its sources quite clearly with its continuous horizontal bands of windows, white painted continuous wall beneath, flat roof and solid flank walls (**6.2.11**). The north side and end elevation indicated many of the problems facing the designers. The uneasy junction of the corridor to the main building, with its slightly decaying detailing demonstrated many of the aspects of this type of approach which subsequently gave the word 'prefabricated' such a bad name. However the interesting thing about these buildings of Stillman's is that they so clearly show the enthusiasm and care of their designers. Even now they carry a certain faded panache that is absent from most school building system work. The care shown in providing ventilation to the north corridor in an effort to combat the typical smell of school is one example. Here the curved asbestos overhang of the roof conceals a simple ventilator operated from the inside, albeit now somewhat difficult to use (**6.2.14**).

6.2.13 School at Sidlesham, West Sussex, England. Interior of classroom showing lightweight steel and window wall (Photo 1969)

6.2.14 School at Sidlesham, West Sussex, England. Corridor showing interior of eaves detail. Sliding ventilator panels located in bottom of void over windows (Photo 1969)

To get these school buildings built, Stillman, it seems, had to engage himself fully. It would appear that he designed them himself and oversaw their execution. To introduce a steel frame far lighter than would normally have been designed or manufactured at that time, Stillman adapted an idea he had worked on with the County Medical officer earlier for using caravans as mobile dental clinics. In this respect it·is interesting how Stillman used local resources to pursue his ideas. In Worthing a garage owner called Gibbs specialized in constructing caravans from cold formed steel channels, angles and the like. It appears that Stillman persuaded Gibbs to engage a structural engineer and to install new machine tools to enable him to make steelwork for the mobile clinics and subsequently for the lightweight frame of the school buildings (White 1965).

The model for building systems was clear: lightweight frame, dry lightweight largely glazed envelope, flexible interior. The limitations of the model, however, suggest that its relevance subsequently was seen in ideal terms rather than as a growing changing reality. The problems of solar gain, heat loss, inadequate heating, poor noise control were well within their architectural context at the time. The schools embody the modern movement's interest in large areas of single glazing, fresh air and sun and in them we see the reformulation of tradition in terms which in many ways produced a worse result than before. However, many of Stillman's schools were better thought out than those that sought to follow him — he used, for example, projecting sunbreakers at transom level on some schools, to deal with solar problems on the extensive glass. Internal flexiblity was another of the underlying aims of the approach. All Stillman's schools used this concept to build in a future change possibility by means of the use of frame and infill. However, White in his book *Prefabrication* can remark:

> 'After 22 years not one of these classroom wings had been extended or modified though the possibility of doing so remained.' (White 1965)

However, when my own visit to Selsey took place in 1969 the classroom wing had been replanned and updated by the introduction of additional toilets and storage. Whilst there had been little radical alteration to the planning concept, since there were still classroom 'boxes' and a corridor as in the original, it had been possible to make internal changes and add new services without interfering with the structure. It may be questioned whether there was any more possibility for flexibility with this type of approach than others, or whether the flexibility investment over 32 years was valuable. White's comment on the work at West Sussex is perhaps of some significance:

> 'At all events the experiment proved highly successful. As it turned out, not so much for its immediate results in initial cost or for its flexibility, but for the great influence it had on post war development, for here, in Sussex, the seeds were planted of all that has since grown to fruition in Herts, Nottingham and elsewhere.' (White 1965)

Here we see clearly the incestuous nature of the argument. It is not for the performance that the experiments are valued but for their influence on later, similar developments. White himself makes no claims for these schools on the grounds of performance but allows their experimental effect and their influence make the claim to validity. Here we see how the repatterning of the school building problem into a new identifiable form, replete with a philosophy tied tightly to a method of construction, carries its own validity. Such validity lies in the claim to the imperatives of the production line — albeit in fact in this case implemented almost on a craft basis — together with the matching of traditional construction costs. To gain credibility such a new set of ideas must have such purifying elements so that it will be clearly seen to have rational authority: indeed, Gropius in his address to the Design and Industries Association in

May 1934 refers to the purifying role of the idea of rationalization together with the satisfaction of the human soul.

In these few, rather neglected buildings, Stillman set the stage for big changes, both in the approach to school building itself, and in what they implied for bringing about a building industry that would more accurately reflect an industry in the service of the people. However, the difference between a handful of carefully thought out schools built with enthusiasm, and that same idea repeatedly multiplied as the only right way of building is vast. Before that happened a number of other dedicated men were to develop and apply Stillman's work in Hertfordshire in 1945, on school building systems and on the post war housing drive.

COMMERCE AND STANDARDS

'The prefabrication systems designed by architects in Britain have concentrated upon giving the individual architect as much planning flexibility as possible within a very limited range of components. This implied, right from the early Hertfordshire experiments, that the commercial systems of prefabricated spans in one direction only, allowing freedom of length only, should be rejected in favour of flexibility in both directions on plan.'

The Architects' Journal
29 June 1961

The establishing of the British client sponsored building systems to the point where they controlled a major part of the market was presaged by a rather different situation. The postwar enthusiasm for prefabrication gave rise not only to hopes amongst architects in local authorities, as at Hertfordshire, but also throughout the whole private sector. Just as there was considerable support for Homes fit for Heroes, so there was support for schools fit for the children of heroes. Indeed this was given greater strength by the intent of the 1944 Education Act and by the several government statements concerning standardization. Such a situation was intended to encourage a suitable climate for bringing about the industrialized nirvana seen by some as long overdue. It certainly did give rise to enormous activity, amongst manufacturers and suppliers, which proved to have little lasting capability. Both for housing and schools there developed a number of commercial building systems, and it is against this climate that the achievements of ARCON and Hertfordshire must be seen.

In housing it is clear that as soon as the direct government support for housing was withdrawn both in the form of directives and finance, the prefabricated house building systems began to wane dramatically. Some local authorities continued to erect small numbers from commercial firms but the market quickly demonstrated that such expensive experiments could not survive without government subsidy, in spite of the apparent logic that had been built up around the myth of the production line.

In school building the situation was very different and, with government encouragement, industry and architects looked closely at appropriate solutions. With the backing of the report of the Wood

Committee on Standard Construction in Schools (Wood 1944) many firms entered the school market with building systems of various sorts. They grew rapidly and were very successful in terms of productivity. The manner in which most of them met their demise offers a curious comment on the period. What is of interest here is how they have managed to disappear from the face of architectural (and technological) history, whilst the Hertfordshire work became so durable. One answer to this must be, however unpalatable, that the architectural profession and its press is an organized body that can, and does, confer durability on certain work. The building industry (or the manufacturers in this case) had no such means of ensuring historical continuity.

Even more fundamental however, is that the Hertfordshire work itself had assured continuity so long as the county were committed to that particular approach. Further, being a part of government they had advantages that no private manufacturer can have. This in the end was crucial, for not only did Hertfordshire people go to other local authorities all over England carrying the word, but several went to the Ministry of Education itself to set up their first development group in 1948. From this vantage point it was possible to press nationally for the component building philosophy and many experiments were carried out with manufacturers and local authorities to test their ideas. With the formation of CLASP in 1955 began the real decline of the proprietary building systems, for now a way was seen to bring back under the control of the architect (and his client) these bounties of industrialization. The commercial systems quickly found it hard to gain orders and, after attempts to refashion them more to architects' tastes they mostly went to the wall.

This situation points up, yet again, one of the great dilemmas that architects have found themselves in with regard to technology. They wish to encourage it, make use of it, but as soon as it ceases to match their view of what it should do, they react. In this case there is no doubt that many of the commercial systems produced very inadequate buildings — particularly environmentally, but even at this distance in time they are not very much worse than the bulk of work produced by architects. Further it was claimed that the commercial systems had cut down the architects' range of choices in favour of the expediencies of production methods. This is particularly ironic in view of the subsequent limitations brought about nationally by the available consortia systems. More importantly architects in local authorities found themselves planning schools in commercial systems, selecting the 'elevational treatment' from the available choices, ensuring that certain standards were met and waiting for the building to appear on site. They themselves began to be concerned that they would in future have little to do if this continued. Such rumblings, however much at the level of the drawing board, also have their effect.

To return to the position towards the end of the war, the Wood Committee, whose findings were published in Post-War Building Study Number 2, Standard Construction for Schools (Wood 1944), had amongst its architect members, C. G. Stillman. The document itself suggests, quite briefly, that the solution to the need for school buildings lay in the standardization of components. Particular attention was given to the use of the frame, and the steel frame in particular, and to the idea of standard bays. This was done against the opinions of the County Architects' Society who expressed doubts concerning the necessity for abandoning traditional materials. The report points out that the procedures used to gain approval and to complete the drawings often led to delays of 'more than twelve months' before the projects could start. It asked the County Architects' Society to make suggestions as to how this could be alleviated since the majority of these concerned the delays in gaining approvals from the Board of Education itself:

> 'The time taken by the Board of Education in approving sketch plans appears to have varied, and the effect in some instances has caused an appreciable delay before progress could be made with the working drawings and quantities, with subsequent delay in securing approval of working drawings.' (Wood 1944)

This, a familiar story to any school architect, was clearly given second place to the whole question of standardization:

> 'Our suggestions, if wisely carried out, need not involve the sacrifice of anything vital. We recognise to the full the constant educational influence exercised by the physical amenities and aesthetic qualities of the school building. There seems, however, no question to suppose that the adoption of pre-planning and standardization, subject to appropriate safeguards, would be incompatible with the production of schools fully satisfactory in both appearance and use.' (Wood 1944)

Their summary enunciated four of the basic rules of prefabrication:

1. 'The structural elements for a steel framework could be produced in large quantity by mass production to standard size or sizes.'
2. 'The adoption of the same basic unit of dimension in either direction — for both length and width — would make it possible to project the layout of the school in any direction desired as a connected whole, and would further make it possible to use prefabricated wall panels or window frames of the same size in either direction.'
3. 'The use of a framework would enable roofing to be taken in hand while walls were still in course of erection, thus saving time on the process of drying out.'
4. '...The framework contemplated could equally well be used for the separate blocks of classrooms or practical rooms, or for separate halls, gymnasia etc., where additions are required to existing accommodation...' (Wood 1944)

The report proposed the 'unit dimension' of 8 ft 3 in (2.52 m), which was based on the standard classroom width of 24 ft 0 in (7.32 m) plus an allowance for walls: three 8 ft 3 in (2.54 m) bays would give 24 ft 9 in (7.54 m). They recognize that it is not applicable to the depth dimension of the normal classroom which was 21-22 ft (6.4-6.7 m), and dealt with this problem by referring to 'practical and special' rooms which could, it said, be planned accordingly. Their proposal, they admitted, gave an excess of space over existing standards — which seems surprising in view of their concern with economy and which they brushed off by saying it would be 'very small in comparison with the school as a whole; and would in our opinion be compensated by the other advantages which this method of standardisation offers'. In short, the economies that they expected to accrue from mass production would pay for extra space, a recurring if unproven view of system designers.

Whilst Stillman, who was by this time County Architect for Middlesex, continued to build schools (Stillman 1949) somewhat similar to those of his West Sussex period before the war, it remained for Hertfordshire to take the initiative offered by the Wood Committee, and it was there that some major advances were made. Before that series of most interesting developments is examined, however, what of the commercial systems? They feature little, as has been suggested, in the history books but they were prolific in their production of school buildings, which in the context of the Wood recommendations for standardization was impressive. That they may lack some quality which may be defined as 'architecture' is possibly true, whether they lack it more significantly than many of the Hertfordshire schools is, I believe, questionable. What is certain is that they are little worse than vast numbers of systems schools which partly owe their existence to these commercial initiatives.

With the cessation of hostilities it seemed an obvious decision to attempt to turn the vast factory resources developed during wartime to better use. As evidenced by the Wood Committee, once again the case was made for buildings to be made in the factory where the machinery and labour were already available. On this occasion the climate was right; probably more right than it had ever been — from the political, commercial and architectural points of view.

The Bristol Aeroplane Company offers a very good example. With the requirement for aircraft and aluminium products now gone they turned their production capabilities in other directions. The AIROH house was their contribution to the prefabricated house demand, although production of this was shortlived. When it ceased, attention turned to the uses of prefabricated structures for other building types. Industrial buildings, pithead baths and school building were all fields where BAC provided structures. As the Bristol Review for Christmas 1948 put it, 'Prefabricated unit installations, sprouting at all points of the compass throughout the British Isles...' (BAC 1948).

By 31st July, 1949, 57 school buildings had been completed by BAC, with 15 in the course of erection and another 49 in the programme. It was not until the end of 1951 that Hertfordshire completed 40 schools. Schools were supplied to Inverness in Scotland and Camborne in Cornwall. An exhibition school was flown to Paris and enquiries came in from all over the world. These were not seen as temporary buildings but as permanent prefabricated structures in which the design of the system was closely bound up with the educational views of the time; finger planning, long corridors, and medium span structures. They attempted to make the maximum use of metals, particularly aluminium, and were designed on a simple repetitive bay system with overhanging eaves and large areas of curtain wall using panel and glass infill (**6.3.1**). Cited as the 'largest aluminium school in Britain', Benarty Primary School in Fifeshire was opened on September 11, 1950 as 'Scotland's unique ultra-modern, educational showpiece'. Excelling even itself the BAC house magazine, Bristol Review for Spring 1951, heralded 'the largest aluminium school yet built' (BAC 1951) when Whitmore Park Primary School was opened by the Minister of Education, George Tomlinson, on 10th January 1951.

This school for 840 children was constructed in 18 months concerning which Mr. Tomlinson was appropriately encouraging. He was also referred to as being 'particularly enthusiastic about the generous window space in the school. Being continually in touch with light and colour', he said 'would not only bring a difference in the attitude, but in the characters of the children' (BAC 1951). Here Mr. Tomlinson was faithfully reflecting the concerns of the recent past in reacting to the rigours of the industrial revolution. This had been reinforced by the hygienic and health vogues of the twenties and thirties, crystallized for architecture in the work of the Bauhaus and the white architecture of the period with its roof terraces and large windows. In many of the schools of this time we can see the overriding obsession with sunlight and ventilation at its strongest. When Mr. Tomlinson referred to light he certainly did not mean artificial light. These lightweight, largely metal clad buildings, embodied the highest

6.3.1. Warblington Secondary School, England. Bristol Aeroplane Company (BAC), early 1950s (Photo 1980)

228

motives of the apostles of industrialization, together with their worst features. Even today a BAC school attracts the attention of some architects because of its clear commitment to a total technology: its honesty to its argument seems clear — warts and all. Whereas more recent work embodies almost the same philosophy, if somewhat more obscured, and perpetuates most of the faults. Of course, heat loss, solar gain, acoustic problems and the lack of variety of surface have given schools of this period an exceedingly hard time, to say nothing of those teachers and children who have had to use them. Sadly the argument that they are buildings produced under stress is still the paramount answer given to such questions: an inappropriate answer since the priorities selected and argued by those setting the standards and by designers have been inadequate in safeguarding basic environmental and sensory needs of users.

Other commercial systems which flourished included the Derwent System (Vic Hallam Ltd) which utilized timber columns and plywood box beams with a range of infill panels all on a 6 ft 4 in (1.93 m) square grid; Uni-seco (**6.3.2**); the Intergrid System (Gilbert-Ash Ltd), using

6.3.2 'Uni-Seco' system of prefabricated construction used for school building. A commercially developed and marketed system of the 1940s. Note the horizontal sun visors and vertical fins to control glare and solar gain. Such devices later disappeared with the Introduction of the client sponsored systems

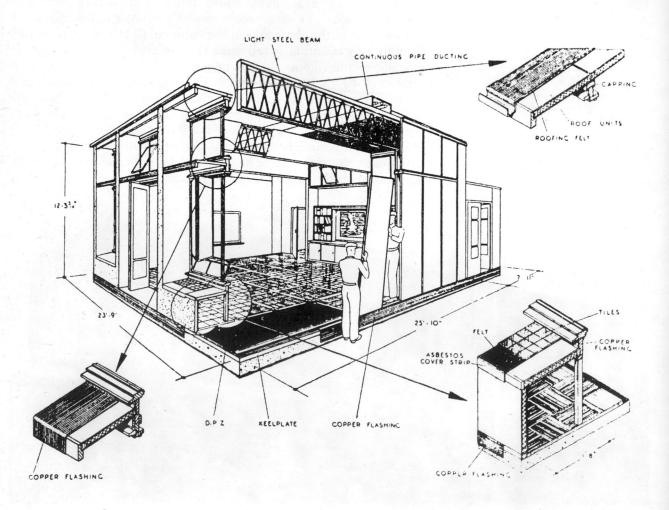

6.3.3 'Presweld' or Hills Dry
Building System: Hills (West
Bromwich) Ltd, 1957. (a) Cross-
section; (b) typical external and
internal corner details

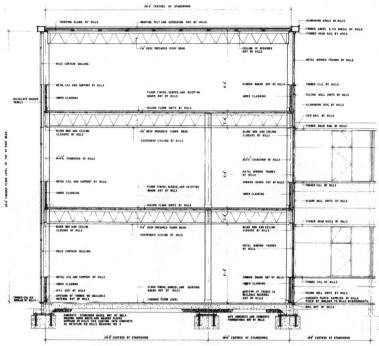

a

TYPICAL CROSS SECTION AND PART ELEVATION ON GRID LINE H 3

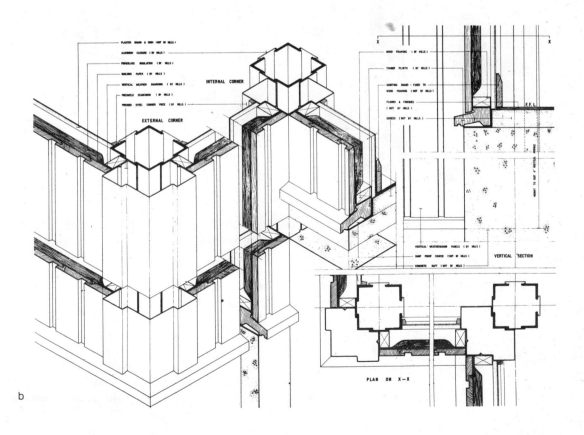

b

230

a

prestressed concrete columns and beams used at Worthing in 1953; the Laingspan System, a concrete system also using prestressing; the Medway System (Medway Buildings and Supplies Ltd) using timber columns and lattice beams on either a 6 ft 4 in (1.93 m) or 6 ft 0 in (1.83 m) grid; and the Hills Presweld System (**6.3.3**). Described as the Hills Dry Building System, the approach developed from their non-traditional housing and with the work at Hertfordshire formed a model for the client sponsored systems that followed. Hills themselves became steel suppliers to Hertfordshire and ultimately went bankrupt. Hills supplied a total system consisting of a rigid light steel frame using angle batten stanchions and lattice beams, 'Hilcon' precast concrete floor 'biscuits', metal curtain walling, windows and concrete

b

1 ERECTION OF STEEL FRAME

2 FIXING CLADDING STEEL & EAVES

3 WALL SLABS & WINDOW SURROUND

4 INSERTION OF WINDOW UNIT

6.3.4 School at Wokingham, Berkshire, England. Ministry of Education Development Group, 1951. (a) From the east; (b) Erection procedure diagrams. Illustrations from *MOE Building Bulletin* No. 8, 1952

cladding panels. The system used a 3 ft 4 in (approx. 1 m) grid with thin concrete panels spanning up to 10 ft 0 in (3.01 m) with other forms of traditional construction designed to junction with the system. The descriptive documentation was very clearly worked out and came with a separate booklet giving component price lists. A prefabricated unit system for school building was also designed by ARCON in 1948, using a series of standard span sections linked by traditional construction. The classroom section 'has been designed to give a minimum daylight factor on the north west side of 5%' and the south east, heavily glazed walls carried horizontal louvres to control the sun (AD 1949).

Through all this the Ministry of Education's (now Department of Education and Science) Architects and Buildings Branch had played a crucial part, first in encouraging development work on the part of commercial firms and subsequently in encouraging the client sponsored systems which replaced them. This group, established in

1949, drew heavily on both Hertfordshire experience and on Hertfordshire staff, with Stirrat Johnson-Marshall and others moving to the Ministry to pursue the work at a national level. The group was also multidisciplinary, including not only architects but quantity surveyors and H.M. Inspectors of Schools. In this way the trio of architecture, cost and client/users were represented. Their task was to draw together work, set up development projects of their own, and disseminate information by means of Building Bulletins. The Development Project was a central part of their work, since it was strongly felt that any advisory or innovative work must go hand in hand with real building to obtain direct results. The first of these projects was a school at Wokingham for Berkshire County Council where many ideas on prefabrication which had developed were brought together (**6.3.4**). Using a modified version of the Hill's frame the designers worked with various manufacturers to develop the parts of the system. A square 3 ft 4 in (991 mm) horizontal grid was used and this project was important in beginning to get the usefulness of this across nationally. In addition a vertical module of 2 ft 0 in (610 mm) was adopted. The school which followed this as a development project was at Belper in Derbyshire, where Brockhouse Steel Structures supplied the frame, and which ultimately led the way to CLASP.

The Building Bulletins which the Branch has produced over the years have formed a useful source of information for the designers of schools. However the structure of government and local government in England is such that far too often these have been taken as edicts from central government with the result that each time a Bulletin is published on a specific building, a rash of similar buildings appears all over the country. This misuse of the work is partly due to the way in which the Bulletins are set out, many of them being a self-contained account of the pursuit of one approach culminating in a particular architectural solution. Few of them demonstrate a real research based background, or draw widely on outside sources. Through its Development Projects the Ministry was able to press the results of its own findings, the early ones encouraging commercial suppliers to develop systems or subsystems for their, and others', use:

'Post-war experience has shown, however, that the incentive of substantial orders based on successive annual building programmes outweighs the difficulties in the eyes of at least the most enterprising manufacturers; others are likely to follow suit if demand seems to justify it.' (MOE 1952)

The next few years were spent in endeavouring to establish such a demand for the several commercial systems which had developed in response to the ministry's suggestion 'that to be efficient and economical the development of a new system must be comprehensive' (MOE 1952). No sooner had this got under way, however, when such encouragement of commercial systems began to dwindle as the Ministry shifted attention to the idea of the client sponsored system.

This combination of both system designers and system users appeared to offer advantages, although the very nature of this relationship has made objective assessment of this difficult. Although some of the commercial systems survived in one form or another (for example Hallam and Medway), some supplying the client sponsored consortia, few could compete with local authorities when the latter backed their own systems. In retrospect it is an odd policy that encourages the considerable investment of time and energy in the development of such systems and then rewards this by removing the market upon which it is based.

HOMES FIT FOR HEROES, AGAIN

'The age we live in will surely be known as the age of invention. This has its dangers and its penalties, but it should also have its rewards and excitements. The skill and ingenuity of our technicians can revolutionise housing as they have revolutionised so many other undertakings.'

Aneurin Bevan, Minister of Health
Foreword to *Homes for the People*,
Association of Building Technicians 1946

During the war of 1939–45 very few houses were built in Britain: in total some 200,000 and these largely for relocated war workers. A similar number were destroyed and some 250,000 made useless. The number of houses and flats built in 1944 was 7000 whereas in 1938 it had been 350,000. The terrors of war had their counterbalance, however, in the renewed opportunities that followed. During and directly after hostilities, committees and individuals gave great attention to ways in which Britain could be reconstructed socially and physically. Endless committees turned their attention to the problems of postwar housing and most of these examined prefabrication as an idea. Many came to the conclusion that the wholesale prefabrication of houses, however desirable, was not likely to be effective with the prevailing situation in the building industry. Inevitably the analogy between consumer products was quoted and requoted by committee reports, it being pointed out that the abundance and low price of consumer goods like refrigerators, cars and so on was a valuable model for the building industry. If only the house was produced in this way, it was said, then it would become cheap and shortage could be eradicated. The number of times this idea recurs is some measure of its seductiveness: it is also a measure of the emphasis put on 'the machine' to solve problems, or perhaps more appropriately, it rests firmly on that dictum of Marx that to change society one must gain control of the means of production.

Of course the irony of this is that Marx saw the division of labour, the alienating effect of factory work and the creation of surplus value as the basis of what he described as the contradictions in capitalist society, whilst many of those architects and others committed both socially and politically to his ideas applied them by encouraging more and more building components to be made in the factory. Thus more and more workers would be exposed to the factory and production line ethic. One might here recall Meyerhold's concept of the beauty of work. However, there is a further layer to the argument in that Marx also held that communism would only be brought about when the contradictions in society caused the capitalist system state to break down. Writing before the Russian and Chinese revolutions, he pointed out that it was the most advanced nations industrially that would first move towards socialism and communism. In summary, the logic of this demanded that industrialization be increased so that its ultimate effects may be hastened. Agrarian or partly industrialized countries, so it was argued, had to go through the process. If social change was to come about through industrialization then industrialization must be brought about. First industrialize, then organize. Gaining control of the means of production in the name of the people was crucial so that a minority of owners of capital would not strip off the 'surplus value' produced by the labour of each worker. Many of those involved in building systems have seen them as a way of controlling parts of the industry and thus turning rapacity to the good of the community.

Marxism had, of course, permeated the intellectual life of the twenties and thirties, and architects and planners whose skills are largely of the planning and predictive kind had consciously and unconsciously absorbed much of this philosophy. The forties and fifties saw attempts to apply this as the overriding, ideal form of action.

If there is amusement in the postwar British situation it may be had from the way in which British government, both Conservative and Labour together with the bureaucratic administrative machine, has absorbed orthodox marxist modes of operation. Of course, the traditional marxist may claim that this is the historical process at work and that it is part of the collapse of capitalism. Seeing the works of society as a superstructure on the production base follows Hegelian logic and is closely bound up with the emphasis on the application of scientific method which gained force during the nineteenth century. Establishing the means of production as primal, has the power and force of a religion: of course production shapes society but it is as dangerous to place too much emphasis on a purely functional argument with regard to society as it is to do this with regard to, say, the actions of individuals or the making of buildings. As recently as 1965, White in his book *Prefabrication* states:

'However, no conservatism in matters of taste and habit can withstand forever the pressure of hard economic facts. When the shoe starts to pinch, a reason will always be found for extolling the aesthetic virtues of wider soles and broader toes! The habits and preferences of the British in housing matters have already undergone a big change: in the cities they are gradually being conditioned to accept life in blocks of flats and a more communal way of living in general. Under the pressure of numbers and the insistent demand for living space with modern conveniences (and inconveniences) the semi-detached house of past suburban development, with its jealously guarded privacy, is becoming an anachronism which should give way to more sociable habits.' (White 1965)

White not only states the primacy of economics, but many another shibboleth relevant to the period in question. People are being 'conditioned' to accept living in flats (it seems) and encouraged to a more communal way of living in keeping with the architectural orthodoxies of the twenties and thirties — and forties and fifties. The semi-detached house is criticized for being concerned with privacy and cited as anachronistic in the face of 'more sociable habits' (White 1965).

The privately sponsored 'Committee for the Industrial and Scientific Provision of Housing', itself a title encapsulating the concerns outlined above, appears to have been connected to a pair of houses designed during 1942/3 on the initiative of Coventry City Architects' Department. It is of interest to the subsequent course of building systems development that the City Architect at Coventry at the time was Donald Gibson (later Sir Donald), and that the man whom he invited to work on the design of the totally prefabricated house in 1942 was a young man named Edric Neel, who later became important in the ARCON group. The City of Coventry was suffering great losses of housing due to the wartime bomb attacks and in June 1942 the Housing Committee agreed to the design and construction of a pair of experimental houses. An offer to collaborate with this work from a firm called Radiation Ltd was accepted by the City Architect. The 'Coventry' house (**6.4.1**), in being constructed of a light steel tubular frame (by Stewart and Lloyds) with small precast concrete panels and metal windows attached, conformed to the most powerful influences in the prefabricators armoury at the time. There was a central heating and prefabricated plumbing installation together with a great deal of steel in the partitions and cupboards. It appears to owe a lot to the Aluminium Bungalow prototype designed by the Bristol Aeroplane Company although never produced in quantity. Apparently satisfactory (notwithstanding the inevitable roof leaks, the bane of prefabricators everywhere), the Coventry house pair were not repeated. As a specific way of employing mass production techniques it failed although it pointed the way to later developments. The Coventry house brought together several strands of the argument: a considerable shortage of housing, an excess of factory capacity due to the expected drop in wartime production and the use of the lightweight precision technology for which the area was famous (Builder 1943; 1945).

The work of the Ministry of Works Standards Committee, set up in 1942, is of some interest in the long history of standardization but only in so far as it reinforces the common arguments on the topic. Their reports of 1944 and 1946 accepted the extension of the scope of prefabrication to a wider range of components and to structural units. They also emphasized that there was no case for alternative methods unless they improved standards without improving cost, and they saw a

6.4.1 Coventry Experimental House. D.E.E. Gibson, City Architect; Edric Neel, Project Architect, 1943-4

nationwide acceptance of standard requirements which should relate to the building regulations. The same might be said 30 years later, in spite of the claimed advances.

The government's temporary housing programme commenced with the passing of the Housing (temporary accommodation) Act in 1944, and allocated £150,000,000. Eleven types of temporary house were used, the three most interesting being the ARCON (41,000 built), the Aluminium Bungalow (55,000 built), and the Uni-Seco houses (29,000 built). Between 1945 and 1948 a total of 156,667 houses were erected under the programme. White describes the temporary housing programme as an:

> 'immense but heterogeneous effort'... 'a disappointing episode of underestimated costs and over estimated potentialities which tended to give prefabrication a bad name among the public, among local authorities, and in parliament.' (White 1965)

Nevertheless, a mere 18 years passed, to 1960 before the same arguments as those heard in 1942 were again being put forward with regard to housing, when the political restyling of the term prefabrication to industrialized building, took place. The demise of the many systems in the mid-sixties could well be described in the same words as White used in connection with the postwar efforts.

The cost of the postwar temporary houses escalated from £375 to £1300 including work on site, and parliament was called upon to provide more and more money. Those involved were, perhaps, less fortunate than other systems designers in that with the costs being directly dealt with by government as they arose there was little that could be hidden, written off to development, or lost in minor works programmes and the like. Certainly the cost of retooling and re-equipping factories was not amortized, a familiar story for housing systems.

However, many of the houses were very interesting indeed from the design point of view and one example will be given here: that of the Arcon house. Although space standards were low and the houses carried the connotations of 'temporary' or 'prefab' they made very comfortable homes which many tenants were very loath to leave. There are a number of reasons for this which are of importance to designers. First, the amount and quality of servicing provided was high with prefabricated kitchen/bathroom units, electric or gas cookers, refrigerators and drying cupboards — luxuries unheard of previously by most council tenants. Second, although small and made of unusual materials, they were detached houses, single storey and standing in their own private gardens. However, such concern with privacy, self-determination and comfort were out of keeping with the social aims of many designers who saw the individual house as a retreat from community and little to do with the collectivization of society. The apartment block seemed more in tune with such views. Further, whilst many architects saw clearly a relation between the industrialization of the building process and change in society at large, this tended to conflict with what people actually wanted. Equally, at this time little real attention had been given to developing useful data for the way people use buildings and space. The technological imperative was here quite powerful and whilst most architects are humanistic in intention, they have been themselves conditioned by their own exclusive culture. This may encourage such empty phraseology as 'giving people what they need rather than what they want' and a certain arrogance which is often necessary to those involved in making a series of complex decisions.

Neel's experience with Gibson at Coventry on the design of a totally prefabricated house using one manufacturing concern quickly led him to the realization that the limitations of this could be overcome by bringing together a number of firms whose combined manufacturing interests covered most of the materials and methods required. Neel spent only six months at Coventry, after which he, Rodney Thomas and Raglan Squire formed a partnership in London which they called ARCON — Architectural Consultants, no doubt following the anonymous tradition of the thirties established by such firms as GATEPAC in Barcelona, TECTON in London.

ARCON was a firm of designers and included, at this time, no manufacturing interests. It concentrated much of its effort on developing ideas about prefabrication. Neel, having established contacts with the Directors of a number of firms during his previous work, opened further discussions with them about continuing to support some of the ideas on the prefabrication of houses and parts of houses. The group of manufacturers, later themselves to take the name The ARCON Group, retained ARCON to act as co-ordinators and designers to the group. It must be emphasized that what the group and the designers were interested in was the prefabrication of parts for permanent buildings. The result of their first collaboration was a pair of prototype houses erected at Coventry.

Their second project was the design of a kitchen/bathroom service unit (**6.4.2**) later to become a significant part of the design of the ARCON house, although in a version by the Ministry of Works. Drawing somewhat upon the proposals of Buckminster Fuller (see earlier) this had all the kitchen and bathroom appliances brought together on a service duct, thus minimizing plumbing runs, and

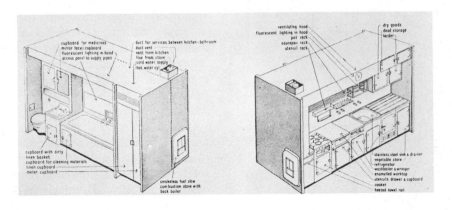

6.4.2 Kitchen/Bathroom unit for Fisher and Ludlow Ltd. ARCON architects, 1945-6

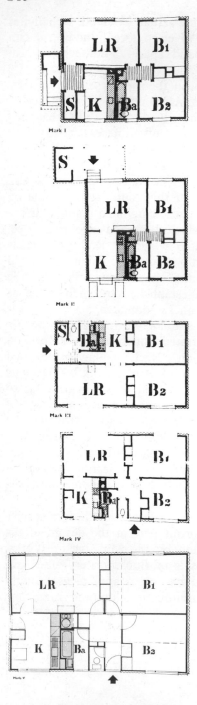

6.4.3 The Arcon house: ARCON architects, 1945-8. Plan development from Mark I to Mark V, the latter being the one used in the built version

producing a compact unit easily manufactured off site and merely connected up when in position. According to Moffett (1955) this work with Fisher and Ludlow encouraged them to try and interest several car manufacturers in their design work, hoping to persuade them to apply the techniques of the assembly line to the manufacture of the kitchen/bathroom unit. But, he reports, they felt that such techniques were out of place in a house. The service unit was exhibited but it was not popular and was not mass produced, although the Ministry of Works in their own kitchen/bathroom unit design were indulging in a popular Ministry activity, that of absorbing and developing good ideas which were looking for markets elsewhere.

It is clear that Neel and ARCON were developing ideas very quickly, and that they had some skill in getting ideas beyond the drawing board stage in that they were able to set up the group of manufacturers and interest government. A subsequent example of this is in the way in which they next designed a complete demountable single storey house (the Mark I ARCON house) to show how their ideas could work and to incorporate the kitchen/bathroom unit. No doubt they were very well aware of the discussions going on at government and manufacturer level regarding possible solutions to the postwar housing project — and indeed of Gibson's part in this. It is of some interest to see how these ideas become, albeit in a modified form, the ARCON Mark 5 house finally erected in large numbers all over England (**6.4.3**).

First the Mark I design was put up to the group of associated manufacturers who discussed it from every angle — materials, construction and cost. Then from this a second proposal, the Mark 2, was evolved. It was of tubular steel frame with asbestos wall cladding on pile foundations, the latter giving it considerable flexibility regarding siting. Since advice on the putting together of such a house was now important, Taylor Woodrow, a firm of contractors, was added to the group so that advice on erection procedures could be built in at the design stage. After public pronouncements by the Prime Minister and Lord Portal in 1944 on the desirability of erecting temporary houses to relieve the housing shortage, Edric Neel on behalf of ARCON and, as Moffett says 'with an eye for business' (Moffett 1955) wrote to Portal pointing out that total prefabrication on a big scale was the only 'inevitable' solution to the housing problem. Again, according to Moffett 'Lord Portal replied asking if Neel and his associates had any experience of designing prefabricated houses using a minimum of steel'. It so happened that the ARCON group had. They then invited him to look at their designs. The result was that erection of the ARCON house (Mark 4) began on a site at the Tate gallery on July 4th, 1944, and was complete two weeks later when the Minister inspected it.

It could be said that the Mark 1 was the designers' proposal; the Mark 2 was the designers' proposal modified by the necessities of manufacturing technique. Subsequent modifications are equally interesting in that they demonstrate the sort of changes that occur in design work and particularly in development work of this sort, and the wide range of people who become involved in making what are, essentially, design decisions.

The Mark 2 house was approved on all but two counts: the use of stressed skin plywood floor and the plan form. It can be seen that it was exceedingly compact with one entrance through the living room and the use of the single living space giving access to the bedrooms and bathroom, which were grouped together. However, the Ministry of Works were insisting that the plans of all temporary houses should conform exactly to their own steel house plan, and although ARCON were unanimous in disliking the Ministry plan they prepared a Mark 3 version. They were concerned that, in the Ministry plan, to get to the bathroom one had to pass through the living room from one of the bedrooms. Clearly the separate WC was better but the size of the kitchen had been considerably reduced since one end was now occupied by three doors.

Many would have been satisfied with such a situation, an approved plan and construction technique, and just concerned themselves with production and profit. ARCON clearly were of sterner stuff, and indeed embodied a very positive attitude to design and problem solving which is all too rare in architects and which resulted in them completely redesigning the plan to satisfy both the Ministry requirements and their own reservations about the Ministry plan. The result was the Mark 4 ARCON house of which a prototype was erected at the Tate Gallery site next to other temporary house types — Pressed Steel, Tarran and Uni-Seco. Local authorities visited the site and, after deciding on the types required, put their requirements to the Ministry. ARCON were asked to proceed with the production of 86,000 houses of which only 41,000 were ever produced.

A number of changes had occurred from Mark 2 to Mark 4: the pile foundations were replaced by a concrete raft with tongued and grooved boarding on wood joists. Some of the tubular steel columns became T sections and structural detailing was simplified. Undoubtedly the size of the order was a considerable compliment to the efforts of the ARCON group but it appears that their relations with the Ministry were to be again frustrating since they were required to incorporate standard fittings and to satisfy various other requirements, which resulted in them deciding to redesign each one of 390 components, thus producing the Mark 5 ARCON house. A number of important features, it was thought, had been lost. The single entrance had been replaced by one back and front; the front entrance was placed in the centre of the house at the insistence of Lord Portal; the large living room window was deleted since it was felt children would break the glass. Economics dictated the deletion of the entrance step at a saving of 5/- (or 25 p) and the living room canopy, an attempt at sun control, at a saving of 15/- (or 75 p).

Most of the implications of total prefabrication and component design, when thoroughly applied, can be seen at work in the process followed by ARCON which, by this time, included architects, structural engineers, several production engineers, window and asbestos experts. The Mark 5 house consisted of about 2500 parts

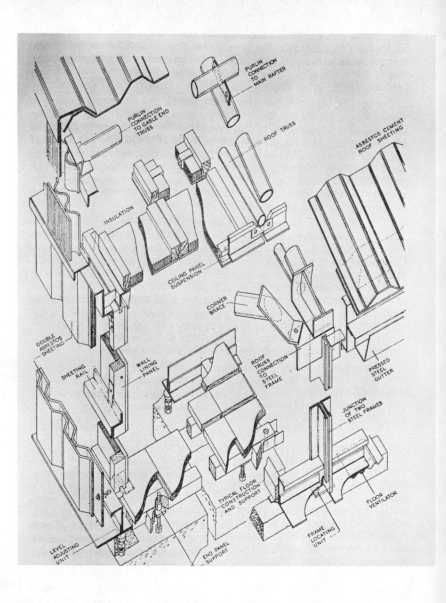

6.4.4 The Arcon house. Exploded axonometric

made by 145 different manufacturers and involving the preparation on the part of ARCON of some 5000 drawings (**6.4.4**). (Ford gives the numbers of parts in a Model T as 5000.) For each component discussions took place with the manufacturers (or their trade associations) to agree production methods and this often further involved discussions with manufacturers and suppliers of raw materials or partly formed products. This led to the formation of two groups of staff, one to deal with the design and drawing of each component, and a separate group whose purpose was to integrate one component with another and to ensure that decisions made in one area could be accommodated within the whole system. For each part there was a specification and these, with the drawings, had to be approved by the Ministry of Works, manufacturers found for production, and contracts set up. Throughout manufacture each component had to be inspected and tested prior to transportation to the point of storage to await shipment to sites. The acceptance of storage by government is an enormously significant point in this process since, part of the point of mass producing components means that there is a ready supply. This means stockpiling which clearly costs money to a manufacturer and ties up capital — something he is usually unlikely to do without firm guarantees. Later groups, like the British Schools Consortia, encountered this problem time and time again, since there was only 'a gentleman's agreement' to take components and this was on an individual contract basis. Authorities or suppliers could not stockpile beyond a certain point without firm guarantees, which they could never be given.

The enormous amount of paper work generated by the process is also typical of most systems and, in the case of ARCON, was produced very quickly in a great burst of enthusiasm and commitment to the idea during the early part of 1945. A group of 100 houses as a pilot run was erected in advance of the main programme so that design, construction techniques and erection could be examined and any modifications made. A number of details were again redesigned for bulk production.

Taylor Woodrow were appointed as managing contractors with the difficult job of co-ordinating transport, production, storage and erection. This involved them in training site erectors capable of putting this rather unusual animal together. Subsequently, after the temporary housing programme, the name ARCON was transferred to the ARCON group of companies and operated as Taylor Woodrow (ARCON) Ltd, retaining A. M. Gear and Associates as consultants but carrying their own architectural and design staff.

The ARCON Mark 5 house incorporated many ideas that only much later were to become standard practice in housing in Britain. Amongst these were ducted warm air heating, modular kitchen fittings, prefabricated electrical wiring harness, prefabricated floor and ceiling panels, and a high standard of insulation in walls and ceilings. It can be seen that quite apart from the efforts to prefabricate all the components the houses were designed with, for the time, a high standard of environmental comfort and convenience. This was undoubtedly primary in endearing them to tenants who, in many cases, were only too aware of the shortcomings of the English house with respect to these qualities. A compact, highly insulated, well serviced box made of factory produced parts transported to sites and erected in two weeks: this formed the stuff from which the dreams of many architects were made. The type of organization that had been set up was in many ways an object lesson since it contained a variety of design and production skills, manufacturers, suppliers, erectors. However, an operation even of this scale, to produce 41,000 houses between 1945 and 1948, required not only the enormous energy of the ARCON designers, but the continuing support of the manufacturers, and most important, the guaranteed captive market back up offered by the government. When housing in 1948 reverted to a local authority responsibility, the government with the escalating costs of the temporary housing programme quickly indicated that the guaranteed market would cease. In addition to this, the prefabricated houses did not conform to the building byelaws (now Building Regulations), and the Minister of Health had to provide local authorities with a special waiver to allow them to go ahead. This offers a very useful comment on such regulations since the 'prefabs' were environmentally better than most houses of the time. Of course they did this by unusual methods and, with byelaws dealing with the way things were made, rather than the standards to be achieved, it is not surprising that they did not conform.

The government, however, continued to encourage experimental house systems of which very few made any significant contribution. White is very critical of the attitudes adopted by government during this period:

> 'but in retrospect it is difficult to understand why it should have elected to concentrate so much effort on inviting the submission of an unlimited number of systems and then rejecting hundreds of them as worthless, a method of approach so patently wasteful that it was already discredited after 1920.' (White 1965)

In view of this evidence, it is surprising that government did a very similar thing during the early sixties, in encouraging industrialized building systems, which produced a rash of some 400 'housing systems' most of which did not get beyond the prototype stage. Strangely, industry has proved more gullible than government with regard to such ideas and they seem to have been repeatedly 'sold' the myth of mass production — not without some hope, of course, that there was truth in it.

However, probably the most important lesson is that, in a country where finance and policy is so highly centralized, the stylistic and conceptual power of such ideas (and they are not only confined to the building industry), once accepted can be thrust upon the country fairly easily. Encouragement, cajolery and sanction can be employed all the way through the hierarchy in the pursuit of a policy. Whilst theoretically local authorities, and indeed industry, have consultative power, their autonomy has in practice been severely limited.

The 1945-8 temporary housing programme is of interest from many points of view, especially that of the application of mass production techniques to housing: nevertheless it was certainly an expensive experiment, although this may perhaps be justified on the grounds of dire need. More important was its commitment to total prefabrication with, it seems, little reference to the effect of this on the whole system, the building industry or on the users. As early as 1944 the Building Industries National Council published a special report on prefabrication, extensively quoted by White (1965). The building industry could be, and was, cited as merely interested in preserving itself, but there was some sound sense in many of the points which it made. These have certainly been borne out since.

The report emphasized the industry's concern to provide a 'long life' product around a standard performance, and the way in which this offers some safeguards in relation to the introduction of new materials and methods. It also pointed out that new methods required much more consideration than conventional methods and probably more time to assimilate the results; also that it was just as important to encourage higher output by new methods within conventional building, as to look for completely new ways of doing things. An important point was also made regarding those employed in the building industry:

> 'the building manpower of the country, similar to the trained manpower of any important industry, was an important national asset. It was based on a long tradition of craft development and organisation.' (White 1965)

The report then proceeds to indicate the importance of this continuity. To most architects at the time this would merely be written off as the traditional responses of a backward industry, afraid that it may lose its position with the rise of new techniques. In many ways however, the sense of understanding the whole system rather than merely emphasizing the technological and artefactual part is a much more likely way to implement change — indeed the many 'building systems' that have failed for various reasons could well have profited from this advice. The pursuit of the industrialization myth has certainly contributed to the difficulties now experienced, for example, in obtaining craftsmen. A building on, and development of, existing or dying skills alongside the introduction of new methods and skills must be more sensible in human, economic and cultural terms.

However, the 1944 committee were not only critical: suggestions were offered for future directions. This involved the development of ranges of internal components standardized throughout by the British Standards Institution and co-ordinated with each other. They further suggested a limited range of building designs which would compromise between the need for flexibility and necessary standardization. This might well be the recipe for the programme of component co-ordination pursued by various ministries, some 20 years later.

To return briefly to ARCON. The government decision to cease supporting the temporary housing programme led them to seek new outlets for their investments and their creative energies, largely overseas. An ingenious roof, the ARCON tropical roof, was designed (**6.4.5**) using a tubular truss and columns. It was designed with an upper roof of corrugated asbestos and a lower roof of termite proof fibreboard, allowing a large gap between to encourage wind passage, and thus cooling. This answered a considerable need in many countries for a lightweight, easily erected structure of spans larger than usually obtainable in local materials. From this a more flexible structural system was developed, the ARCON Roof (**6.4.6**), a three-pin frame roof (**6.4.7**), sawtooth roof (**6.4.8**), and monitor roof (**6.4.9**).

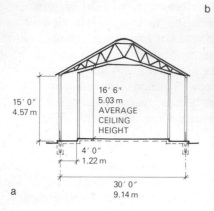

15' 0"
4.57 m

16' 6"
5.03 m
AVERAGE
CEILING
HEIGHT

4' 0"
1.22 m

30' 0"
9.14 m

a

b

6.4.5 Tropical roof (Troproof).
ARCON, early 1950s. Double roof to
permit free circulation of air in the
intervening space. Structure tubular
steel. (a) Cross-section; (b) view

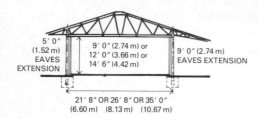

5' 0"
(1.52 m)
EAVES
EXTENSION

9' 0" (2.74 m) or
12' 0" (3.66 m) or
14' 6" (4.42 m)

9' 0" (2.74 m)
EAVES EXTENSION

21' 8" OR 26' 8" OR 35' 0"
(6.60 m)   (8.13 m)   (10.67 m)

6.4.6 Arcon roof: ARCON: early
1950s. Section, two half trusses of
welded tubular steel

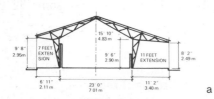

9' 8"
2.95m

7 FEET
EXTEN
SION

15' 10"
4.83 m

9' 6"
2.90 m

11 FEET
EXTENSION

8' 2"
2.49 m

6' 11"
2.11 m

23' 0"
7.01 m

11' 2"
3.40 m

a

6.4.7 Three-pin frame structure.
ARCON, early 1950s The design
transferred no bending moment to the
foundations and allowed for wind
loadings of up to 75 m.p.h. (121
km/h) (a) Cross-section; (b) view of
prototype at Ruislip, West London

b

6.4.8 Sawtooth roof. ARCON, early 1950s. Factory under construction at Bournemouth, England, 1958. (a) The steel structure before cladding. An elegant use of tubular steel. Note the steel valley gutters and sheeting rails. (b) Roof sheeting and cladding in progress. (c) Detail of steel connections

248

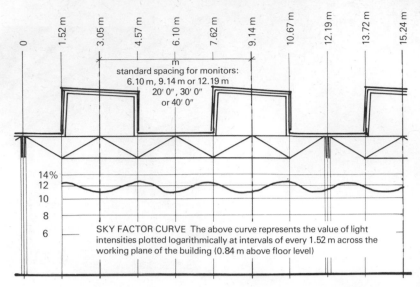

standard spacing for monitors:
6.10 m, 9.14 m or 12.19 m
20' 0", 30' 0"
or 40' 0"

14%
12
10
8
6

SKY FACTOR CURVE The above curve represents the value of light intensities plotted logarithmically at intervals of every 1.52 m across the working plane of the building (0.84 m above floor level)

6.4.9 Monitor roof. ARCON, *c.*1955. Section of typical bay. A well worked out solution that proved popular in the face of the crude structures often offered for work environments

These, with a number of other designs, and optional ranges of components, allowed the ARCON tradition to continue in the open market situation. All these early structures used 4 in (approx. 10 cm) as the basic module and 3 ft 4 in (approx. 1 m) as a grid multiple. These were strictly adhered to in the positioning of purlins and other members, and this bore fruit as the components multiplied. The sets of dimensional and detailing rules built up over the years meant that the components could retain a measure of open-endedness necessary in the market situation. Taylor Woodrow (Building Exports) Ltd continued to market the components and complete buildings for the ARCON group, although the building industry's own assimilation of many of the ideas pioneered by ARCON gradually produced a number of companies competing with similar ideas, almost all of which lacked the thoroughness and elegance of the precursors, ARCON.

To back up their development work ARCON made a policy of carrying out research into specific problems as they seemed to require attention — one of these was some highly original work on rainwater disposal on roofs, which resulted in their use of the rainwater sump, and another was concerned with methods for calculating the sun angles in any part of the world. This work was refined into a series of simple design aids which could be employed when problems arose. One of the most significant pieces of research received almost no publicity although Moffett (1955) gives it considerable attention. This concerns the research into component interchangeability, jointing, and dimensional co-ordination carried out by Rodney Thomas. The work on the ARCON house, coupled with the need to see marketable components in a broader context than that of a closed system, caused the group to allocate a sum of money to allow research into these areas in 1951. The original idea was to produce information on how the ARCON export house might be made of almost infinitely interchangeable components, rather than as it originally was when merely based upon the modular size.

Having examined the problems involved, Thomas realized that any module only permitted interchangeability if all components were of the same thickness, had the same edge shape and were manufactured to the same tolerances. Of course all components and materials on the open market varied in these respects, and mere attention to dimensional co-ordination or attempts at limiting components was not likely to suffice as long term solutions by themselves.

This soon involved Thomas in some very basic research which was not completed until 1955, when he presented the findings to the ARCON group with the aid of a colour film. He claimed to have developed an approach which would enable any architect to build accurately to any dimension with differing ranges of components which would fit together without scribing and regardless of their surface dimensions, thickness and edge shape. This would allow any new component or material to be absorbed by the system without change and further, would provide a solution to the difficulties created by each component having its own manufacturing tolerance, this being compounded by erection and movement tolerances. Although they had previously been involved in the development of flexible joints ARCON now realized that the joint was much more important than the component – a significant breakthrough.

The concept rests on the use of a series of 'weights' and the way Thomas came upon it is a further illustration of how solutions may come to the prepared mind. He realized that a shopkeeper can weigh any commodity no matter what its shape, size and colour by using a few weights. He then reasoned that if a limited number of dimensional 'weights' were used in varying combinations almost any range of sizes could be built up and any component introduced (**6.4.10**).

6.4.10 Proposals made to achieve component interchangeability. Rodney Thomas, ARCON, 1955. (a) Plan showing how the edge conversion pieces would make the transitions from any component to any other. (b) Vertical section showing the use of the same approach

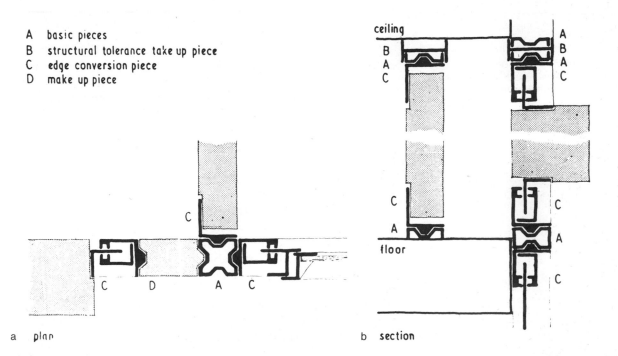

A basic pieces
B structural tolerance take up piece
C edge conversion piece
D make up piece

a plan

b section

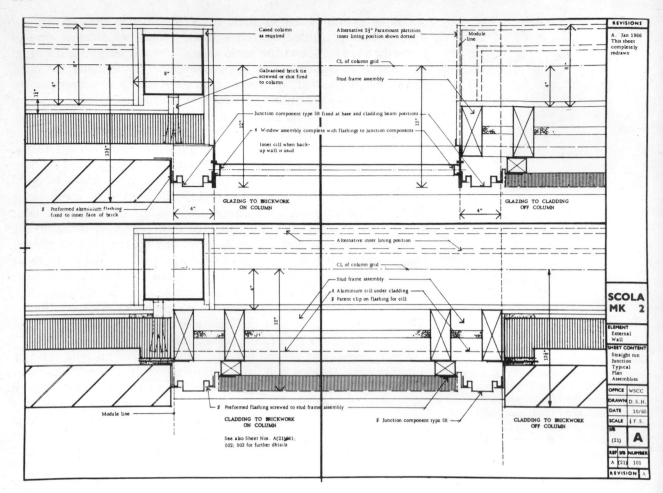

6.4.11 Second Consortium of Local Authorities (SCOLA), Mark 2 system junction components, 1965, using an idea similar to that proposed by Neel in 1955

Five 'weights' of 1, 2, 5, 8 and 9 inches (25, 51, 127, 203 and 229 mm) were selected from which dimensional units could be made up to make inches or centimetres (a large amount of ARCON work was in metricated countries), and to deal with fractions and tolerances. With a 'foot' weight in the components themselves almost any combination of components was possible together with the introduction of non-standard components. It also allowed four-way tolerance with very few basic pieces and really opened the way for component interchangeability to occur. However, it does not appear to have been acted upon by the ARCON group, and appears to have been almost completely ignored by those working in the field in spite of its publication in the *Architectural Review* for 1955. This in some ways is not surprising for, at the time and indeed since, most work in the field of component interchangeability has been concentrated on the co-ordination of dimensions of components around the 100 mm (4 in) module and 900 mm (1 ft 0 in) preference. Ultimately, in the late sixties, the Department of Education and Science (having resisted the idea from several external quarters) proposed, in a limited circulation document concerning performance standards for external walls, that the introduction of a 'third member' at the component junctions would simplify many problems. Subsequently the DOE Component Co-ordination Group proposed a similar idea. Unfortunately neither of these seemed to have much effect. However, one of the schools consortia, SCOLA, briefly flirted with the ideas put forward by Thomas when their Mark 2 system introduced the concept of joint components and converters which would take up tolerances and movement, and encourage future interchangeability by being capable of accommodating any thickness component (**6.4.11**). Although the components perhaps lacked sophistication they did manage to give the Mark 2 buildings using them a precision and a component flexibility they sadly needed. The pressures of cost plus a lack of understanding of the implications led to its early rejection from the system. However, this is looking ahead. Before the institutional systems are examined the place where many of the ground rules were worked out must be studied. This is the Architects' Department at Hertfordshire.

HALCYON DAYS AT HERTFORDSHIRE

'We went to a meeting at the Ministry and everyone sitting around the table, had at one time, worked in the Herts Architects Department.'

Overheard at a meeting, *c.* 1965

Although the intervention of World War II between 1939 and 1945 caused a cessation of all normal building activity in Britain, it had the effect that most such wars have: that of focusing the minds of men around the solutions to social problems. This then released a great flood of energy with few immediately available resources with which it could be satisfied. At the polls in 1945 the country voted in the Labour Government, and the great social ideals of the thirties were seen to be even more powerful now that fascism had been defeated. Britain would be made a better place to live, grabbing capitalist industry would be nationalized and thus turned from the production of private wealth and be harnessed for the public good. Social services would be created to deal with health, unemployment, housing, schools. In building terms homes fit for heroes, and schools fit for their offspring, were to be built without delay and it was to be done by using the industrial resources now freed from the manufacture of weapons. The 1944 report of the Wood committee on *Standard Construction in Schools* encouraged new approaches.

In this climate those architects imbued with both social conscience and the ideals of the modern movement saw their chance. Frederick Winslow Taylor, the pioneer of the use of scientific method in his studies of productivity and the man who had impressively motivated the Constructivists amongst others was, in spirit, alive and well again. The concept inherent in the marriage of social good and the production line surged again in the architectural breast. Methods of building more suited to the new age about to dawn would be developed. In housing and in schools architects turned this into their own call for a new order. Utopias were again to be put to work. Various types of prefabricated houses were developed and in school building the halcyon days were commencing in Hertfordshire County Council.

In a series of articles in the *Architects' Journal* during 1955 Lacey and Swain set out the arguments which had motivated those at Hertfordshire:

'With this early work the process started of establishing a system of "non-traditional" building and of developing the system over successive building programmes in order to achieve better schools.' (Lacey and Swain 1955)

Not only do we again hear echoes of Pickett who, in 1845, was calling for a new system of architecture 'developing the properties of metals' and Daniel Badger with his Badger fronts where lightness of structure, facility of erection, beauty and economy were emphasized, but also of the masters of the modern movement with their cries for new methods to fit the new social aims brought about by machine technology. In fact Le Corbusier's later recipe for 'a completely new method of construction' (Boesiger 1960) was firmly embedded in the Hertfordshire work. Unlike Le Corbusier, those at the Bauhaus and others, Hertfordshire had the support of a great wave of social concern. However, in purporting to set aside not only traditional building but the architectural high stylists, Hertfordshire were drawing directly on the great mythologies of the modern movement. They were also working in a context where new methods and materials were being actively pursued both by commercial systems and a handful of other local authorities. For example, Sidney Loweth the Kent County Architect, pointed out in a lecture to the Chadwick Trust April 1949 that in new schools they were aiming at maximum flexibility, rapidity of erection and the use of materials that are readily available (Loweth 1949).

Hertfordshire was faced with having to provide a considerable number of school places and the pressures on the County's resources were heavy. The County Architect, C. H. Aslin, put the case clearly in his report of 1947:

'Before embarking on the construction of the Schools in the programme it was necessary to make an analysis of existing building conditions because they had changed so much since 1939. It became clear that the shortage of skilled site labour and traditional building materials such as timber and brickwork would call for an entirely new approach to buildings. Moreover labour conditions are such that men are being attracted from the site to the factory. In recognising this trend it is felt that the architect should design in such a way that his buildings will consist of standardised factory-made units capable of simple assembly on the site by a small number of semi-skilled men.' (Lacey and Swain 1955)

In addition the raising of the school leaving age to 15 created an extra demand for school places. This resulted in school sites in most counties being dotted with HORSA huts, an acronym derived from Hutted Operation for the Raising of the School Leaving Age. These were supplied and erected by the Ministry of Works. Hertfordshire, well aware of the dangers of accepting, for however 'brief' a time period, temporary and substandard accommodation, suggested to the

Ministry of Education that it should investigate its ideas on prefabrication instead of utilizing the HORSA huts. Hertfordshire here demonstrated how such a situation could be turned to advantage and that their concern was with the quality of the result rather than with the convenient expediency that caused many local authorities to subsequently rush into the embrace of the Schools Consortia. There was, however, a closer relation at this time between reality and the current architectural preferred image than had previously been the case. Nevertheless it can be seen that, far from setting aside traditional methods, Hertfordshire was calling up the great machine age gods. The *Architectural Review* went so far as to introduce a study with:

> 'in them (the Herts schools) we can now see what sort of architecture does in fact result from that long-expected revolution, the impact of industrial production on building. This revolution is none the less important because it is taking place rather gradually and not in the violent manner prophesied by the visionary architects of the twenties.' (Davies and Weeks 1952)

After a study of the requirements of educationalists, specialists and teachers a method of building was proposed embodying the following principles:

1. rapid erection: thus saving on the hiring of alternative accommodation and transport costs.
2. economical, but not cheap, building: the equitable distribution of costs within the overall costs allowance was the aim.
3. repair and maintenance costs comparable to those of traditional building.
4. a flexible system: this was not interpreted as the ability to make frequent or rapid changes within the building envelope but much more it was seen as removing one of the main obstacles to planning freedom and allowing each building to be individually tailored to its site. The Dom-ino House of Le Corbusier (1914) with its frame and platforms freeing the planning of the building can be seen in the comments of Lacey and Swain:

> 'A flexible system... which would not unduly restrict the planning of rooms and spaces by structural limitations'. (Lacey and Swain 1955)

> Like Stillman, Hertfordshire embodied the idea that change should be allowed for but very much in terms of architect controlled change, and this was emphasized by the very specific forms of the buildings.

5. The schools produced should be: 'pleasing to look at and to work in' and 'the architects had in mind the need for designing buildings that are acceptable to the largest number of people' (Lacey and Swain 1955)

The people's architecture was to arise from this setting aside of traditional building and from the use of totally new methods. We are here faced with architects proposing on the one hand to remake buildings in formal terms so that it would be more in keeping with their view of society and yet expecting that this in some way would accord with the commonly understood perceptions of 'a school', or 'a building'.

However, those at Hertfordshire did have several advantages that others before did not have in this respect. The most important was that the ideas were not new, but had permeated the architectural profession from the twenties onwards and thence to the general public. Although at this time it was fashionable to deride the 'prima donnas' of architecture their ideas had been avidly sought by many a student, then practising. This, together with the fact that the county was faced with many real problems requiring imaginative solutions and a county administrative structure that was notably progressive, provided fertile ground for many of the ideas that had been postulated over the previous forty years. The county's education department, with John Newsom as Director of Education, was very concerned to change the image of school building from the institutional one that had hitherto prevailed. Great emphasis was also laid on establishing a good understanding of what went on in schools and, in an article in 1949 entitled 'User Requirements in School Design' (ABN 1949) an 'Educationist' pointed out that architects should spend as much time as possible in a wide range of schools. The view was also held that schools should be as small as possible — no more than 500 for seniors, 360 for juniors, and 240 for infants. The architect contributor to the same article pointed out that 'at least three experiments in teaching spaces' were being introduced after discussions with the education department. These were subsequently to become of central importance in the development of school planning throughout Britain. The first was the creation of a series of alcoves for small groups instead of the normal single classroom; the second was the reduction of the height of teaching spaces 'to suit the smaller client' and the introduction of a special rooflight with louvres; the third was the introduction of folding doors between classrooms and corridor so that the latter spaces could also be used for teaching.

It was decided that one system of construction should be used for the schools and that the architect would work in close liaison with manufacturers to produce components for the system. The wartime uses of an interdisciplinary development group to solve problems was of some significance here although it was clearly part of the ideal of many architects to take architecture into the factory. Stirrat Johnson-Marshall and David Medd had served in the army together, assisting in the fabrication of a false D-day attack, and employing the techniques of operational research (Merriman 1978). Unlike later building system development groups, which have been separate groups created to do that work, the development work in Hertfordshire was done (right up until the creation of the Consortium, SEAC) by distributing it amongst those actually designing the schools.

6.5.1 Junior Mixed Infants school, Cheshunt, Herts, England. Hertfordshire County Council Architects Department, 1946-7. One of the prototype buildings for the initial Herts programme. View of classroom

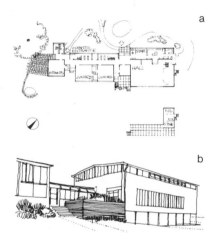

6.5.2 Junior Mixed Infants school, Essendon, Herts, England. Hertfordshire County Council Architects Department, 1946-7. One of the prototypes erected prior to the main programmes. (a) Plan; (b) view of hall and link to classrooms

6.5.3 Steel frame houses. Hills Patent Glazing Co Ltd. Northolt, London, c.1945. The houses were clad with precast concrete panels

Whilst this creates all sorts of organizational problems it has some very healthy advantages over the separate development group method, since it ensures that development and design are closely related and that loyalties are not split. Further, designers can see the results of their developments with a more critical eye when it has been carried out in the context of a job that actually gets built.

The building which was to test these ideas first was a prototype infant section erected at Cheshunt, and stage two was the addition of a junior section also at Cheshunt (**6.5.1**). A two class Junior Mixed Infant School followed at Essendon in 1946 and 1947 (**6.5.2**). The Cheshunt infants section was designed to a finger plan consisting of three square classrooms with cloakrooms, each with its outside teaching area, and each connected to a corridor link. The second, junior section, consisted of a double loaded corridor with five classrooms one side and cloaks and changing rooms on the other. Both sections were linked to a central block containing hall, dining room and administration which was not in the prototype construction. In principle these classroom blocks followed the same principle of construction put forward by Stillman at Sidlesham and elsewhere in West Sussex, and embodied the stylistic qualities of the thirties with one whole wall single glazed above waist height and metal windows which rolled right back to admit nature. The steel frame at Cheshunt was pinjointed at the foot and fixed in one direction at the cap with diagonal cross braces. This was developed by Hills Patent Glazing Company who already had a house building system of their own on the market using the frame together with precast concrete wall and roof units (**6.5.3**). This was, initially, their 'Presweld' system but they had already demonstrated a new system at Birmingham which had steel beams consisting of a steel flat top and bottom with a ¼ in (6.4 mm) diameter rod lacing spot welded to it and this was used

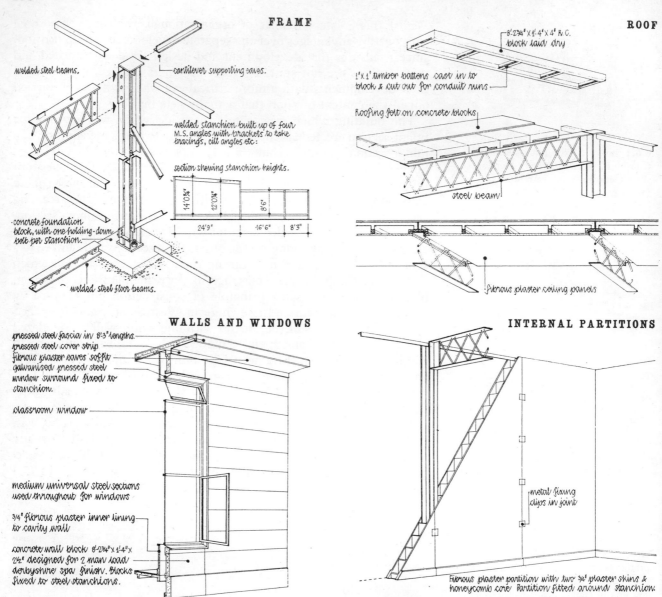

FRAME

welded steel beams.

cantilever supporting eaves.

welded stanchion built up of four M.S. angles with brackets to take bracings, cill angles etc:

section shewing stanchion heights.

14'0¾"   12'0"   8'6"

24'9"   16'6"   8'3"

concrete foundation block, with one holding-down bolt per stanchion.

welded steel floor beams.

ROOF

8'-2¾" x 1'-4" x 4" P.C. block laid dry

1' x 1' timber battens cast in to block & cut out for conduit runs

Roofing felt on concrete blocks

steel beam

fibrous plaster ceiling panels

WALLS AND WINDOWS

pressed steel fascia in 8'3" lengths.
pressed steel cover strip
fibrous plaster eaves soffit
galvanised pressed steel window surround fixed to stanchion.

classroom window

medium universal steel sections used throughout for windows

¾" fibrous plaster inner lining to cavity wall

concrete wall block 8'-2¾" x 1'-4" x 2½" designed for 2 man load derbyshire spa finish. Blocks fixed to steel stanchions.

INTERNAL PARTITIONS

metal fixing clips in joint

fibrous plaster partition with two ¾" plaster skins & honeycomb core. Partition fitted around stanchion.

a

6.5.4 (a) Cheshunt JMI, Infants section, 1946. Design of steel frame, walling, floor and roof all by Hills Patent Glazing Co. Sketch design and drawings checking by Hertfordshire CC architects. (b) Cheshunt JMI, junior section and Essendon JMI, 1947. Shows modifications made to prototype design

on the Cheshunt prototype building, with Hills designing walling, floor and roof (**6.5.4**). The Cheshunt prototype indicates the direction for later development throughout the school building consortia: the light pinjointed steel frame with its angle batten stanchion; the concrete 'biscuit' roof laid dry; the honeycomb partitions; and the external wall with horizontal precast concrete units 'designed for a two man load' anchored back to the stanchions. Metal windows, pressed metal trim and overhanging eaves unit complete the industrialized package.

The account of the development work indicates the sort of problems that arose: lining and plumbing of such a light steel frame, and the condition of the steel beams arriving on site. A very high standard of

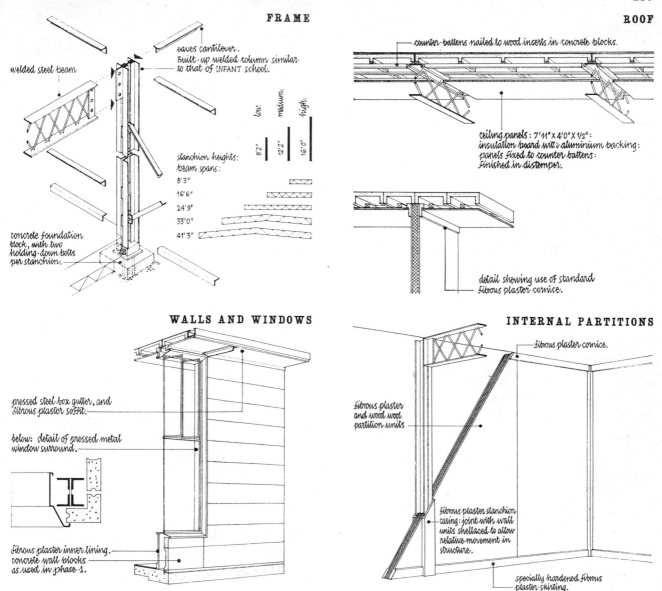

FRAME

- eaves cantilever. Built-up welded column similar to that of INFANT school.
- welded steel beam
- concrete foundation block, with two holding-down bolts per stanchion.

stanchion heights:/beam spans:
- 8'3"
- 16'6"
- 24'9"
- 33'0"
- 41'3"

low. 8'2"
medium. 12'2"
high. 16'0"

ROOF

- counter-battens nailed to wood inserts in concrete blocks.
- ceiling panels: 7'11" X 4'0" X 1/2": insulation board with aluminium backing: panels fixed to counter-battens: finished in distemper.
- detail shewing use of standard fibrous plaster cornice.

WALLS AND WINDOWS

- pressed steel box gutter, and fibrous plaster soffit.
- below: detail of pressed metal window surround.
- fibrous plaster inner lining.
- concrete wall blocks as used in phase 1.

INTERNAL PARTITIONS

- fibrous plaster cornice.
- fibrous plaster and wood wool partition units
- fibrous plaster stanchion casing: joint with wall units shellaced to allow relative movement in structure.
- specially hardened fibrous plaster skirting.

b

skill was required for many of the operations such as laying the concrete roof slabs to gain a good flat surface and the casting of the slabs themselves. The difficulties in eliminating such casting irregularities caused the dry concrete roof to be quickly discarded. The control of production tolerances on concrete wall units also proved difficult with ultimately one of +0 in and −¼ in (6.4 mm) being agreed. The size and weight of the partition units also created problems. The two prototype schools were designed on an 8 ft 3 in (2.5 m) structural and planning grid, the use of such grids being a key part of the standardization philosophy in the county. The 1947 building programme consisted of eight junior and/or infants schools, but design work had to commence on these before the prototype

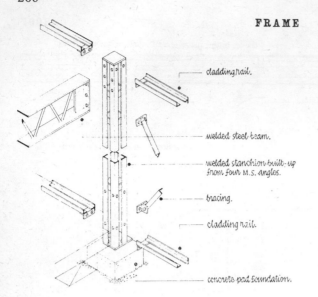

cladding rail.

welded steel beam.

welded stanchion built-up from four M.S. angles.

bracing.

cladding rail.

concrete pad foundation.

8'2¼" X 1'2" X 4' precast concrete roof unit, with timber fillets cast in for fixing ceiling battens.

open ends and webs to pass conduit.

roofing felt.
vermiculite screed.
concrete roof units.

galvanised M.S. grille stops.

insulation board expansion joint.

insulation board ceiling on counter-battens.

6.5.5 1947 Schools programme, Hertfordshire CC Architects department. A number of detail changes were made including the simplifying of the beams, and the stanchions. These latter became angle and batten stanchions of the type subsequently used by other systems in the sixties. At this time also the external cladding became vertical spanning and the overhanging eaves eliminated

schools were completed and, although it was possible to build in some of the experiene of the first schools, this was obviously a far from ideal situation. Nevertheless it is a problem occurring again and again in systems development. As can be seen from **6.5.5** a number of changes occurred in the detail construction although the principle remained the same. One of the most important things to be established was that of a standard stanchion range, thus simplifying erection manufacture and ordering. The external skin appeared to go through the most changes and this is well in keeping with the history of building systems where an inordinate amount of time appears to have been spent on this part of the whole system with surprisingly little return. Indeed there is in many ways a negative return since the abandonment of what were called traditional methods of building meant also the abandonment of the environmental qualities of such methods which had developed over long periods.

However, it is clear that the new methods had more to do with the supposition that a new social order could be paralleled by a new visual order than with a real examination of environmental qualities. That new social order, as with others before, was to be directly expressed through the technology with mass production being seen to be brought under the control of the publicly employed architect and not left to the vagaries of the capitalist market place.

For the 1947 Hertfordshire building programme the concrete external wall slabs were redesigned so that instead of running horizontally they ran vertically (**6.5.5**), thus necessitating the introduction of horizontal cladding rails to support them. It may be of some significance that at this point Hills withdrew from the manufacture of these units. The eaves overhang, which echoed certain stylistic concerns of the time, was dropped in favour of a completely flush external face utilizing an eaves block to make the junction.

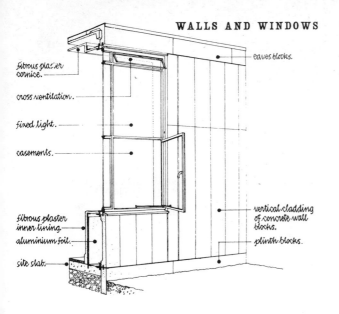

fibrous plaster cornice.

cross ventilation.

fixed light.

casements.

fibrous plaster inner lining.

aluminium foil.

site slab.

eaves blocks.

vertical cladding of concrete wall blocks.

plinth blocks.

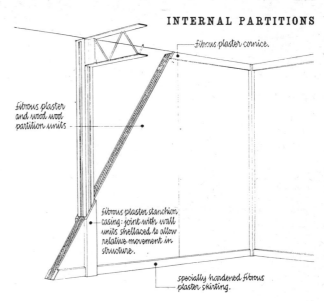

fibrous plaster cornice.

fibrous plaster and wood wool partition units

fibrous plaster stanchion casing: joint with wall units shellaced to allow relative movement in structure.

specially hardened fibrous plaster skirting.

6.5.6 1948-9 Schools programme Hertfordshire CC Architects department. A return to the overhanging eaves of the first buildings, in which internal corner and block junctions posed particular design problems because of their complexity in components and assembly

Unfortunately no tolerance had been allowed between the walls and the roof and this resulted in haircracking. Further, the finish was exposed aggregate and this was considered unsatisfactory. Whether due to this latter or to its stylistic deficiencies the flush eaves was dropped and the overhanging eaves was brought back in the 1948 and 1949 primary programme. The simple flush concrete exterior of the 1947 programme had seemed to cause some disturbance. It is clear that the technical problems could have been solved but it is also clear that its stylistic rigour fell outside those norms acceptable at that time, a mere two years before the Festival of Britain in 1951. A few years later, in the early 1960s, such a facade would have been thought to rightly reflect the pristine virtues of the machine. Davies and Weeks (1952) go so far as to say:

> 'Technically, the provision of projecting eaves cuts right across the principles of modular design, creating almost insoluble problems at the inner angles (**6.5.6**) and greatly increasing the number of units required to form a roof. It is the only instance in the programme of purely aesthetic considerations being allowed to override efficiency.'

With the approval of the county's first postwar secondary school building programme for 1949/50, of five schools (**6.5.7**), it was felt that the work so far developed in primary schools could be usefully extended. Design work had commenced on a new 3 ft 4 in (1.016 m) grid system but this was abandoned in view of the difficulties of designing a new system from scratch and the 8 ft 3 in (2.5 m) grid system was then looked at for this programme. These schools were expected to be more robust, to be capable of going up to three storeys and to fit to a cost limit of £290 per pupil place initially, reducing to

1946—1947: PROTOTYPE

Cheshunt, *Burleigh*	J.M.I.
Stage 1: Infants' section	
Stage 2: Junior section	
Essendon	J.M.I.

1947 PROGRAMME

Letchworth, *Wilbury*	J.M.I.
Hitchin, *Strathmore Avenue*	Infants.
Hemel Hempstead, *Belswains*	J.M.I.
Croxley Green, *Little Green Lanes*	Junior.
Croxley Green, *Malvern Way*	Infants.
Watford, Oxhey, *Warren Dell*	Infants.
Watford, Oxhey, *Warren Dell*	Junior.
Bushey, *Highwood*	J.M.I.

1948—1949 PROGRAMME

St. Albans, *Spencer*	Junior.
Hertford, *Morgan's Walk*	J.M.I.
East Barnet, *Monkfrith*	Infants.
Borehamwood, *Cowley Hill*	Junior.
Watford, *Leavesden Green*	Junior.
Ware, *St. Mary's*	Infants.
Welwyn Garden City, *Templewood*	J.M.I.
Hitchin, *Highover*	J.M.I.
Harpenden, *Batford*	J.M.I.
Hatfield, *Gascoyne Cecil*	J.M.I.
St. Alban's, *Aboyne Lodge*	Infants.
Watford, Oxhey, *Oxhey Wood*	Junior.
Watford, Oxhey, *Oxhey Wood*	Infants.
Watford, *Cassiobury*	J.M.I.
Hemel Hempstead, *Maylands*	Junior.
Hemel Hempstead, *Maylands*	Infants.
Barnet, *Whitings Hill*	J.M.I.
Hemel Hempstead, *South Hill*	J.M.I.
St. Albans, *Mandeville*	J.M.I.
Stevenage, *Fairlands*	J.M.I.
Watford, *St. Meryl*	J.M.I.

1950 PROGRAMME

PRIMARY SCHOOLS PROGRAMME

East Barnet, *Oaklands*	Infants.
Baldock, *St. Mary's*	Infants.
Welwyn Garden City, *Blackthorn*	Junior.
Letchworth, *Grange*	J.M.I.
Hatfield, *Brookmans Park*	J.M.I.
Abbots Langley, *Hazelwood*	Infants.
Watford, Oxhey, *Little Furze*	Junior.
Watford, Oxhey, *Little Furze*	Infants.
London Colney, *Bowmans Green*	J.M.I.
ickmansworth, *Mill End*	Junior.

SECONDARY SCHOOLS PROGRAMME

Hoddesdon, *Stanstead Road*	Sec. Modern.
St. Albans, *St. Julians*	Sec. Modern.
Hemel Hempstead, *Adeyfield*	Sec. Modern.
St. Albans, *Sandridgebury Lane*	Grammar.
Welwyn Garden City, *The Howard*	Sec. Modern.

1951 PROGRAMME

PRIMARY SCHOOLS PROGRAMME

Watford, Oxhey, *Greenfields*	J.M.I.
Borehamwood, *Kenilworth Drive*	J.M.I.
Stevenage, *Broom Barns*	Infants.
Hatfield, *Cranborne*	Infants.
Borehamwood, *Brookfield*	Junior.
Borehamwood, *Merydene*	Infants.

SECONDARY SCHOOLS PROGRAMME

Rickmansworth, *Scots Hill*	Grammar.

1952 PROGRAMME

PRIMARY SCHOOLS PROGRAMME

Watford, Garston, *Lea Farm*	Junior.
Waltham Cross, *Park Lane*	J.M.I.
Letchworth, *Icknield*	Infants.
Bishops Stortford, *Havers Lane*	J.M.I.
East Barnet, *Livingstone*	J.M.I.
Hemel Hempstead, *Hobbs Hill*	Junior.
Hemel Hempstead, *Hobbs Hill*	Infants.
Borehamwood, *Saffron Green*	J.M.I.
Watford, Oxhey, *Site 7*	Junior.

SECONDARY SCHOOLS PROGRAMME

Barnet, *Barnet Lane*	Sec. Modern.
Hitchin, *Old Hale Way*	Sec. Modern.

1953 PROGRAMME

PRIMARY SCHOOLS PROGRAMME

Watford, Oxhey, *Site 7*	Infants.
St. Albans, *New Green Farm*	J.M.I.
Hemel Hempstead, *Bennetts End*	J.M.I.
Hemel Hempstead, *Chaulden*	Junior.
Hemel Hempstead, *Chaulden*	Infants.
Abbots Langley, *Hillside*	Junior.
Stevenage, *Bedwell East*	Junior.
Stevenage, *Bedwell East*	Infants.
Stevenage, *Broom Barns*	Junior.
Borehamwood, *Cowley Hill*	Infants.
Welwyn Garden City, *Blackthorn*	Infants.

SECONDARY SCHOOLS PROGRAMME

Stevenage, *West Shephall*	Sec. Modern.
Hemel Hempstead, *Bennetts End*	Sec. Modern.
Hemel Hempstead, *Bennetts End*	Grammar.
Borehamwood, *Leggatts Farm*	Sec. Modern.
Borehamwood, *Potters Lane*	Grammar.

6.5.7 Hertfordshire CC Schools Programme 1946 to 1953, using 8ft 3in (2.5 m) system. List of schools completed

£240 per place. Apparently Hills were already developing an 8 ft 3 in (2.5 m) frame system and it is of interest that the county was using the Hills Presweld system of construction — it is not entirely clear why they then designed their own version. However, this pattern will again become a familiar one in systems work: the manipulating of commercial systems and firms so that the control of design and, to some extent, production is in the hands of architects. Techniques which for the architect in private practice designing a single building are a necessary element of control, can become weapons of stultifying power when used with government and local government backing over a series of contractss or over a whole building system.

A key part of the Hertfordshire philosophy, and a development from Stillman's ideas, was the use of the planning grid. Stillman, and even the first Hertfordshire schools, utilized frame construction on the bay principle. The use of a two-way grid enables a plan to change size and direction at any point on the grid which, if it is of an appropriate

a

WHAT SEEMS TO
BE REQUIRED OF
THE STRUCTURE
IS SOMETHING
ANALOGOUS TO
THE FREEDOM
OF MOVEMENT OF
A QUEEN ON A
CHESSBOARD AS
OPPOSED TO THE
LIMITATIONS OF
A PAWN

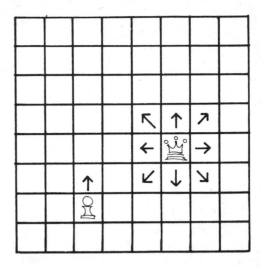

b

THIS CAN BE TRANSLATED INTO A STRUCTURE GIVING
THIS FREEDOM AND NOT THIS RESTRICTION

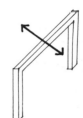

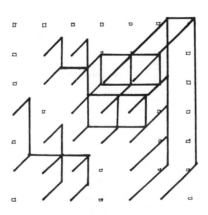

ALL COMPONENTS ARE
GRID MULTIPLES.
PLAN OUTLINE TO CONFORM
TO ANY PATTERN OF GRIDLINES

6.5.8 (a) Bays and grids: the chess-board analogy. The Queen has flexi-bility of movement whereas the pawn may only move forwards, one or two squares at a time. (b) Using a regular grid, columns may stand in any position and have beam connections from any or all of four sides thus offering considerable planning flexibility. Using the bay system for the design of components there has to be a given range of spans and the system can most easily be extended by adding more bays

dimension, can create much more flexibility than can be achieved using the bay system (**6.5.8**). In an article signed 'The County Architects' Dept' in the *Architect and Building News* in 1948 (Hertfordshire Architectural Department 1948) the advantages of the grid are simply set out, after making an analogy between the

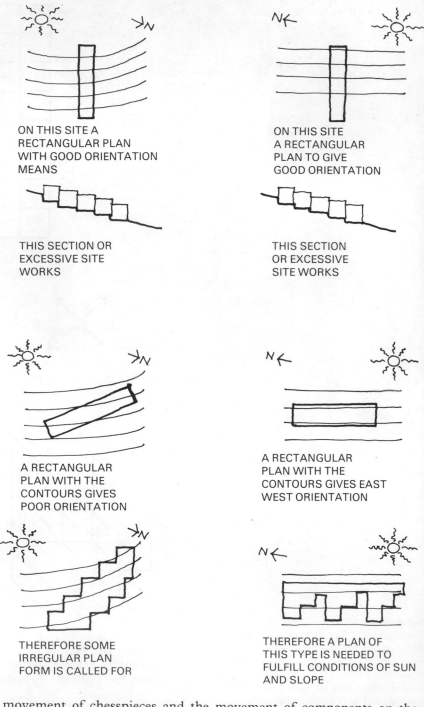

ON THIS SITE A
RECTANGULAR PLAN
WITH GOOD ORIENTATION
MEANS

ON THIS SITE
A RECTANGULAR
PLAN TO GIVE
GOOD ORIENTATION

THIS SECTION OR
EXCESSIVE SITE
WORKS

THIS SECTION
OR EXCESSIVE
SITE WORKS

A RECTANGULAR
PLAN WITH THE
CONTOURS GIVES
POOR ORIENTATION

A RECTANGULAR
PLAN WITH THE
CONTOURS GIVES EAST
WEST ORIENTATION

THEREFORE SOME
IRREGULAR PLAN
FORM IS CALLED FOR

THEREFORE A PLAN OF
THIS TYPE IS NEEDED TO
FULFILL CONDITIONS OF SUN
AND SLOPE

6.5.9 Site arrangements showing how the ability to create buildings with flexibility at small increments can assist in dealing with orientation and site problems

movement of chesspieces and the movement of components on the planning grid. The diagrams show how such a method assists in dealing with orientation, site slope and shape and are direct and useful aids to designers (**6.5.9**).

Such a use of grids was significant since it not only offered planning flexibility but suggested the way to a vision of component interchangeability which would embrace the whole building industry. The manner in which such a charmingly direct and useful idea has been transformed into the complex ineffectiveness of the endless series of reports, recommendations and directives surrounding the various Ministries' attempts to co-ordinate components nationally is an important lesson for those creative souls altruistically feeding their wares into the machinery that now exists in Britain. It is a small but significant side issue that the 1949/50 development programme included not only the steel frame version referred to above but a timber-framed system which was to take advantage of steel shortages. That this is not even mentioned in the *Architects' Journal* account is also significant: many system builders at this time felt that some sort of betrayal was involved in using such traditional materials as timber. However, the fact that Hertfordshire developed the timber system did reflect that they had a much more pragmatic approach than many of the later proponents of building systems or indeed than of their own publicists. An examination of their early schools will also find the use of brick, block and timber to be considerable and considered. Design was still the driving force at Hertfordshire and only a few signs were to be seen at this time of the organizational and administrative forces that were later to become significant, particularly in the schools consortia. As Wallis rightly observes:

> 'An industrialised technique is proved when the units developed gain eventually wide application elsewhere, resulting in continuous production.'

> '...it is worth recording that even at its most standardised a great deal of construction was in fact produced on site in more or less conventional way [*sic*] and also that many factory units were specials, though part of the system, in the sense that the production quantity might be very small.' (Wallis 1966)

The Gropius house of 1927, for the Stuttgart exhibition, is an early example completely made of factory produced components and designed on a module, or dimensional unit, of one metre (**6.5.10**). Hertfordshire quickly discovered the limitations of a grid as coarse as 8 ft 3 in (2.5 m) and moved to one of 40 inches (almost a metre; 3 ft 4 in = 1.016 m), as recommended by the *Working Party Report on School Construction* of 1948. Their first attempts with this grid were Oxhey Clarendon Secondary School and a crisply designed nursery school at Garston in 1950. Davies and Weeks point out that Gropius (with Wachsmann) used the 3 ft 4 in (approx. 1 m) unit for his 'General Panel System' of 1947 (Davies and Weeks 1952) and it was used for their own work on the LMS 'Unit' railway stations of 1946-8. The advantages of such a dimension are many. The structure may be of any multiple of 40 inches (or approx. 1 m) and can be freed from the external wall, where the doors, windows and cladding units can all be of 40 inches or a multiple of it. This dimensionally controlled division of external wall and structure was a more thorough expression of the

6.5.10 Experimental House, Weissenhofsiedlung, Stüttgart, Germany: Walter Gropius 1927. Built by 'dry mounting' process in an effort to eliminate the moisture problems associated with traditional wet forms of construction

Dom-ino idea and brought together the possibility of an architecture uniting those concepts with the industrialized component building philosophy and its techniques.

'No longer need the rhythm be restricted to a single, steady pulse. The 40-inch units may be combined to give intricate counterpoint both horizontally and vertically.' (Davies and Weeks 1952)

By 1956 Hertfordshire had three structural systems available to designers: brick, concrete and steel with a set of components that was interchangeable on each. Within its own organization at least, the designers were trying to practise what they preached: a bank of dimensionally related components which could be taken off the shelf and used as required. Ironically the very success of the group at Hertfordshire caused this movement to grind to a halt, for the very transplanting of the component building seed from Hertfordshire to other places created a number of autonomous organizations each with its own 'best' system: CLASP, SCOLA and many others. The situation subsequently hardened to a point where such component or subsystem change became very nearly impossible. This, however, is an important topic in its own right and will be dealt with separately.

The Hertfordshire approach had two aspects which seemed to elude protagonists of total system building. The first was that relatively few items were bulk purchased; the second was that 'wet' construction was used wherever economy, time and architectural requirements considered it sensible. The size of the building programme and the autonomy enjoyed by the authority meant that experimental work could be pursued where necessary — indeed in 1961 a totally brick cross wall school was designed at Cheshunt, close to the original school built there using dry construction (AJ 1961a). The success of the Hertfordshire work in establishing the credibility of the factory mass production ideal was considerable, and its proponents and its ideas spread throughout Britain and gave rise to an enormous reputation for school building throughout the world. What then, in summary, were its key innovations?

1. The rationalization of dimensions using grids of various types and modular (or dimensionally related) components. The grid first used was 8 ft 3 in × 8 ft 3 in (2.5 m × 2.5 m) as with Stillman's schools in West Sussex. Then one based on 3 ft 4 in (1.016 m) was introduced which has particular significance since it was strongly felt that, being almost a metre size, it would facilitate rationalization on a European scale. Later a grid of 2 ft 8 in (0.813 m) was also employed. This work on grids laid the foundation for later interest and growth in modular planning and components, preferred dimensions, tolerances and jointing.
2. With grid planning and the standardization of components went the idea of bulk purchase of components. If the design of the

components was known and their numbers assessed by establishing a programme of building for a year or more in advance, manufacturers could be informed accordingly. This in turn would encourage them to plan ahead and thus construction schedules and delivery times could be brought into line, thus eliminating the many delays from this source.

3. Grid planning and standardization of components also led to changes in the way in which drawings were prepared, stored and issued. Component drawings had to be prepared that were unencumbered with information specific to a particular project, so that manufacturers could readily see what was required and this could subsequently be checked. This predesign meant that the possible conditions that a component would have to satisfy would have to be predetermined. In other words not only did components have to be able to fit any one of a number of situations in different schools but there had to be assembly drawings covering all these possible combinations of components. Thus the idea of standardized details which could be repeated on many projects was developed and rationalized. Architects, when faced with producing a new detail, pull a previous job drawing out and revise it to suit. In many ways this offers the worst of both worlds since there is usually no way of accumulating feedback on whether it had worked out or not (let alone recording it) and further such a method saves very little time since it has either to be drawn or traced afresh and maybe adapted as well. True standardization of drawings for details means that these are quite separate from job drawings (the drawings for each project) and do not carry job specific information on them. They can be worked out, checked with manufacturers, contractors and others and when built the deficiencies can be fed into a redesign of the detail. The model here was again the car industry. It will readily be appreciated that drawings of this nature not only need to be drawn in a particular way but also their manner of use needs to be examined: the way in which they relate to job specific drawings, their storage, their issue to manufacturers and contractors.

Just as grid planning and the component approach later developed into the establishing of the modular idea, and its partial incorporation into legislation, the ideas here growing around standard details and thus standard drawings were quickly to develop into a lusty child in its own right — that of project documentation, in the following sequence of events: the standardizing of details and components around a particular mode of building led to the rationalization of drawings. The classification of the building into parts was closely tied to the classification of the drawings into groups. Thus developed separate drawings for the steel frame, drawings for the external wall, roof and so on. The need to divide and yet to relate gave rise to numbering systems of various sorts, and subsequently laid the foundations for the acceptance of the Swedish SfB classification system later to become the CI/SfB system in Britain.

4. The existence of sets of components autonomous from specific jobs meant liaison with manufacturers could be on a more continuing basis, and free of job conflicts, than was the case in the traditional contract situation. Thus the architect could 'get into the factory', and, hopefully, the manufacturer could bring his special expertise to bear on the design at a more useful stage than was usual.

5. All these various developments were closely related to sets of new procedures for tendering and cost control, for the supply and fixing, or the supply only, of the components purchased under bulk arrangements and for which special sorts of tender documents were required. These would specify materials and workmanship together with the approximate numbers required over a given period and indicate roughly to which areas of the county delivery was required. Some guide was given as to expected job start dates when deliveries would be required. Standardization could again benefit since, if the costs of many of the components were known in advance, it was a lot easier to see whether a given building design was likely to fall within the cost limit or not.

Thus the County Architect could show intent and credibility by endeavouring to be less at the mercy of the vagaries of the market than many of his colleagues. In thus reducing the areas of unpredictability in costing and programming, Hertfordshire not only controlled its own internal environment it also had salutary effects on its external environment. The Ministry of Education (now the Department of Education and Science) was faced with an authority which knew more about the control of building costs than it did itself. A part at least of the social ideal had been achieved: how to control and gain maximum use from the available money. The built floor area per child in Hertfordshire schools was above the national average, and they could ensure that not only did they not overspend on a school but, of more importance, that they did not underspend since unused money was not transferable to another school.

However, what commenced as an experiment and a significant contribution must, for all its successes, be looked at in the light of subsequent events. It is true that the Hertfordshire architects themselves resisted for many years the formation of a consortium, maintaining that they had no evidence for assuming that more advances could be made by being bigger. The ideas, so messianically developed there, seeded in the Ministry of Education, in Nottinghamshire (CLASP) and the London County Council, ultimately gave rise to the institutional buildings systems.

Many of the architects who worked at Hertfordshire during this formative period went on to develop the ideas in many other places, thus spreading and transforming those ideas throughout the building industry. Stirrat Johnson-Marshall founded the Ministry of Education development group, subsequently entering private practice with Robert Matthew, building York and Bath Universities using the CLASP system. Swain moved to Coventry with Gibson, and then to Nottinghamshire where he chaired the CLASP working party, subsequently becoming County Architect. Lacey became Chief Architect to the Ministry of Education (now DES), and ultimately moved to the Property Services Agency as Chief Architect. Wallis draws attention to the special conditions under which the Hertfordshire work had been achieved:

> '...there was what might be called a "war footing" where individuals submerged their personality to the common task freely; and secondly, the members of the group responsible in turn for the design were also responsible for the working out of the system.' (Wallis 1966)

In this lies much of the difference between what happened at Hertfordshire and what was to follow. In spite of their 'social realism' and team approach, they were prima donnas in some sense since they were imprinting themselves (and their values), not on one building, but on year after year of school building programmes. In the end they have also, in no small way, imprinted themselves on almost all school buildings in Britain since. The success of the Hertfordshire approach inevitably meant that it was taken up elsewhere, and in a variety of forms. Also inevitably, much was lost in the various transformations with the ideas frequently applied with little understanding of their true context. The concepts of cost control and predictability, in the hands of some architects and many politicians, suffered this fate with many of the schools resulting from this exhibiting more the norms of cost accountancy than the humane concern that had inspired the early Hertfordshire work.

ALL AU FAIT
AT THE GLC

'I found myself ringing up Lord Esher (my old friend who will be in the chair at the meeting) and when I asked him about my speech he said "Oh, why not put in a paragraph about the superiority of public sector housing, because you know, it is superior".'

Richard Crossman entry for Sunday 6th December 1964 in his Diary, prior to speaking at the RIBA on the presentation of prices for the best designed housing of the year, in *Sunday Times* 2 February 1975

It would be surprising if, in the development of building systems, the largest local authority in Britain had not played an important part. Equally surprising is that although it has had many flirtations with methods of off-site industrialization, the Greater London Council (formerly LCC) has not managed to demonstrate decisively those economies of scale often talked of. Indeed at the GLC there have been many serious attempts since Robert Matthew (appointed Architect to the Council in 1946) took over the housing programme in 1950, but these have never assumed the proportions of a total in-house design approach. In view of the difficulties encountered in many of these attempts this is perhaps just as well. An early scheme in Picton Street (now the Elmington Estate), Camberwell, completed 1957, with slab blocks using a great deal of prefabrication was an experimental project (Jackson 1970). It was partly designed under the hand of Cleeve Barr, who was then heading the LCC Development Group, having previously worked at Hertfordshire. This project not only suffered many of the usual difficulties associated with prefabrication — tolerance, fit and so on — but more serious oversights such as the double handling of all material to one block made necessary by the positioning of one of the two parallel blocks just that much too far apart for the tracked crane jib to reach.

Since the scheme was designed to study site rationalization and the team included the contractor (at design stage) as well as the Building Research Station (now Building Research Establishment) much of the confidence the scheme was intended to instil in such methods was lost by difficulties of this sort. Further, great problems in the working of the shunt flues (flats had coal fires in Britain at this time) kept one block empty for some 18 months. Such problems caused questions within the department concerning the benefits that were supposed to accrue, but no satisfactory answers appeared.

There were many skirmishes of this nature during the fifties and various attempts at prefabrication or industrialization were made throughout the organization, particularly in the Council's Housing Division. The newly established development group, first under Cleeve Barr and later Oliver Cox, were very concerned to implement standardization at various levels. This took a variety of forms, from standard 'type plans' for flats and maisonettes to standard details for components like windows, doors and their assembly.

This work is not without significance since it formed a platform for many of the later attempts at total standardization culminating in the metrication programme. It grew out of a strong commitment to the rationalization and progress associated with the Scandinavian social democracies. Sweden in particular was held in high esteem by those many in Britain during the postwar period who were trying to create some sort of social equality. This concern had its counterpart in architectural values which were socially inclined, humanist in intent. That Sweden had managed to create many of the things that those in the thirties in Britain had yearned for was, it seemed, self-evident. Building types such as schools, housing, clinics — in fact the whole paraphernalia of the welfare state — were the desired objects upon which many architects wished to lavish their care and attention.

The forms, the layout and the materials used by the Swedes were seen as a humane solution to the problems posed in Britain. Many an architect during this period made the pilgrimage to Sweden, to Stockholm and the surrounding new developments. Many went to work there, some staying permanently. That the architects' department at the LCC was imbued with this spirit is now a matter of history, to the extent that many of the buildings produced during this period acquired the epithet mock-Swedish or, more rudely, Scandinavian folksy. The Festival of Britain Exhibition in 1951 effectively sealed the fate of this idea as a stylistically respectable one, nevertheless its effects persisted for many years in many a local authority and private office.

It must not be forgotten that the vast tide of social concern, and such attempts to translate this into architectural terms, were sincerely held and often brilliantly carried out. Men such as Robert Matthew, Leslie Martin, Michael Powell and Whitfield Lewis were not only professionally highly competent, but their attitudes permeated the whole way in which their departments were run. Whitfield Lewis was trying to push through a major housing programme; he had also to find time to fend off pressures from that curious animal the LCC Establishment Officer whose concerns not only extended to the hiring of staff and the mechanics of their employment but, apparently, to such matters as how they dressed. Lewis and others were constantly defending staff who, although not out of place in any self-respecting architect's office, were seen by the grey suits of County Hall as bohemian in the extreme. Who were all these people in sandals and open necked shirts starting at 10 am and finishing at 10 pm or later? An example may serve to give the atmosphere — the most well known is probably that of the so-called 'Late Book' in which late arrivals must inscribe their name below a red line and give the reason for the late arrival: as might be imagined the possibilities inherent in such an idea for creative people were vast. The complications of endless interviews, the pasting over of scurrilous entries and the vetting of excuses provided a situation to match a Kafka novel.

This digression, to attempt to establish the strength of the zeal and genuine humanist content which existed at the time, is important since a strong strand of it was concerned with rationalization and standardization seen in humanist terms. This may seem somewhat quaint now that the juggernaut of technology is seen to be a much more ferocious beast. Within this humanist attitude standardization had an important place and Sweden could offer a number of useful examples. They were well advanced with the idea of modular co-ordination and the uses of grids in planning and in relating components. Further, they had extended this to the standardization of specification clauses into an industry wide master specification upon which architects could draw. The seeds planted in Britain by this approach took a long time to germinate into the National Building Specification of 1973. Both the Schools Division, under Michael

6.6.1 Part of standard range of wooden windows. London County Council Development Group, early 1950s

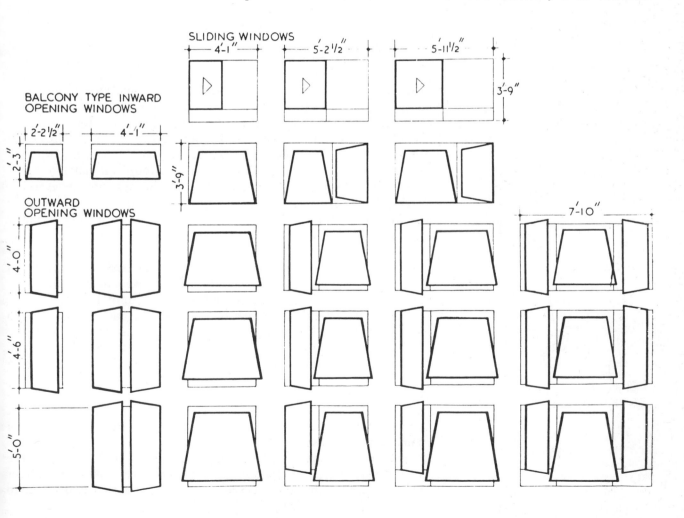

274

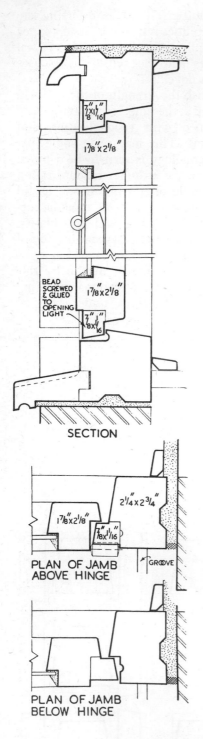

SECTION

PLAN OF JAMB
ABOVE HINGE

GROOVE

PLAN OF JAMB
BELOW HINGE

6.6.2 Standard timber window details.
LCC Development Group, early
1950s. The detail was modified in the
early 1960s and this improved version
was in use until 1978

Powell and the Housing Division, under Whitfield Lewis, had groups of people or individuals endeavouring to introduce such ideas. Of course there were strong connections with what was going on in the Hertfordshire County Architects' Department: indeed many architects came from Hertfordshire to work at the LCC, Cleeve Barr being one, Oliver Cox another.

The development group produced a range of standard type plans to which the Housing Division would work; further, they produced a bank of details which were tried and tested by feedback to eliminate error and build on experience. The work of the group made a number of significant attempts to encourage a co-ordinated approach on the part of the many architects working in the department. Crosswall construction is dealt with elsewhere but other work included the introduction of the now commonplace hexagonal grassed paving blocks to facilitate fire vehicle access, a range of 25 timber window types (**6.6.1**) using EJMA sections (**6.6.2**) (including a pivot hung window which caused complaints about cleaning and ventilation control), ironmongery, staircases and — most important of all — shunt flues and single stack plumbing. The latter two innovations became the norm in all multi-storey work. The design approach also showed itself in practice terms in that each group of 15 or so architects was set up to act as a more or less autonomous unit within the overall structure to ensure maximum freedom and to encourage individual responsibility. This tended to work against the entrenched methods of the bureaucratic machine. The search for new solutions often led far away from the type plans, which were sometimes found to be incorrect anyway, and the complexities of many sites with the particular problems posed made their application difficult. Project architects were not slow to find any excuse to avoid such attempts at standardization, and the Development Group ultimately gained something of the role of court jester to the department. Another example was standard details — these were often very thoroughly worked out and, for example, components such as windows certainly did need rationalization. Even such attempts were met with hostility by project architects, who disliked their proportions, detailing or performance characteristics. Nevertheless both these areas provided fertile ground in which seeds, later to blossom, were to grow and these window details were used (in a modified form) until 1978.

Cleeve Barr moved to the National Building Agency and Cox left to form Shankland Cox Associates with Graeme Shankland late of the LCC Planning Department. Barr, as is discussed elsewhere, attempted to apply many of these ideas on a more grandiose scale and played a key part in the systems work of the sixties through his role at the NBA. All this is by way of setting the scene for what was next to happen, and to make clear that not only are later developments in systems building in Britain traceable to few sources, but to throw some light on those sources.

Most of the pioneering work in standardization attempted at the LCC during this phase has not reached the history books for a number of reasons: first the social basis upon which it rested is difficult to

6.6.3 Elmington Estate (Picton Street) Camberwell, London. LCC, Architect to the Council J.L. Martin, 1955. Two slab blocks intended to be served by one single track crane during construction. (Photo 1977)

comprehend in the face of the subsequent disenchantment which all institutions faced; second it was very diffuse, even if strong within the organization. That is to say there were few total experiments, but many partial ones: thus the idea was rarely comprehended as a totality outside those closely engaged upon it. Third, and probably of most importance, many of the results somehow lacked the architectural panache which would have established them as continuing source material. Picton Street (Elmington Estate 1955), Camberwell was such a result — once seen instantly forgotten (**6.6.3**). Its architectural qualities were insufficient to override its functional difficulties, and make clearly explicit any qualities which it did possess (**6.6.4**). Adjoining the new development, delightful streets from an earlier period offer a bitter comment on the humanity of the new world of public housing (**6.6.5**).

To see the opposite of this one only has to take the example of the Roehampton Estate. Here, many similar ideas were followed through but with an architectural literacy which has, in spite of endless problems and press attacks allowed it to remain as a valid example. The estate, originally Alton East (1954-6) (**6.6.6**) and Alton West (1956-61 (**6.6.7**) covered an area of some 130 acres (52.61 ha) between Putney Heath and Richmond Park. Pevsner established its pedigree in

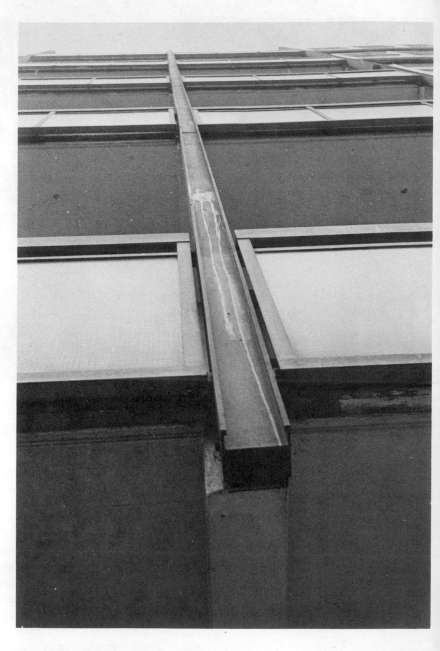

6.6.4 Elmington Estate. Junction of balcony support grid to concrete frame. Clearly shows the difficulty of achieving the precision demanded by the aesthetic. (Photo 1977)

his article in the *Architectural Review* (Pevsner 1959) which, in short, is: Le Corbusier's Ville Contemporaine of 1922, his Plan Voisin of 1925, Gropius and Fry's Isokon project near Windsor 1935, Beaudouin and Lods at Drancy 1932–3 using the Mopin system ('bleak estate') together with Gropius' recommendations concerning high slabs in parallel lines to the CIAM congress of 1930. Pevsner's two other precursors, apart from a reference to Chambord, are Gibberd's 'The Lawn' at Harlow New Town 1950 and the LCC's own Ackroydon Estate close to Roehampton at Wimbledon 1951. Roehampton marks a change in the traditions of the LCC, which by

6.6.5 Directly to the north of the
Elmington Estate lies housing from an
earlier period which uses repetitive
elements in a manner which has stood
up well over time and which has a
human face

6.6.6 Alton East, Roehampton,
London. LCC Housing Division,
1954-6. Eleven-storey point blocks, *in
situ* concrete and brick. (Photo 1978)

6.6.7 Alton West, Roehampton, London. LCC Housing Division, Group leader Colin Lucas. (1956-61). Slab block (Photo 1968)

this time had 500 qualified architects of whom 250 were in the Housing Division. It begins to show a clear move from the 'softer' intentions of the earlier buildings with their emphasis on humane materials, to a tougher, more pristine logic which is strongly Corbusian. Pevsner notes this within the two stages of the scheme itself:

> 'But although humanism and variety are the hallmark of the whole scheme they are not active equally in all parts. In fact there is a very noticeable change between Alton East and Alton West.' (Pevsner 1959)

The first stage he describes as 'architecture at ease', the second as 'exacting' and that:

> 'It is highly intelligent, concentrated, of great integrity, crisp and precise. The point blocks are completely flat in their elevation. Nothing must stick out. . . . In fact, if the foreign source of Alton East was Sweden, the foreign source of Alton West is the Unité at Marseilles.' (Pevsner 1959)

Pevsner gives some attention to naming the teams who worked on each stage — Alton East: Michael Powell (later head of Schools Division), Cleeve Barr (subsequently LCC development group and Principal Housing Architect and Ministry of Housing and Local Government Architect, before heading the National Building Agency), Oliver Cox (subsequently Shankland Cox Associates), Mrs. Rosemary Sternstedt, J. N. Wall and H. P. Harrison. On Alton West the group was headed by Colin Lucas (formerly of Connell, Ward and Lucas) with John Partridge, W. G. Howell, John Killick, S. Amis

6.6.8 Alton West, Roehampton,
London. Point blocks showing the
totally flush detailing of the facades.
(Photo 1968)

(later of Howell Killick Partridge and Amis), J. R. Galley and Roy
Stout (later of Lichfield and Stout).

The architectural rigour implicit in Alton West was particularly
important in the building systems argument. If this seems perverse it
is only necessary to remind ourselves of the very close relation
between a thoroughly visually consistent piece of architecture and its
implicit or explicit supporting arguments. In this case it embodies a
powerful social idea liberally adapted from Le Corbusier and others,
together with a rigorous use of precasting and standard type plans. Let
us note that Pevsner points out that the elevations are thoroughly flat,
nothing sticks out. Let us also note that this is evident not only at the
Unité at Marseilles, but also often in the aesthetic of Mies van der
Rohe. Completely flush, supposedly undemonstrative facades were
felt to be somehow more morally correct than lots of decorative
projections and folksy attempts at reticulation (**6.6.8**). All flats shall be
equal and they shall look equal, although, as Fleetwood (1977) argued
in a powerful critique some 20 years after completion, 'all that
Roehampton has done, all that the modern movement has done, is to
change the appearance of inequality'. Clearly Lucas' link with the
thirties is of crucial importance as is his experience with concrete and

the consistency with which the architectural idea is sustained — something very difficult to do on a large scheme. Equally the concern with large scale standardization of concrete components is a precursor for much that was to follow, first in systems and then throughout architectural endeavour.

In offering a convenient model in which a number of myths were combined it offered a strong clue to the way things would develop. The flush facades referred to and the very detailing itself was to appear in system after system, some using concrete and some not. Even a lightweight school building system, SCOLA, first developed by ex-LCC men in Shropshire, imported a concrete detail aesthetic into a timber technology, thus raising considerable technical and maintenance problems (**6.6.9**). In awarding Roehampton the accolade of the best produced scheme in England in the ten years previous to its completion, Pevsner was perhaps, not saying a great deal. The basis upon which that judgement rested, its authority, was however of some importance.

'There has been as everyone knows, a great longing among architects after the war to get away from rationalism and to recover fantasy. Ronchamp is as much a sign of this as the Neoliberty of Italy, the chunky concrete of the English brutalists as much as the chequerboard patterns of fenestration of the English non-brutalists or the arbitrary patterns in screen walls inside and outside the Americas. Roehampton demonstrates the possibility of up to date sanity. It is not necessary to cover the surfaces of buildings with arbitrary bits. Planning and siting ought to serve the purpose of creating that sense of variety, contrast and relief which everyone wants. Buildings can remain rational, as they ought to, if they are sensitively grouped and if they are placed in

6.6.9 Library, Ferring, West Sussex. F.R. Steele County Architect, M. Heffer Project architect, SCOLA Mark 1a 1964. Plywood fascias, timber horizontal rails all bolted to stanchions. An aesthetic owing much to Le Corbusier's use of heavy concrete horizontals, simulated in the slab blocks at Roehampton by the LCC

6.6.10 Elmington Estate, Camberwell, London. London County Council, Housing Division, 1955. Narrow fronted maisonettes in eleven storey slab blocks

juxtaposition with lawn and trees. That is what distinguishes the best mid-twentieth century schemes from those of the twenties and thirties.' (Pevsner 1959)

Building, and system building particularly, certainly remained rational in the sense in which Pevsner meant, although usually without the lawns and trees and that ingredient which he undoubtedly recognized as architecture. It is that which commended Roehampton to his attention over, say, Picton Street (Elmington Estate) Camberwell (**6.6.10**) which, after all, also consisted of slab blocks and contained a number of technical innovations. It was that which caused him to intone the 'ought'. Lucas' team in building Alton West were working out a version of architectural thought which certainly had a rationale in it, but whether such rationalizations are permanently linked in the way Pevsner suggests is open to great doubt. Once again it seems that a well argued simple rationale that directly links specific visualizations and social purposes is open to serious doubt or worse, when the subsequent history of estates such as Roehampton are examined.

During September 1973 an exhibition entitled '85 years of housing by LCC and GLC Architects' was mounted. It consisted of a series of photographs showing the various in-county and out-county developments since the creation of the LCC in 1888. It also clearly showed the swings and roundabouts of pressures and tastes acting on, and within, the department: and as a social document the handout at the exhibition is of particular interest to us. Our particular concern here, systems building of various sorts, is dealt with somewhat curiously: it starts, under a subheading industrialized building:

'From 1960 the government, expecting an expansion of the economy and fearing a shortage of building labour, pressed local authorities to use systems of industrialised building. The LCC in the first instance decided to use the Larsen and Nielsen system.' (GLC 1973)

This statement in itself is interesting, not only in view of the industrialization used in part in Roehampton between 1955 and 1959, but in view of the schemes already carried out by the LCC to try such methods. Neither the large experiment at the Elmington Estate in 1957, the Reema Construction involvement in a large scheme some time before 1960, and many other ventures were mentioned. It is of significance that the date of 1960, given by the GLC for the inception of such methods, is the date when the political seal of approval was given to such work by central government, thus partially absolving the GLC's own previous decision to use such methods. The publication goes on to say:

'Morris Walk was the first major scheme carried out on this basis... The flats are spacious and well-equipped but the layout suffers from a rigidity of alignment imposed by this system.' (GLC 1973)

The publication then points out that by 1966 a dozen different schemes had been tried but all with less success than Morris Walk. Then:

'By 1967 the shortage of manpower and materials had not materialised and industrialised methods disappeared almost as quickly as they had appeared. By 1970 the Council was using only 3 systems.' (GLC 1973)

Here in a few glib sentences we see an official version of events that at the same time both reduces the GLC's own very real contributions to naught and also gives us a clue to the way in which such trends are perceived. The GLC are here gently pointing out that it was political pressure that caused them to use industrialized methods 'from 1960' and that when political predictions concerning manpower were proved wrong, they carried on with their sensible methods of work. This is also the view presented by the book based upon the exhibition, *Home Sweet Home* published in 1976. The truth is that much of the impetus for the political pronouncements of 1960 came from the work at the GLC (LCC) and at Hertfordshire. The ideas then in good currency suddenly found themselves with support at government level. In more human terms it also meant that many of the ideas of the thirties and forties had reached the ears of politicians via the simple route of many of their advisers, who were imbued with these philosophies.

By this time, the brave new world of industrialization was no longer crying in the wilderness; architects had been very convincing, they had also been successful and the policies were implemented. That British facility for assimilating all ideas was all too effective and architects only gradually began to realize that these ideas, when absorbed into the total political system, then became something other than they had intended.

Architectural ideals, when developed from a social (or socialist) concern were one thing, but those architectural ideals absorbed into the Britain of the sixties was quite another. Of course many large contractors and suppliers were eager to comply with this 'drive to industrialization', of course they agreed with guaranteed orders, of course they agreed with cost limits which reflected agreed minima more than maxima. All this would give them stability and confidence. It also gave them considerable power in eliminating smaller firms. With architects pouring scorn on the considerable number of small firms in England able to build houses, what large firm needs further encouragement? Here was an opportunity to get more work, to reduce all this diversity and competition and many large firms made the most of it. For a brief moment architects in fact found themselves as close to the machine ideal as they are likely to get. It was they who were the spokesmen to government, encouraging legislation and pressures which would give industry the situation and the even flow of orders which they thought it ought to have. Some of the Ministries and the Treasury remained sceptical, the latter no doubt for its own reasons, and ultimately, as the above brochure says, it came to naught, at least as far as the GLC was concerned.

Although subsequently Howell, Killick Partridge and Amis were to leave Lucas' group at the LCC, and not involve themselves in the institutional systems, it is in Roehampton that the combined threads of visually and visibly ordered structure, attitudes to the machine and the Corbusian aesthetic are to be seen most clearly. We should bear in mind that the double height through-maisonette developed by Le Corbusier at Marseilles, when subsequently employed at Bentham Road Hackney in 1954 (AJ 1954) (**6.6.11**), and at Roehampton turned into a mere visual representation of the idea. The *brut* concrete turned

6.6.11 Bentham Road, Hackney, London. LCC Housing Division. Architect to the Council Robert Matthew succeeded by Leslie Martin, project architect Colin St Wilson 1954. Narrow fronted maisonettes in eleven storey slab blocks using precast and *in situ* concrete. The first scheme in Britain to clearly draw on the Corbusian hypothesis of the village in the air. (Photo 1959)

into the precision of the precast concrete cladding, and the greensward under the piloti into drab expanses of concrete. It was the Bentham Road scheme, published as early as 1954 but in design in 1952 when a mock-up maisonette was constructed, that established a new, more rigorous approach. The design group, headed by C. G. Weald, included Colin St. J. Wilson, P. J. Carter, A. H. Colquhoun, J. F. Metcalfe, A. Weitzel and I. Young with Felix Samuely as consulting engineer. The two blocks made extensive use of precast concrete, as did Roehampton subsequently, on a podium slab raised on piloti. Something of the ideas at work behind Roehampton can be seen much later in Howell's lecture at the RIBA in the 'Architects Approach to Architecture' Series. Howell says:

> 'When my partners and I were students after the Second World War, the first buildings to fire our imagination were the Hertfordshire schools.' (Howell 1970)

Once again Le Corbusier's Dom-ino House drawing is shown which Howell describes as:

> 'the most perfect ideogram of structure as a basis for the new architecture. But turn it from an ideogram into a building, from a structure into an enclosure, and what do you get?... Usually an interior enclosed by plain walls, uncommunicative of whether they were support elements or not, a plain flat lid...' and later,

> 'The wonders of technology have brought us to a world of wallpaper' (Howell 1970).

Although Howell here placed great emphasis on structure he was careful to point out that it was only one sort of architecture. Nevertheless, the link from Stonehenge (which he makes) through the Crystal Palace, Le Corbusier and Hertfordshire schools is the one that many architects of his generation would have made: it had a seemingly indisputable logic which only later drew criticism. This argument again elevates the structure, if not structural exhibitionism, to a point undreamt of by Paxton, and is evident in Howell's comments concerning structure as a means of articulating space:

> 'We are often asked if it is not inconsistent that we do not feel the same urge to communicate over services. We flog our poor structural engineering friends into designing things which must all be seen, and at the same time drive our services engineering friends mad by insisting that all their efforts must be invisible.' (Howell 1970)

and the coup de grace:

> 'Displaying structure and concealing plumbing is not really all that inconsistent — after all, there's a world of difference between a chap who likes displaying his biceps and one who walks round with his fly zip undone.' (Howell 1970)

6.6.12 Prefabricated timber houses used to provide accommodation on sites awaiting development. Greater London Council, Architect to the Council, Hubert Bennett, 1960s. (Photo 1976)

In this telling and discreet analogy muscular displays of strength are approved whereas the source of sexual strength must, it seems, be kept hidden!

The scale of the work carried out by the Greater London Council has always meant that it can experiment more confidently and there are a long series of projects which would give an interesting sidelight on industrialization in Britain. Roehampton stands slightly aside from this in that it also draws in many other strands of architectural thought and this in turn has tended to obscure these propositions. What is of particular interest is that, with the vast GLC building programmes and their many attempts at industrialization, the results seem to have been less than successful, as they themselves admit in the document referred to earlier. Not only Elmington, the Reema experiments and Roehampton but the later timber house (**6.6.12**) and the use of Balency at Thamesmead all seem to show that such total approaches are doomed. Curiously they passed almost nothing into the organization, or to later schemes, and they have done little to affect the mainstream approach. If the GLC cannot create suitable conditions for these idealizations of the mass production line to take root then, the thought must be faced, the approach must be invalid. There is indeed something about a building that makes it distinctively unlike a motor car.

7

THE INDUSTRIALIZED VERNACULAR

THE AUTHORIZED VERSION

'Everyone knows that it is much harder to turn word into deed than deed into word.'

Maxim Gorky in
USSR in Construction April, 1937

'Modular Co-ordination is a system devised to co-ordinate the sizes of factory-made building parts with the designs of buildings. It is a method by which the dimensions of standard building components and of the buildings which incorporate them are related to each other by a common unit of size. This unit, which is both a common denominator of all the sizes, a dimensional factor and an increment of sizes, is called "the Basic Module", which derives from the Latin word "modulus", meaning a small measure.'

Modular Co-ordination
European Productivity Agency, 1961

Components on Catalogue

Vernacular methods of building, says the *Shorter Oxford English Dictionary*, are those 'native or peculiar to a particular country or locality', and this appears to have been first applied to cottage building in 1857, having grown from its previous usage in connection with custom and literature. The idea of a building language which is 'naturally spoken' and comes directly from the cultural situation is not only embedded in the ideals of many an architect but it is, in a different form, favoured by the public at large. The idea that 'unselfconscious' (Alexander 1964) building has advantages which selfconscious building does not have is a recurring motif to architects. This is perhaps because it is something that, by definition, they cannot achieve. With the development of industrialized society has come the notion that architecture and building should be a direct reflection of these pressures. It is but a short step to the concept of an industrialized vernacular of building components. Implicit in this is the view that the advent of mechanization and mass production will make available ranges of building products which will, through use and time, become more and more refined providing a 'bank' of components upon which designers, or indeed anyone can draw.

In Europe during the inter-war period this notion existed but, as we have seen, more as an ideal than a reality. However, the great cast iron catalogues, the development of the steel frame and the American timber-frame house have demonstrated, in very real terms, the possibilities of such an idea. But at no time had the ideal become abstracted and applied to the building industry of a whole country. The sort of development that industry had shown it was capable of in the United States was already evidence that an industrialized vernacular existed. Indeed many of the European apologists of mass production drew their inspiration from these sources. John Gloag, after a visit to the United States in 1938, gave voice to what was to become a frequent traveller's tale:

'Some of their experiments with industrial materials have an adventurous flavour; but they are always adjusted to some practical end.' (Gloag 1938)

Since industrialization has developed integral with North American culture itself, it has proved particularly powerful to European eyes. Impatient with the apparent unresponsiveness evident upon their return to Britain, such travellers frequently find themselves falling into the more European habit of elaborate theory or concept building around the realities they have experienced. As we have seen, the United States was not without its own apostles of the machine from the propositions of Ford and Fuller, to the dimensional practicalities of Bemis.

7.1.1 Lustron Home, Muncie, Ind, USA, c.1949. View of gable end showing timber addition on left

The period following World War II saw another fascinating attempt to follow these lines of development. This was the Lustron Home, an all steel house built on a production line basis, by a company created by engineer and inventor Carl Strandlund, with the aid of a loan from the Reconstruction Finance Corporation (RFC) (7.1.1). The latter had been set up in 1946 and was authorized to offer government loans to encourage factory built housing in the United States. Strandlund had been Vice-President of the Chicago Vitreous Enamel Company and, after originally proposing to use porcelain enamel panels on a series of petrol stations for Standard Oil of Indiana, he was convinced by the government to apply the idea to housing. Lustron was given the Curtiss-Wright aircraft factory at Columbus, Ohio, and a series of loans starting in 1946 with 15½ million dollars. After a number of marketing and political difficulties Lustron was closed in March 1950, having only commenced production in 1949.

The original house, built totally of mass produced parts, had a plan area of 1025 square feet (95.22 sq m) and a prototype was erected in 1947 at Hinsdale, Illinois. The external panels, in a variety of colours, were of 2 ft 0 in × 2 ft 0 in (610 × 610 mm) matt finished porcelain enamel on a steel stud structure (7.1.2). The interior panels were also of porcelain enamel steel but of storey height (7.1.3). The panels employed rubber gaskets and glass fibre insulation, and all the internal fittings including bathroom, kitchen, and bedroom cupboards were all built in steel (7.1.4). The ceiling contained a radiant panel heating system and the pitched roof was covered with wide porcelain enamel sheets formed to look like roof tiles (7.1.5).

The factory used a range of automotive assembly techniques, producing four houses an hour, nearly 100 each day. Whilst Lustron encountered few problems with unions, local building regulations and financing posed continuing difficulties. Individuals or builders paid 6000 dollars before the single trailer containing all the components left the factory and this, together with traditional loan constraints, made cash flow difficulties for purchasers.

a

7.1.2 Lustron Home. (a) Gable end showing flush window type and porch. (b) *next page* Bay window type

7.1.2 b

According to an interview conducted by the author and Dan Gagen with a young couple living happily in a Lustron Home some 30 years later, in Muncie, Indiana in 1979, two of the chief factors contributing to its original purchase by one of the parents was the lack of maintenance necessary and the fireproof qualities. I was very surprised indeed to find that, inside and out, the house was still in a remarkably good condition. The surfaces of the panels, the joints, and the gaskets were still in a pristine state, with none of the scratches and dents one has come to associate with this technology. The only complaint the occupiers could think of concerned the difficulty of putting pictures on the walls, which could only be done by the use of magnets.

7.1.3 Lustron Home. Living room showing vertical panels, grooved to simulate joints

7.1.4 Lustron Home. (a) Bathroom. (b) Divider unit between kitchen and dining area. (c) *Next page* Bedroom cupboards

a

b

7.1.4 c

7.1.5 Lustron Home. Steel roofing units pressed to tile forms

The Lustron Home constituted one of the few thorough attempts to create a complete industrialized package which could be sold on the market in the same way as an automobile, refrigerator or washing machine and this was summed up in the small plate screwed to the bathroom wall (**7.1.6**) and which stated: 'The Lustron Home. A new standard for living. Call your dealer for service'. The brief account of the history of Lustron by Floyd E. Barwig (in Bender 1973) points out that although technologically sophisticated such an approach must consider complete changes in the whole system of home provision to

295

7.1.6 Lustron Home. Manufacturer's plate in bathroom, complete with serial number

7.1.7 Eames House, Santa Monica, Cal, USA. Charles and Ray Eames, 1949. Reproduced from Smithson (1966) by permission of *Architectural Design*

be successful. Such changes could have negative results which outweigh the advantages. It may be that all those who see houses being produced like cars should heed the words of Senator Ralph Flanders, commenting on the enormous effort that had gone into Strandlund's project: 'If Lustron doesn't work, let us forever quit talking abut the mass-produced house' (*AF* 1949).

The attempt by Lustron to create an industrialized vernacular package house was mirrored in post-1945 Britain, where the implications of such an approach were being worked out in terms of

modular co-ordination, the use of grids, and dimensionally related components at Hertfordshire, by ARCON and later by the Modular Society. However, the possibilities were given renewed architectural status by a single building: the Eames house at Santa Monica completed in 1949 (**7.1.7**). This was coincident with the end of the first group of Hertfordshire schools and the running down of the postwar prefabricated housing programme in Britain.

The Eames house demonstrated all the characteristics of the argument for the industrialized vernacular with its vision of the 'catalogue availability' of building components. A description of the house and the way it was designed will make clear why Peter Smithson said that 'the whole design climate was permanently changed by the work of Charles and Ray Eames. By a few chairs and a house' (Smithson 1966), and why their work holds particular interest in the development of the idea of an industrialized vernacular. First, they embody an approach to design and to technology that is inclusive, highly professional and original. Second, this has led to them working across a very wide range of design activities: film making, industrial design and the design of buildings. Third, they have designed only three buildings, one of which managed to capture the essence of what many architects had been trying to do for years: to build a good piece of architecture from standard industrialized parts.

In an issue of *Architectural Design* devoted to the work of Charles and Ray Eames, Michael Brawne described the significance of the Santa Monica House in this way:

> 'It also posed again all the questions we had hesitantly but still persistently been asking; in particular, whether a system of factory-made parts could be devised in which the components were small and variable enough to make them equally useful and valid for all the buildings within the village, town, city; and then at what point, if any, some new order was necessary among this deliberate diversity of units.' (Brawne, 1966)

The strength of the Eames aesthetic on others of the same, and subsequent generation, may be seen again in the remarks made by Hugh Morris in a talk in the series Architects' Approach to Architecture at the RIBA on 18th January, 1966:

> 'Eames Santa Monica house; an early and sophisticated example of the kind of component building we will have to spend the next twenty five years developing.' (Morris, 1966)

Morris, a student in the radical years at the Architectural Association during the late forties, and later a member of the LCC Hook new town team, was to design Bath University employing a special version of the CLASP school building system.

To those who yearned for the reordering of society on a more equitable basis and for an ethical use of the machine, the Eames house

7.1.8 Showroom from Herman Miller, Beverly Hills, Cal, USA. Charles and Ray Eames, 1949. Reproduced from Browne (1966) by permission of *Architectural Design*

was clearly a powerful symbol. It had been designed from 'off the shelf' industrialized components put together in such a way that the life of the occupants and their multi-styled objects could go on changing without affecting the enclosure. The buildings of Mies van der Rohe have some superficial similarities to the Eames house but they dominate and seem to formalize the very activities within, whereas Charles and Ray Eames had created an 'anonymous' architecture that was at the same time significant and cool.

Charles Eames, born in 1908, studied architecture at Washington University in the mid-twenties and later became Head of Experimental Design at Cranbrook whilst working on projects with Eliel and Eero Saarinen. Eames had been asked to leave the Beaux Arts atmosphere of Washington University at St. Louis because of 'an over enthusiastic interest in Frank Lloyd Wright' (McCallum 1959). The pioneering work on chairs which began in 1940, using moulded plywood techniques developed from experiments with plywood splints, has led to the many designs for chairs and furniture, most of them produced for Herman Miller. In 1947 Charles and Ray Eames designed the Herman Miller showroom in Beverly Hills, which used a glass wall similar to that subsequently used in their own house (**7.1.8**). In 1950 they started making a series of short films which have become classics, some of these, 'Two Baroque Churches in Germany' (1955) and 'House' (1955), using still frames as the 'components' of the film. These were then photographed onto cine film and music added to form a movie film. Later, in 1964, multiscreen experiments like 'Think', for the IBM pavilion at the New York World Fair, extended the film experiences.

Continuous exhibition and furniture design all demonstrated an interest and competence in many media of communication, and in this Charles and Ray Eames embody that Californian ease with technology: an acceptance of the products of current technology similar to the attitude of craftsmen of former days with their confident and loving use of available tools.

The utterances of the Eameses do not contain the polemic beloved of those who see in such tools either a threat to the present order or a means for reordering society. They may of course be criticized for this, or perhaps for their involvement in the selling of the industrial complex. To accept this criticism would be to misunderstand them and their approach which is highly selective but at the same time inclusive. To this must be added the enormous richness which they have made available, particularly in the visual world. They have taken the puritanical rigours of the modern movement and elegantly combined this with the breath of the Orient and high California technology. It seems fitting that the west coast of the USA should so literally provide an example of what happens when the philosophies of Europe and the East merge in a benign clime.

The west coast magazine, *Arts and Architecture,* under the editorship of John Entenza, instigated a series of house designs under the title of 'Case Study Houses' in an endeavour to show what could be done with the single house. These designs were not only published but also built,

7.1.9 Eames House. Drawing of first scheme, 1947, showing the two blocks at right angles. Reproduced from Browne (1966) by permission of *Architectural Design*

thus offering real live conditions for seeing what happened and what the region could learn from such work.

Charles and Ray Eames produced their first design for Entenza's series in 1947. A house for themselves on a steep hillside in Santa Monica, it was basically a steel framed and braced box suspended over the site with a studio at right angles to it (**7.1.9**). Lower down the slope they also designed, with Saarinen, a house for John Entenza; the third of their three buildings. After their initial proposal for their own house they designed a second, and this is of interest for a number of reasons. First it utilized the materials ordered and already on site for the first design, but used them in a different way. Second, it used those materials to enclose the maximum volume of space. The final plan consisted basically of two rectangular boxes backed onto a reinforced concrete retaining wall and with a courtyard between. Both the studio and the living areas have a double height space. Particularly significant in the developing systems argument is the nature of the materials used and the precise way in which they have been put together (**7.1.10**). Here lay some of the great goals of the system builders: off the peg components being put together in any number of configurations and a maximum of enclosed space for least materials.

The structure is a steel frame with 4 in (101.6 mm) by 4 in (101.6 mm) columns on a bay size of 7 ft 4 in (2.24 m), with open web steel trusses 12 in (305 mm) deep spanning the 20 ft 0 in (6.01 m) width of the building. This was erected in 16 hours by five men with the roof being completed three days later by one man. The labour cost using steel was 33% of the total whereas with timber it would have been

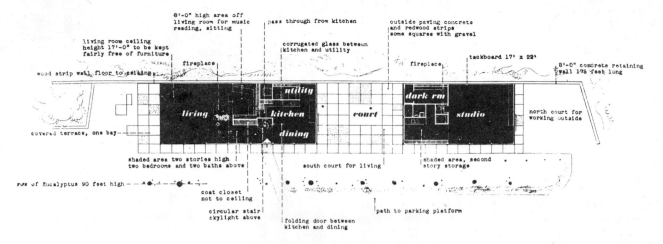

8'-0" high area off living room for music reading, sitting

pass through from kitchen

outside paving concrete and redwood strips some squares with gravel

living room ceiling height 17'-0" to be kept fairly free of furniture

corrugated glass between kitchen and utility

fireplace

fireplace

tackboard 17' x 22'

8'-0" concrete retaining wall 175 feet long

wood strip wall, floor to ceiling

utility

dark rm

living

kitchen

court

studio

north court for working outside

dining

covered terrace, one bay

shaded area two stories high two bedrooms and two baths above

south court for living

shaded area, second story storage

row of Eucalyptus 90 feet high

coat closet not to ceiling

circular stair skylight above

path to parking platform

folding door between kitchen and dining

7.1.10 Eames House, 1949. Plan showing design as built

7.1.11 Eames House: Elevation showing use of standard windows and cross bracing to steel frame. Reproduced from Holroyd (1966) by permission of *Architectural Design*

50% of the overall cost. The joists are omitted where the mullions of the external skin can carry the roof load, and wind loads are taken by means of diagonal bracing which is expressed on the outside of the structure (**7.1.11**). The upper floors and the roof structure are of industrial steel decking which is fixed to the structure to form horizontal wind-bracing to the structure. The external skin, a mixture of glazed and solid panels, is used as an envelope and is a continuous flush surface using colour like a Mondrian painting (**7.1.12**). The skin terminates at the roof edge in a thin line, with no eaves overhang and no more capping than is necessary to make the junction. The wall is freed from its major role as structure and is seen as an array of components that can be added to or subtracted from as required: it does not terminate in massive corners or a positive pediment. It is seen

7.1.12 Composition with Red, Yellow and Blue. Piet Mondrian, 1939-42. One of the long series of neoplastic paintings using a grid of lines. These experiments commenced with 'Composition: colour planes with gray contours' painted in 1918. Copyright by SPADEM, Paris 1980. Courtesy the Tate Gallery, London

7.1.13 *Opposite* First Church of Christ Scientist, Berkeley, Cal, USA. Bernard Maybeck, 1909-11. (a) West elevation. Factory steel windows subdivided vertically with lead cames, pale pink glass. Asbestos sheet panels above with diamond cover pieces at junctions. (b) Side elevation, showing windows and panels

almost literally by many as a piece of the machine age concept, 'cut' like a pattern from the assembly line and clipped in place: a prototype endless architecture.

The importance of this one house can perhaps now be more clearly sensed in the whole pattern of concepts and building which have given identity to the building systems idea. In many ways, perhaps surprisingly, it is similar to Bernard Maybeck's First Church of Christ Scientist in Berkeley built 1909-11 (**7.1.13**). Here Maybeck, working in the Californian tradition, used existing technology in a new way in a single building of great confidence. Here, no rhetoric is necessary concerning the right use of the machine: no clarion calls to use built form to indicate a new vision of society. Maybeck with his use of exposed concrete, industrial sash and asbestos sheet and later, in 1949, Charles and Ray Eames with their quietly confident use of the industrial vernacular, are part of the same tradition.

The lessons derived by the building systems designers from the Eames house concept are clear. To the light steel frame, already developed by Hertfordshire, is added the rolled steel roof deck, although it did not see implementation until used in SCOLA Mk la in 1964 and in 1972 in Mark 5 CLASP. The pinjointed braced frame is evident in CLASP, inaugurated in 1955. The lightweight galvanized steel window wall, used by Stillman and developed by Hertfordshire

a

b

with the Crittall Manufacturing Company was subsequently adopted by SCOLA in their Mk 2 system in 1966.

The lightweight wall as an anonymous unit of mass produced parts is the recurring theme of many of the school building systems. Unlike the Eameses, none of them has really utilized the idea behind it, adopting the imagery rather than the concept. Hence what was a practical possibility in the California of the late forties was, for many reasons, an impossibility in Britain in the fifties, sixties and seventies. Charles and Ray Eames really were creatively looking around at the industry and using components which they saw as fit for their purpose. In Britain and much of Europe, the idea of the building process itself as a system was not considered by the closed systems, and thought by many architects to be outside their concern.

What the British building systems designers saw illustrated of the Eames house seemed to confirm that line of development involving the light steel frame, lightweight skin and components all rolling off the production lines of factories. They saw the house through European eyes — and European architectural eyes at that. How to do it? The prescription was: change the industry, discard tradition, encourage mass-production by grouping large orders and long runs.

Only much later do we see any real attempt to use the industry in the manner indicated by the Eames house. An example, dealt with later, can be seen in the work of Norman Foster, who starts a project by surveying what is available and, after a process of elimination, uses the results of his analysis to its limit. This is the real Eames lesson. Ironically it is Foster's IBM building at Portsmouth, England (1971), which most thoroughly embodies some of the basic ideas in physical form; right down to the external skin which has virtually no edge.

However, the impact of the Eames house on those architects of the 1950s and 1960s cannot really be underrated. It brought into sharp conjunction the act of design and the realities of production. As we have seen, for most, these realities have been seen largely through Henry Ford spectacles: that standardization and high productivity on a repetitive basis equals distribution of the wealth of the machine across the mass of society. Also with buildings: make shelter cheap, standardized, and there will be a home for everyone. Such simple minded suppositions, however repetitive, have not as yet significantly changed the problem of lack of housing or poor housing. Poor housing is only poor housing because there is better housing.

It was not the style of Charles and Ray Eames to make large and grandiose claims for their work, nor to attempt to invest it with high moral purpose as a panacea for all. Their use of 'the industrialized vernacular' was right in keeping with their ability to competently draw on developing technologies whilst at the same time using intelligently their skill as designers. In this respect their attitudes to information and communication are as important as, if nor more than, their house. Even here the system builders could have profited had they understood. However, to most architects, the artefact is all, the ideas written and otherwise communicated are all filtered through the object, through the building. Architects' 'saw' the house, they 'saw' its component

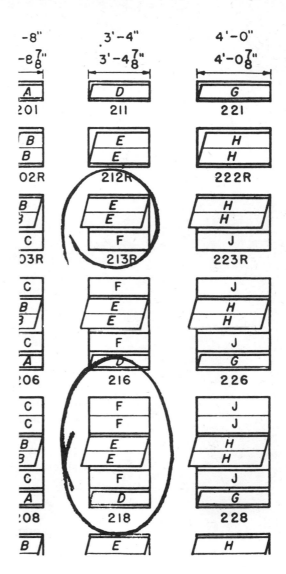

7.1.14 Eames House. Window components used straight from the manufacturer's catalogue (McCallum 1959)

vernacular, complete with component assembly drawing in McCallum's (1959) persuasively glossy book (**7.1.14**). Most of those who became involved in system building nevertheless found that they had become much more involved in information and communication than in designing a building. The amount of paper, administration and the difficulties created by the 'invention' of the building systems in Britain are still with us both on site and in the office. Charles Eames himself had some interesting things to say about communication in his RIBA Annual Discourse of 1959 in London:

'One of the most penetrating of the changes is that change which makes our society almost completely dependent on current information that is, information current, and as contrasted to information that is accumulated. In a traditional society, that is a traditionally orientated society, information is mainly accumulated, and almost any action within this pattern calls for a specific reaction. Today there are very few isolated pockets where this could be said to be true.' (Eames 1966)

In language close to that which Marshall Mcluhan was to use five years later in *Understanding Media* (McLuhan 1964), Eames also gave his view of the effects of multimedia communication:

'Ours is a world so threaded with high frequency, inter-dependence, that it acts as one great nervous system. It requires all the feedback controls man has devised to keep from oscillating itself out of existence.' (Eames 1966)

The very broad range covered by the work of Ray and Charles and Ray Eames — films, toys, exhibitions, furniture and building, is testimony to the belief held by many designers that if one can design in one field one can design in any. However, this does only hold perhaps when the appropriate attitude of mind is adopted. A training in designing buildings does not, automatically, make a person able to act in all spheres. Often quite the reverse: the quite specific ordering processes necessary to the act of building design can indeed close the mind of the architect to other forms of order, other modes of communication. In seeing the Eames Santa Monica house 'cold', as a model, many of those involved in building systems design have missed its most significant fact: that the house itself is embedded in a whole, larger way of working that is very open. Furthermore, it happened in California. Good wine, it is said, travels badly.

Of Modules and Methods

If the propositions of the twenties and thirties had presented a possible view of an industrialized building industry, postwar Britain possessed the climate for encouraging a belief that such a possibility had now arrived. The impetus derived from the postwar prefabricated housing and school building programmes led many to see this as the only way forward. The convenient conjunction of popular social concern, welfare state, available philosophies of industrialization and the practical tools now available through the Hertfordshire and other work, engendered enormous enthusiasm amongst architects that an industrialized vernacular was not only possible but at hand. This view was furthered by the emergence of a number of forces all concerned to gain political support for these ideas.

Foremost in this was the Modular Society, founded in 1953, which sought to bring about the idea of a dimensionally related building industry vernacular, through a series of ground rules, the most crucial one being the use of the 'basic module' of 4 in, or 100 mm, for all component sizing. Clearly the work of Bemis was of significance here for, as we have seen, he had proposed a three-dimensional cubical module based upon the 4 in dimension in the United States in 1936

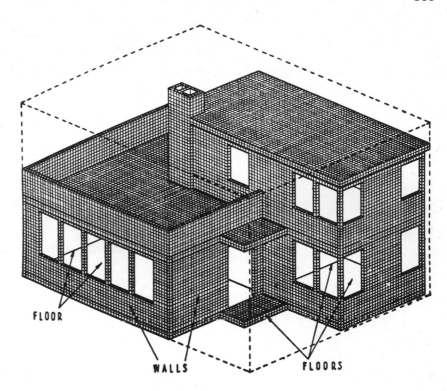

FLOOR

WALLS

FLOORS

7.1.15 Drawing from Bemis's book *The Evolving House* volume 3, 1936 showing the 4 inch (100 mm) module applied in all three dimensions

(7.1.15) (Bemis 1936). In Europe, Sweden led the way in developing prefabrication and rationalized approaches to building and these came to be viewed as a necessary part of the programmes of social engineering now sought. However, it was The European Productivity Agency, of the Organization for European Economic Co-operation, which set out the case for industrialization and dimensional standardization in their first report on Project No. 174, published in 1956, and entitled *Modular Co-ordination in Building* (EPA 1956). In a more political sense the Agency was tied to the whole concept of productivity as the concomitant of higher standards of living in Europe as is evidenced by its own opening statement:

> 'The European Productivity Agency, which is responsible for the publication of the present report, was set up as a new branch of the OEEC in May, 1953. Its task is to stimulate productivity, and thereby raise European standards of living, by influencing not only Governments but also industrial, agricultural and research organisations, private and collective enterprises and public services. One of its primary aims is to convince management and workers alike of the benefits of productivity and to enlist their co-operation.' (EPA 1956)

Many of the protagonists of the modular idea saw it as one way of drawing Britain and the rest of Europe closer together. This was no doubt partly because of the inter-war architectural influences but it also had a strong appeal to architects who were grappling with a building industry that could well use any injection of organization — particularly if this had to do with prediction and control of time and cost. The first EPA report makes out a forceful case when introducing the concept:

'It appears that the only solution to this urgent problem (housing in Europe) lies in the use of modern industrialised production methods and techniques in building, which until now has been carried out more or less on a "handicraft" basis. With industrialised production of standardised building components it should be possible to achieve an appreciable decrease in building cost and consequently in rent.' (EPA 1956)

From this the deduction that follows is:

'One fundamental condition for such industrialised building production is the adoption of a modular system as a basis for standardisation of building components.'

ONE OR MORE COMPONENTS OF THE SAME SIZE WILL FIT THE MODULAR REFERENCE GRID IF THEIR SIZE IS BASED ON A MODULE

GROUPS OF COMPONENTS HAVING THE SAME SIZE WILL NORMALLY FIT THE MODULAR REFERENCE GRID

COMPONENTS OF DIFFERENT SIZES WILL FIT THE REFERENCE GRID

GROUPS OF COMPONENTS HAVING DIFFERENT SIZES WILL NORMALLY FIT THE MODULAR REFERENCE GRID

7.1.16 Ranges of dimensions possible using modular components. With a single component size of 2M or 1M only dimensions which are multiples of the basic size can be achieved. With 3 components of 1M, 2M, and 4M many choices of overall dimension are possible. From the European Productivity Agency's book of 1956 *Modular Co-ordination in Building*

7.1.17 (a) United Kingdom modular test building. Offices and laboratories at Hemel Hempstead, Hertfordshire. Entrance elevation. (b) United Kingdom modular test building: plan

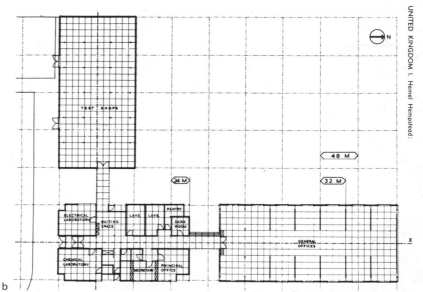

The report was divided into three sections, Design, Manufacture and Building. The first dealt with reference grids, sizes and components, the relation of different dimensional systems, the use of modules and concluding with the proposition that 10 cm (or 4 inches) was to be the preferred module from which resulted ranges of sizes (**7.1.16**). The manufacturing section dealt with component sizing, ranges and tolerances of components whilst the section on building dealt largely with the way in which these ideas were to be applied on site using grids. It pointed out the importance of controlling the 'station' (or position) of components and establishing rules for tolerances. The extensive appendices dealt with the ranges of sizes of different component types available in each country and examples of dimensionally co-ordinated building from a number of countries. For example the Norwegians showed the 'Modern Frame House', Sweden the Elementhus system whereas Britain, submitted the 'PUNT' system — basically only a roof system and one that has not survived.

The second report of the EPA on Project 174, *Modular Co-ordination* published in 1961, was less general in that it was based upon the construction of a number of test buildings in each of the participating countries. The countries involved, Austria, Belgium, Denmark, Germany, Greece, Italy, Norway, Sweden and the United Kingdom submitted widely differing constructional approaches, although all used some form of modular co-ordination. Norway and Sweden both offered versions of the timber-frame house with Sweden also offering a three-storey block of flats of *in situ* and precast concrete. The United Kingdom offered two buildings, the first, offices and laboratories at Hemel Hempstead (**7.1.17**) which was a thoroughgoing essay in lightweight, dry construction in the Hertfordshire or British Aluminium Company, school style. Also shown were two terraces of brick houses:

> 'Terrace 1 demonstrates the use of conventional brick in this context, the system of sizes being extended to work down to 1½ inch multiples, in order to accommodate the brick size of 4½ inches'

whilst

> 'Terrace 2 minimises the use of the 4½ inch brick dimension, and more use is made of foot multiples, and complementary sizes based on 3 inches.' (EPA 1961)

All this reflected the conflicting interests of those modular co-ordination enthusiasts intent on introducing the 4 inch (100 cm) module, and the British brick industry with its heavy investment in plant for traditional sized bricks of 9 in × 4½ in × 3 in (nominal) (229 × 114.3 × 76 mm). The British position was stated quite clearly

> 'The emphasis throughout the scheme is on the common multiples of 3″ and 4″, i.e. foot intervals, as offering the best possibility of linking the national and international positions.' (EPA 1961)

A compromise in the best British tradition, and one which was more than a little confusing to apply. It was also one that would continue to recur, right into the introduction of metric dimensions into Britain, with 300 mm (or the 'metric foot') being retained as a preferred grid dimension. The EPA report itself in 1961, in its Recommendations, asked that:

> 'the responsible authorities and institutions should consider steps to be taken in order to encourage and speed up measures to eliminate the discrepancies between the foot/inch and the metric systems of measures.' (EPA 1961)

The most interesting of all the Test Projects in the second EPA Project is the Greek submission: the reconstruction of 600 houses on the Island of Thera after an earthquake. Modular concrete blocks were used to produce houses that were Greek in style (**7.1.18**), and with a diversity lacked by the other entrant countries. However, this approach significantly did not meet with the entire approval of the Report compilers who went so far as to say that they had

a

b

7.1.18 Greece: modular test buildings. Houses on Thera. (a) Side elevation. (b) End elevation showing vaults. These, and the rest of the houses were constructed using hollow concrete blocks and reinforcement. In the report it was pointed out that the vaults were a common traditional element in the area

'not sufficient breadth to be able to draw from them general conclusions on modular co-ordination. We believe that only the adoption, and application of methods of co-ordination by an industry of a country or even better by a group of neighbouring countries can bring out the advantages of the system.' (EPA 1961)

Here we see the philosophy in its most all embracing form, and it is such a view that gave encouragement to many British architects endeavouring to introduce modular co-ordination into the United Kingdom. Many of these ideas had already been well worked out at Hertfordshire and there was a widespread commitment to the philosophy amongst the majority of architects. The formation of the Modular Society in 1953 was an important step in the attempt to gain national acceptance of the ideas. It focused discussion and gave practical support and publicity to the growing pressure on government to introduce legislation related to the idea of modular co-ordination.

The founder, first secretary and driving force of the Modular Society was Mark Hartland Thomas, who committed a large part of his enormous energy and his life to selling and developing the idea he believed in so passionately. Curiously enough, immediately prior to Thomas' embracing of the 4 inch (100 mm) module he had written a book, *Building is Your Business* in 1947, which categorically came down against such a dimensional basis. In a brief section on dimensional standardization he had pressed the advantages of a German system based upon a plan grid of 1.25 metres (about 4 ft 1 in). He was also very sceptical of attempts to co-ordinate heights around a module:

'This German system has very interesting possibilities, and it has a better chance of success than the American and British researches which are aiming at the establishment of a small unit (3 or 4 inches) applied in all three directions, heights as well. Such a small unit would not realise economies sufficient to offset the disturbance caused to production by adopting it: and the standardisation of heights, as well as horizontal dimensions, would put a severe encumbrance upon architectural design.' (Thomas 1947)

However, driven by the zeal of Thomas and others, the Modular Society became exceedingly influential in making the case for modular co-ordination at meetings, through articles and publications including the *Modular Quarterly* and the *Modular Primer*. The latter, first published in the *Architects' Journal* for 1st August, 1962, became the handbook for a whole generation of architects intent on bringing some rationalization into the tortuous British building industry. In many ways it was not only a board-side tool but, as one architect put it, an act of faith. This was particularly true since, by the time of its publication in 1962, there was already a great deal of evidence to suggest that the mass production theme which had so strongly underpinned the postwar efforts at Hertfordshire and elsewhere was already at breaking point. Hertfordshire itself had already demonstrated to their own satisfaction that the cut off point at which quantity production meant significant saving was, for most components, at a much lower level than most architects had imagined. Whilst it varied greatly from component to component, even with absolutely standard components, batch production was generally the mode used by manufacturers, where they continued to match production runs against single orders and not to stockpile. This, combined with the methods actually used in production, suggested that changes in design caused much less difficulty and cost than the great mass production myths had led many architects to believe. Nevertheless although modular theory was firmly based in these great myths of mass production, in practical everyday terms, particularly on the drawing board, there are a great many things to be said in favour of the methods it proposed.

Alexander had, in an early paper in 1959, made out an elegant and mathematically supported case for the 4 inch (100 mm) module. Entitled 'Perception and Modular Co-ordination' it examined the whole question of visual order and proportional systems, such as the use of the golden section. The paper demonstrated how these systems were 'closely connected with modular co-ordination — that technology, in fact, is developing a system that is intimately related to ours' (Alexander 1959). Corker and Diprose in the *Modular Primer* put it thus:

> 'Modular co-ordination is a universal system for relating industrial techniques to the building process by linking the essential requirements of the designer to those of the manufacturer and constructor by co-ordinating the size of components with a 4″ basic module; specifying a system of tolerance; and establishing a modular reference system by which all components are located.
>
> 'It will facilitate the maximum variety of design use for factory-made components, preplanning, and bulk purchase, and the organisation of building consortia.
>
> 'It will simplify design, manufacturing, construction and general procedure and it will achieve economies; by speeding up the building process and increasing productivity.' (Corker and Diprose 1963)

The *Modular Primer* points out the advantages for each section of the building industry:

1. The manufacturer, in having foreknowledge of component sizes can make long term plans for production and thus reduce the risks associated with storage.
2. The architect will save time in the preparation of drawings which will also be simpler since one will not have to continuously repeat complex dimensional information. Further such dimensioning,

with a tolerance system, will ease specification and eliminate wastage and cutting on site.

3. The quantity surveyor will have a greater ability to forecast and control cost in view of known factory programmes and this will encourage more efficient tendering. This will in turn simplify Bills of Quantity, standardize the preparation of information and lead the way to the mechanizing and computerizing of information.

4. Structural and mechanical engineers will benefit from a common basis for the integration of their work and a standardization of services will be possible: this being particularly important with the increasing amount of a building's cost being devoted to such servicing.

5. The contractor will gain advantage because the approach will be rationalized and comprehensible to him and will assist with the difficulties of site setting out and the positioning of components. More importantly he will be able to programme more effectively, use factory made components, which will be more easily put together by mechanical means.

In a section on aesthetics it points out that the sizes produced by use of the 4 inch (100 mm) module are 'not significantly different' from those produced by using Le Corbusier's Modulor (1961) but more interesting is the

'fact that modular co-ordination contributes a positive aesthetic advantage as it facilitates the design of standard components which, when used in various combinations, lead to "open" systems of construction as opposed to the "closed" systems derived from a limited application of dimensional co-ordination.' (Corker and Diprose 1963).

This important reiteration in the *Modular Primer* of the basic point of modular co-ordination is interesting in view of the great increase in the number of closed systems in the two years previous to its publication. After CLASP in 1955, there had followed SCOLA in 1961 and Method Building in 1963. Most of the original protagonists of the modular idea had firmly held to the view that it was wrong to develop closed, institutional systems of the sort which subsequently emerged in Britain. Such evidence as now exists suggests that their view was well-founded. Apart from the basic module itself other related aspects of the modular idea are these:

1. Reference grids: a grid of lines which assists the location of buildings and components, similar to the grid used on a map.
2. Basic modular grid: this is a grid spaced at 4 inches (100 mm) in each direction and used for component sizing and component assembly. It is emphasized that components must 'keep station' on the grid. That is they must always be located within grid lines so that their perimeter faces always end up relating to a grid line (**7.1.19**).

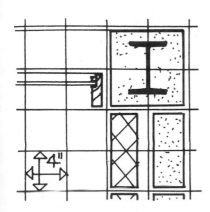

7.1.19 Basic module grid, from *The Modular Primer,* (Corker and Diprose, 1963)

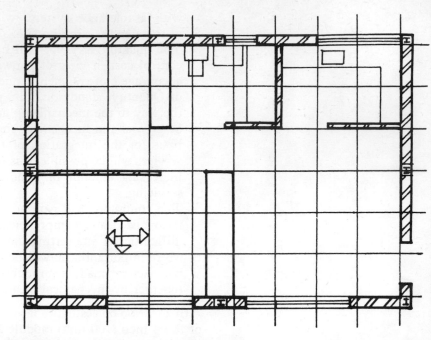

7.1.20 Modular planning grid, from *The Modular Primer*

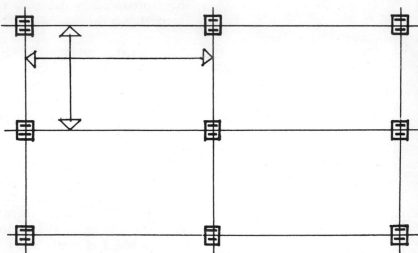

7.1.21 Modular structural grid, from *The Modular Primer*

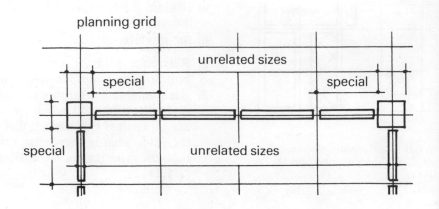

7.1.22 Dimensional co-ordination by centre lines, from *The Modular Primer*

3. Modular planning grid: such grids are used for the planning of spaces and for layout purposes and sizing components. Such grids must always be in multiples of the basic module (**7.1.20**).

4. Modular Structural grid (**7.1.21**): this is a grid for the location of main structural components — columns, walls, beams — and although it should be a multiple of the basic module it is not necessary for it to be a multiple of the planning grid. In this way components may be 'freed' from the tyranny imposed by the more normal use of structural grids as in the Hertfordshire 8 ft 3 in (2.5 m) system, or as in the buildings of Mies van der Rohe. One of the most crucial, and overlooked, points made by Corker and Diprose (1963) is the distinction between the use of a centre line grid and a modular reference grid:

> 'The conventional method of co-ordinating dimensions is to relate the centre lines of all components to the planning grid, using this also to determine component sizes but taking no account of their thickness' (**7.1.22**)

and

> 'It is impossible to maintain a constant factor in the size of the spaces left over and "specials", dimensionally unrelated to the grid, must be provided. The end result is the "closed" system applicable to only one type of structure.'

The modular reference grid, on the other hand (**7.1.23**):

> 'allows for the interchangeability of different types of structure and the location of components in a variety of positions — hence an "open" system.'

The section on tolerances points out the implications of manufacturing, erection and movement tolerances (**7.1.24**), whilst that on components spells out in some detail the way in which specific types of components are dealt with by the use of modular grids. The short section on drafting techniques indicates how the use of the modular system, backed up by appropriate drafting techniques, can simplify the production of drawn information and thus aid manufacturing and erection processes (**7.1.25**).

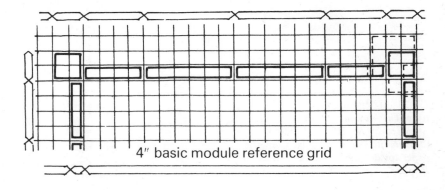

7.1.23 Modular co-ordination using the 4 inch (100 mm approx) basic module reference grid, from *The Modular Primer*

4" basic module reference grid

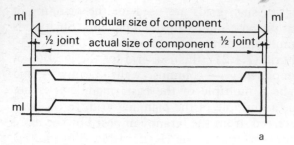

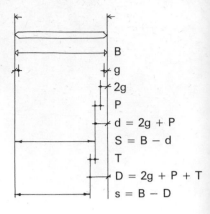

1 modular planes
2 modular space
3 modular size (basic)
4 minimum gaps
4a two minimum gaps brought together for convenience of diagram
5 position tolerance
6 minimum deduction
7 maximum size (upper limit of component)
8 manufacturing tolerance
9 maximum deduction
10 minimum size (lower limit of component)

B
g
2g
P
$d = 2g + P$
$S = B - d$
T
$D = 2g + P + T$
$s = B - D$

b

7.1.24 (a) Tolerance for a modular component, from *The Modular Primer*. (b) Modular tolerance diagram, from *The Modular Primer*

7.1.25 (a) Modular drafting paper, from *The Modular Primer*. Initially proposed by M. Hartland Thomas. (b) *Opposite* Use of modular isometric paper for detailing, from *The Modular Primer*

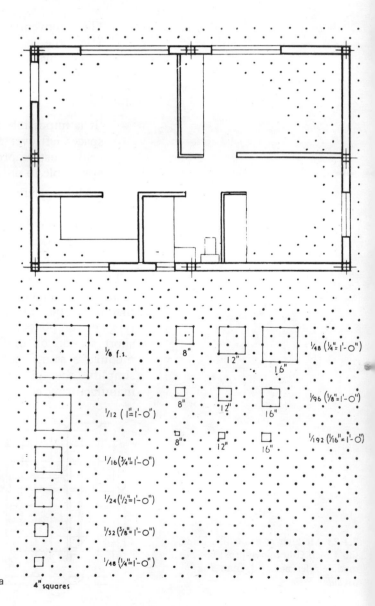

a

Although the *Modular Primer* establishes in documentary form the basis of the modular idea in Britain it was not published in the *Architects' Journal* until 1962 and in book form until 1963. By this time the ideas so vigorously championed by the Modular Society were already in good currency. Hertfordshire was already a legend and CLASP was well established. Nevertheless the beginning of the consortia and systems boom was only just commencing, with modular co-ordination becoming part of legislation in the ensuing years. In this process the ideas were dissected, twisted this way and that by endless committees until they became institutionalized in a form some way removed from the pioneering and basically simple notion that had fired the early enthusiasts. The process of metrication began in Britain in 1965 in the building industry. The programme allowed ten years to 1975 for the changeover and made necessary the rewriting of British standards and related documents. This gave an opportunity for modular ideas to be incorporated and to become mandatory. Although the 100 mm module has gained widespread use the British building industry as a whole is still, at time of writing, far from committed to metrication and the continued use of feet and inches still demonstrates the strength of continuous tradition in the face of attempts at rationalization.

b

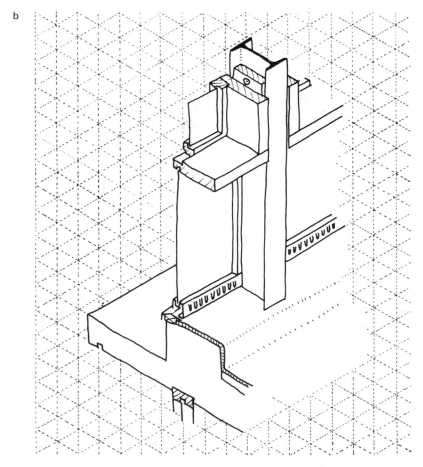

1961 saw the publication in English of another, even more powerful document than the European Productivity Agency report with its dry, civil service mode of presentation: this was Konrad Wachsmann's *The Turning Point of Building* (Wachsmann 1961). This latter volume contains one of the clearest statements of the case for mass production and industrialized building. It does this by the time honoured means of selecting from the nineteenth century a very specific series of examples such as the Crystal Palace, Eiffel Tower and bicycle frame. This is followed by a section on industrialized methods from the idea of modules and standardization to environmental control systems, and culminates in Part 3, with specific examples taken from projects carried out by students working with Wachsmann. It concludes with the case for anonymous teamwork, and more research.

Although 1961 saw the beginnings in Britain of the drive to establish national school building consortia and housing systems, and of the publication of the documents on modular co-ordination, the timing of the Wachsmann volume is curious, although it can be seen as summing up a life devoted to industrialization. In many ways it seems more a product of the twenties or thirties than of the postwar period. Clearly the commitment to prefabrication of the forties and fifties gave the volume its context, but it came just at the point when it was becoming clear to many systems designers and others that such rigorous adherence to the machine doctrine had to be revalued.

Wachsmann's 'Turning Point' carries a strong flavour of those arguments of the inter-war years, with their emphasis on the necessity for architecture to recognize the spirit of the age manifest in the machine and machine production, and is explained by his connections and commitment to the ideas of the period. Wachsmann, born in 1901, was first apprenticed to a cabinetmaker and carpenter and, after a period at the Dresden and Berlin Academies of Arts was, for three years, the Chief Architect to Europe's largest prefabricator of timber components. Although never translated into English his book on timber-frame building techniques was an influential publication in the thirties (Wachsmann 1930). After a period in Rome and France (where he worked with Le Corbusier) he emigrated to the United States in 1941. There, with Walter Gropius, he designed the Packaged House System and in 1942 founded the General Panel Corporation (**7.1.26**). Le Corbusier commented rather pointedly on the Wachsmann and Gropius use of a square grid referring to the Japanese traditional houses 'on a modulus which is certainly much subtler than this: the plait (the tatami)' (Le Corbusier, 1961). Wachsmann's work in tubular steel was shown to advantage in the Mobilar Structure (**7.1.27**) of 1944 at the Museum of Modern Art. In 1950 he became a Professor at the Institute of Design at the Illinois Institute of Technology in Chicago and directed the Department of Advanced Building Research. From this it can be seen that Wachsmann's vision of *The Turning Point of Building* in 1961 relates very closely to his own concerns in the twenties and after, and this explains the somewhat anachronistic

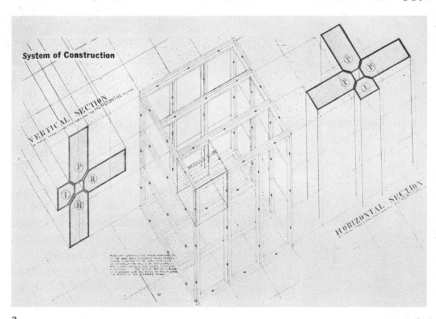

a

b

7.1.26 House by General Panel
Corporation. Walter Gropius and
Konrad Wachsmann, 1941–2.
Designed to standards set up by the
National Housing Administration for
Temporary Dwelling Unit No 1
(TDU-I) the system used a series of
metal clips and wooden wedges for
assembly, which could be carried out
merely using a hammer. (a) The
construction was based on a system of
timber module frames which could be
used in any one of three directions.
These were based in size on a 'module
cube' of 3ft 4in (1.016 m).
(b) Assembly showing use of 10ft 0in
(3.048 m) high panels with vertical
siding

flavour of the book and its arguments, many of which had become
familiar ones concerning the expansion of technology and the new
possibilities thereby offered.

Although there have been changes throughout the building industries
in both Britain and America, the surprising thing to those closely
involved in it is that it is still very much a craft based industry. Many
of the arguments surrounding the shortage of craftsmen that we have
heard from those at Hertfordshire, and subsequently from the
building systems apologists, have always been very selective.
Hertfordshire themselves were very aware that the real shortages of
certain skills at particular times, like bricklayers, did not necessarily

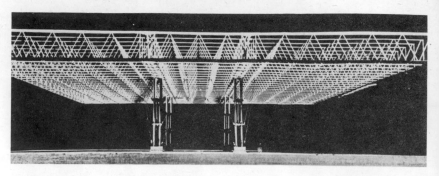

7.1.27 Open hangar structure using the Mobilar System. Cantilever on four sides of 55ft 0in (16.764 m). Konrad Wachsmann, *c.*1950

mean that they would be short on other occasions. Many of the schools consortia found, not that there were shortages of craftsmen, but that often there were complaints that contractors could not use the crafts skills which were available but had to learn new ones to go with the nut and bolt technology.

In Wachsmann we can see the reasons for many of the strange contradictions encountered in recent developments in building systems in this connection. It is a commitment to a philosophy of machine production, which however attractive, bore little relation to what was actually happening in the industry. Of course machinery was being introduced, of course there were and are examples of mass production machinery being used, but this demand for a 'comprehensive idea of technology' is in reality a demand for an architecture that is symbolic of the machine. One sees in it that predilection on the part of architects for a 'total architecture'. This is to be an architecture which embraces the philosophic suggestions of the zeitgeist in the way it is designed, in its actual construction, in its formal appearance and in the details of its making. It is an approach which, I believe, results partly from an education which strongly emphasizes the necessity for the design of buildings to have a total, holistic and visually apparent structure. The building must be a unified Gestalt, it must have at the very least a visual consistency which makes this plain. Architects often then find it necessary to order their whole world in such a way as to establish a similar consistency. Since mechanization has affected large areas of other industries in specific ways its effects must, runs the argument, be similar in building: and if they are not, it is because of the backwardness of the industry. In this way the argument can actually ignore real craft skills (and real craftsmen) which actually do exist and are available, in favour of symbolic gestures to the machine which demand different skills which in the building industry are not actually available. By this means, the disappearance of the craftsmen is actually encouraged by architects, and those involved in building systems have consistently taken such a view. For Wachsmann, industrialization is central:

> 'As one of the great virtues of industrialisation is the ability to turn out products of uniform, peak quality, meeting the requirements of all to the same degree, while using the most suitable materials in the best possible form and achieving the

highest standards in the most economical way, the industrial process can only have its full effect within a system of all-pervasive order and standardisation.' (Wachsmann 1961)

Wachsmann proceeds to establish the link between this argument, dimensional standards and performance standards. He applies the latter not only to the structure and fabric of buildings but also to the integration of their environmental control systems. Finally, in this opening part of his book, Wachsmann shows that moving fabrication from the site to the factory will give improved environmental performance and lead to building being seen as a process rather than as merely a series of products and through this will develop a research capability which will be applied by 'anonymous teamwork':

> 'The mounting and justifiable demands for perfect environmental control can only be met by simultaneously integrating all the complex mechanical and electrical services and the equipment with the structure, the factory-made element and the entire assembled buildings. For, in the technical sense, the modular, static, dynamic and mechanical problems have now become a universal whole.' (Wachsmann 1961)

Wachsmann's case is a thorough one, positing a materialistic view, with materials, methods and technology paramount. Nowhere in his book does he talk about the use of buildings, the psychological or sociological effects. There is no discussion whatsoever of performance in relation to human comfort and certainly no references to those immaterial aspects of places and spaces, their symbolic effects, that are an important part of any humanistic architecture. But then the book is called *The Turning Point of Building* and clearly it is a title that has been chosen with care. If the distinction between architecture and building still has any significance it is clearly in evidence here, where Wachsmann quite explicitly rejects the architectural past in favour of 'the new laws'.

In Abel's words this set of propositions is indeed a 'dinosaur' (Abel 1969), since it stands as a marker of the end of a commitment to the machine. Nevertheless, the very coherence of the Wachsmann argument, has been tremendously influential both on building systems and on architecture at large. The longspan, large shed, flexible interior, environmentally controlled spaces of Ehrenkrantz, Rogers and Foster all owe much to these propositions. The elegance of the space frames Wachsmann developed quickly encouraged further commercial development in Europe and the United States. The British NENK system itself appears chronologically to emerge from this injection, and later in a more complex way the influence can be seen in the MACE system. Both NENK and MACE used a form of spacedeck. It is also relevant that Wachsmann lectured at the Architectural Association School in 1950 and again in 1957, the latter lecture being published in the Association's journal for April, 1957,

7.1.28 Alexander Graham Bell
demonstrating structural systems
based on the tetrahedron, *c.* 1900

complete with illustrations of his work. 'The Crystal Palace' says
Wachsmann 'was a work of art'. Like Giedion before him (Giedion
1967, 1941) Wachsmann uses Paxton's Crystal Palace as an example
of 'the new spirit of the times.' He thus makes the same
oversimplification, by extracting one object from Paxton's work and
approach. Thus Paxton's much more interesting characteristics are
suppressed in favour of the single symbolic object. For Paxton, each
problem had its unique solution, each situation demanded a new
response, which we might call holistic eclecticism. He was quite
prepared to produce, say, rusticated Gothic if that was appropriate.

Wachsmann also draws on Alexander Graham Bell's work with the
tetrahedron showing how, long before Fuller in 1907, he was using
this idea in space frame structures (**7.1.28**). However his most succinct
example is that of the bicycle frame which, he claims, proves there is
'such a thing as anonymous and enduringly perfect form' even though
the pace of technical development might suggest otherwise. The

bicycle frame is the example of a norm as called for by Gropius in the thirties. Because its design had remained static for some 70 years Wachsman makes the assumption that it is the perfect norm and form which he seeks. Unfortunately, Alex Moulton with his Moulton Bicycle led a reappraisal of the bicycle frame at the moment Wachsmann had discovered its durability. Wachsmann's apparent need for a stable, a 'perfect' and enduring form is in itself of significance and possibly relates to his rejection of forms which had shown themselves to already have an enduring quality in architectural terms. Wachsmann's most significant section, on the influence of industrialization, rightly draws attention to the difference between machines such as saws, lathes, and so on which do specific tasks, and those that are multipurpose and automated.

> 'This means that the toolmakers, the machine builders and the mechanics, who make these tools, are the most important craftsmen of our time, with whom and for whom we must learn to work.' (Wachsmann 1961)

Rather strangely his insight into the implications of automation led to a chapter on modular co-ordination which opens with a version of Leonardo da Vinci's universal man (**7.1.29**): from this he then develops the concept of a three-dimensional module for the co-ordination of components. At this point another aspect of Wachsmann

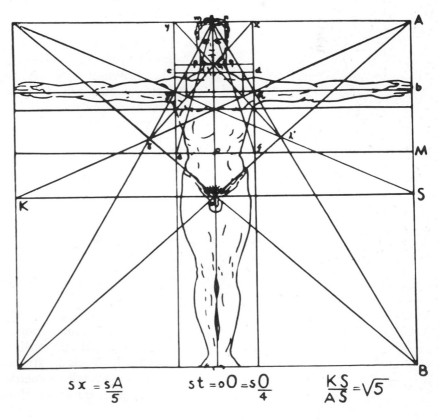

$$s\,x = \frac{sA}{5} \qquad s\,t = oO = \frac{sO}{4} \qquad \frac{KS}{AS} = \sqrt{5}$$

7.1.29 'Man, the measure'. Wachsmann, 1961

emerges, and one that is not entirely consistent with the machine rationale we have come to expect. The relevance of some of these diagrams seems to owe more to developments in painting than to the clear presentation of his component philosophy. Wachsmann cites twelve different types of module:

1. The material module: these are the related dimensions of raw materials. Here Wachsmann makes the statement that has probably caused, when put into practice by the system buildings, the most difficulty.

> 'Therefore it does not seem logical that the development of the material module of industrial production should be influenced by existing standard sizes for raw materials, which have quite different origins.'

This is one of the basic tenets of those proponents of 'the authorized version' of industrialized building: that the dimensions of all building materials must relate together and that to this end machinery will have to be replaced to accomplish this. The metrication and dimensional co-ordination programme undertaken in Britain since 1965 is the somewhat tardy attempt to establish this.

2. The performance module: this is a somewhat difficult notion but basically means the multiplication of the basic modular sizes to give module dimensions in which the particular material can be used to best advantage.

3. The geometry module: this is 'the proportional system governing the structure, the individual element and the overall planning.'

4. The handling module: this is straightforward and governs questions of transport, storage and erection. Wachsmann makes a particular point that if a component is just too heavy for one man, or for two, then by means of time and motion study it should be redesigned to suit such handling.

5. The structural module: this concerns the relationships between the structural elements.

6. The element module: here is introduced the idea of the 'universal surface' which, says Wachsmann 'requires a re-appraisal of the concepts walls, window, door, ceiling and floor'. Such elements are space defining ones such as load-bearing elements or transparent elements, and this allows their dimensional and jointing characteristics to be looked at in isolation from their specific function as support or light conveying elements. One of his examples here is that of the Inland Steel building in Chicago by Skidmore, Owings and Merrill, which illustrates the Miesian concept of a modular anonymous space and a separated structure.

7. The joint module: gives the position of every interface between the system of load-bearing structure and other elements and it must not disrupt the overall geometric module being employed.

8. The component module: this covers those parts not included by the

two divisions structure and element, for example, stairs or mechanical systems.

9. The tolerance module: this is self-explanatory and is concerned with accommodating small inaccuracies.

10. The installation module: this is not, as might seem to be the case, the installation of components. It is the modular relationship of the cables, ducts and outlets relating to the servicing systems and as such emphasizes the need for these aspects to be properly integrated in the total building.

11. The fixture module: this concerns built in and loose fittings and emphasizes that although such products have quite a highly developed modular order of their own, these rarely relate to that of the building itself.

Wachsmann again calls for all the parts of the building to relate in modular terms and it is this, he says, that 'gives the whole building its form and effectiveness': total architecture with a vengeance.

12. The planning module: this is seen as a theoretical device which controls and relates the other modules into a coherent tool for disposing the parts in the planning of buildings.

Wachsmann's views on joints are of particular significance and his statements embody completely the philosophy related to the use of components, whether in a building system as such or elsewhere:

> 'The joint is not a necessary evil. Accordingly, it does not need to be concealed with seal strips and so on, like an object of shame. It stands out as a formative element, which has evolved with progress in technology... These joints not only indicate zones of contact but scrupulously define any object they enclose. They not only reflect processes of aesthetic importance but represent the results of technical functions and are to be understood as such... In the perfect relationship of object, function and separation the joint communicates a new visual attitude.' (Wachsmann 1961)

Here we see again how machine processes have been invested with a morality and that there is moral virtue in demonstrating by visual means what are the underlying forces of the age. This view means that traditionally sound methods such as a cover strip over a joint, although of proven efficiency, must be set aside as must be the covering of joints by plaster and other techniques — even if such techniques are industrialized. This philosophy and its implications have been swallowed almost whole by the architectural profession, with enormous time and effort being expended on designing joints that have the desired visual integrity and at the same time keep out wind and rain, whilst at another level it has imbued much modern architecture, for all types of construction, with a symbolism thought appropriate to the first machine age.

Wachsmann develops his arguments on mass production and standardization into a thorough examination of joints and jointing and

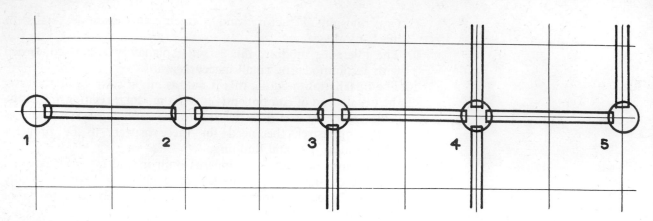

7.1.30 Wachsmann's 'Five fundamental ways in which surfaces can be related at a joint'. The 5 junction conditions for partition design

this analysis remains a fundamental one to design using preformed components. His five joint conditions (**7.1.30**) for components using rectilinear grids shows clearly that components of constant dimension and profile can be used provided each of these joint conditions is solved. Wachsmann describes the joint line as 'a manifestation of energy' (Wachsmann 1961).

However, the innovativeness of Wachsmann's approach is most evident in his development of joints which will take a multiplicity of components and his proposition that the joint and the jointing device, in industrial processes, merits considerable attention. The joint or connector should, he says, be capable of taking all components of no matter what cross-section. The joint is likely, then, to become a complicated device in its own right, which is justified by its universality. In this way industrialized components are connected by an equally sophisticated highly tooled joint. This approach led to the ingenious multi-way connectors of his space frames (**7.1.31**), some of these using wedge fixings, or interlocking methods of considerable complexity (**7.1.32**). Nevertheless, at all times the integrity of the components was maintained to facilitate mass production (**7.1.33**). Inevitably with such jointing systems, the cost would be high, because of their highly tooled nature and the multiplicity of their component parts. Here Wachsmann was suggesting that building was following the pattern of the motor industry with its multitude of highly tooled components: this commitment to a total 'high technology' is, however, most apparent in his comments on environmental control:

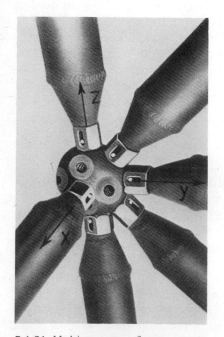

7.1.31 Multi-way space frame connector. Wachsmann, 1950s. 'The three principal axes of the node x, y, z enable up to 8 members to be connected in any of three planes at angles of 45 degrees'

'While the production of synthetic building materials is already providing us with insulation capable of smoothing out local climatic conditions so effectively that is equally useful in the face of both extreme heat and extreme cold, complex mechanical air-conditioning equipment is making it possible to ignore the degree of latitude, and the local climate in general, as a direct influence on construction. Mechanical equipment of this kind helps to create autonomous space that manufactures its own climate. Accordingly, no design need necessarily be determined by regional climatic conditions. The anonymous, universal room thus becomes a reality.' (Wachsmann 1961)

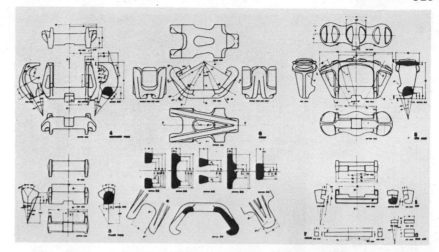

7.1.32 Five element joint. Wachsmann, 1951. Part of a 3-dimensional system developed for US Air Force hangars and an important advance on the earlier linear system

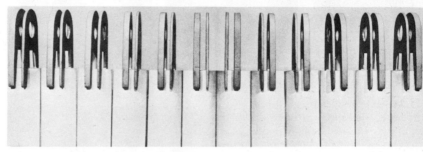

7.1.33 'The uniform, standardized, mass produced product. The eye plates are exactly the same at both ends of the pipe' (Wachsmann 1961)

This accords with Fuller's propositions (Fuller 1946) that modern transport and technology has evened out regional differences and made localized building methods obsolete. Those arguments set up in the inter-war years have taken a further turn, for not only does the machine ideology give a uniformity of product but it also provides a universality of internal environment:

> 'These divergent facts, the technically determined, completely dust-free room, sealed off from the outside world and ventilated by air-conditioning apparatus, and the extension of environmental control to the open air with the possibility of taking in outdoor space, will have a strong, liberalising effect on planning.' (Wachsmann 1961)

This is best summarized in his model for the Mobilar Hangar (**7.1.34**) (of the mid-fifties), with its longspan steel roof and moveable partitions. The Salzburg project of 1959 shows the idea in a simpler form (**7.1.35**) and one which is readily recognizable as that developed by Ehrenkrantz for SCSD beginning in 1961 (**7.1.36**).

It can be seen that Wachsmann's argument is total. In drawing together the threads of mass production, component building, jointing techniques, environmental control and performance standards he provides the best introduction to the ideological propositions which under-pin the work of the building systems. The presentation of that ideology as a bland, matter of fact explanation of technological 'facts'

7.1.34 Mobilar open hangar structure. Wachsmann, 1950s. 'A completely enclosed building can be transformed in a few minutes into one open on all sides, all the wall elements moving away under their own power' (1961)

7.1.35 'A space structure resting on two pairs of columns forming a bay 72ft 0in (21.946 m) × 36ft 0in (10.973 m)' Wachsmann (1961)

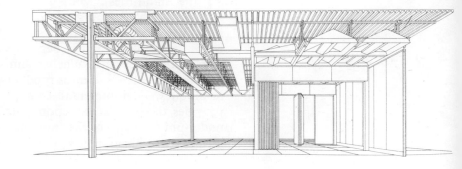

7.1.36 Schools Construction Systems Development (SCSD). Ezra Ehrenkrantz, 1961. Cross-sectional perspective

7.1.37 Double page spread from *The Turning Point of Building* Wachsmann, 1961 with the curious caption: 'Perspective of a structure built with a single standard structural element; the skeleton of a five-storey building is suggested'

should be noted. There is the same heavily deterministic view of the course of events that we have encountered before. However, Wachsmann includes some curious, almost mystic, comments and drawings which are difficult to explain within the terms of his argument. The concept of the energy inherent in the joint is one such aside: another is contained in his drawings of 'dynamic structures' which, one suspects, indicates the brief emergence of Waschsmann's otherwise concealed sensuous and artistic side (**7.1.37**). Appearing at Art Net in London in July, 1976, Wachsmann demonstrated his adherence to the promises of high technology claiming that 'man's great enemy is nature' and adding 'I reject this word creative completely'.

This section has attempted to show some of the factors which have contributed to certain working assumptions which lay behind much building systems work. Within the idea of an 'industrialized vernacular' for building these assumptions have been accorded considerable credibility and have become separated from a number of other developments. These postulates have been forged into what I have called 'The Authorized Version': a set of views which have become professionally and politically linked into a coherent and acceptable philosophy for the pursuit of specific aims. The idea of industrialized components on catalogue availability (as in the Eames house), the development of modular theory and Wachsmann's 'universal room' that is independent of local climatic conditions, add up to a picture that most of us will recognize in many building systems. In such a form, the philosophy must be questioned since it is clearly based upon a number of oversimplifications. However, it is still clear that the proposition put forward by Charles and Ray Eames is a valid one: it is only when we see their benign pragmatic approach against the full rigour of Wachsmann's idealized, total view that it becomes clear that the 'authorized version' gave authority to many buildings which contravene basic comfort levels, to say nothing of their architectural quality.

SYSTEMATIZATION WITHOUT A SYSTEM

'A generating system is not a view of a single thing. It is a kit of parts, with rules about the way these parts may be combined.'

Christopher Alexander,
Systems Generating Systems,
1968

Any discussion of building systems inevitably carries an image with it of that which is permissible as a building system and that which is not: whatever falls within the boundary shows a measure of agreement as to what might constitute a building system. There is a constant interplay between concepts developed by designers, such as Wright, Le Corbusier, or the Eameses, and the developments and capabilities of the process of making buildings. In spite of all this, as we have seen, the power of the machine analogy and the imagery developed from it have formed a very particular boundary, or set of criteria, within which system builders see themselves working. Within this boundary only heretics will talk of the brick as being the first prefabricated unit and the most flexible one. As a material it had been cast out from within the boundary described by the hard line building systems view because of its heavy reliance on skilled labour and wet jointing on site. This argument persists as a support to systems ventures, with the purists of CLASP only 'considering' the introduction of brickwork in 1972 (CLASP 1972). Whole ranges of ideas, processes, methods, and materials thus become acceptable or unacceptable within the concept of a building system.

The developments relating to timber housing constitute a good example of this process for, although timber building has been with us for a long time and offers all manner of possibilities, it has seldom featured strongly in building systems work in Britain. A great wealth of ideas and results have been ignored because the process of timber housing, in encouraging choice, variety, and user participation automatically sets it outside of the self-described 'building system' boundary. In the brick based British building industry, the lessons to be learnt from such approaches have only slowly been recognized, although by 1979 a fifth of all new housing starts employed a timber frame inner skin.

From the Scandinavian timber tradition (**7.2.1**) through to the growth of the American timber family home, project or 'tract' house,

a

b

7.2.1 (a) Norwegian traditional timber construction. Part of the Bjørnstad farm group, late 18th century, re-erected at Maihaugen Open Air Museum, Lillehammer, Norway. (Photo 1975)

(b) Re-use of traditional Norwegian construction. Lidartunet, a group of timber farm buildings collected from the Øystre Slidre in Valdres, re-erected and used by the Koxvold family over a period of thirty years

7.2.2 *Opposite* Tunhus system. Housing development at Ramstad, Oslo, Norway. (a) A two storey house. (b) External corner and eaves detail. (Photos 1975)

to the work of Walter Segal in Britain, there is certainly a system at work. That such developments have not conformed to the somewhat narrow definition of the system approach used in Britain is significant.

Nevertheless both Sweden and Norway have a number of timber based building systems — some, like the Norwegian Tunhus system, are particularly well thought out and elegant (**7.2.2**). Apart from the availability of timber, Scandinavia has had the advantage of not having its craft traditions destroyed by the industrial revolution. If we turn to the United States of America and Canada we find that there the use of the timber frame has grown with their industrialization.

The development of house-building methods in North America is of particular significance in two ways: first for how much the study of it

a

b

has to offer those interested in prefabrication, industrialization, rationalization, or whatever euphemism currently is in use; secondly, for just how little notice has been taken of such methods by those avowedly interested in building systems design.

Some of the lessons to be learned from the timber frame, and the balloon frame in particular, are set down by Giedion in *Space, Time and Architecture* (1967, 1941) where he devotes some eight pages to discussing its merits and describing its inventor, who was most likely George Washington Snow (1797-1870). Snow was a Chicago civil engineer, surveyor, contractor, financier and 'jack of all trades'. Konrad Wachsmann, before Giedion, however, wrote his little known work, *Holzhausbau* (1930), published in Germany in which he described in great detail this form of construction — this work was clearly influential on many youthful European architects of the thirties.

Giedion speaks only of the balloon frame and not at all of the platform frame, which has now largely superseded it as a form of construction. He cites St. Mary's Church Chicago, 1833 as the first balloon frame building and draws attention to the significance of the invention of the mass produced machine made nail which superseded the wrought iron nail and eliminated complex jointing. Machines for cutting and heading nails were patented in both England and the United States around 1790. The case of the nail is a dramatic example of the type of economies that the initial mass producing of a product can have, as well as showing that it can cause far reaching changes in methods.

The introduction of cut nails early in the nineteenth century quickly undercut the high price of wrought iron nails and by 1842 they were 3 cents a pound (Bolles, in Geidion (1967, 1941)). The variety of nails increased to cover every specialized task in timber-frame construction until, by 1968 there were 270 different types being used in residential and farm buildings in the United States (Stern 1968). Clearly the development of the timber frame, first in its early form of balloon frame, and later in its platform frame form was a new approach, albeit developed from former timber-frame ideas. The latter, however, required a degree of skill with their mortice and tenon and other joints which meant that those who could exercise it also controlled it to a degree. The skills involved and their initiation procedures described the craft that its practitioners practised. With the new methods, which used lighter timbers, nailed together to a very simple pattern, it was possible for almost anybody to build a reasonable house quite quickly. In this, the change in technology had its repercussions on the very basis of the originating craft itself and, ultimately, upon the whole of society. The ubiquitous timber-framed house is a fundamental part of the North American way of life. Thus the reasons for the very considerable and gradual move by the North American house building industry towards rationalization involves many of the ideas, not to say ideals, of the building systems propagandists. This evolution to systematization without a 'system' is worthy of close examination for it embodies not only a view of techniques and processes but also a view about such 'architectural' matters as environmental performance, the use of space, reflection on the user's needs and form and meaning.

It is first necessary to say something of the traditions of house building in North America. Clearly the plentiful supply of timber and the ease with which it can be worked have formed a basis for the developments of this mode of construction. Not only houses, however, but all forms of building have been constructed out of timber, often rising several storeys high. Apartment houses and hotels of five or six storeys in timber construction were quite common.

The unification of the developing technologies of the mid-nineteenth century, with the constructional advantages of the timber frame, were crystallized by Catherine Beecher in *The American Woman's Home,* published in 1869 (**7.2.3**). Banham (1969) draws particular attention to the significance of her proposals in introducing 'for the first time the conception of a unified central core of services'. He goes on to argue that this is a precursor of Fuller's dymaxion house of 1927 (**7.2.4**). Indeed, as he rightly points out, the subsequent development of the typical tract builders house in the United States and Canada owes a great deal to this concept. In this we see the full application of the idea of a highly serviced, well insulated lightweight box. Wright and others particularly contributed to the extension

7.2.3 The American Woman's Home. Catharine E. Beecher and Harriet Beecher Stowe, 1869. 'When "the wise woman buildeth her house" the first consideration will be the health of the inmates. The first and most indispensable requisite for health is pure air, both by day and night' (Beecher and Stowe, 1869). Key: I, hot air stove; 2, Franklin stove; 3, cooking range; 4, fresh air intake; 5, hot air outlet; 6, foul air extracts; 7, central flue; 8, foul air chimney; 9, moveable wardrobe

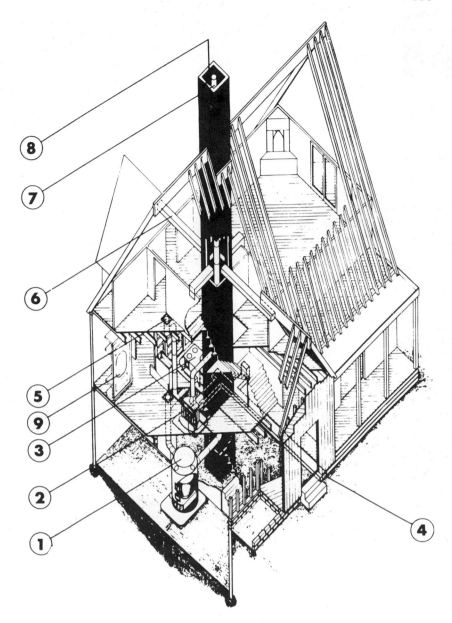

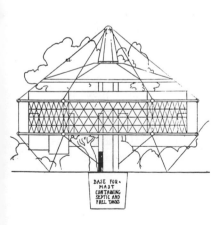

7.2.4 4D Dymaxion house. R. Buckminster Fuller, 1927. As with The American Woman's Home of 1889, this proposal uses a central services core and the design emphasizes the importance of an environmentally controlled interior.

of Catherine Beecher's ideas and their rephrasing in more high style terms. A number of these threads, emanating from high and low style sources (from architecture and from the vernacular) have woven together to give us first the emphasis on environmental control and servicing flexibility as evidenced in Ehrenkrantz's SCSD, (Schools Construction Systems Development) and then in the work of the 'environmentalists' — Richard Rogers, Norman Foster (IBM), John Wright (Ilford), and Peters and Paterson at Chichester.

Banham refers to the significance of the emphasis on roof, not walls, in the work of Wright and the California school: and we can see the same view still in evidence in the above examples. SCSD considered the wall element unworthy of development time since it constituted only 7% of the total cost; John Wright at Ilford, claiming not to be interested in style reduced the external wall to its simplest level at the same time elevating (literally) the air handling plant on the roof to symbolic status, and Foster visually eliminates the wall almost entirely by making it of bronze glass. Lance Wright described it 'the true disappearing wall' (Wright 1972).

The tradition of building in timber, using the balloon and the platform frame has, in North America, been developed to an operationally efficient method of providing homes. Further it should be noted that, unlike Britain and Europe it is there called home-building, not house building, or some such bureaucratic phrase as 'the provision of dwelling units'. Since the 1930s the United States has developed a mode of financing home-building which had brought home ownership to 63% in 1963 as opposed to 43% in Britain. Through such agencies as the Department of Housing and Urban Development (HUD) and the Veterans Administration the federal government insures or guarantees finance obtained through insurance companies or mortgage houses so that the lenders do not lose out in the event of foreclosures. Although a 0.5% charge is made for the insurance cover so provided this is offset by the lower interest rates for FHA mortgages. This and other financing arrangements bring about very low initial downpayments, and thus home ownership is brought within the reach of a far broader section of the community. Downpayments have been as little as 3% over a 30-35 year term. Home builders offer additional arrangements which can also reduce down payments dramatically. Whereas in Britain, even for young professionals it is often difficult to raise the capital necessary for the downpayments of 5-20%, depending on the type of property. In the United States all such government approved loans have the progress of construction inspected by government inspectors so that standards are assured. There have been similar arrangements throughout Canada for a number of years where such loans are made, or insured by the Central Mortgage and Housing Company and construction is similarly inspected. CMHC has branch offices throughout Canada to which purchasers can turn for advice, finance or purchase difficulties.

CMHC publishes a book of 100 designs for small houses for which working drawings are available for both home owners or builders for a nominal sum. All the houses are designed by architects working basically in the frame tradition. In addition CMHC have published a number of guides for prospective purchasers that are couched in non-professional language. 'Choosing a House Design' and 'The Principles of Small House Groups' are typical titles that have been made available over the counter in any CMHC office. 'A Lot to be Proud of' deals with soil fertility, mowing, watering, turf disorders, the planting, cultivation and feeding of trees. With leaflets concerning the inspection of your new house, urban renewal, home building and loan provision freely available, it is possible to get some sense of the contrast between North American practice and European practice. The latter seems often more aimed at discouraging home purchasing and building. Indeed there is some truth in this. Certainly in Britain until recently, the pressures (elsewhere described) to provide housing as a social service via central and local government has led to frequent discouragement of the private sector, thus creating the enormous institutional housing providers that we now have: e.g. the GLC and all large local authorities. This in turn has led to powerful criticism from a later generation, who now see that the agglomeration of design power and administrative control into such few (and bureaucratic) hands, gives the individual much less self-determination than owning his own home. He is not allowed to personalize it overmuch inside, nor allowed to add to it outside, even if he is so inclined. In addition, far from becoming a freely available cheap public service as hoped, the very attempts to plan and control how people can make choices has led to shortage after shortage (Pawley, 1971a,b).

The American methods of financing the individual rather than financial control through the intermediaries, such as building societies and banks, has created a situation where an increasing number of people can become home owners. It has also given the industry some stability to develop viable home building methods to meet the demand.

There is a further aspect to this in connection with the homebuilders themselves, numbering some 50,000 (MHLG 1966a). The vast majority of these are small firms with an average output per firm of 60 units annually, and the means used has encouraged them to develop. This is in contradiction to the approach of the system builders in many parts of Europe where methods have been encouraged that favour the large contractor. In Britain particularly the avowed intention of many of those encouraging the use of industrialized methods was to reduce the number of small firms of house builders, apparently because they were not efficient. This was a practical result of the embrace of the mass production philosophy that economy and efficiency only follow when there are long identical runs of items. This view resulted in government employed architects becoming involved in putting together the large markets that adherence to such a philosophy predicated. Hence local authorities were encouraged to join their requirements together and this meant that methods and means had to be standardized. This happened both in house building and school building, and quickly led to severe repercussions on the market situation.

In housing it can be said that apart from the single large contract (which is like one building anyway) the attempt on the part of housing systems to penetrate the small authority housing market has not proved successful. Where it has been adopted the results have often put the local authority and the architects at the mercy of the large contractor who virtually had backing from central government because he was 'using industrialized' methods. In this way the smaller contractors with their, often painfully, slowly evolving techniques were frequently cut off from such work because of the apparent construction mystique involved. Further, as is elsewhere described, the process of very real breaks with the traditions of the industry were rapidly accelerated. In the case of the client sponsored consortia the continued placing of orders with larger and larger firms has meant that all sorts of options closed. The big firms became bigger and the smaller, sometimes more go ahead and responsive firms, went to the wall. Of course, in many areas this has been positively encouraged by large firms, sometimes through their trade associations. The logic of all this is simple: a client/architect body had a large order for components or a subsystem; it casts around for firms capable of manufacture and reliable delivery; this predicates that the firms must have a good administration and adequate financial backing since authorities investigate this aspect. This usually eliminates all but the most well established and large firms; negotiations commence and the firms discover (if they didn't know already) that they have considerable bargaining power. They recognize, or learn quickly, that working with consortia and development groups, in addition to the usual complexities of working with local authorities, is time consuming and subject to considerable bureaucratic intervention. All of this has to be included somehow in the costing, either initially or as 'planned' extras; thus such firms gain more work and on occasion encourage cost increases because of this interdependent nature of the machinery. This is gradually reflected in the rise in house or school building costs generally with competition from inexperienced, outside firms becoming less and less possible.

The situation in the United States is markedly different since the government and the industry have responded to a large extent to support the small home builder. It is possible for the builder to do everything for himself: design, materials purchase, and hiring of labour. Alternatively, he can obtain a completely designed and prepackaged house from a home manufacturer and employ all the necessary labour through subcontracting (**7.2.5**). There are also many variants between these two poles. Tindale (MHLG 1966a) sets out the complex relationships between these different elements in the industry in a useful diagram.

The Federal Housing Administration (later HUD), set up in 1934 during the depression, gave some encouragement to continuity of market which ultimately gave rise to a mainstream development largely consisting of prefabricated versions of the traditional timber-frame house. Manufacturers of prefabricated homes gradually increased the number of items they supplied at the same time enlarging their likely market. Commencing by manufacturing external wall panels and roof elements they gradually took in more and more of the work that had formerly been done on site. Preglazed windows fitted to panels, claddings fixed in the factory, partitions, flooring and ultimately, servicing elements and furnishings were brought within their embrace. Houses prepackaged in this way could then be loaded onto a couple of trailers and trucked to the site where they could be rapidly assembled. During this period of development many innovatory prefabricated systems were produced — Fuller's Dymaxion house, plywood and steel panel houses and Koch's Techbuilt experiments. But although some of these undoubtedly fed the mainstream of development, its basis lay in a financially well supported and clearly explicit relation between provider and consumer. This meant that such development was not distorted beyond the market capability at any one time by a 'one shot' specific and total innovation. It is thus not surprising that, in this context, the Dymaxion house was not marketed.

This mainstream approach to home building is again contrary to many developments in Britain where many of the closed commercial housing systems that failed in the early sixties, including the MHLG's own 5m housing system, had to be discarded. The fact that 'innovation' was itself coming from the MPBW/MHLG development groups meant that, apart from exhortation, its only other recourse to gaining acceptance for the ideas and techniques was by co-operation with willing authorities or by gentle persuasion of the less willing ones. Ultimately many such attempts at change, some good, some bad, have to be enforced by sanctions of various sorts — in the case of the encouragement of the use of 'industrialized' methods this was done by careful control of loan sanctioning — which just as quickly reversed after the political abandonment of industrialized building at the time of the Ronan Point collapse. Apart from the rationalizations brought about in the use of timber construction and related financing, the North American house has a number of other distinctive features. Often houses will be single storey with a full basement, half or completely excavated below ground. This will be unfinished, the poured concrete or block wall being left for the owner to complete.

The provision of such basement space is very cheap since the cost of excavation using mechanical excavators is very low. In addition to extra space, which can be completed when the owner can afford to finish and fit it out, there are other benefits. All the servicing, furnace, warm air ducting, plumbing and drainage runs can be housed there and these are kept weather free and immediately accessible for maintenance or additions or replacement. The standards of servicing are usually high, often with full ducted warm air heating to perimeter floor grilles, and where required, air conditioning. There is a high provision of electrical outlets to cater for the many machines used in such homes and the flexibility required by a family. Very well fitted bathrooms, showers and kitchens are the norm, and to encourage sales, home builders will often include cookers, refrigerators and dishwashers.

The methods of construction developed by American and Canadian homebuilders is of particular importance: this is so because it has evolved some aspects of rationalization and standardization whilst ignoring others. Amongst the latter are many thought to be significant by those system designers laden with a set of preconceptions concerning the nature of industrialized building. The fact that such rationalizations have occurred around a direct consumer supplier link is doubly important, reinforced as it is by a set of federal government standards and adequate financial provision. Of course such a market oriented situation is open to the charge that producers are manipulating the consumers to buy what they want to sell: those 'Ticky Tacky Boxes' of the popular song by Pete Seeger, produced in

a

b

c

7.2.5 The development of the timber house 'package'. (a) 1812 log home later used as one room schoolhouse. Richmond, Ind, USA. (Photo 1979). (b) Sears Readycut houses shipped to buyers as a complete package, 1904-5. Cost $1700-1900. These examples in a row of 6 in Madison, Ind, USA. (Photo 1979). (c) Package timber homes, Lafayette, Ind, USA. National Homes, 1979. National Homes offer a large range of options — Bi-levels, Tri-levels, Split levels, Two Storeys and Ranches. (d) American timber home building. The structure of manufacture, supply, and distribution. MHLG 1966a. Whether building a single house, a small group or a large estate, the US home builder can choose to buy complete packages in a range of designs or sets of parts which he can assemble as required

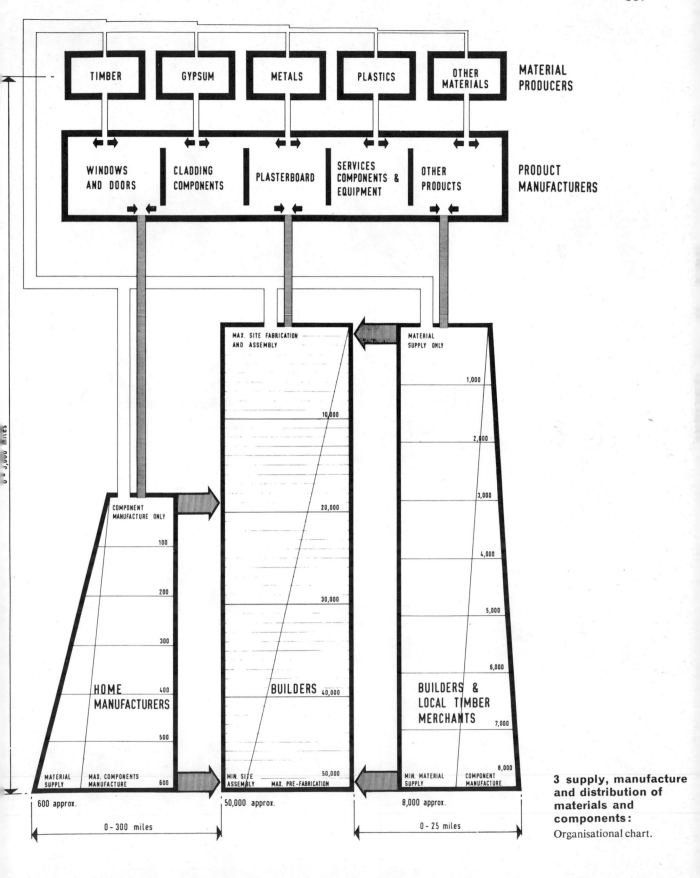

3 supply, manufacture and distribution of materials and components:
Organisational chart.

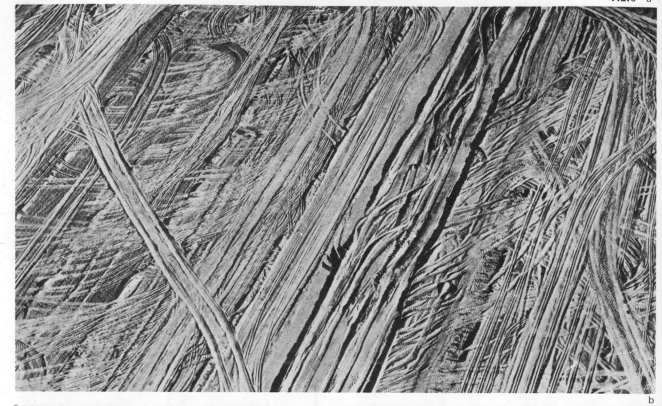

c

d

e

f

7.2.6 Project, or Tract, housing: a subdivision under construction, near Los Angeles, USA. (a) Site after clearance. (b) Foundation walls and garage site slab laid, timber stacked ready for construction. (c) Timber framing in progress. (d) The shells of houses completed. A typical layout for suburban development in the USA: a broad street, open front gardens, houses of various designs, backyards and service alley. (e) and (f) After the passage of time the personalized houses with lush planting can create an interesting and pleasant environment. Streets in Muncie, Ind. (Photos (e) and (f) 1979)

endless conforming lines (**7.2.6**). The rise of the ecology lobby has added further force to this view, drawing as it does on the wastage of energy inherent in the provision of a structure which provides little of its own environmental control and relies heavily on incoming energy systems like electricity and natural gas, for its survival. Whilst there is some truth in this, frame construction is so flexible that it can be adapted easily and, furthermore, many such critics end up using it themselves when it comes to enacting their theories. Such timber-frame housing developments quickly mature into desirable and humane environments, well planted and easily personalized. Gans (1967) has shown, contrary to popular belief amongst many design

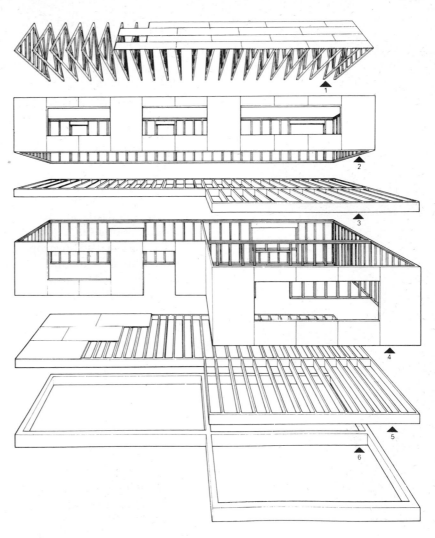

7.2.7 Platform frame construction.
Exploded perspective

professionals, that these developments quickly grow strong networks
of interaction and community concern.

The platform frame system used has become highly standardized
around a few dimensions, and this capability plus its simplicity has
made it far and away more used than the balloon frame (**7.2.7**). It
consists of timber studs 4 in × 2 in (102 × 51 mm) at 16 in (406 mm)
centres in 8 ft 0 in (2.42 m) high wall panels of varying widths based
upon the stud centres. Rafters and ceiling joints have long since been
replaced by trussed rafters, very light section timbers plate nailed and
erected at two foot centres to a range of roof pitches. Floor joists of 8
in × 2 in (203 × 51 mm) at 16 in (406 mm) centres is normal. The
walls, roofs and floors then have a series of layers, added like clothing.
There is vapour barrier and plaster board and insulation inside,
plywood sheathing, weatherproof membrane and a wide range of
finishes. All this has been facilitated by the nationwide use of
presurfaced timber, already machined to a finished surface.

This gives rise to another aspect of the approach, and one viewed with suspicion by many architects who see a close relation between moral value and built form. To these designers a building should 'express its functions' usually by means of its structure. Decoration, or expression that cannot be so described, is frowned upon and a building, runs the argument, thus loses its integrity. The tract houses of the United States have the ability to have almost any appearance clipped, pasted, stapled or nailed on. This allows an expressive freedom, or 'untidyness', that is anathema to many architects. The instant styling of suburban America is enough to set many architects reaching for their Design Centre glasses; until they have lived in it, when it is usually found to be part of a very congenial lifestyle.

The incorporation of broad dimensional standards into government property standards in the United States encouraged the use of basic planning modules of 2 ft or 4 ft (610 mm or 1.22 m). Bemis it will be recalled had in 1933-6 proposed his theory of modular co-ordination based on the 4 inch cube and this in turn was developed from studies of practice, largely in timber housing. The trade associations also carry more responsibility in this co-ordination of effort than they do in Britain. Referring to the National Lumber Manufacturers Association Unicom handbook Tindale says:

> 'The differences between the Unicom publication and technical data on British systems of construction are interesting and significant. . . . Unicom is more akin to a code of practice than to a system handbook. The method of construction is specified in sufficient detail of prefabricated components to be obtained from any manufacturer by simple reference to the Unicom manual.' (MHLG 1966a)

Tindale goes on to describe how the manual deals with rules for dimensioning and coding of non-modular parts, which as components are outside the range of the handbook. Nevertheless this does not preclude the system from working because all other manufacturers of doors, windows and cladding are working to the same basic set of standards. An important aspect of the whole approach is the growth of nailing techniques which are advanced and well understood apart from being well covered in official publications.

The standards, plus the handbooks produced both in Canada and the United States, have eliminated mystique rather than encouraged it as 'systems documentation' has done in Britain. The timber construction manuals produced by the Canadian Institute of Timber Construction give the properties of timber sections of various sizes, design charts and typical detail guidance, which shows the designer or builder how to make the necessary calculations for timber construction. Such commercially produced documents as 'Canadian Wood Frame House Construction', 'Wood Frame Construction' and 'Post and Beam Construction' give an indication of attitude to systematization which is supportive and encouraging rather than censorious. All this has led to a gradual change where more and more

sections have been precut, then prefabricated in a factory and trucked to the site. It is of special interest that, as Tindale says:

> 'the US homebuilding industry has been fortunate in comparison with its British counterpart, because it has not had in a few short years to evolve a new technology while assimilating the processes and effects of industrialisation.'

Then she goes on:

> 'It may be argued that this initial advantage may prove to have considerable limitations in the long run since the industry may find it difficult to switch from what are basically developments of craft processes to more sophisticated and essentially industrial methods of production.' (MHLG 1966a)

In these few lines we can see a total misunderstanding of what had been seen because it was still being viewed from a narrow view of what constituted a system. It is full of contradictions. First we might question why in Britain it seemed necessary to 'evolve a technology'? What is meant here is to evolve a NEW technology and one in keeping with the myths of the machine age apostles. It is elsewhere described how such attempts to implement this has meant that existing methods, often already industrialized ones, have to be discredited. Second, it does not really state why Britain has 'in a few short years' to assimilate the 'processes of industrialization' nor does it question the validity of that assumption. It is also implicitly accepted that it is 'a good thing'.

Third, a distinction is made between craft processes and 'essentially industrial methods of production' which are more 'sophisticated'. At the very least this statement says an enormous amount about the author's approach and is representative of the whole building systems ethos in Britain. This dictates that craft processes are not sophisticated and that industrialized methods are sophisticated.

Fourth, is the automatic assumption that the methods described so eloquently in the book are not industrialized and are craft methods which will be one day surpassed. Did the author not see the true 'industrialization' inherent in the process? Did the prefabricated dimensionally co-ordinated dry construction with flexibility produced in a factory not constitute some part of the concept of industrialization? The methods are certainly not labour intensive as is British housebuilding nor are they inefficient as she herself so clearly indicates. The key of course rests in the valuation of craft. This view creates a boundary between craft processes and industrialized processes. Although the standards of the former may be valued, its existence is not in accord with the concept of the machine age held by many architects. This view of what constitutes a craft process (a hammer and a power saw are in their way part of the process of industrialization) has been instrumental in creating the difficulties that many system builders have encountered. It also partially accounts for

7.2.8 (a) House sections showing modular size range. Redrawn from MHLG, 1966a. (b) *Opposite*. A complete package house erected in under a week by National Homes, Lafayette, Ind, USA (1979). The components arrive in the two trailers seen at each end of the house

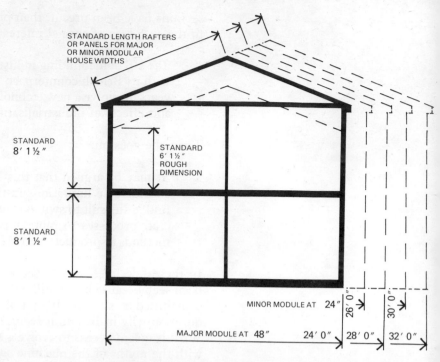

a

STANDARD LENGTH RAFTERS OR PANELS FOR MAJOR OR MINOR MODULAR HOUSE WIDTHS

STANDARD 8′ 1½″

STANDARD 6′ 1½″ ROUGH DIMENSION

STANDARD 8′ 1½″

MINOR MODULE AT 24″   26′ 0″   30′ 0″

MAJOR MODULE AT 48″   24′ 0″   28′ 0″   32′ 0″

the difficulties (especially in housing) where a recognition of the successes of the US timber-frame tract house experience would mean a radical change of basic assumptions, an acceptance that housing is built in small numbers at a time, in many different places by small groups of people. This is what British systems designers failed to recognize and act on. Even more ironically a continuing catalogue of

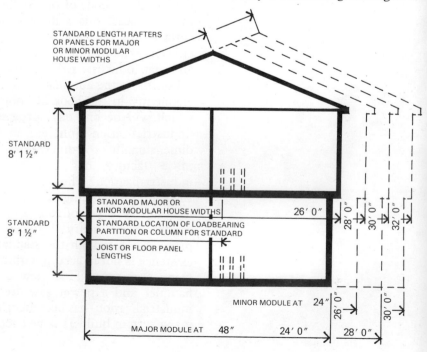

STANDARD LENGTH RAFTERS OR PANELS FOR MAJOR OR MINOR MODULAR HOUSE WIDTHS

STANDARD 8′ 1½″

STANDARD MAJOR OR MINOR MODULAR HOUSE WIDTHS   26′ 0″

STANDARD LOCATION OF LOADBEARING PARTITION OR COLUMN FOR STANDARD

STANDARD 8′ 1½″

JOIST OR FLOOR PANEL LENGTHS

28′ 0″   30′ 0″   32′ 0″

MINOR MODULE AT 24″   26′ 0″   30′ 0″

MAJOR MODULE AT 48″   24′ 0″   28′ 0″

b

the methods developed around the US home building industry reads like the Utopia of the machine aided cornucopia so often described by Europe's architects:

1. The increasing move of work into the factory has led to a reduction in the labour content of the house to match the high cost of skilled labour.

2. Construction times are very short which is a great advantage in the adverse weather conditions often encountered. House shells can be put up in one day and in good conditions completed in as little as three weeks. (The author has seen a number of sites where the completion of a house from start to finish, including lawns and trees was completed in some six weeks.)

3. The use of prefabricated elements cuts down the time on site measuring, cutting and fitting as well as in the contractor's office (**7.2.8**).

4. Factory production has created higher quality and this in turn has increased standards on site.

5. Components and materials, often trucked some distance, are very well packaged since this saves replacement due to damage. Kraft paper, polythene and card containers are common. On site materials are almost always protected.

6. Components such as doors, windows and siding have all been designed to fit easily into the frame after erection and self-weathering junctions have evolved which do not let the water in (**7.2.9**).

7. Wall construction, usually consisting of: a siding or cladding, waterproof membrane, sheathing, stud frame, insulation, vapour barrier and an internal skin, gives a weather and airtight container with a minimum of complication (**7.2.10**).

8. Joints have been reduced to a minimum and straightforward solutions found.

9. New methods and materials are introduced as they prove themselves: plastic film coatings to claddings and interiors, new fittings and finishes.

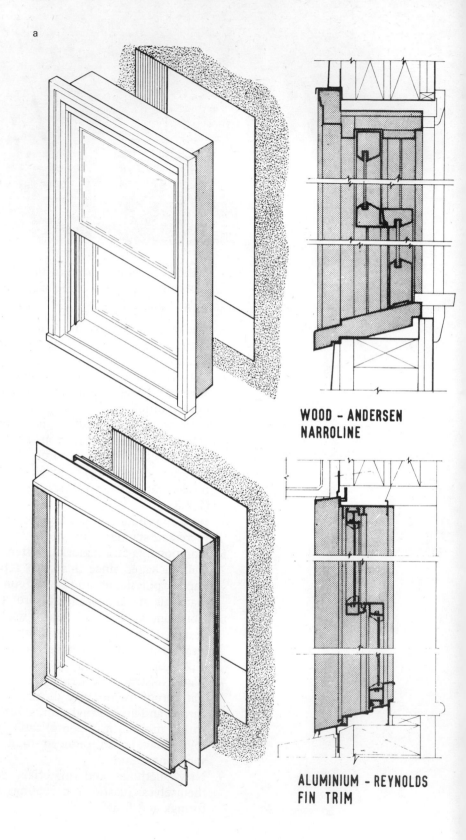

WOOD – ANDERSEN NARROLINE

ALUMINIUM – REYNOLDS FIN TRIM

7.2.9 Window to wall junctions. (a) Showing windows manufactured with an integral surround which acts as a fixing and provides a good basis for a water and draught proof joint. (MHLG 1966a) (b) *Opposite.* Window set in wall. National Homes house, 1979

b

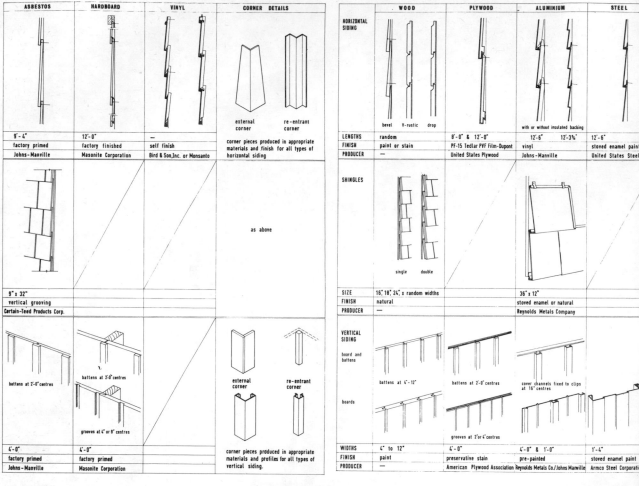

	ASBESTOS	HARDBOARD	VINYL	CORNER DETAILS
	9'-4"	12'-0"	—	external corner / re-entrant corner
	factory primed	factory finished	self finish	corner pieces produced in appropriate materials and finish for all types of horizontal siding
	Johns-Manville	Masonite Corporation	Bird & Son,Inc. or Monsanto	
	9" x 32"			as above
	vertical grooving			
	Certain-Teed Products Corp.			
	battens at 2'-0" centres / 4'-0"	battens at 2'-0" centres / grooves at 4" or 8" centres / 4'-0"		external corner / re-entrant corner
	factory primed	factory primed		corner pieces produced in appropriate materials and profiles for all types of vertical siding.
	Johns-Manville	Masonite Corporation		

	WOOD	PLYWOOD	ALUMINIUM	STEEL
HORIZONTAL SIDING	bevel V-rustic drop		with or without insulated backing	
LENGTHS	random	8'-0" & 12'-0"	12'-6" 12'-3¾"	12'-6"
FINISH	paint or stain	PF-15 Tedlar PVF Film-Dupont	vinyl	stoved enamel paint
PRODUCER	—	United States Plywood	Johns-Manville	United States Steel
SHINGLES	single double			
SIZE	16", 18", 24", x random widths		36" x 12"	
FINISH	natural		stoved enamel or natural	
PRODUCER	—		Reynolds Metals Company	
VERTICAL SIDING — board and battens	battens at 4"-12"	battens at 2'-0" centres	cover channels fixed to clips at 16" centres	
boards		grooves at 2" or 4" centres		
WIDTHS	4" to 12"	4'-0"	4'-0" & 1'-0"	1'-4"
FINISH	paint	preservative stain	pre-painted	stoved enamel paint
PRODUCER	—	American Plywood Association	Reynolds Metals Co./Johns Manville	Armco Steel Corporation

7.2.10 Materials available as claddings to timber-frame houses. MHLG 1966a

10. Since much of the work is done by small subcontractors the procedures adopted have created a situation where overlapping of trades has been gradually reduced so that continual return visits are not necessary. Further, the clear programming of work leaves little doubt that when a contractor says he is ready for a trade he really means it. This reliability plus the competitive nature of the market has encouraged the subcontracting groups to develop new methods to 'get in and out in a day'. The Ames tape machine for jointing plasterboard was one of these. Similarly, those fixing ceiling finishes invented short adjustable stilts (**7.2.11**) to enable them to fix the sheets and tiles thus giving them a mobility and cost saving over elaborate scaffoldings and plankings: whilst other work in a room can continue simultaneously. The use of staple guns to fix polythene, weatherproofings, ceiling tiles and sheatings has been a commonplace for many years without any loss of quality.

11. Power tools are readily used on site, drills, saws, power floats and so on. Fibreglass has been formed into paper-covered 'batts' for many a year (**7.1.12**) whilst the glued asphalt shingle on ply sheathing for roof construction gives speed, flexibility and variety

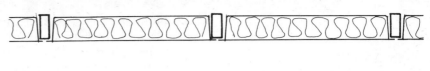

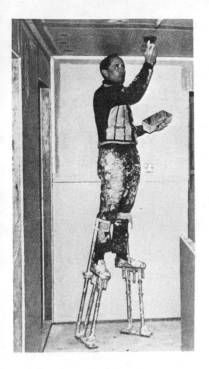

7.2.11 *Above.* Operative using stilts for ceiling work

7.2.12 *Above right.* Glass fibre insulation batts, paper covered with edge lip for fixing to timber frame

7.2.13 Plasterboard nosings, trim, and clips. US Gypsum and National Gypsum company. MHLG (1966a)

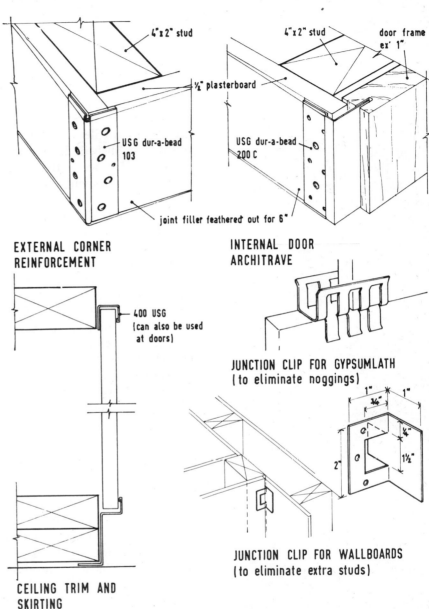

4"x 2" stud

½" plasterboard

USG dur-a-bead 103

joint filler feathered out for 6"

EXTERNAL CORNER REINFORCEMENT

4"x 2" stud

door frame ex' 1"

USG dur-a-bead 200 C

INTERNAL DOOR ARCHITRAVE

400 USG (can also be used at doors)

JUNCTION CLIP FOR GYPSUMLATH (to eliminate noggings)

CEILING TRIM AND SKIRTING

JUNCTION CLIP FOR WALLBOARDS (to eliminate extra studs)

to the house types. Prehung doors are widely used: each has three hinges to forestall any tendency to bow at the lock position and the knobset which requires only the drilling of two holes is an object lesson of real as opposed to imposed factory production. These knobsets are simply catalogue coded and arrive complete with assembly instructions in neatly packaged containers. Plasterboard edgings and nosings have long been standard practice and considerably simplify internal partition works (**7.2.13**).

12. Plumbing walls, prefabricated wiring harnesses, and a solution to the difficult bath wall junction are all examples of a responsive rather than a dictative technology. Ordering schedules and methods are simple and direct (**7.2.14**).

'Housebuilding in the USA' is a model of clarity concerning, in my view, a very well developed system indeed. However, the work appears to have been ignored in Britain. During this period, of course, Sir Donald Gibson and a party did go to Canada to look at house construction and certain other explorations were once again made, but by and large the architect system designers found difficulties in assimilating these lessons. Commercially developed systems such as that of James Riley's 'Riley form' Guildway and Colt were however already established. Indeed, the document seems curiously reticent about its relevance to the British scene. The Ministry's involvement with the development of their own architect designed 5M system from 1961 until its public burial at the RIBA in 1969 (Tindale and O'Toole 1969) not only calls into question the political advisability of a central government department inventing a nationwide system, but also juxtaposes this with the success of US timber housing as reported in 'Housebuilding in the USA' in 1966.

Many reports have conveyed the good news of the North American timber house, not least Giedion (1967, 1941) in *Space, Time and Architecture* in 1941. The British Timber Development Association, whose 1947 publication, *Prefabricated Timber Houses,* also described American developments and 'one American firm (which) has sold 110,000 precut houses during the past 40 years' (TDA 1947). The sad fact is that there was so little effect upon the British building industry during the 30 postwar years, although more recently the acceptance of timber-frame construction has grown in Britain and there are a number of examples of its use in both public and private developments: one projection suggests that by 1982 over a half of new houses will use such methods. Where it has been successful it has been used much as in North America with external cladding of traditional type, often responding very directly to the market. A particularly thoughtful and elegant use (**7.2.15**) of platform timber frame has been made by Farrell and Grimshaw in a series of infill housing schemes for London housing associations. In a carefully worked out programme, design and price tenders were invited against a performance specification. Construction times were good by normal standards and on one scheme tenders came in up to 15% under the very tight cost yardstick allowed by the government (Farrell and Grimshaw 1978).

Much of the reawakened interest in the use of timber-frame method, particularly amongst architects, has been due to the work of Walter Segal. Segal built his first timber house in 1932 near Locarno in Switzerland, and it appeared in Yorke's book *The Modern House* (1943, 1934). Segal, who subsequently settled in England, was brought up amongst artists, architects, writers and mystics, his father being a painter. Near his home, one Henri Oedenkoven from Antwerp had founded a utopian community in 1900. Oedenkoven was

7.2.14 (a) Order form for homes manufactured by Great Lakes Homes Inc, Sheboygan Falls, Wisconsin, USA (MHLG 1966a). (b) *p. 354-5.* Order form used by National Homes, Lafayette, Ind, USA. Completed by buyer/builder 1979

GREAT LAKES HOMES, INC. | Quote No. | Production No.

Date 10/7/63   Send Quote To

~~XXXXX~~ New American Classics   Address

Builder   City & State

Plan Description: Newport - 28'x34' #1   (Drywall) - Plaster -- Basement -- Crawl -- (Slab)

Living Room on Left   Living Room on Right   Basement Height

Lower level ext. walls No Insulation No		Med. cabinets numbers #7462	
Lower level windows No		10" Ceiling fan, Kitchen - 3 speed - Damper No	
Lower level storms & screens No		4" bath fan 8" wall fan - kitchen No	
Lower level int. walls No Doors No		Bath accessories Yes	
Lower level trim - Base ☐No Shoe ☐No		Shelf material Metal	
1x2 Wall furring ☒ ☉ Ledge cap - 1x6 ☐, 1x8 ☐, No ☒		Special front lock No	
Lower level siding No		Sills ☒ Oak — ☐ Pine	
Floor joist size None		Metal threshold Yes Oak threshold No	
Deck No Block for single floor No		Nail package Yes Weatherstrip Applied	
Adjustable steel posts No Length		Wall insulation 2" Garage insulation No	
Wood basement beam (stock only) No		Ceiling Insulation 2"	
Mudsill (or sill plate) by No		Stairs Basement Main	
House sheathing ½ - ¾ - None		Treads No No	
Gable sheathing ½ - ¾ (None)		Risers No No	
Garage sheathing ½ - ¾ (None)		Stringers No No	
Common wall sheathing ½ - ¾ (None)		Number of handrails No Floor stripping - (1x2 - 16" o.c.) No	
Sill (or box) sheathing --		Oak floors in No	
Garage size No		⅜" Underlay in No	
O. H. door size No By		Variations No	
Garage mudsill by No		**BELOW WILL NOT BE QUOTED WITHOUT DETAIL**	
House siding Horizontal masonite		Kitchen cabinets No Kohler sink No	
Gable siding Horizontal masonite		Counter Top No Color	
Garage siding No		Vanity cabinets No Vanity top No	
Siding variations No		End Splash No Hood Model No	
		Oven Model No Burner Model No	
Roof - Hip ☐ - Gable ☒ Roof Pitch 4½/12		Primed exterior trim Yes Flower box Per plan	
Trusses 2x4 Size 24'o.c.		Plaster grounds by GLH Yes ☐ No ☒	
Ceiling joist size -- Rafter size --		Mas. chim. ☐ (Gas-Oil) Metal Chimney by Bldr.☒ by GLH ☐	
Main overhang Per Elev. Gable overhang Per Elev.		Planters No Divider No	
Build gable out for brick - Roof sheathing 3/8" ply w/clips		~~If metal windows - ¼" round sweet bend only ☐~~	
Shingle type 235# Color		~~1x4 casing sill & stool ☐ full trim ☐~~	
No. lineal ft. metal drip edge Yes		~~1x4 jambs~~	
Window header size Std.		Additional 2x4's for - Drop Ceiling ☒ - Duct Work ☒ No	
Window Type Wood D.H.		**MISCELLANEOUS**	
No. Lites Per plan		Options:	
Storms & Screens Alum. Comb.		Elev. #7	
Picture window description Per plan		Elev. #8	
		Elev. #9	
Window sizes as shown --		Elev. #10	
Window sizes by Great Lakes Yes		Elev. #11	
Glass Sliding Doors No Size Glass Type		Elev. #12	
No. pair window shutters Per plan		12'x22' garage	
Door shutters size No		22'x22' garage	
Trim (Pine) Oak Mahogany		Deck material-crawl	
2¼" Base ☒ 3¼" Base and base shoe ☐		Deck material-base. -includes	
Type of heating by bldr. --		Basement stair material	
Bath room base trim by GLH ☒ by Bldr. ☐		Not Included: Flooring Nails - Sheetrock W. I. Rail	
Doors - Birch Oak (Mahogany)		Other: Paneling, pre-finishing of	
Sliding doors No		doors	
Bifold doors Metal - Classic style			
Number of Pocket doors No		NET F.O.B. JOB SITE + Tax	
Front doors Per plan (Side lights) no		OPTIONS ~~DEDUCT~~	
Rear doors Per plan		Col. #2	
House to garage door No		Col. #3	
Garage service door No		Col. #4	
Hang exterior doors Yes		Col. #5	
Comb. door - Type Alum. Number 2		#6	

The manufacturer's representative will go
through this form with the builder to ensure
that full details are given to the factory.
The circle on the plan opposite refers to
details which show alternative arrangements for houses with basements.

7.2.14 a

NATIONAL HOMES MANUFACTURING COMPANY
Post Office Box 7680
Lafayette, Indiana 47903
(317) 448-2000

SSR No. _____

☐ — Mark if New Builder's First House

☐ — Erection Supervisor Required

☐ — No Supervisor Required

1979 HOUSE ORDER FORM

Current Date _____ Builder Order # _____ Delivery Date _____ Time _____ Zone _____

FINANCIAL CLEARANCE BY: Letter of Credit From _____

Check in Advance: Check # _____ Bank _____

NHAC Loan No. _____ Branch Office _____

Sight Draft From _____

Other _____

THE INFORMATION REQUESTED BELOW IS NECESSARY TO AVOID DELAYED OR LATE DELIVERY OF THIS SHIPMENT IF ADDITIONAL INFORMATION IS NECESSARY, PLEASE ATTACH A MAP.

JOB SITE:

No. and Street _____ City _____ County _____ State _____

Lot # _____ Block # _____ Subdivision _____ Purchaser _____

PLEASE GIVE ANY NECESSARY DIRECTIONS TO JOB SITE IN SPACE BELOW:
(Indicate here if you have attached a map.) (If rural delivery, list nearest town.)

_____

_____

_____

Model No. _____ Design No. _____ Marked Up Plans Attached _____

Hand of House: RH _____ LH _____

Foundation: Slab Floor _____ Crawl _____ Basement _____
(Refer to Floor Plan)

Below For Office Use Only

Hand

Floor

Trim

( ) SPLIT DELIVERY: DATE TRUCK 1 _____ TRUCK 2 _____

DELAYED TRIM SYSTEM:

( ) (a) House to be processed for delayed trim (2 stage delivery)

( ) (b) Ship with this order delayed trim from sequence # _____

1. **BASIC ERECTION OPTIONS**
( ) A. Manual Panel Erection
036 _____ ( ) B. Crane erection w/NHMC crane
046 _____ ( ) C. Crane erection w/Local crane

2. **FRAMING** Basic: 2 X 4 Exterior and Interior Studs — 16" O.C.
(With metal gyp clips; sheetrock by Builder)
344 _____ ( ) A. No Garage
344-310 _____ ( ) B. With 1 Car attached Garage
344-090 _____ ( ) C. With 2 Car attached Garage
430-344 _____ ( ) D. One Car Garage Under per plan
440-344 _____ ( ) E. Two Car Garage Under per plan.

3. **OTHER FRAMING**
050 _____ ( ) A. Wood Wall Blocking in Lieu of Metal Gyp Clips
043 _____ ( ) B. Door Rough Opening Height Framed for Carpet or Oak Flooring
060 _____ ( ) C. Lower Level of Upper Level Finishing Package Per Plan Option No. _____
-040 _____ ( ) D. Double Jack Studs.
057-058 _____ ( ) E. Heavy Headers . . . . . Interior ( )   Exterior ( )
059 _____ ( ) F. 1 X 4 Corner Bracing (Basic w/Styrofoam)

4. **STANDARD 2' ADDITIONS** (Ordering Additions will affect loading)
_____ ( ) A. Bedroom End (Indicated on plan in bedroom by open arrow)
_____ ( ) B. Living Room End (Indicated on plan by closed arrow)
_____ ( ) C. Garage (Indicated on plan in garage by closed arrow)

_____ Net (Base package and Options   Firm: _____

_____ Transportation _____ Trailer @ _____ Address: _____

_____ Sales Tax. Signature & Title _____

(Must be signed prior to delivery)

_____ Total Amount   Emergency Phone No. _____

ORDER SUBJECT TO CONDITIONS OF SALE ON REVERSE SIDE.

Order Completed By: _____

Transportation is charged in accordance with actual number of trailers used. Price sheet estimates are based on 40' trailers and basic house packages only. Options, extensions, unitubs, ceiling insulation or extra material will result in an additional trailer.

5. | PRECISION FLOOR | Basic: Upper level of 1½ story and 2 story models only, per spec. sheet.

610 _____ ( )   A. Precision Floor w/5/8" T&G floor plywood (16" O.C. floor joists)
Ground level ( )   Upper level ( )

600 _____ ( )   B. Precision Floor w/3/4" T&G floor plywood (16" O.C. floor joists)
Ground level ( )   Upper level ( )

_____ ( )   C. Basement KD Stair package (When precision floor not selected)
_____ ( )   D. Optional kneewalls ( )   F.A. kneewall coverage ( )
_____ ( )   E. Laminated Wood Garage under header (with NHMC Kneewalls)

6. | EXTERIOR WALL AND GABLE COVERAGE |

736 _____ ( )   A. Basic: Front coverage per design for field application; basic wall sheathing is 1/2" fiberboard, offwall fiberboard gables, front gables per plan, shutters, porches and other design material per plan. Note: two story houses have basic flush fiberboard gables for field applied siding.

| Front |

730-753 _____ ( )   1. Deduct basic front coverage
735 _____ ( )   2. Optional front coverage (see product manual for pricing)
Type_____ Color_____

| SIDES , REAR AND GABLE COVERAGE OPTIONS |
B.   Field Applied Coverages _____

| Horiz. |

737-114 _____ ( )   1. 8" Horizontal aluminum shipped loose for field application over fiber, w/ offwall gables shop applied over fiber.
Colors: White, Sandtone, Desert Gold, Colonial Blue, Fern Green, Willow Green, Musket Brown, Primrose Yellow & Sierra Tan.
State Color_____

| Gable |

738-119 _____ ( )   2. Double 4" aluminum shipped loose for field application over fiber, W/off wall gables shop applied over fiber.
Colors: White, Sandtone, Desert Gold, Colonial Blue, Fern Green, Willow Green, Musket Brown, Primrose Yellow & Sierra Tan.
State Color_____

| 3 Side |

739-146 _____ ( )   3. Double 5" aluminum Horizontal shipped loose for field application over fiber, w/offwall gables shop applied over fiber.
Color: White, Colonial Blue, Sandtone & Sierra Tan.
State Color_____

| Panel # | Type |

4. Wood and Hardboard Coverage
Specify type_____ Color_____
(See product manual for pricing.)

C.   Shop Applied Coverage Where Available Per Plan _____

727 _____ ( )   1. Viking aluminum shop applied over fiberboard, offwall gables w/viking shop applied direct to frame.
Viking Colors: White, Sandtone, Desert Gold, Colonial Blue, Fern Green
State Color_____

773-143 _____ ( )   2. Simpson 5/8" Ruf-Sawn T-1-11 Pattern 4" O.C. Shop Applied over fiberboard, w/offwall gables shop applied direct to frame.
047-049 _____ ( )   1. T-1-11 applied over styrofoam
Colors: Pinecone Brown, Mauvewood Gray, Prairie Sage.
State Color_____

774-144 _____ ( )   3. Simpson 5/8" Inverted Batten 12" O.C. shop applied over fiberboard, w/ offwall gables shop applied direct to frame
047-049 _____ ( )   1. Inverted Batten 12" O.C. 5/8" over styrofoam
Colors: Pinecone Brown, Mauvewood Grey, Prairie Sage.
State Color_____

775-117 _____ ( )   4. 3/8" Ruf-Sawn shop applied over fiberboard, w/offwall gables shop applied direct to frame, w/field applied batts.
047-049 _____ ( )   1. 3/8" Ruf-Sawn applied over styrofoam
Colors: Pinecone Brown, Mauvewood Grey, Prairie Sage

775-776 _____ ( )   5. 7/16" Hardboard Sawtooth Pattern Shop applied over fiberboard w/off wall gables shop applied over fiberboard.
145
047-049   1. 7/16" Hardboard over Styrofoam. Select Color
094   Bark Brown w/Musket Brown trim
093   Grey w/Mauvewood Trim
093-094   Grey w/Musket Brown Trim

D.   Optional Coverage _____

118-006 _____ ( )   1. Flush gable in lieu of offwall (1/2" fiberboard sheathing) (coverage is
004-008   shipped loose for field application when sidewall coverage is ordered.)
047-049 _____ ( )   2. Styrofoam in lieu of wall fiberboard for field applied coverage.
_____ ( )   3. Add fiberboard under Shop Applied gable coverages
085 _____ ( )   4. 4" offset gables for brick with Shop Applied coverage.
State coverage type_____ Color_____
733 _____ ( )   5. Vinyl Flashing (for Field Applied Brick sides and rear.

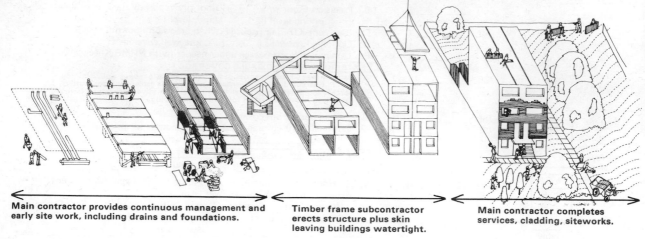

Main contractor provides continuous management and early site work, including drains and foundations.

Timber frame subcontractor erects structure plus skin leaving buildings watertight.

Main contractor completes services, cladding, siteworks.

a

b

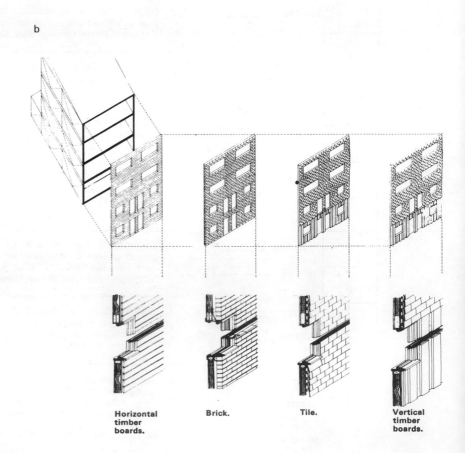

Horizontal timber boards.

Brick.

Tile.

Vertical timber boards.

7.2.15 Timber-frame housing. Farrell, Grimshaw Partnership (1978).
(a) Erection sequence diagram which illustrates the use of timber-frame construction site and organizational procedures. (b) Cladding options.
(c) *Opposite.* Range of home types

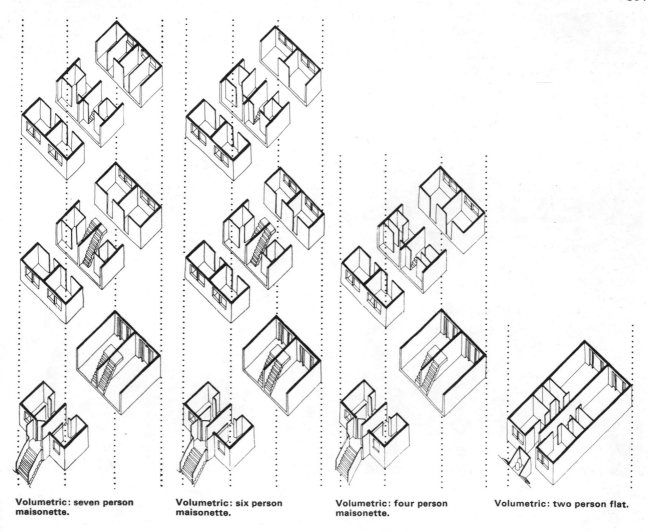

Volumetric: seven person maisonette.

Volumetric: six person maisonette.

Volumetric: four person maisonette.

Volumetric: two person flat.

c

'a practical man with considerable talent for architecture, one of the unrecognised innovators and who used the flat roof, roof terraces, sliding windows and untreated materials long before others.' (Segal 1974)

According to Segal, surrounded by such a rich diet caused him to seek out ordinariness, and this is a quality which has imbued his later work with timber. For many years Segal designed housing in Britain using methods deriving from traditional processes and contracting procedures, although the results were very much mainstream modern in approach.

In 1964 Segal was about to take down his Victorian house in Highgate, London and replace it with a new design of his own. To house his family in the interim period he designed a small single storey timber house in the garden, and it was this house that led him to design many more and to his discarding the traditional form of practice.

7.2.16 Addition to Children's Home, Singleton, Sussex, UK. Walter Segal, 1973. (a) View from west. (b) Showing construction

The house had an area of 66.15 sq m (712 sq ft) and cost the very low figure of £853 in 1965 (AJ 1966a). By 1970 Segal had built seven such houses (**7.2.16**), going on to receive as many commissions as he could handle and building at about half the normal cost in 1976 (AJ 1970). The basic difference between Segal's houses and the American timber-frame technique is that he uses a calculated frame which is raised off the ground and the slender columns supported on paving slabs set on pad foundations. The wind bracing for such a structure is placed below floor level. Where most of the building systems designers used grids dimensioned in accordance with the demands of modular theory, and therefore requiring components of special sizes to be made, Segal recognized that there already existed an effective supply industry making many components in standard sizes.

Rather like the traditional Japanese house, the dimensional discipline employed arises from the materials used. The roof, external walls and partitions comprise woodwool slabs on a constant 2 ft 0 in (610 mm) width, 2 in (51 mm) thick and in 3 lengths, 6 ft 0 in (1.83 m), 6 ft 8 in (2.03 m) and 7 ft 0 in (2.13 m). He simplified all details of panels and roofs, endeavouring to eliminate screws, nails and other such fixings and relying upon pressure and friction fixings. Products are used as directly as possible with as little cutting and fitting for each job as possible. Windows and doors have only linings, not frames, and sliding windows are made up on site very simply. He describes the envelope of the building as being like layers of clothing, each doing a different job. His roof of loose laid felt which is not stuck to the substructure is an example of this and unike many felt flat roofs they have given little trouble. Such an approach permits air to circulate between layers in both roof and walls thus eliminating problems of condensation.

Segal's drawings are simple and direct and usually drawn freehand on A4 paper, permitting them to be easily photocopied (**7.2.17**). He employs no main contractor but manages the building for the client, ordering materials and appointing the few trades necessary. The major part of the work is easily carried out by two carpenters, some of whom have moved from job to job with him, living on site. Segal refuses to be drawn into letterwriting and deals with all the queries on one site visit per week. All this has worked exceedingly well and his buildings are amongst the few where one hears nothing but praise both during and after the contract. By carefully developing timber construction Segal has created a system of his own involving existing materials and new processes. He produces economic and quickly built buildings which seem a pleasure for the builders and for the users — on occasion these two have been one and the same since, as Segal points out, they are so simple that almost anyone can build them.

Segal has been successful in producing low cost housing largely because he has seen through much of the familiar ideological argument of the system builders, and recognizes that the building industry already produces quality materials. Segal shows us, says Kainrath, that 'so many architects dream of something which is in fact

b

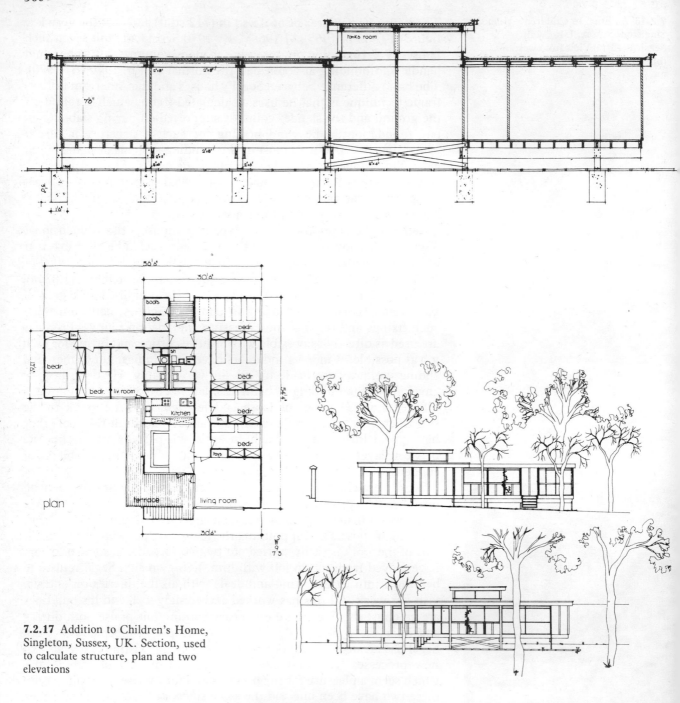

7.2.17 Addition to Children's Home, Singleton, Sussex, UK. Section, used to calculate structure, plan and two elevations

already at their disposal' (Kainrath 1970). This is not to say that materials must be accepted uncritically. In fact quite the reverse is true, for such an approach requires a close knowledge of the market and availability so that quality may be assessed against cost.

A number of quite basic assumptions underly Segal's work. One of these is the use of presurfaced timber. Significantly the use of presurfaced timber is the norm in North American home building and

Segal has realized how this simplifies design, construction and finishing. The British tradition of using largely rough, or 'unwrot' (unworked), timber encourages a lack of precision and requires to be covered in the final building. Arising from the use of presurfaced timber is an increased use of light, powered hand tools. Again this is rare on British building sites, usually due to the reluctance of general contractors to install the necessary temporary electricity supply. The frame and roof of such a structure are erected very quickly and operatives are thus working in the dry. With his simple drawings there is a list of materials right down to screws and nails, which acts as a bill of quantities, a specification and an order list.

Segal develops a very personal relationship with his clients and the craftsmen who work with him. He usually produces many alternative schemes and encourages his clients to suggest plans. On one of his schemes, an extension to a children's home at Singleton in Sussex, he submitted 10 proposals. He has also long been interested in low cost high density housing, publishing a book, *Home and Environment*, on the subject in 1948. Work has begun on applying his ideas to small self-build groups, one of which is in the London Borough of Lewisham. Segal's studies of density are extensive and he has demonstrated that it is possible to achieve 54-60 persons per hectare using two storey houses.

Most important of all perhaps is the realization that Segal is not against standardization, he points out that he has endeavoured to implement it all his life. But his attitude to it is pragmatic and carefully thought out. In comparing his approach with that of CLASP he

'gives the motto: "The second handling and the second shaping is not a gain." This is why CLASP is not significantly cheaper. CLASP, he believes, has lacked incentive to refine itself, being too tied to Brockhouse, its steel manufacturer. But, more importantly, it has the traditional European three state approach: one, the raw materials; two, change their shape in a ('CLASP') factory; three, put them together and finish on site. For me there is no intermediate state, no components. I don't change a given shape into another, I use it in the shape given.' (McKean 1976a)

The quantity surveyor who commented on the phenomenally low cost of Segal's own first house in 1966 (AJ 1966a) pointed out that

'The nature of this particular building has, however, conditioned the cost by excluding any luxury finishes that might normally be expected' (AJ 1966a)

This thrifty aesthetic permeates all his work, embodying some of the ideals of Henri Oedenkoven—flat roofs, large sliding windows and untreated materials. Oedenkoven's colony rejected private property, adhered to a strict code of morality, were against conventions in marriage, dress and politics as well as espousing vegetarianism and nudism. Along with this went strict views about building which, suggest Segal (1974), may have influenced Giedion.

Segal's buildings are in many ways very basic and, although comfortable, heat loss problems are not unknown. Most significantly, however, they are also very much in the mainstream modern movement tradition with the vertical panel aesthetic frequently seen in Bauhaus proposals. Segal points out that he had been early influenced by Konrad Wachsmann's book on American timber framing, *Holzhausbau,* published in the thirties.

At a time when the modern movement is being subjected to constant criticism it is surely of interest that such a valuable line of approach can be developed from it, and can fire the interest of many involved in that criticism. Segal's example has a great deal to offer us in understanding the application of a technology within a humane working framework.

This brief account shows the variety of possibilities that has developed through the use of timber-frame techniques for housing and it demonstrates that, if one is prepared to set aside dogma, it can deal with most of the issues raised by those pressing for industrialized building systems. It is both an advantage and a disadvantage that it has not been perceived as a system worthy of theoretical attention. As Christopher Arnold's (in Ehrenkrantz *et al.* 1977) elegant account shows, the timber-frame approach really is 'The Apple Pie Building System'.

COMPONENTS IN CONTEXT

'Perhaps there is hope for CLASP if used out of context.'

James Stirling,
quoted in Jencks 1973; 1968/9

The Open and the Closed

The preceding two sections have been concerned with the concept of an industrialized vernacular and its place in relation to the development of building systems themselves. By the use of examples and architectural propositions it has been shown how such ideas have grown with both modern architecture and with building systems. Charles and Ray Eames in a house, Wachsmann in a book, the development of the timber-frame house, all tell us something about how industrialization is or how it might develop. It is one of the arguments of this book that many building systems' proponents have so circumscribed their field and what is possible that the real relevance of some of these ideas has become confused. We now turn to the idea of the industrialized vernacular in terms of viewing the building industry as an open component system which it is the designers' responsibility to utilize, and we set this against the building systems idea as it now stands. This, it seems, is necessary because the very pervasiveness of the notion of closed building systems, which appeal to political, financial and professional power groupings.

The Modular Society's vision of sets of inter-related components freely available on the market seems unlikely to occur against such pressures. Unlike the United States, where public preferences can to a degree be demonstrated through the market, Britain (with, perhaps, the disadvantages of both capitalism and socialism and the advantages of neither) has solidified certain approaches into institutional forms. In the case of the building systems their institutional form itself embodied a narrow set of approaches which made any change difficult. In becoming predictive, the building systems became both proscriptive and prescriptive: alternatives were then ruled out and the filling of their prescription was the closed systems themselves.

It is this severe reduction of alternatives and the exclusive nature of the building systems that has caused such disquiet amongst both the public at large and professionals. Of course, the design of any building is a progressive reduction of alternatives merely to make it sufficiently comprehensible for designer and constructor: however, until recently the decisions concerning what selection to make from the available choices, what system to employ, has been carried out by the designer albeit in consultation with his users, clients and other consultants. The idea of an industrialized vernacular, or an open component system relates well to this function and indeed would probably enhance its operation as it appears to have done in the United States. With building systems and their component parts being put on the same basis as National Health Service spectacles, or unemployment benefit, a certain utility may be effected but choice and self-determination have all but disappeared. The open component view of the building industry does prevail, however, although in unlikely places. We shall examine developments whose impact has been obscured both by the short time that has elapsed since their appearance, and by the subsequent careers of those involved. The first is Eric Lyons, who with crosswall construction changed the face of speculative housing in Britain in the early fifties; the other is James Stirling who, according to Jencks, observed that 'perhaps there is hope for CLASP if used out of context' (1973; 1968/69).

The Resurgence of the Crosswall

The development of crosswall construction forms an important part of postwar moves to rationalize and industrialize building. The success of the work done by Eric Lyons, and at the London County Council (now GLC) were important stepping stones for later developments. The crosswall system uses those internal walls which are at right angles to the main axis of the building as vertical loadbearing members. They can be in any material, although most of the early work used brick. During this period, work in Denmark and Sweden also began to utilize concrete in this way, thus paving the way for its future use in a more industrialized manner.

For many architects of the early fifties one of the most potent architectural objects in England was Parkleys, the housing scheme at Ham Common, London designed by Eric Lyons for Span. Completed in 1953 Parkleys contained three flat types and, says Lyons:

'The whole matter of standardisation was taken very seriously: one type of kitchen, one type of bathroom, one type of everything, and maximisation of production. It was a fallacy in a way, but it was a discipline and an interesting one.' (Lyons 1968)

Parkleys was Lyons' first use of crosswall construction (**7.3.1**) and it was used on a standard bay system into which could be arranged various combinations of rooms. The crosswall gave fire and acoustic separation as well as load-bearing properties, the floors took their bearing on the walls and the 'open' ends were infilled with a variety of materials — block, tilehanging and lightweight panels. The approach was close to that of Gropius' serial houses at Torten (**7.3.2**) of 1926, although a good deal more human in appearance. Although Lyons had clear attitudes towards standardization at this period he saw them in the context of the whole design — unlike much of the closed systems work. He 'designed for production' but did not place that before his

7.3.1 Parkleys, Ham Common, London. Eric Lyons and Span, 1953. (Photos 1980). (a) A highly controlled vocabulary of brick crosswalls, infill panels, tile hanging and lavish planting. (b) The basic repetitive crosswall bay

main task, which he amply achieved in the Span work, of designing livable humane environment. In 1968 he did not 'think progress is inevitable; productivity is not all — not that is, if we believe in a man-orientated society' (Lyons 1968) and his views on much of the work of the industrialized closed systems showed his attitudes, since to become also the cry of the public:

> 'Think of all those dreary schools, smugly called "our success story" — in numbers, I suppose — and the bulk purchase myth. . . .' (Lyons 1968)

Few others have had the long experience of Lyons in designing for standardized production, and so perhaps his works should be considered rather carefully — all the more so if one returns to those

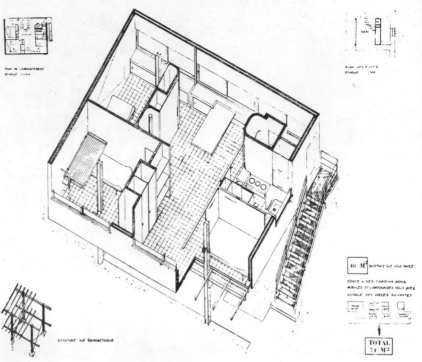

7.3.2 Serial houses, first stage of Bauhaus development at Törten, Dessau. Walter Gropius (1926). Party crosswalls, concrete slab, concrete beam at front and back

early span housing schemes at Ham and Blackheath, for with their careful attention to layout and landscaping (**7.3.3**), Lyons demonstrated how a real open component system could work in the hands of a good designer. It has often been said that his Span schemes were for the private middle class owner and barely confronted major housing problems. In fact, Lyons' original intention was to provide very low cost houses indeed, for first time buyers — not an easy task in the postwar period. That his design approach proved attractive to other architects and designers and then a larger public is surely a sign of success if choice is to mean anything other than that of five different local authority alternatives. And Lyons' contribution to both public and private housing has been considerable: in the latter he caused a minor revolution, although it took time to cover the country. The Span work popularized communal gardens again, the use of a residents' association to create self-governance, the introduction to the mass of speculative builders of the possibilities of mixing different sized units, the use of terrace housing again, and of course the now ubiquitous cross wall with its vertical slices of panel and brick. His reintroduction of tile hanging has swept the land and, although it upset the more rigorous proponents of the machine aesthetic, it is a 'subsystem' that embodies many of the properties of an open component system — light, mass produced, dry, easily handled, but at the same time it can maintain a character completely human (**7.3.4**).

Close to this time the newly reorganized London County Council Architects Department had been experimenting with crosswall construction, and the *Architects' Journal* summarized their work in this field in 1955, opening with the words:

7.3.3 Parkleys. (a) Careful attention to layout and planting. (b) Segregation of pedestrians and vehicles by an intricate use of paths and small courtyards (Photos 1980)

'When cross walls are used economy is produced not by mere mass production but by a nice balance between structural strength, ease of assembly and practical convenience.' (Dunican and Barr, 1955)

Prior to this the LCC Research and Development group under Cleeve Barr had made cost analyses of the advantages of different forms of structure and had come to the conclusion that:

'Calculated load-bearing brickwork still is cheaper than any form of reinforced concrete or steel structure for blocks of flats or maisonettes at least up to five storeys high, and this fact has inhibited the development of new techniques.' (RIBAJ 1955)

7.3.4 Parkleys. (Photo 1980)

7.3.5 Terrace houses, South Hill Park, London. Stanley Amis, William and Gillian Howell, 1956. Narrow fronted crosswall houses utilizing a double height at one end of the section. 'Our exemplar for the use of the double height room was, of course, Marseilles' (Amis and Howell, 1956)

An influential example of what could be done with crosswall construction was a small group of terrace houses with 12 ft 0 in (3.66 m) frontages built in South Hill Park Hampstead in 1956 (**7.3.5**). These were designed by Bill Howell, and Stanley Amis, of the LCC's Roehampton team, together with Gillian Howell (AR 1956).

This reintroduction of the crosswall was affected by the initial non-acceptance by local authority building inspectors and it was partly due to the campaign waged by the LCC and the AJ to use the system that it ultimately became an acceptable form of construction. Basically doubts were expressed about overall stability, with the cross walls seen as a row of playing cards in parallel which would fall over in a puff of wind. The main issue was longitudinal bracing and in the crosswall system this is provided by the rigidity of the junctions between walls and floor, where the floors are concrete. Alternatively, or where the floors are timber, bracing is introduced into the structure by means of sections of brick or block at right angles and or by means of edge beams across the perimeter (**7.3.6**). The main point is that this type of structure can be calculated as a three-dimensional framework. Many building inspectors tended to look upon each part of the construction as a separate entity, and thus architects initially had to persuade their local authority that the crosswall system 'deemed to satisfy' the conditions for good building. Gradually regulations began to permit the use of crosswall construction generally and it became the accepted model for much local authority and private house building.

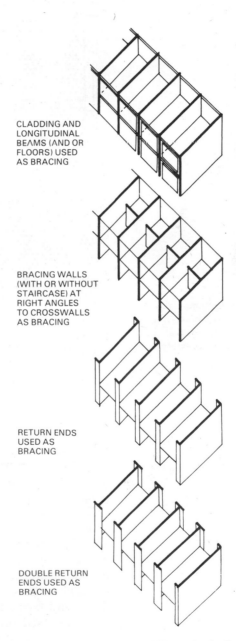

CLADDING AND LONGITUDINAL BEAMS (AND OR FLOORS) USED AS BRACING

BRACING WALLS (WITH OR WITHOUT STAIRCASE) AT RIGHT ANGLES TO CROSSWALLS AS BRACING

RETURN ENDS USED AS BRACING

DOUBLE RETURN ENDS USED AS BRACING

7.3.6 Crosswall construction. Principal methods of bracing cross walls

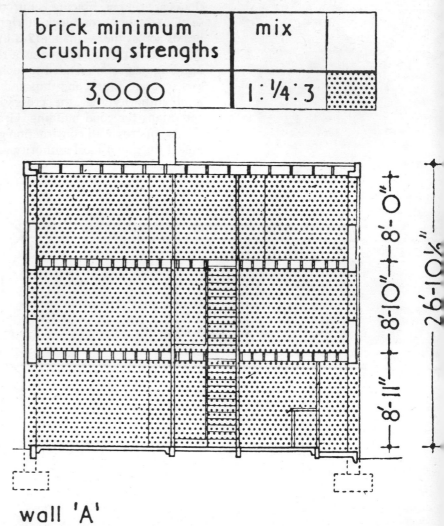

brick minimum crushing strengths	mix	
3,000	1 : ¼ : 3	

wall 'A'

7.3.7 Crosswall construction. Up to 3 storeys using bricks of same crushing strength. This method emloyed at the Ashburton Estate, London County Council, early 1950s

However, its use was not without problems, and for many builders the greater precision of construction it demanded was a constant source of difficulty. Not only did position tolerances have to be good but deviations in verticality and horizontality had severe repercussions on other components — cladding panels for example. Further, up to three storeys the same sort of brick could be used (**7.3.7**) but as heights increased so did the complexity of brick types in the crosswalls (**7.3.8**). This introduced further materials and site complications in controlling construction and there came a point where the law of diminishing returns operated. At a detail level the waterproofing of the ends of the crosswalls became crucial since with the tolerance questions already referred to unexpected differences between wall and cladding occurred. With the crosswall aesthetic operating at its most direct and favoured, that is with the 9 inch (229 mm) walls and the floors exposed on the elevation these problems were at their most severe. A variety of means were developed to overcome these (**7.3.9**)

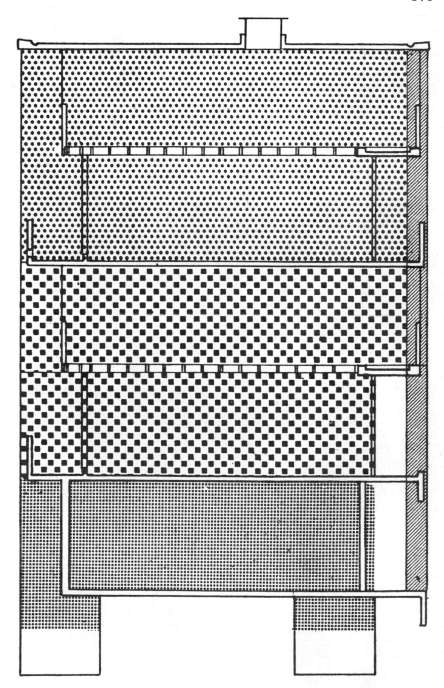

7.3.8 Crosswall construction. As height increases so does the variety of brick type and mortar mix. As used on 5 storey maisonettes at Golden Lane, London. Chamberlin, Powell and Bon, 1956-7. Ground floor: bricks of minimum crushing strength 4000 lb/sq in and 1:1:6 mortar mix; first and second floors: bricks of minimum crushing strength 5000 lb/ sq in and mortar mix of 1:2:9; top two floors: bricks at 3000 with a mortar mix of 1:1:6

but clearly the minimal aesthetic had its price when set against the traditional use of brick frontages which contained certain inbuilt margins which dealt with such problems.

Nevertheless considerable savings could be made, not least in the foundations which need only occur directly under the crosswalls and such savings, even if more complex to design and construct, could be shown (**7.3.10**). One of the most severe limitations was that the repetition required tended to favour 'a block of all one type of

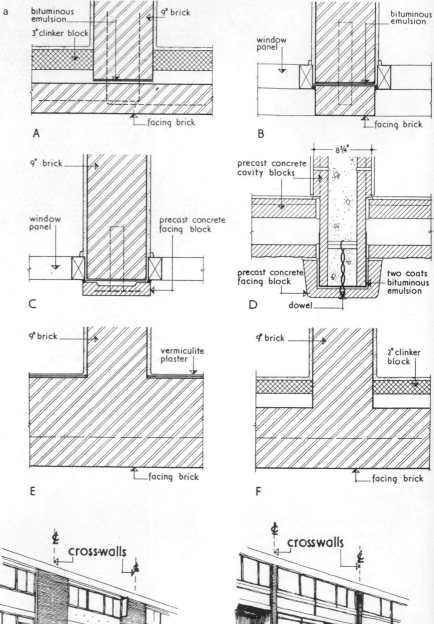

a

A
bituminous emulsion
3" clinker block
9" brick
facing brick

B
window panel
bituminous emulsion
facing brick

C
9" brick
window panel
precast concrete facing block

D
8¾"
precast concrete cavity blocks
precast concrete facing block
dowel
two coats bituminous emulsion

E
9" brick
vermiculite plaster
facing brick

F
9" brick
2" clinker block
facing brick

7.3.9 (a) Crosswall construction: alternative end treatments, 1955.
(b) Cross wall construction: differing elevational solutions, 1955

b

crosswalls

crosswalls

crosswalls

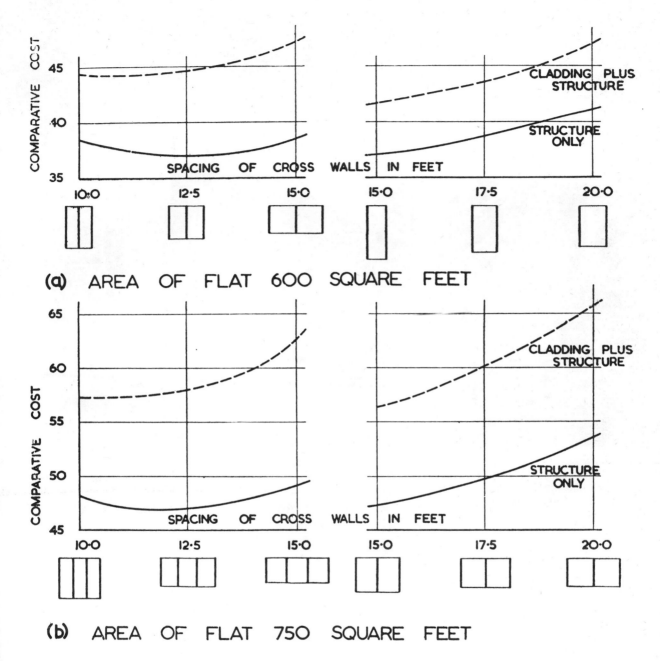

(a) AREA OF FLAT 600 SQUARE FEET

(b) AREA OF FLAT 750 SQUARE FEET

7.3.10 Crosswall construction: cost effect of differing spacing of crosswalls (in feet). Theoretical flats of assumed constant area of (a) 600 sq ft (55.74 sq m) and (b) 750 sq ft (69.98 sq m). The spacing of walls may be between 10ft 0in (3.05 m) and 18ft 0in (5.5 m) without varying the overall cost of the structure by more than 3%. Outside these limits structure cost increases rapidly

accommodation or at most with a minor variation' (Dunican and Barr, 1955) so that vertical services and construction could be kept economic. This mitigated against those authorities who had a policy of mixing different accommodation in one block — in principle good social policy in that it avoided ghettoes of single type groupings such as all elderly people, or all no-children families.

Early work was at the Ackroydon Estate in Wimbledon and the Ashburton Estate, Putney, whilst one of the most publicized of

7.3.11 Housing at Golden Lane, London. Chamberlin, Powell and Bon, competition won 1952. Maisonettes using crosswall construction. (Photo 1957)

London County Council schemes using crosswall construction was that at Loughborough Road. The Building Research Station designed its own prototype at Abbotts Langley were the cross walls were of 6 inch (152 mm) *in situ* concrete unreinforced. Chamberlain Powell and Bon used the system for their maisonette blocks (*c* 1955) at the prize winning Golden Lane housing scheme in the City of London, won in 1952 (**7.3.11**) and Powell and Moya employed it at Mayfield Comprehensive School in Putney, 1955.

Gradually crosswall construction became the norm for housing. Private developers took on the system and the aesthetic language of projecting cross walls, with panels or alternating bands of brickwork, window and tile hanging. The organization of the appearance of such housing into simple rectangles of brick, tile, boarding and window created the basis for an adaptable aesthetic which, in a perverse way, brought the modern movement within the language of the speculative builder large and small (**7.3.12**).

For housing in general both the style and the rationalizations of technique inherent in crosswall construction were a prelude to the high rise systems. The three-dimensional box construction, and the standardization lent itelf to use with concrete *in situ* and precast, and thus formed an important link in the development of such systems.

7.3.12 2 storey, 3 bedroom, single family terrace houses using crosswall construction with brick (in this case), tile hanging or timber spandrel panels. Chidham Park, Havant, England, Page-Johnson Ltd, 1962. (Photo 1968)

Clearly questions of rigidity and stability are fundamental to the success of the brick crosswall system; the development of the idea of the three-dimensional structure to one composed of precast components with *in situ* junctions as with the Larsen-Nielsen Anglian System at Ronan Point in the mid-sixties is not such a far cry in concept, but in practice was the difference of an open flexible and basically human system to a closed inflexible inhuman environment. An examination of that, however, must await a later section.

Constructivist Components

The introduction of the work of Stirling and Gowan into a study of building systems will no doubt create some conceptual difficulty. Nevertheless, it is an important inclusion and not to be avoided merely because the conventions dictate otherwise. By this I mean that stream of thought which has been a strong one in the development of systems ideas and which places the rationalizations of those who see themselves engaged in real social problems against those whom they describe as the 'prima donnas' of architecture. The questions raised by this are discussed elsewhere: the most relevant point to this section is to counter that myopic view which suggests that problems or solutions are not central, because the architects involved can demonstrate their personal abilities (or otherwise) in the form of positive stylistic qualities.

Stirling, in 1945 and with no examination qualifications, entered the Liverpool School of Architecture on a service grant, at a time when there was great debate concerning the ideals of the modern movement which were taking hold, belatedly, in Britain after World War II. One of his teachers was Colin Rowe. About his post-qualification period in the early fifties he says: 'I still felt the last significant architecture was the white masterpeices of the 20's and 30's' (Stirling 1965). This,

7.3.13 Engineering building, Leicester University, England. James Stirling and James Gowan, 1959-63. (Photo 1971)

together with the Bauhaus-like plan form for the Poole Technical College competition and the use of Hertfordshire prefabricated concrete panels on part of this building indicate his concerns at that time. From these early concerns he has consistently shown how the whole building industry can be seen as a system which, with the right inventiveness, can be made to yield the sort of riches which system builders can conceive of as only being available by means of a limited set of rules. Time and time again Stirling has demonstrated an attitude to industrialization that is far more flexible in terms of what might be possible but, because of the set of rules he has imposed on himself has even more extreme limitations in other directions. Let us take

7.3.14 History Faculty Library, Cambridge, England. James Stirling, 1968. Each component subsystem given a clear identity. (Photo 1968)

materials: Stirling's attitude, in the classic tradition of the modern movement is that visually, few materials good, many materials bad. His aesthetic at this level centres around simplifying the materials used in almost a caricature of those earlier simplifications by the modern movement. Both Leicester Engineering Building (with James Gowan) (1963) (**7.3.13**) and the Cambridge History Faculty (1968) (**7.3.14**) use the concept of clean volumes with planes of red brickwork. Between these are stretched, bent and wrapped glass, often in the form of patent glazing. Like many system builders, having decided upon a skin material it is used consistently, as a unifying device. Even to the extent that at Leicester tiles to match the brickwork are fixed onto external doors. However, unlike most system builders his concern then turns to the joints and, having dealt with the junction between his brick and his glazing, the dictates of the industrialized vernacular being employed are then permitted to work. For example, the patent glazing, having had its edges defined is then filled in by the subcontractor using the techniques then current. In this Stirling is able to take advantage of the available open component vernacular and is not endeavouring each time, as many systems have done, to create their own set of components from scratch. Nevertheless, Stirling's continuous concern with technology and its use has a certain mythic quality:

> 'My greatest concern is to produce a building with a high degree of environmental quality (one off or not) which somehow improves the human condition.'

> 'I think architecture at the moment is rather static because I think architects are cynical about the society which they have got. It seems to me that in the 'twenties and thirties Corb, the Constructivists, Futurists, and others, had an intense vision of a society which was about to arrive, and now that it has come we are all somewhat disillusioned...'

> ...'I think the vision which they had gave them a consistent plastic inventiveness, something which is lacking now.' (Stirling 1965)

This gives some guide as to his particular mode of selection, his system of operation. His comments on the use of glass further accord with the qualities invested in that material by the modern movement in the twenties and thirties for when asked about his emotional attachment to that material he answered:

> 'You are quite wrong if you think I choose to use brick and glass on an entirely emotional level. I use glass skins because they are lightweight, rapidly applied, low in cost, keep out the rain, self-cleaning, let the light in, let you see out.' (Stirling 1965)

This might almost be one of the systems builders talking: the selection of certain properties of glass whilst insisting on its great functional virtues and making no reference to problems of heat loss, solar gain, acoustics and security. Jencks has described Stirling and Gowan's method as:

> 'pragmatic "one-upmanship", skimming the trade magazines for new industrial products which they could then use in an unforeseen way... The basic idea was ingenious stealing to save money, the use of systems building out of context.' (Jencks 1973)

For example, in the Preston housing (1961) glazed stoneware block traps are used as roof drainage outlets, sills are of standard splayed or bull nosed bricks. The vigour of the approach at Preston not only reintroduced into architecture a proper regard for vernacular methods, but stylistically moved the flaccid architecture of the late fifties in a new direction. The use of the bright red Accrington brick (local to Preston) has ensured that such a brick has now been plastered indiscriminately over town centres up and down Britain. Nevertheless this was not the first set of ideas about industrialization that Stirling and Gowan were to crystallize. The first building was a single house on the Isle of Wight (**7.3.15**) which, in the directness of what it had to say about rationalization, was probably far more potent. It is certainly far less documented. Its external appearance, in bearing so closely on its philosophy, commended it first to architects and ultimately (for less exalted reasons) to the public and private house-building fraternity.

The house, completed in 1958 for an art teacher was small and economic, having a plan area of 1000 sq ft (304.8 sq m) and being built for £2800. As can be seen (**7.3.16**) all the serviced rooms are concentrated in a core which was also the link between a living and a bedroom wing. This plan form not only clearly defined functions but also created enclosed and sheltered courtyards almost within the house. The idea relates closely to The Mechanical Core H-shaped house (**7.3.17**) designed by J. and J. Fletcher in 1945 and which was illustrated in Giedion's *Mechanisation Takes Command* (1969, 1948). This house was for the returning soldier, who:

> 'First (he) goes to the factory to get the "mechanicore" which has all the latest conveniences, and then to the mill for lumber...' (*Pencil Points* May 1945: in Giedion 1969, 1948)

Although the Stirling and Gowan house followed this plan it carried clear statements about the way in which traditional construction could be viewed and rationalized in the light of the assembly line argument. These statements were as follows.

7.3.15 House, Cowes, Isle of Wight, England. James Stirling and James Gowan, 1958. Elevation showing repetitive window units 5ft 0in (1.5 m) and 10ft 0in (3.048 m) wide. (Photo *c*.1972)

7.3.16 House, Isle of Wight. Perspectivized plan showing service core and modular nature of the design

1. bedroom.
2. studio bedroom.
3. bathroom.
4. kitchen.
5. living room.
6. dining room.
7. entrance hall.
8. store.

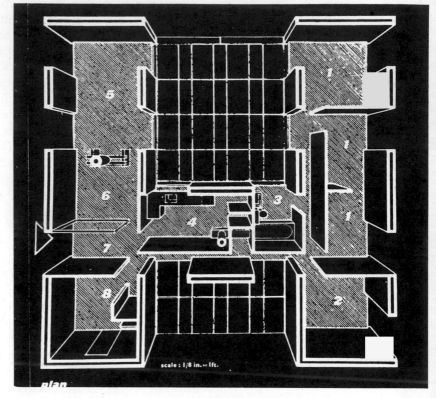

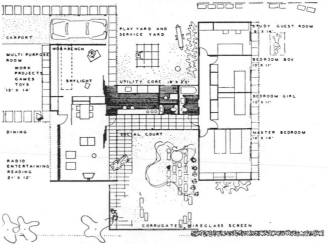

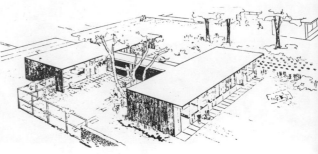

7.3.17 Mechanical Core H-shaped house. J. and N. Fletcher, 1945

7.3.18 House, Isle of Wight. Stirling and Gowan, 1958. Detail of elevation showing storey height panels. (Photo *c.*1972)

All spans were 10 ft 0 in (3.048 m) so that all timber joists could be cut to the same length. The external walls were simplified to two basic alternatives; the first a brick panel which was either 5 ft 0 in (1.5 m) or 10 ft 0 in (3.048 m) in length, the second a full height timber-frame insert with identical divisions for all windows and all doors (**7.3.18**), with the grid ensuring a dimensional consistency over the entire building. The eaves detail was a true systems solution in that it produced the same detail over both solid and void, thus simplifying drawings, construction and appearance. The building was, in itself, a complete thesis concerning rationalization. It had demonstrated that, by the co-ordinating influence of a grid, the use of a limited range of component sizes and the consequent limitation of alternative junction details, even quite traditional building processes could be viewed from the 'industrialized' stance. It was a stance which was to change architectural style throughout the land and pave the way for 'Welfare State Style' architecture itself (**7.3.19**) or what was sometimes subsequently called 'ratrad' — rationalized traditional construction. The lessons the house had to offer concerning drawing techniques were also clear in their relation to the attempts of the systems designers: that by simplifying dimensions, numbers of components and details, working drawing production could be rationalized and drawing time reduced.

Curiously, this house was neither referred to by Stirling in his talk at the RIBA in 1965 nor included in the exhibition of Stirling's work at the Heinz Gallery in London in 1974 although it is briefly referred to in *Buildings and Projects* (Stirling 1974; 1975). It is not mentioned by Jencks (1973), who is well tuned to Stirling and Gowan's attitudes to industrialization during this period and who coined the phrase 'industrial adhocism' for their approach:

> 'The extract unit, the lighting fixtures, the glazing and engineering brick are all standard parts which are resolved together through the axonometric technique... This method of drawing really is a method of designing for it allows the architect to work out the space, structure, geometry, function and detail altogether and without distortion. It was a necessary tool for the Constructivists to analyse their meccano-like joinery;...' (**7.3.20**). (Jencks 1973)

7.3.19 Gosport House, Havant, England. City Architect, Portsmouth, mid-1960s. A three-storey example of the type of rationalized traditional construction that became common in the late 1950s and 1960s — Welfare State Style (Photo 1979)

382

7.3.20 Constructivist table. Designed by Vkhutemas students working with Alexander Rodchenko, *c.*1920

Jencks points out that the axonometric drawing of Leicester Engineering Building (1959-63) is the first fully worked out example of this technique by Stirling and Gowan. An early axonometric was 'Stiff Dom-ino Houses' drawn in 1951, and other projects including the Ham Common Flats of 1955-8 (completed 1958), the Preston Housing of 1957-9 (completed 1961) and the Churchill College competition of 1958 all used this approach in various ways. The appropriateness of drawing to building technology is most clearly apparent in a later project, the additions to the Olivetti Training School of 1969/72 (**7.3.21**). Discarding his usual vocabulary of brick, concrete and glass Stirling used moulded glass reinforced plastic wall and roof units on a repetitive basis, styling the building rather as Olivetti designers style their business machines. The intention to make each segment self-supporting in two pieces, each extending from ridge to podium was not possible due to transportation problems and so each unit consists of four jointed components. This, fire regulations and economic reasons also caused the introduction of a precast and *in situ* reinforced concrete portal frame internally. Since self-colouring of the GRP units would have resulted in variations during production, the units were coated with a two part polyurethane paint. So repetitive units suggesting the consistencies of mass production have been laid up by hand, supported by reinforced concrete and painted. The paradox, to casual observers, is that it appears more industrialized than Stirling's other work, and this is supported by the elegant axonometric with its suggested infinite extendability. The building itself is exciting, if somewhat clinical.

7.3.21 Olivetti Training School, Haslemere, England. Addition and link to existing house by James Stirling (alterations to existing house by Edward Cullinan), 1969-72.
(a) Axonometric of addition and link.
(b) View of elevation through glass link, showing GRP panels

a

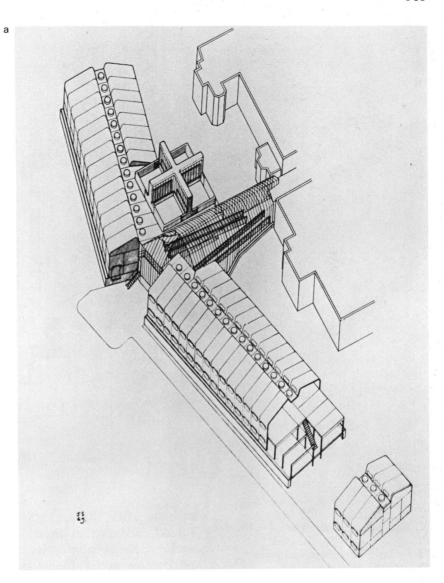

b

Jencks is correct in drawing attention to the important relationship between the method of drawing and the designs of architects. The building systems advocates had constantly pressed the techniques of engineering on to architects. This was not surprising since their view of the building industry was that of a finely tuned machine. The axonometric and exploded axonometric were crucial parts of the drawings made for machines, machine tools and machine products such as the motor car. Clearly then, a building industry which was to progress from its 'backward' craft basis to the precision assembly of many parts made in the factory could look to such methods. There is no doubt that to describe the complex sets of component assemblies required something better than the traditional plans and sections. The whole point of rationalization was to make everything absolutely explicit to the designer, to the manufacturers, and to the man on site.

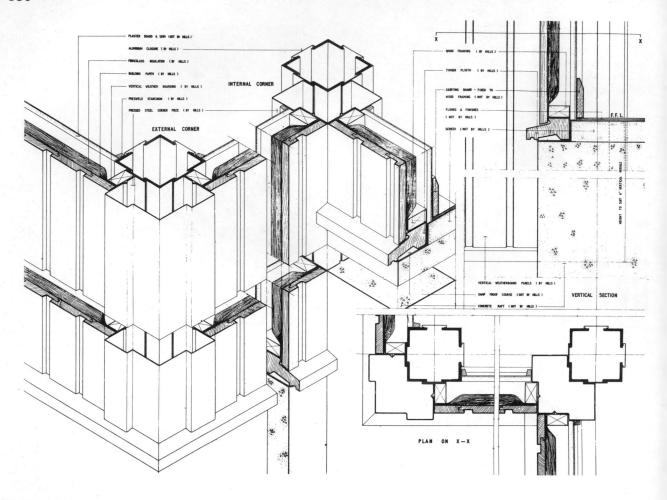

PLASTER BOARD & SKIM (NOT BY HILLS)
ALUMINUM CLOSURE (BY HILLS)
FIBREGLASS INSULATION (BY HILLS)
BUILDING PAPER (BY HILLS)
VERTICAL WEATHER BOARDING (BY HILLS)
PRESWELD STANCHION (BY HILLS)
PRESSED STEEL CORNER PIECE (BY HILLS)

INTERNAL CORNER

EXTERNAL CORNER

WOOD FRAMING (BY HILLS)
TIMBER PLINTH (BY HILLS)
SKIRTING BOARD - FIXED TO
WOOD FRAMING (NOT BY HILLS)
FLOORS & FINISHES
(NOT BY HILLS)
SCREED (NOT BY HILLS)

X   X

F.F.L.

HEIGHT TO SUIT 4' VERTICAL MODULE

VERTICAL WEATHERBOARD PANELS (BY HILLS)
DAMP PROOF COURSE (NOT BY HILLS)
CONCRETE RAFT (NOT BY HILLS)

VERTICAL SECTION

PLAN ON X—X

7.3.22 'Presweld' or Hills Dry Building System. Hills (West Bromwich) Ltd, 1957. The use of axonometric drawings, in addition to plan and section, to illustrate the assembly of components

An early Hills, Presweld System manual (*c* 1959) illustrates this (**7.3.22**) as does the special axonometric drafting paper especially developed for this approach and shown in the *Modular Primer* (Corker and Diprose 1963). In his book of 1947, Mark Hartland Thomas (founder of the Modular Society) had used some elegant perspectives by Hugh Casson (**7.3.23**), showing the erection sequence of the LMS prefabricated Unit Station (Thomas 1947) together with an axonometric showing construction.

As Jencks indicates, drawing in this fashion actually alters the way it is possible to think about design and is particularly relevant to an observation of the various parts of a building as fragments in space. Equally, another device used by Stirling and Gowan, in their Isle of Wight house and for their Ham Common scheme, has had an effect on representation techniques: this is the cross-sectional perspective (**7.3.24**). Such a drawing can show not only the whole construction across a building together with its servicing, but also gives a good impression of its interior as well. It is well suited to the concept of a building as being more than the sum of its parts: constructional system, environmental control system, furniture and people can all be

7.3.23 Unit station. London, Midland and Scottish Railway Architect's Department (under W.H. Hamlyn, c.1946. (a) Erection procedure drawings (by Hugh Casson). (b) Axonometric

Stage 1. Excavation and preparation. Placing of pre-cast foundation blocks and site concreting. Erection of six main columns.

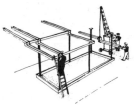

Stage 2. Erection of cranked cantilever roof beams.

Stage 3. Plywood roof units slide into position.

Stage 4. Steel-frame wall-posts erected. Concrete dado blocks and plinth mould inserted.

Stage 5. Enamel wall-panels fitted. Door and window frames.

Stage 6. Internal plywood unit partitions, glazing, decorating, etc.

a

b

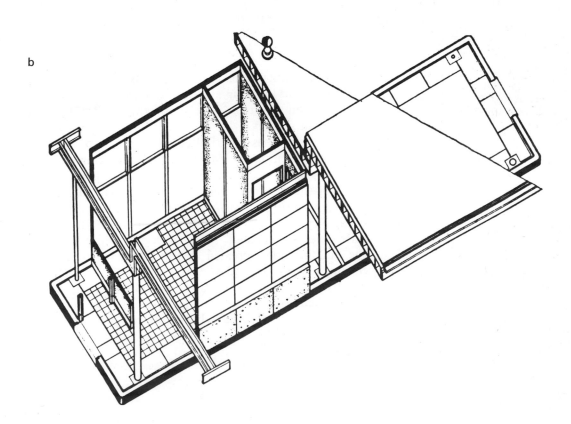

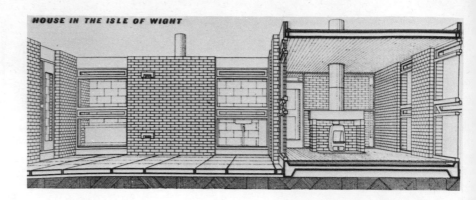

7.3.24 House, Isle of wight. Stirling and Gowan, 1958. Cross-sectional perspective

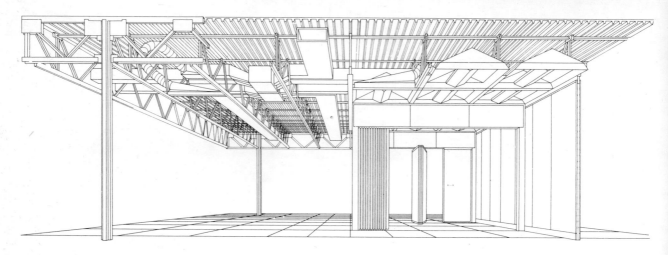

7.3.25 Schools Construction Systems Development (SCSD). Ezra D. Ehrenkrantz, project architect, 1961. Cross-sectional perspective

related in dimensional space. It has been re-invented to heroic proportions in building systems by Ezra Ehrenkrantz for his Schools Construction Systems Development (**7.3.25**), described in a subsequent section.

That Stirling has contributed enormously to architectural attitudes to industrialization is clear, and it is interesting that a number of his buildings have been criticized on similar performance grounds to those of the systems designers — as shown, for example, by Stamp (1976) in *Stirling's Worth: The History Faculty Building* where post occupancy leaks, sound insulation, temperature control and other difficulties are detailed. Many of the building system designers saw the buildings they created as functional machines, and in much of Stirling's work a similar view can be seen, albeit expressed more overtly and with a great deal more style. It is salutary, and somewhat ironic then, that these two important strands of twentieth century architectural thought have both suffered performance difficulties.

8

'SOME SORT OF ARCHITECTURE'

SYSTEMS GALORE

'It is quite likely that prefabrication will arrive, on a large scale, before we are practically and aesthetically "ready", that prefabricated houses will be designed, huckstered, and sold, not for the advantage they can offer, but on the basis of what people are used to, prejudiced in favour of, or, can be titillated by...'

Karl Koch
At Home with Tomorrow, 1958

The success of the work at Hertfordshire and elsewhere encouraged the feeling that the idea of a componentized industry was realizable. A series of events then conspired to create, first, a plethora of separate building systems, and then disenchantment. It appeared that the industrialized vernacular was not at hand, and that many of the attempts made to create it via the 'closed' building system actively mitigated against it, at the same time destroying many of the vestiges of the traditional system in the process. However disorganized the latter its resilience ensured that, with the disappearance of many of the systems, it was able to continue (given the finance) to build and, as architects were again to discover, could do this in a way far more sympathetic to places and people than the many inadequacies that had been thrust upon them by the systems proponents. When in 1960 CLASP (Consortium of Local Authorities Special Programme) showed, with their Milan Triennale exhibit, how it was possible to use the idea of systems building to bring some efficiency and sense of purpose into the rather staid world of local authorities, its political implications did not go unnoticed.

Not only was it holding costs, but in becoming predictive on a larger scale it could much more effectively bargain with the providers of resources, than had hitherto been the case. It was normal practice for the Ministry of Education to circularize authorities immediately prior to the end of each financial year offering them extra money if they could make 'token' starts on site before that financial year ended on 31st March. One of the crucial advantages of the schools consortia approach was that with standard drawings and procedures they could make such starts very quickly indeed. Indeed it would not be going too far to say that the initial purposes of the architects involved in the creation of building systems was to make maximal use of their share of money obtained from tax revenue for the purposes of providing much needed facilities. Many socially conscious architects often felt that, nationally, the distribution of the financial cake was inadequate and, more particularly, that a greater proportion of the available finance should be used for housing, schools and so on rather than on grandiose gestures of some dubious sort.

By becoming efficient themselves architects were able to demonstrate credibility to the financing system at large and to draw out more resources than would otherwise have been available. Inevitably this ability was ultimately turned against them, when their own techniques gradually became used by central government and many of their innovations became political means for reducing areas, standards and costs. This resulted, ultimately, in disenchantment from the point of view of the architect, his client and the many users of the provided accommodation. The creation of such a convincing animal as CLASP in turn produced smudged carbon copies, many of which were contrived for the aggrandisement of particular authorities, who seemed to understand little of the underlying architectural and social intent.

The rampant institutionalization which has characterized postwar Britain has seen an extreme example in the building systems world. The unusual confluence of social intent, creative energy and expertise put over half Britain's architects into central or local government. It was here that it was felt that socially responsible work could be done, and if it was done, this would influence both industry and clients in an appropriate direction. Unfortunately these views were also tied to a particular architectural philosophy, moreover one with a deterministic aesthetic so that architects were ultimately attacked for being responsible for buildings looking the same, for monotonous repetition and buildings that looked 'like prisons'. Such reactions also reflect a real public disgust at the abdication of responsibility that is often inherent in the sheltering umbrella of local government and its building systems.

This section deals with the period during which the idea of an open component building industry turned into the era of the institutionalized system. These latter can claim many credits in their attempts to grapple with the pressures upon them. Equally however, many seemed unable to recognize or act upon what was happening to them. The erosion of both personal and professional responsibility has been slow but inexorable with many committed system builders now looking around and wondering what sort of animal it is that they have created.

THE ADVENT
OF **CLASP**

'Viewing the operation of the system as a whole, it cannot avoid being some sort of architecture, since it is conceived, directed, developed and executed by architects. . . . This is not, in any visible sense, architecture considered as one of the accepted fine arts; not architecture as the expressed will of a highly developed personality. And yet it carries its own visual conviction, the air of being the expressed will of something or some body of things, the product of some sort of highly developed creative force.'

Reyner Banham,
'Ill Mct by Clip-Joint',
Architectural Review (1962b)

The creation of CLASP, the Consortium of Local Authorities Special Programme, in 1957 is of particular significance in the process of the translation of a number of key threads into a practical reality. It is important in relation to what has gone before and particularly important to what is to follow.

One of the more surprising aspects of the growth of building systems in Britain is how little published material there is on them. This is even more interesting in the case of CLASP in view of its size and importance. The fact is that other than the Ministry of Education's own Building Bulletin No. 19 'The Story of CLASP' (1961), Bulletin 45 CLASP/JDP (1970) — which deals with the extension of CLASP to Further Education, and a number of articles in the technical press, one of the most potent forces in postwar school building in Britain has not been properly documented and has certainly received far less than its due share of considered comment. This is in part due to the nature of the animal and those associated with it. White (1965) in his brief comments on CLASP makes one of his many apologies regarding the introduction of individuals and their importance in (as he says) an 'official publication':

'The development of the CLASP method is another striking illustration of the way in which methods of building involving prefabrication have come to fruition as the result of a combination of circumstances and — above all — of personalities. It is not usual in an official publication to acknowledge the names of officers in public service whose particular contributions have been outstanding, but it must be said here that the teamwork of a dozen or two of energetic individuals at the Ministry of Education, at Hertford and at Nottingham have sufficed to revolutionise the

8.2.1 Mining subsidence and its effects. Effect on the ground surface of underground working. Sectional diagrams show the wavelike formation of the strata curving into the void left by the coal extraction using longwall mining. The angle between the vertical at the working face and the limit of subsidence effect is called the angle of influence, being a near constant angle for any particular area. (a) A building at P will be affected by subsidence if coal is extracted in zone of influence below. (b) Section parallel to direction of mining working. (c) Section looking towards coal face. (d) Horizontal strain on ground surface. (e) Effect of differential vertical subsidence. Building has tendency to break its back as the ground curves away beneath it (curvature exaggerated in diagram). (f) Effect of horizontal tension in the ground. Two points A and B on the surface may stretch apart by up to 2 inch in 100 ft (50 mm in 30 metres). The building will have a tendency to tear apart (curvature ignored). (g) Movement of a pinjointed frame with spring bracing on subsiding ground (h) Building on horizontal ground: all frames square. (i) Ground begins to subside below A. The weight of structure on stanchion A overcomes the resistance of springs in the wind brace, allowing panel AB to 'lozenge'. All stanchions remain vertical because the majority of spring bracing units are still on horizontal ground. (j) Bracing in panel ED tilts because springs in GF are approximately balancing springs in AB. Hence GF is 'lozenged' and the lozenging of AB is relatively reduced. (k) Building almost completely subsided and squaring on horizontal lower level. FG is still affected by the subsidence wave and is 'lozenged'. DE and AB are both on horizontal ground and, overcoming the resistance of the springs in FG, have brought all stanchions vertical. (After Swain, 1974)

(a)

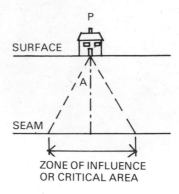

(b)

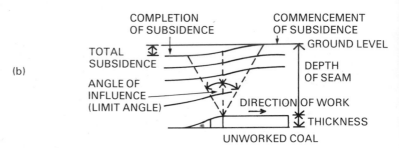

(c)

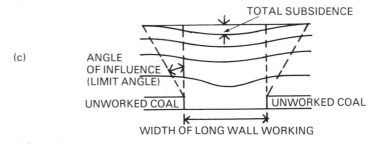

(d)

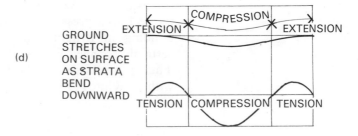

(e)

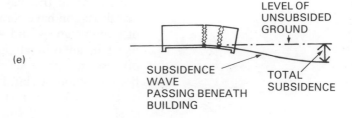

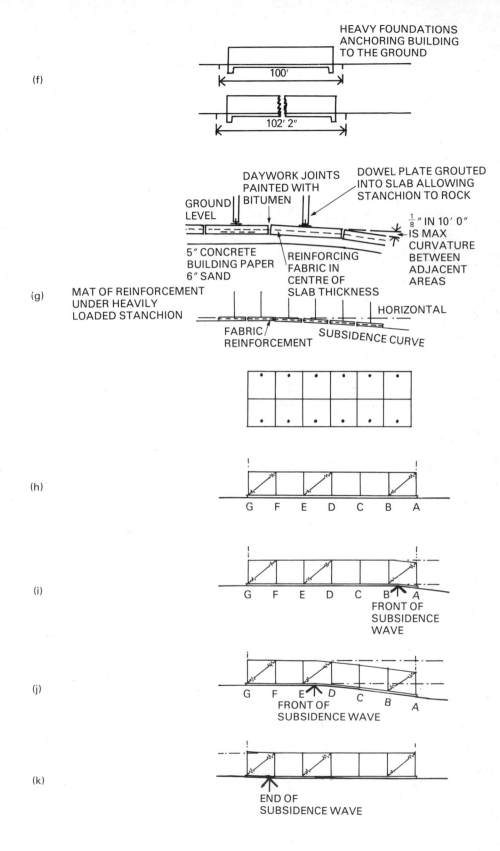

(f)

HEAVY FOUNDATIONS
ANCHORING BUILDING
TO THE GROUND

100'

102' 2"

(g)

DAYWORK JOINTS
PAINTED WITH
BITUMEN

GROUND
LEVEL

DOWEL PLATE GROUTED
INTO SLAB ALLOWING
STANCHION TO ROCK

$\frac{1}{8}$" IN 10' 0"
IS MAX
CURVATURE
BETWEEN
ADJACENT
AREAS

5" CONCRETE
BUILDING PAPER
6" SAND

REINFORCING
FABRIC IN
CENTRE OF
SLAB THICKNESS

MAT OF REINFORCEMENT
UNDER HEAVILY
LOADED STANCHION

HORIZONTAL

FABRIC
REINFORCEMENT

SUBSIDENCE CURVE

(h)

G F E D C B A

(i)

G F E D C B A

FRONT OF
SUBSIDENCE
WAVE

(j)

G F E D C B A

FRONT OF
SUBSIDENCE WAVE

(k)

END OF
SUBSIDENCE WAVE

school building methods over a large part of the country, and at the same time to hold the cost of building by these methods within the limits imposed.' (White 1965)

It is interesting that the two criteria mentioned above are the revolution in building methods and the holding of cost within required limits. There is no mention of the quality of school environment or its relation to educational requirements, an omission that is also evident in the account not only of CLASP, but throughout the book. The discussion of grids, for example, concentrates on their relation to the building materials market and little on their relation to anthropometrics, planning and the school environment. However, no apologia should be necessary for the naming of names because in spite of its seeming anonymity the development of building systems is dependent upon people just as is the rest of the history of architecture. A very small group of people have caused the assimilation of the idea of building systems into the way of life in Britain, some bridging the gap caused by the 1939-45 war. One of the links was Donald (later Sir Donald) Gibson, another is C.G. Stillman. Gibson's work was referred to earlier when, in 1942, as City Architect to Coventry he and Edric Neel (later to be one of the founders of ARCON, designers of one of the better postwar prefabs) designed an experimental house. In 1954 Gibson became County Architect to Nottinghamshire. The problem that he faced initially was how to build over areas where mining subsidence was likely to occur and how to do this economically. There were discussions with the Building Research Station and others, with the result that a pinjoined braced light steel frame was proposed on a 5 inch (127 mm) slab cast in sections with only a binding layer of reinforcement through the middle (**8.2.1**). This was the reverse of the solution previously used for such conditions which was to put down large concrete foundations and build as heavily as possible in an endeavour to withstand movement. With this development the light frame and skin had acquired, in its long history, another valuable functional supporting argument. Some years later Henry Swain was to comment:

> 'So we went ahead on the theory (and I can hardly believe this) for six years without proof. It was only in 1963 with 213 CLASP jobs complete all over England that five buildings proved it. They had survived moderate and extremely severe subsidence without danger or anything more than superficial damage.' (Swain 1974)

The survey reported on in 1974 shows that the repair bills on schools affected by mining subsidence had been very small. Furthermore, Nottinghamshire had never claimed any of the additional 7% cost available for buildings in areas of mining subsidence.

The Ministry of Education in 1953/4 had constructed a development project at Belper, Derbyshire, using a light steel frame manufactured by Brockhouse, and Gibson (then Nottinghamshire

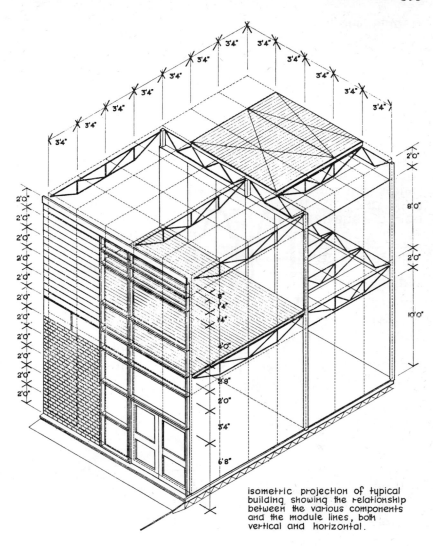

isometric projection of typical
building showing the relationship
between the various components
and the module lines, both
vertical and horizontal.

8.2.2 Consortium of Local
Authorities Special Programme
(CLASP), 1957 onwards. Isometric
showing dimensional system. MOE
1961)

County Architect) was not only interested in this aspect but also in the
prefabrication argument. He was joined by two young men from
Hertfordshire, W.D. Lacey (later to become Chief Architect to the
Department of Education and Science and then to the Property
Services Agency) and H.T. Swain.

One of the problems facing them was that the Nottinghamshire
building programme was not of sufficient size to warrant the ideas in
which they were interested. The result was an association of local
authorities. Initially, Nottinghamshire, Coventry and Derby combined,
thus creating a more sizeable market and proliferating the proposals
amongst a number of local authorities. The ideas developed at
Hertford could now be introduced across a broader field:

1. Bulk purchase of components with agreed and known prices to
   enable forecasting.
2. Standardization of components, methods and drawings.
3. Development feedback.

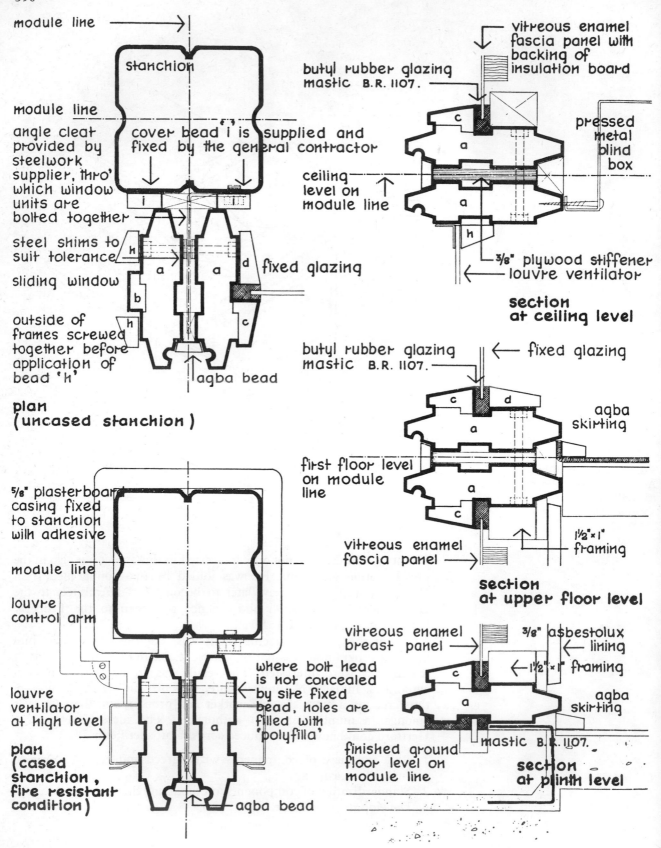

module line →

stanchion

module line

angle cleat provided by steelwork supplier, thro' which window units are bolted together

cover bead 'i' is supplied and fixed by the general contractor

i

steel shims to suit tolerance

h

a   a   d   fixed glazing

sliding window

b

h

c

outside of frames screwed together before application of bead 'h'

agba bead

plan (uncased stanchion)

⅝" plasterboard casing fixed to stanchion with adhesive

module line

louvre control arm

louvre ventilator at high level

where bolt head is not concealed by site fixed bead, holes are filled with 'polyfilla'

a   a

plan (cased stanchion, fire resistant condition)

agba bead

vitreous enamel fascia panel with backing of insulation board

butyl rubber glazing mastic B.R. 1107.

c

a

pressed metal blind box

ceiling level on module line

a

h

⅜" plywood stiffener
louvre ventilator

section at ceiling level

butyl rubber glazing mastic B.R. 1107.

← fixed glazing

c   d

a

agba skirting

first floor level on module line

a

c

vitreous enamel fascia panel →

1½"×1" framing

section at upper floor level

vitreous enamel breast panel →

⅜" asbestolux lining

← 1½"×1" framing

c

a

agba skirting

finished ground floor level on module line

mastic B.R. 1107.

section at plinth level

8.2.3 CLASP: Window details showing relationship to frame. (MOE 1961)

4. The development of a relationship with manufacturers leading to quality and programming control.
5. Reduction of professional time spent on repetitive items by the architect, quantity surveyor and contractor.
6. Through serial tendering develop a relationship with contractors so that experience may build up.

The method of construction used initially (**8.2.2**), in addition to the cold rolled frame and concrete slab, consisted of a precast concrete kerb that took the lightweight external skin of two types, solid and glazed. The former were timber stud frames with shiplap boarding or clay tiles nailed on. The window wall units were timber-framed units either 3 ft 4 in (991 mm) or 6 ft 8 in (2.03 m) wide with a constant section (**8.2.3**). Various infill was used in addition to glass, including vitreous enamel sheet and boarding. The external corners drew upon the Miesian aesthetic (**8.2.4**) in a pragmatic, perhaps whimsical, way with a softwood version of the steel frame re-entrant corner (**8.2.5**).

8.2.4 Illinois Institute of Technology, Chicago, USA. Ludwig Mies van der Rohe: one of a series of structures 1942-58. The 'negative' corner (Photo 1980)

8.2.5 CLASP. Timber external corner detail, 1957 on

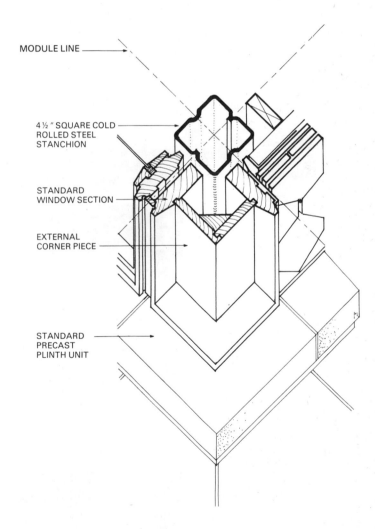

MODULE LINE

4½″ SQUARE COLD ROLLED STEEL STANCHION

STANDARD WINDOW SECTION

EXTERNAL CORNER PIECE

STANDARD PRECAST PLINTH UNIT

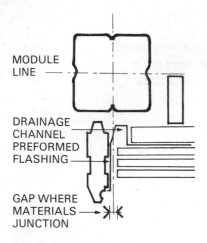

MODULE LINE

DRAINAGE CHANNEL PREFORMED FLASHING

GAP WHERE MATERIALS JUNCTION

8.2.6 CLASP. Detail of cladding and frame junction in plan showing the 'flash gap' principle

8.2.7 (a) Building system designers have attached great importance to component legibility, with each unit shown as discrete, joints firmly expressed. Cladding detail. Brockhouse CLASP, 1967

(b) York University. CLASP: Robert Matthew, Johnson-Marshall and partners, 1965 onwards. Details of concrete cladding panels. (Photos 1976)

b

Considerable attnetion was given to the way in which the components junctioned with the use of an anticapillary joint (**8.2.6**) which relied on receiving the water and draining it rather than trying to seal it directly. The Building Research Station studies of such an approach as it applied to joints in concrete panels is set out in BRS Digest 85 of 1967. This technical development has very respectable philosophic precedents and rests in that moral attitude which has always favoured the view that if the building is made up of separate parts this separateness must be made visually evident. The basis for this may lie in Platonic Idealism, that truth equals beauty, which in modern architectural terms may also be seen in the puritanism of the functionalist school. Joints must look like joints; components must look like components.

The stylistic interests of the twenties and thirties placed particular emphasis on the right use of the machine. Crystallized in the rectangularities of the Bauhaus, the way in which the components of the new machine architecture were to be articulated has continued to obsess architects and designers. The development of the 'flash gap' joint is evidence of this at both a practical and a philosophic level. A joint was required that would accommodate manufacturing and erection tolerances, the movements of materials under heat and cold, and the exclusion of wind and water. There were of course, already many such methods of making joints, but none of them really satisfied one of the main design criteria in this discussion. This, usually

a

implicit, rarely expressed, is that each component must still maintain its component legibility (**8.2.7**) and not be 'blurred' by lesser considerations. Thus, such a well tried method of making a joint as an applied cover strip, thoroughly established by tradition, was often considered unsatisfactory. The way in which the 'flash gap' joint has dominated architecture in recent years, finding its way into building systems and buildings generally, is another example of the power of such moral arguments. These make it particularly difficult for many architects and designers to cast off this early legacy of the machine age — for example in the face of the impact of the environmental control argument. More seriously, it has actually discouraged a cold look at the very real changes in technology and the way machines are used.

There is a further important point, concerning the way in which educational requirements were turned to technological solution. White (1965) talks of CLASP having 'improved (the) total environment' of schools, mentioning in passing daylighting, artificial lighting and decorative schemes. In many ways of course, there were improvements but it is doubtful if there was a real improvement in overall environmental standards. The emphasis given to some aspects, like daylighting, (see Broadbent 1973, Manning 1967) at the expense of others, is largely the result of the stylistic concerns of immediate historical precedent. The emphasis on hygiene, large amounts of sunlight, and pristine cleanliness can now be seen as the residue of the stylists of the twenties and thirties. From Stillman's refashioning of the Bauhaus image, Hertfordshire and CLASP, to the present situation, we can see the development of this idea. The necessities of style demanded large transparent areas backed by the great moral argument of the value of fresh air and sun as an antidote to the previous images of the industrial revolution. These were firmly implanted in the architectural mind. Giedion did much to fix such an image in the caption to his photograph (**8.2.8**) of the Dessau Bauhaus (1926):

> 'In this case it is the interior and exterior of a building which are presented simultaneously. The extensive transparent areas, by dematerialising the corners, permit the hovering relations of planes and kind of "overlapping" which appears in contemporary painting.' (1967, 1941)

A close look at most building systems indicates that one thing they have been particularly unable to do is to integrate the total environment, except around a limited series of visual norms. This is perhaps the opposite of a view popular amongst some system builders which is that in integrating the total environment they are responding to needs and not concerned with creating a piece of architecture. In endeavouring to confine themselves to only 'facts and realities', a stylistic straightjacket has been developed, so powerful that the refashioning of its products to serve more embracing goals is severely inhibited. That it became supported by administrative and organizational machinery of an exceedingly complex kind is also

8.2.8 Bauhaus at Dessau. Walter Gropius, 1925-6. Corner of the workshops wing. From Gropius' *The New Architecture and the Bauhaus* published in England in 1935, and subsequently used by Giedion in 'Space, Time and Architecture' 1941

relevant; the architect who designs his one off building around a number of stylistic and functional concerns is prone to be a figure of fun to systems designers, and described as a prima donna. How much more arrogant is it to fashion a building system around a series of stylistic and functional concerns and then have it used by architects throughout England and the world? The fact is that the lightweight construction and large glass areas of most school building systems have been very poorly integrated with concepts of heating, ventilation and lighting and often give a less than comfortable environment for school children to work in. High heat loss through glass and lightweight panels together with heat leak has been a considerable problem, whilst solar gain in south or near south facing classrooms has created unnecessary difficulties for staff and children.

It is of interest here to refer to schools Stillman built in Middlesex in the 1940s after his experience with lightweight construction in West Sussex in the thirties. These schools (**8.2.9**), whilst continuing to use the principles of lightweight construction, took particular care to introduce solar shading devices in the form of vertical and horizontal projections, as had his earlier work. Not so with the later systems work. The continued reliance on the technology of the structure (in the form of a frame) and the external envelope of the building meant that the environmental quality of many schools suffered in holistic terms as opposed to conventional construction with its heavier walls creating a 'flywheel' effect. More significantly, these ideas became formalized into administrative, and subsequently, ministerial terms. Thus change becomes doubly difficult. This is particularly evident in the total environment which systems builders were supposed to be considering, for it codified certain views about what was important, how the money should be spent and upon what. A typical cost analysis indicates this. The emphasis on structure and fabric served only to depress the scales further and further from a proper consideration of

8.2.9 'A light prefabricated classroom'. C.G. Stillman: 1943. The clearstorey permitted cross-ventilation and natural light on two sides. The large areas of glass partially protected from solar gain by vertical and horizontal projections

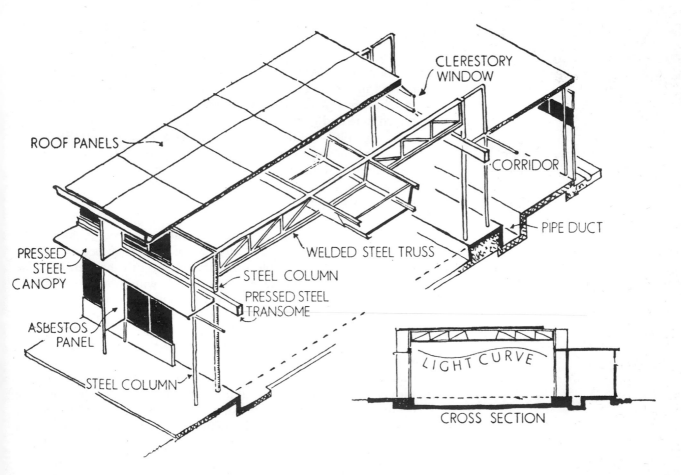

ROOF PANELS

CLERESTORY WINDOW

CORRIDOR

PIPE DUCT

WELDED STEEL TRUSS

PRESSED STEEL CANOPY

STEEL COLUMN

PRESSED STEEL TRANSOME

ASBESTOS PANEL

STEEL COLUMN

LIGHT CURVE

CROSS SECTION

heating standards, ventilation, lighting and acoustics. This is the reason why neither CLASP, SCOLA nor many of the other systems have been able to deal adequately with the integration of services. It has remained for those outside the building systems field to do this in a fashion which can again act as a critique of the conventions of the building systems.

In one of the few critical articles on CLASP, in the *Architectural Review,* Banham (1962b) indicated that it was exceedingly difficult to make any useful comment on CLASP because of the lack of published architectural evidence:

> 'Incredible though it may seem, after five years' operation and a Special Grand Prize at the 1960 Triennale di Milano, the detailed architectural appearance of CLASP buildings remains unknown to most readers of architectural magazines, including — be it admitted — the AR, in spite of the undoubted success of the operation.' (Banham 1962b)

He might have added that the creators of CLASP had a healthy disregard for most architectural magazines and what they had to say. This was tied up with the belief that a new architecture was being created (or new non-architecture) which was based on considerations other than those which interested the 'glossies'. Such an architecture, it was thought, would be styleless, anonymous, a background. Its appearance, in the event, is an all too obvious welfare state style. However, Banham's comment is still largely true: that since the foundation of CLASP there has been a very limited amount of published material. True, CLASP buildings have been cost analysed in the *Architects' Journal,* brief technical and explanatory articles have appeared here and there, Bulletin No. 19 (*The Story of CLASP*) was published in 1961 (MOE 1961), and Bulletin 45 in 1970 (DES 1970), and CLASP features inevitably in those endless catalogues of industrialized building systems. But it has still managed to avoid the sort of critique given to most buildings of merit: it is almost as if there were no tools by which it could be examined. Banham rightly draws attention to one reason for this — the systems designers' lament. This states that because the system is under constant change and development, with (of course) each Mark of the system better than the last, no comment is possible on what is currently happening. The failure and replacement, for example, of 60 windows on a Mark II CLASP Computer Centre at Worthy Down, Winchester after only six years can be blamed on the weakness of the earlier version. Thus, change has been built into the critical field as well as the building system and judgement is suspended forever. No-one must mention its shortcomings. In many ways what has happened is thoroughly laudable, in that quietly and with little architectural fuss, an idea has become a reality, well intentioned people have been getting on with the good work of making school buildings, with modest reward and little recognition. The failures of earlier marks being fed back, in a proper systems manner, to the next revision of the system.

8.2.10 CLASP school at the Triennale di Milano, 1960

Another view might have it that, partly because of the nature of local and central government, the growth of CLASP has been protected from proper public scrutiny since such scrutiny can only be through those organizations committed to it.

The award of the 1960 Special Grand Prize at the Triennale di Milano to the full sized CLASP primary school (**8.2.10**) erected there was, in spite of its limited publicity, of considerable import to the turn that buildings systems development was to take in Britain. Apart from the truism that heroes are always unsung in their own land, the interest in CLASP until this time was largely from architects and educators and particularly those with a commitment to some sort of innovation. Having discovered that the then Board of Trade were indifferent to the Triennale and that it was unlikely that a submission would be put in, CLASP decided that it would enter a complete primary school. In retrospect the occurrence of a group of British local authorities making such an effort to show their worth at the Milan Triennale sounds exceedingly unlikely. However, it displays just another facet of the drive exhibited by CLASP. Most of the components for the school were given by the manufacturers and it was set in well landscaped surroundings and supported by a display of slides and other material concerning the British efforts in school building and design. It was also fully equipped with furniture and fittings.

The British school (designed by W.D. Lacey, L.H. Blockley and Rex Goodwin) was an enormous success and, apart from gaining the Grand Prize, aroused a very great deal of interest from Europe in the British approach to school design and in particular to CLASP. Prominence was given to the oft quoted reductions in cost achieved by such methods: the 1948 cost per school place was £320: the increased costs of materials would have made this £550 in 1960 (the year of the exhibition) whereas in fact the actual cost was £260, whilst the 1961 Estimates 'Committee Report' pointed out that 'Nottinghamshire have been able to build schools with quality finishes, teaching areas considerably above minimum standards and facilities superior to those previously achieved, while still keeping within the cost limits' and that CLASP schools were 'eight per cent cheaper than all the schools in England and Wales built in other ways' (AJ 1961b). These powerful figures can brook no critical comment it seems. However, it should be pointed out that whilst (apparently) CLASP costs were held or, in the case of a number of items, reduced, this had to do with a complex of interacting design factors and was not merely due to the introduction of prefabrication techniques. For example, most younger architects held the view that school buildings at the time were particularly wasteful and inefficient. Indeed many schools still embodied the prewar spirit, out of keeping with the 1944 Education Act and more progressive educational thought. Brought up on the efficiency argument of the modern movement many schools architects wanted to see a number of changes. One of these was in the way that space was treated, and here the legacy of the modern movement was all powerful. Giedion had crystallized the argument around the effects of the cubist painters in the way in which they had endeavoured to break with Renaissance perspective:

'It views objects relatively: that is, from several points of view, no-one of which has exclusive authority. And in so dissecting objects it sees them simultaneously from all sides — from above and below, from inside and outside. It goes around and into its objects.

'The advancing and retreating planes of cubism, interpenetrating, hovering, often transparent, without anything to fix them in realistic position, are in fundamental contrast to the lines of perspective, which converge to a single point.' (Giedion 1967, 1941)

The interlocking of these ideas with those evident in the architecture of Le Corbusier, Gropius, Van Doesburg and the De Stijl movement, laid the foundation for its subsequent assimilation into the general architectural subculture. The conceptual addition of the fourth element, time, to the three dimensions of Renaissance perspective and its connection to the idea of the inter-relationship of space is deeply embedded in the growth of modern architecture. Much of this had been explored in the cubist work of Braque, Picasso, Duchamp and the Constructivist sculptor, Naum Gabo, who had distinguished between carving and construction (8.2.11).

'The first represents a volume of mass; the second represents the space in which the mass exists made visible.' (Martin, Nicholson, and Gabo 1937)

There is irony in the way in which many architectural users of these ideas are very concerned to disassociate themselves from such 'art' influences. Architecture as a series of largely flat rectangular planes intersecting cleanly with each other around a flowing conception of space has thus been highly valued. It is of course, very bound up with

8.2.11 Head No. 2 Naum Gabo: 1916. (Enlarged version 1964). Space defined by plane and line rather than mass. This version of the Gabo head, in Cor-Ten steel, is in the Tate Gallery, London. Courtesy The Tate Gallery, London

the concept of freedom and flexibility in planning together with constructional ways of achieving this. As we have seen, Le Corbusier's Dom-ino House posits one model for this. Mies van der Rohe's Barcelona pavilion, at the International Exhibition of 1929 is a further one about which Giedion says:

> 'with unsurpassed precision he (Mies van der Rohe) used pure surfaces of precious materials as elements of the new space conception.' (Giedion 1967, 1941)

However, this has all been well documented by Bonta (1975) and needs no further detailed exposition here. The relevance of such ideas of space to the development of building systems is, however, considerable since many of the philosophies, arguments and artefacts have been fashioned around them. Plan flexibility, together with the idea of the interpenetration of space mean that formerly finite spaces around finite functions have now had their edges blurred. Functions and spaces flow into each other, are interchangeable. This allowed the 'order of things' to change, to be fluid, and for architects to see themselves as relevant to the changes they felt were taking place in society. With space seen as non-finite, divisions can be broken down, rooms and activities flow into each other, the outside can flow into the building and the building can flow into the landscape. With the erosion of the idea of specific spaces the concept of the multi-use of space is architecturally strengthened.

This slight excursion into the antecedents of the multi-use of space may serve as a useful illustration of one aspect affecting the way school building systems developed. For, when reductions in the costs of school building are claimed, as they are in the previously quoted figures, it must not necessarily be assumed that one is comparing like with like. The example referred to shows how the concept of the multi-use of space has had a powerful effect on the economics of school design in that it gave 'authority' for changing all the criteria. Architects themselves, for the architectural reasons outlined, were prime movers in trying to establish a different concept of space in school building which saw its expression, and some functional validation, in the multi-use of space argument. In practice this included the elimination of corridors and other circulation space wherever possible and the combination of functions within one space: as with the assembly hall, dining hall and gymnasium. It is then but a short step to the argument that, in a primary school, the assembly, gymnasium, eating, music and drama functions can all take place in the one space and then that this space can also act as circulation for the classrooms. In this way one can see that from the more traditional school planning it was possible to make enormous savings merely in floor area using these and similar arguments.

There were, of course, educational arguments for integration of this sort and it is upon these that most school designers rely. However, the power of the architectural argument must not be under-rated and amongst themselves architects still discuss such innovation as if there is almost a causal relationship between the building and its activities. Many believe that patterns of behaviour can be drastically changed by the 'patterns' on the ground. There is certainly little evidence for this although some for the argument that the built form can inhibit and encourage the way people act. Certainly good teachers have not felt that a particular sort of building is necessary before good teaching and learning can take place and very many of them are very vocal concerning the way in which, for example, the multi-use assembly-hall-all-purpose-space has eroded certain approaches considered important. Environmentally, acoustics are often poor, solar gain high, glass areas too large for comfort and difficult to black out, heating if adequate for gym is not adequate for sitting and eating, residual aromas from lunch interfere with drama, residual odours from gym interfere with lunch and so on. It is interesting that the very same Estimates Committee Report of 1961, which had applauded CLASP methods, had also received two criticisms from teachers. The first was the lack of separate dining rooms and the second was concerned with the large areas of glass used and which resulted in glare and heat problems. Whilst the teaching profession is not notable for its innovatory ability it is fair to ask, then, in what context was the multi-use of space launched? To this there are two main answers: the power of an architectural argument partially concealed within an educational one and the tremendous savings in cost due to savings in floor area. Hence the cost reductions, whilst spectacular must be viewed in their total setting to ascertain whether or not they were due to prefabrication, or due, in part, to a number of other changes in the approach to the design problem.

Not only could such planning changes save money in themselves but changes in specification and detail could be made which would also make savings. Since there were almost no criteria of a definitive nature laid down in the School Premises Regulations it was always difficult to assess the real effect of such changes. Therefore when the cost of a component item is claimed as held or reduced it is necessary to know the specification before and after the event. It is also necessary to know what cost increases it may reflect in other parts of the system and, often ignored, the hidden costs that will appear in tenders which all contractors put against novel or changed methods. It is necessary to know the answers to questions such as this when intepreting claims to cost effectiveness such as those referred to by Oddie from the CLASP report for 1974 which:

> 'shows that in the fifteen months up to 1st July 1974, when the Royal Institute of Chartered Surveyors Index of general building costs rose by 30 per cent, the cost of CLASP components rose by only 18.87 per cent. For the year 1972/73 the equivalent figures were 25 per cent and 9.89 per cent.' (Oddie 1975)

The same study also points out that the basic aim of CLASP was not to reduce costs but to maintain the output of school buildings whilst endeavouring to maintain standards. In the event it is perhaps an ironic comment on these good intentions that much system building is seen by the public to have been brought about by the primary pressures of cost control and to embody a general diminution of standards. A thorough analysis to examine the merits of the two sides in this argument has yet to be undertaken. An important part of such a study would be the influence of the 'serial tendering' methods employed by the Nottinghamshire office. This was an arrangement where contractors price either a typical job or a hypothetical Bill of Quantities and a contract is then awarded on this basis for a series of buildings. This has the benefit of reducing administrative costs both for the contractors and for the clients and architects. Clearly it encourages contractors to forward plan and permits continuity of workforce and experience. Although the standardization of CLASP was an aid to such an approach, serial tendering has been used on non-system jobs by other local authorities.

The political impact of the Triennale Grand Prize for the CLASP school had considerable effect in Britain. Although the mass production ideal had already gained many adherents within and without the building industry, public and political recognition was needed and this the Grand Prize provided. Whereas before administrators and government had to be encouraged, cajoled and convinced by all sorts of means to embark on the next moves in the industrialization game now being played by architects, almost overnight the whole climate began to change. The success of CLASP subsequently saw the ministerial coining of the phrase 'Industrialised Building'. It led to a push towards the adoption of similar sets of ideas throughout the public design sector and, hopefully it was thought, beyond. The *Architects' Journal* in an issue relating to the 1961 IUA Congress in London laid out the case all too clearly, in an article entitled 'The Way Forward':

> 'This (the impact of industrialization) suggests that the way forward for the architect is by full participation in the large-scale activities of the building industry *full enough to begin to direct them*; (AJ italics) in the closest attention to the human needs of the occupants of buildings, as the prime basis for any advance in design; and in the reorganisation of the large office to give the individual architect freedom for action and initiative.' (AJ 1961b)

The article then went on to set out the way this might be achieved:

1. Programmed building: this implies the agreeing of forward programmes as developed by the Ministry of Education rather than the approval of each single project as it came up.
2. Development work with the pattern of spending planned: development groups could then be set up which would look at the whole process from the client's brief to the erection of the building on site. It would ideally be a multidisciplinary team on the operations group models developed during World War II.
3. Development work on a national scale: this posited the need to set up on a national scale in each government department, groups of a nature similar to that at the Ministry of Education, and the article referred to recently created groups in the Ministry of Housing and Local Government, Ministry of Health, University Grants Committee and the War Office. Donald Gibson, late of Nottingham, who had in 1958 been appointed as Director-General to the Directorate of Works at the War Office provides an interesting example of the manner in which an idea can be carried by a man from small scale experimental work, as at Coventry, to its implementation at a national level.

Again the *Architects' Journal* in citing the example of CLASP with its annual programme of £7,000,000 in 1961 puts the case for the architect and manufacturer:

'Not only does this reduce building costs but provides the architects with a vocabulary of interrelated components which *they themselves have designed.*' (author's italics). (AJ 1961b)

The implication is obvious: the architect is able to control the manufacturer in a way which echoes the traditional architect/craftsman

8.2.12 (a) First Engineering building, Bath University, England. CLASP. Architects: Robert Matthew, Johnson-Marshall and partners, 1969. (Photo 1971)

8.2.12 (b) University of York, England. CLASP. Architects: Robert Matthew, Johnson-Marshall and partners (Andrew Derbyshire in charge), 1962 onwards. (Photo 1976)

relationship and which reaps the expected full benefits of mass production. The article goes on to point out that CLASP was still quite crude in terms of the use of the industrialization argument, and that further development was likely to make it more industrialized and less craft based and to simultaneously widen the 'vocabulary' of design.

Three points are of interest here: the restating of the argument implying the morality of the machine, and the ignoring of the social and human implications of this; the recurring pious hope of building systems designers that although the necessities of the argument imply that the range of choice of components is initially few, one magic day it will be possible to give great variety; and finally the way in which the architect is seen as controlling design of components and the manufacturer. Not only will the machine be used for the good of society, runs this view, but architects will control that in design terms. The realities of the British political scene had created a situation undreamt of by the Bauhaus, hoped for by the Constructivists, made scientific by the work study methods of Frederick Winslow Taylor and strongly recalling Marx's dictum that to change society one had to gain control of the means of production.

Although CLASP was initially designed for school construction it gradually developed to include buildings in almost all building types: health centres, community buildings, railway buildings, geriatric units, old persons homes, universities (**8.2.12**) and so on. It was of course used at York (1962 onwards) and Bath (1969 onwards) universities after considerable modification of the system (DES/UGC 1970). Further, Brockhouse Steel Structures marketed their own version of CLASP in Britain and abroad, paying the Consortium a royalty.

Nottinghamshire itself has pioneered a unique attempt to unite design and production with their Research into Site Management Programme (RSM), which started in 1967. The purpose was to set up a combined design and build group which would not only design, but manage the construction of jobs on site. By 1976 work handled by RSM constituted 7% of the Nottinghamshire building programme and the RSM staff amounted to 48, broken down as follows: 7 architects, 1 quantity surveyor, 2 QS technicians, 3 administrative staff and 35 site operatives. The original intention was to discover, at first hand, more about the implications of building with the CLASP system and to feed back the findings to the whole building programme. Rabeneck (1976b) points out that the most important thing learned by the RSM team is a greater awareness of the value of resources. As an example is cited the discovery that it costs more to prepare a detailed site layout drawing with quantities than the cost of hire for an excavator/grader to carry out the work itself: as a result only sketch drawings are prepared and detailed instructions given to the excavator driver on site:

> 'It is one of the unusual advantages of RSM that we can equate the cost of an architect with that of a D9 tractor and scraper.' (Rabeneck 1976b)

Clearly Nottinghamshire has learned a great deal about the implications of using CLASP within their own programme including how to save and control cost. However, as Rabeneck points out:

> 'it is debatable as to whether RSM has made systematic findings that would be of much use to more conventional organisations using CLASP' (Rabeneck 1976b)

Further many of the earlier 'findings' of the RSM project team merely confirmed the experience of countless job architects using CLASP and similar school building systems:

> 'Experience on site showed that ordinary plan and section details, particularly of complex junctions, were difficult to penetrate. . . .'

> 'they were dismayed at the variety of small timber sections required. . . .'

> 'Experience on site yielded many lessons... delivery straight from the factory, which applied to most firms, runs the gauntlet of too many hazards. . . . The nominal skills available... do not fit the actual task to be done. . . . The team sees that the bill (of

quantities) in places embodies a degree of detail and accuracy that bears little relation to realities of cost on site.' (Carter 1968)

Such comments were common currency, long before 1968, in many offices using such systems. However, it does amply demonstrate the impasse created by the client sponsored systems with some types of feedback information. In short, it just did not gain attention. Clearly even with such seasoned systems exponents as at Nottinghamshire it was necessary to create the special situation of RSM for such feedback to be codified into a communication which the system operators could 'hear' and act upon. Many of the client sponsored systems, in recreating technology and the building process in the new image of the production line, also created discontinuity. The initial findings of the RSM team unwittingly show just how strong this image must have been for it to obscure many of the real difficulties created by the system. The RSM project has provided a wealth of useful and continuing information for the Nottinghamshire office and in addition brought many architects to a realization of the complications of their decisions on site.

'This association of actual with site jobs makes one think of the operatives as if one was doing the job oneself: if jobs are de-mystified, design is improved.' (Peter Johnson-Marshall in Rabeneck 1976b)

It cannot be doubted that such a direct experience is a valuable one for designers; however, even this may result in a continuous simplification, or working down to available skills. It is often the demanding and complex problem that obtains the best from a craftsman and causes him to extend his normal standards to meet those being asked. Rabeneck cites three shortcomings to the otherwise successful Research into Site Management Project: the first is the lack of worker involvement in decision making; the second are the difficulties of control arising from the fact that 36% of a contract is still carried out by specialist subcontractors; third, since there is no sponsorship for the 'research' aspects of the project all the data collection and processing is done within the normal fee structure. This accounts for the limited amount of useful data that has been co-ordinated and disseminated and, says Rabeneck, this is 'ironic in a group that has carried the label "Research" for the last 8 years'. It is clear that well formulated and continuous appraisal reports could form a valuable function for CLASP users and for the users of systems in general, and it is difficult to understand why such studies have not been forthcoming. Part of the explanation may lay in Rabeneck's claim that it 'seemed to fall on deaf ears in London, mainly because of the NIH ("not invented here") mentality of much central government' (Rabeneck 1976b). Another part of the explanation may lay within the approach itself which is not to court publicity but to proceed with implementing its ideas. One of these central ideas has to do with control, first, the attempt to gain control of the manufacture and delivery of components and then with RSM an attempt to gain control of the whole on-site process. It is doubtful whether many commentators have explored the implications of a local authority gaining this degree of control, where the user, the designer, the producers and the builders are virtually all one organization. For many years it has been talked of in some architectural circles as a desirable end but its broad sociopolitical effects need to be carefully weighed. RSM has clearly shown an interesting alternative route: is it one to be followed?

In recent years CLASP has become sensitive to the professional and public reaction to years of system building. In its Annual Report for 1975, for example, it says:

> 'Some elements of the construction industry criticise system building on the grounds that it is short cut technology, a bureaucratic convenience, and a struggle to achieve the cheapest building regardless of cost and regardless of environmental consequences'

and answers this with:

> 'Do we attempt to use the technology of the 20th Century or do we go on doing things in the way they were done 50 years ago? System building is successfully reducing costs by applying the lessons learnt in other industries. This is not cheap building but value for money. The environmental consequences of what we build are the responsibility of our clients and the professional designers.' (CLASP 1975)

Three observations only need be made here: first the technology is very much that of almost 50 years ago, in that it is still close to Stillman's Sussex Schools; second, the same report gives precedence to a number of brick clad CLASP buildings; third, the 'consequences of what we build' can only be the responsibility of those clients and designers within the rules of the available system.

GO FORTH AND MULTIPLY...

'By 1965, there were 224 industrialised building systems available in Britain from 163 developers; 138 of them specifically recommended for housing.'

David Crawford
A Decade of British Housing 1963-73, 1975

Political Weight

The early 1960s were important years in the endeavour to establish building systems in Britain. The cumulative effects of the development work that had gone on in various places since the war, coupled with the political realization that the success of CLASP at the 1960 Triennale di Milano could be turned to good use, gave the arguments great force. Due to the passage of time, many of the architects imbued with a social (not to say socialist) mission to harness the productive machine for the public good were now in positions of relative authority and thus were able to put their views at a high level. Further, in 1961 there commenced yet another drive by government to end the housing problem. The Conservative Government, returned in 1959, saw the political potential inherent in the idea of building systems as a positive explanation of how the industry might be encouraged to produce more houses per annum and this was reinforced by the Labour Government returned in 1964. The Bauhaus doctrine seems to have been once again politically swallowed, this time with the shining light of CLASP and the Hertfordshire schools as exemplars, and supported by the very many architects and convinced administrators who seemed to be offering confident and enthusiastic knowhow.

In the public mind the whole idea was still very closely associated with those postwar asbestos and metal 'prefab' bungalows dotting many large towns. Although the occupants were often very attached to them because of their compact planning and (for the period) high standards of environmental control and equipment, in most people's minds they were seen as temporary houses that had been around too long. The image of stained asbestos and poorly maintained sheet metals used in much postwar housing and schools was still very strong, and not seen to be as good as permanent building. Prefabricated building, as it was then called, was thought to be expedient, quick and cheap to build. In the public eye this may have just been acceptable for a postwar situation: something else was needed in more peaceful and affluent times. Of course many of the criticisms were entirely accurate since, as has already been discussed, the architect's approach to the industrialization of building has been heavily weighted towards a total new view of what the building is and the way in which it uses materials. It had set aside, quite deliberately, many existing methods (however industrialized) where these did not visibly exhibit the new order, did not match the ideals of the machine aesthetic. That new order was composed of lightweight materials with strong emphasis on expendability. The way in which the feelings of the public might be assuaged was to emphasize that using machine made components and methods did not necessarily mean shoddy, poorly finished products, and to establish a new public image and a new name for the methods. The new term, a masterstroke if rather a short-lived one, was Industrialized Building or IB as it quickly became known. Of course the phrase had been in use before, but in 1961 it began to be used politically and publicly to encapsulate yet another brave new world.

Both Wachsmann (1961) and Mumford (1946, 1930) had used similar phrases to carry the concept of a fully machine oriented industry. Indeed Mumford in 1930 had not only referred to 'the industrialisation of architecture' that was already apparent, but he also clearly stated the reasons why such an approach by itself would solve few problems, indicating that housing is more a financial problem than a technological one (**8.3.1**). The result of the post-1961 drive was the creation of one school building consortium after another together with strong attempts to apply similar methods to the construction of housing in an endeavour to increase the rate of building by local authorities. There was more than a grain of truth in the thoughts that Richard Crossman (Minister of Housing) confided to his diary on Wednesday, 24th August 1966:

> 'the recognition that most Local Authorities don't want to build houses and, if they do, are grossly incompetent and drive any contractor crazy by the arbitrary methods of their committees and their sullenness and incompetence.' (Crossman 1975)

Political pressure, public encouragement, backed by zealous but often inexperienced, untried architects, quickly led to a vast proliferation of housing 'systems'. By the early sixties a great number were on the market, although very few were viable. The loss of confidence that has been suffered by the British construction industry ever since the early 1930s is nowhere more clearly demonstrated than by their wholesale and often blind acceptance of the IB philosophy so persuasively put during this period. Even in 1945 they had been more realistic, reports White (1965), in questioning attempts to change the whole industry overnight rather than to use the idea in an evolutionary way.

However, the successive blows of the architectural modern movement have not been without their effect. From being rank

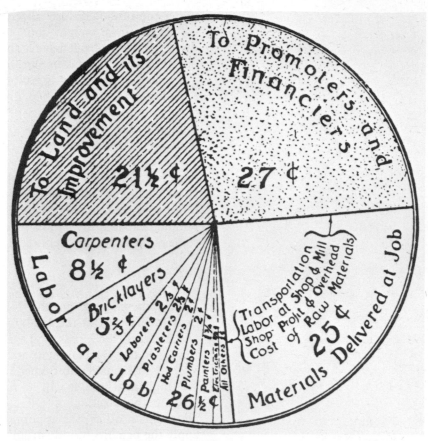

8.3.1 'A large saving in the shell is a small saving in the final house.' In this chart from the *Primer of Housing* (Holden *et al.*) something over half of building cost is shown to be in the construction. The balance, 48½%, covers land and finance costs, neither of which are affected overmuch by mass production techniques

outsiders in the twenties and thirties, architects themselves have proliferated in Britain at a time when more and more spending has been controlled by government. Over a half of the architects in Britain work in publicly financed employment. The powerful message of the modern movement, and in particular of the Bauhaus and Le Corbusier, has permeated the profession and grown consonant with the British version of the welfare state. The building industry has therefore been under mounting pressure to accord with these influences although it has seldom had the literacy or awareness to understand properly what was going on. Under successive attacks both at a conceptual and a practical level, the industry appears to have lost confidence in itself.

Having been told time and time again that conventional methods must be discarded and that much of what they did was of little relevance, the building industry began to believe the arguments. Contributory in itself of course, has been the building industry's own slothfulness in developing managerial expertise (as in the United States) or indeed any clear public evidence of its own efficiencies and strengths. As with all myths, those particular to the modern movement and its view of the building industry contained such elements of reality: such elements could be used in any reordering of ideas. In this way a commonly perceivable pattern was crystallized from the many strands of the building systems argument. These were drawn together under the title of Industrialized Building. From 1961 the movement gathered official weight and powerful allies, mushroomed and then just as rapidly subsided as feedback increased. Its symbolic watershed occurred with the gas explosion which blew out one side of Ronan Point and called into question the philosophy and many of the techniques to which it had given support. Nevertheless its legacy deeply permeates the practice of architecture in Britain.

During this period a number of events occurred in rapid succession, which put both architects and the building systems idea in the centre of the stage. The resultant searching public scrutiny found both of these wanting, but this did not stop many of the ideas from, so to speak, passing into the architect's repertoire. One of the most influential areas of activity was that of public works, with its vast building expenditure and its many employees. This chapter deals with some of these developments as they pertain to building systems.

NENK Method of Building

Development work on the NENK method of building began in September, 1961 at the Directorate of Works in the War Office but was, in April 1963, passed to the newly formed Directorate General of Research and Development at the Ministry of Public Buildings and Works. Sir Donald Gibson had moved from being County Architect at Nottingham where, it will be recalled he had established CLASP, to take up the position of Director General. Ralph Iredale from the Nottingham office was appointed to head a team whose brief was to apply the benefits of industrialized building to the very wide range of building types needed by the army. The unusual name given to the method is also of significance since it reflects the success of a particular idea. David Nenk, who was an administrator, had held the post of joint head, (with an Architect) of the Architects and Buildings Branch of the Ministry of Education from 1950 until 1955 and had been responsible for establishing the success of the work of the Ministry's development group. This had been politically auspicious since it demonstrated the importance of a government department being headed by both an administrative civil servant and a professional technical adviser with equal authority. This was in the face of the tradition in Britain where invariably professional administrators, most of whom had usually spent most of their life in the civil service, were given the most responsible posts.

The success of the Ministry of Education's Architects' and Building Branch (established in 1949) was seen as a powerful example, at least to architects, of how changes could be made in the structure of a government department to encourage collaborative working and innovation. The acclaim received by CLASP, and the establishing politically of 'Industrialized Building' as the major solution to building problems, was a strong factor in the reorganization of the War Office department as a civilian one. A further important strand is that David Nenk had been a member of the Weeks Committee which recommended the setting up of a civilian Directorate of Works at the

War Office. This contrasted with the previous position where almost all army work had been carried out by the army themselves, with the result that they were not as open to change and new ideas as other less contained parts of the building professions. The quality of design, and particularly the way in which it related to the countryside (often the most scenic areas), was frequently bad and they were constantly under attack for taking a less than responsible attitude to the environment. Further, barracks, married quarters and other buildings still lagged behind the sort of standards expected elsewhere and this again reflected in recruitment difficulties.

The opportunity to rethink the building systems argument offered by the army's building programme of £10,000,000 per annum in the United Kingdom alone was clearly an attractive one to the NENK team, and they were at pains to describe their approach as a 'method' of building rather than a system of construction which they said:

'might be defined as that where every component and condition has been anticipated and uniquely solved within a particular selection of disciplines' (MQ 1963)

8.3.2 NENK system. Ministry of Public Building and Works: development work started 1961. Isometric showing hypothetical assembly

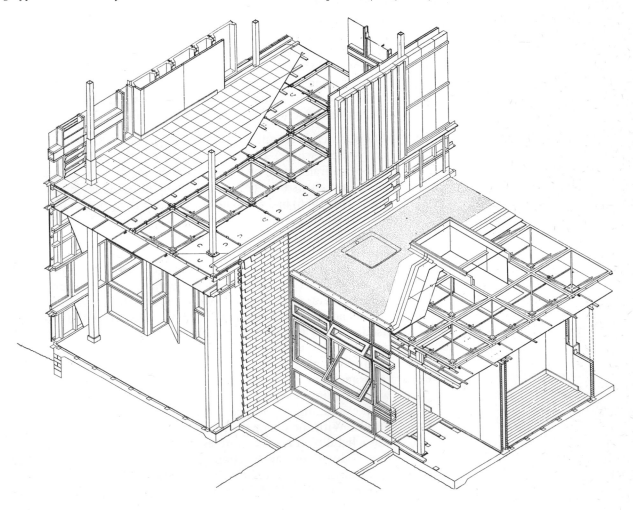

The NENK approach aimed to establish principles which would allow new materials and techniques to be included as they became available. Unlike a number of other development groups many of the NENK team already had experience of the problems posed by the application of the mass production idea to building, and consequently, there are a number of aspects to the NENK idea which show interesting developments. That they were not in themselves developed further elsewhere is perhaps an instructive lesson in the capacity of the industry to absorb change of the sort proposed (**8.3.2**). Walters and Iredale (1964) enumerate the basic decisions surrounding the NENK concept as follows:

1. Dimensions of all spaces and thickness of walls, partitions, floors and roofs would be multiples of the basic module (M) which was 4 inches or 10 cm (approx).
2. Submodular thicknesses would then be considered and preferred sizes for components decided.
3. Structure based on the use of a space frame.
4. The carcassing of internal and external walls, floors and roofs would be considered independently of their finishes.
5. External walls and partitions would be in vertical panels spanning between floors and ceilings.
6. External walls and partitions to be made up of two independent leaves thus allowing differing combinations to achieve differing performance requirements.
7. Services to be housed in roofs and floors and in wall cavities.
8. Dry construction to be used wherever practicable.

These criteria recognize a number of problems that had already been exposed elsewhere. The first of these was the inflexibility of the post and beam frame used by CLASP and others. Economic spans tended to be around 30 ft 0 in (9.12 m) for all the schools consortia: CLASP mark 4 (in 1966) had a maximum span for floor beams of 27 ft 0 in (8.2 m) and single piece roof beams of 27 ft 0 in (8.2 m). Two part roof girders went up to 42 ft 0 in (12.8 m) and long span girders, usually only used for assembly halls and gymnasia up to 60 ft 0 in (18.3 m). SCOLA Mk 1, Mk 1a and Mk 2 had ordinary floor and roof beams up to 30 ft 0 in (9.12 m): these were identical in design but used at differing centres, that is more widely spaced for roofs than floors.

The fact that spans around 30 ft 0 in (9.12 m) are the economic ones, itself reflects the way in which the problem had been set up and derives from the 'classroom box' type of school, which generally required spans between 22 ft 0 in (6.7 m) and 30 ft 0 in (9.2 m) with the stanchions lost in the wall thicknesses. Clearly at the time Stillman was designing his steel frame schools in the 1930s such spans resulted from examining what actually went on in schools with classes of 30–40 pupils. With the changes in approach in education such a restriction was a liability. This is an interesting example which shows the roots of the 'realities' of construction. If a thorough analysis of educational requirements suggested that the most suitable spans were, say, in the region of 50 ft 0 in (15.2 m) to 60 ft 0 in (18.3 m) then presumably the quantity equals economy argument backing up the schools consortia would have made such spans the 'economic ones'.

The limited spans do reflect a particular approach to school planning which has been touched on earlier for not only the classroom box type of school, often with a 'finger' plan, but the type of school embodied most clearly by the work of the Ministry of Education (now DES) development group demands considerable flexibility in beam size under 30 ft 0 in (8.2 m). In this way the structure and external skin can produce the type of 'informal' planning and inside–outside relationship embodied in their philosophy. The counter argument to this is to drastically reduce the numbers of components needed and to simplify the external envelope thus saving money which can be put into longer spans giving internal flexibility, or more floor space, better environmental control — or all three (**8.3.3**). However, this particular argument emerged most clearly in the work of Ehrenkrantz and will be dealt with in a subsequent section. The basic issue here is that what is economic in building has much to do with the way in which you ask the question: in the case of CLASP and similar school building systems the questions had been asked in quite specific ways. This is an important point for building systems development since, to be viable it should not make major component changes at too frequent intervals, yet if it does not do so it is incarcerated in its own initiating logic.

The NENK development group were clearly aware of these issues and could already see the disadvantages of organizing the industry around such narrow criteria with their implications for users. Subsequent to CLASP, building systems proliferated: within this proliferation there were a number of attempts to introduce new possibilities into the ground rules being set up. The efforts of the NENK group must be seen in this light. To escape the inflexibility and difficulties imposed by the post and beam frame, NENK introduced the double layer flat grid space frame, with the minimum of supports. The advantages claimed were:

1. The design and location of structural members more directly related to stress distribution giving greater economy of material.
2. Wide distribution of stresses means that location of supports becomes less critical.
3. Longer spans possible for any given depth.

The result were constant 2 ft 0 in (610 mm) deep floor and roof structures made up of prefabricated inverted tetrahedra giving 26 ft 0 in (7.9 m) one-way and 44 ft 0 in (13.4 m) two-way floor spans with 44 ft 0 in (13.4 m) one-way and 88 ft 0 in (26.8 m) two-way roof spans. What it gained in internal flexibility is perhaps lost in external flexibility since the overall sizes of buildings had to be in multiples of 4 ft 0 in (1.22 m) as did changes in direction of the external wall.

It was, no doubt, inevitable that the attractive concept of the space deck should sooner or later find its way into building systems since this would reflect the conceptual place it had, fairly recently, acquired in architects' minds. Although Fuller's work involving tetrahedra was the most intellectually tough it was the work of Konrad Wachsmann

422

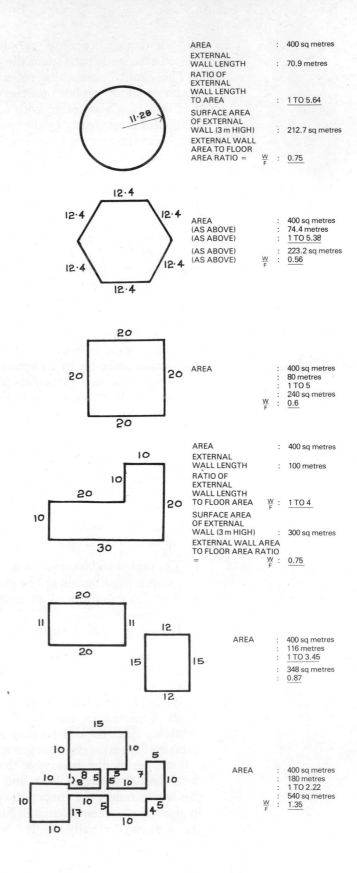

AREA	:	400 sq metres
EXTERNAL WALL LENGTH	:	70.9 metres
RATIO OF EXTERNAL WALL LENGTH TO AREA	:	1 TO 5.64
SURFACE AREA OF EXTERNAL WALL (3 m HIGH)	:	212.7 sq metres
EXTERNAL WALL AREA TO FLOOR AREA RATIO = $\frac{W}{F}$	:	0.75

AREA	:	400 sq metres
(AS ABOVE)	:	74.4 metres
(AS ABOVE)	:	1 TO 5.38
(AS ABOVE)	:	223.2 sq metres
(AS ABOVE) $\frac{W}{F}$	:	0.56

AREA	:	400 sq metres
	:	80 metres
	:	1 TO 5
	:	240 sq metres
$\frac{W}{F}$	:	0.6

AREA	:	400 sq metres
EXTERNAL WALL LENGTH	:	100 metres
RATIO OF EXTERNAL WALL LENGTH TO FLOOR AREA $\frac{W}{F}$	:	1 TO 4
SURFACE AREA OF EXTERNAL WALL (3 m HIGH)	:	300 sq metres
EXTERNAL WALL AREA TO FLOOR AREA RATIO = $\frac{W}{F}$	:	0.75

AREA	:	400 sq metres
	:	116 metres
	:	1 TO 3.45
	:	348 sq metres
	:	0.87

AREA	:	400 sq metres
	:	180 metres
	:	1 TO 2.22
	:	540 sq metres
$\frac{W}{F}$	:	1.35

8.3.3 Plan forms and external wall area. All the plans shown have the same floor area: the circle encloses more area per given length of wall than any of the others. The last, highly articulated plan has over twice the length of external wall than the circle or the square

8.3.4 In his 1957 lectures at the Architectural Association in London Konrad Wachsmann showed his open hangar structure using the Mobilar system

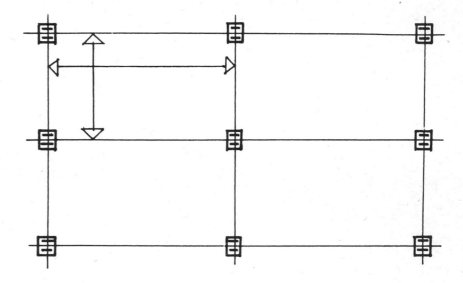

which encapsulated the concept in a form most attractive to designers. Wachsmann's work has been referred to earlier. His book of 1961, *The Turning Point of Building* and particularly his lectures at the AA in 1957 (**8.3.4**) had a profound effect on those looking for support for those arguments surrounding the idea of the 'right use of the machine'.

The developers of NENK were also aware that structural grids tend to dictate planning grids and there was a concern to free the two in order to give greater flexibility. This led to the decision to use the 4 inch (10 cm) module as the basic sizing and positioning dimension, and Walters and Iredale (1964) point out that this was in accordance with the Second Report on Modular Co-ordination produced by the European Productivity Agency (EPA 1961). They also drew attention to the fact that Chief Architects in central and local government had accepted the 4 in/10 cm recommendation in June 1962 and that the RIBA had accepted it as 'policy' for industrialized building since July 1963 — a rising acceptance curve. The NENK team had examined many proposals regarding the combination of numbers for building components and put forward the idea of a number trio. For example, with only three panel widths of 5M, 6M and 7M it was shown to be possible to produce every modular dimension from 10 m (3 ft 4 in or 1.02 m) upwards in an increasing number of different ways. Computer produced tables of combinations showed the ranges available. It can be seen again that a central part of their ideas lay in a co-ordinated system of dimensions, producing, in this case, considerable flexibility in overall sizing but based on relatively few components.

The way in which the external walls were seen to relate to the structure is also of significance: especially since so many attempts to produce industrialized systems have managed to ignore much previous experience in this respect. External walls and partitions were to span

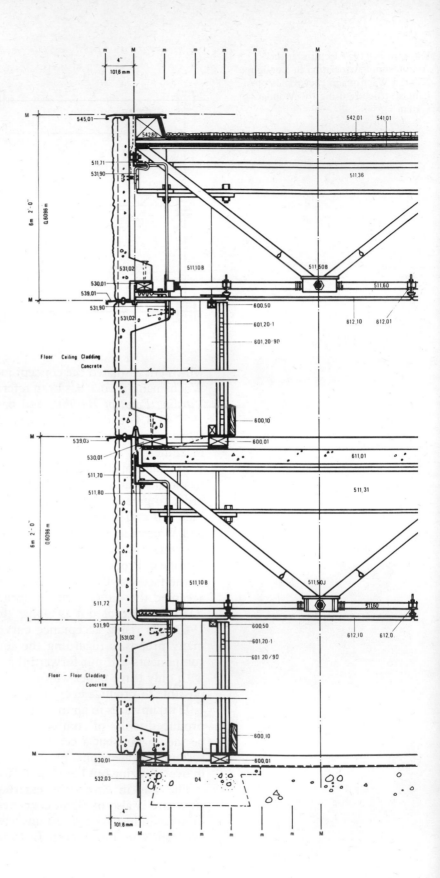

8.3.5 NENK system: Ministry of Public Building and Works. Wall section showing one type of cladding, spacedeck junctions, and coding of components

vertically floor to floor, or floor to roof, thus allowing them to be independent of columns and, incidentally, reducing the difficulty of horizontal joints. The logic of this was carried further with the decision to divorce finish from structure in walls and partitions, thus giving the possibility for the basic construction of these to remain the same whilst introducing flexibility in cladding and finishing materials (**8.3.5**). At least there is a recognition here that the curious moralities of the machine age argument as it had applied to the use of materials, and the 'honest' expression of function and means, were more a hindrance than a help if 'Industrialized Building' was to begin to match the choice and flexibility of conventional building and also to remain economically viable.

The final important recognition by the NENK team, perhaps, is that some positive effort initially was made to accommodate increasing servicing requirements. This reflects in the use of the space deck which, on paper at least, gave more freedom for floor and roof service runs and in the way in which the external walls were designed in two discrete skins. When Iredale and Walters gave their presentation on NENK at the RIBA on 17th March, 1964, it was described by one questioner as a 'half industrialised development' and further that the decisions regarding the external wall were 'a concession to architecture' (Walters and Iredale 1964). The attempt in NENK to offer the opportunity for the use of conventional materials and/or industrially produced materials can here be seen against the commonly held view that to be industrialized a system has certainly to look industrialized. Iredale in his reply emphasized that to attain true flexibility, a system must be able to absorb components both new and existing. This was a necessary and welcome approach after the puritanical rigours of other attempts at building systems design.

NENK has a further interest and that is in the area of documentation. Considerable attention was given to endeavouring to develop ideas concerning standard drawings and their automation. An examination was made of the types of document needed by different persons involved in the design, construction and manufacture (**8.3.6**). The result was six types of document:

1. information sheets giving the building method;
2. data sheets directed at the designer giving the function and use of each element;
3. component drawings for all elements with details and ranges;
4. assembly drawings;
5. specifications for components, materials and assemblies;
6. cost data in the form of analyses and schedules of rates.

Each component and junction was drawn separately and given a discrete code number and all drawings were reduced to A3 size to form a basic manual for the method. The coding system developed was a decimal method related to possible computerization and microfilm recording onto aperture punched cards. Thus all the information could be coded and retrieved automatically. Further the scheduling of components, cost comparisons and so on could be generated. The

426

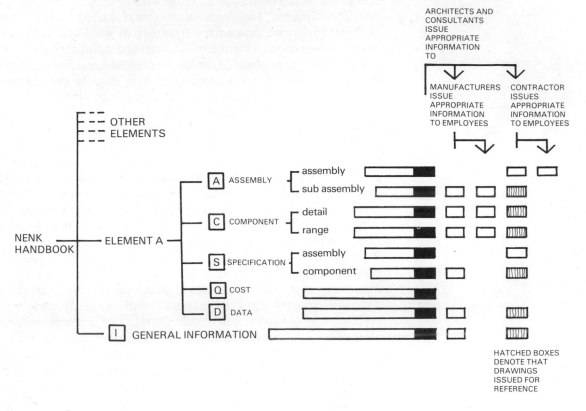

ARCHITECTS AND CONSULTANTS ISSUE APPROPRIATE INFORMATION TO

MANUFACTURERS ISSUE APPROPRIATE INFORMATION TO EMPLOYEES

CONTRACTOR ISSUES APPROPRIATE INFORMATION TO EMPLOYEES

OTHER ELEMENTS

NENK HANDBOOK — ELEMENT A

A ASSEMBLY — assembly / sub assembly

C COMPONENT — detail / range

S SPECIFICATION — assembly / component

Q COST

D DATA

I GENERAL INFORMATION

HATCHED BOXES DENOTE THAT DRAWINGS ISSUED FOR REFERENCE

8.3.6 NENK system. Types of document and their distribution: 1964. (After Walters and Iredale, 1964)

complexity, however, had its effect on site where some confusion was caused by the ten NENK handbooks necessary.

Certainly the NENK idea had many interesting aspects, more as a model for future lines of action than for its own effect on the ground. Its recognition of possible next steps in terms of structural flexibility, envelope flexibility and documentation was far-sighted. However, it probably still held a number of sacred cows in too high an esteem for it to break out of the difficulties imposed both by the system building tradition in Britain and the organizational context of its operation. For example, it continued to place great emphasis on certain modular and dimensional rules, when the jointing problem was the one needing attention if the component vernacular was to develop. The NENK structural engineer, in views expressed at the RIBA lecture, certainly felt that economy and structural efficiency had been compromised to maintain, for example, the constant 2 ft 0 in (610 mm) deck depth:

'Because of the programme which had been laid down for development, quite rightly certain decisions were made. One would not say they were "imposed", but they were made. One had in mind for instance the depth of the floor structure which was 2 ft 0 in overall. The engineers with 12½% of the total responsibility felt that it was too shallow....' (Walters and Iredale 1964)

427

Similarly with problems of jointing and tolerances, and the degree of precision required: these were not, it seems, entirely alleviated by the use of the Building Research Station (now BRE) dry joint technique. Perhaps one of the more significant comments came from Mr. K.H. Price representing the National Federation of Building Trade Operatives. Having praised the way in which facilities had been provided for district officials and the unions, he criticized the comment that had been made concerning such methods lessening the need for the use of craft skills:

> '... the tolerances which were required and the high degree of finish to the base slab indicated that there was no lessening of craft skills. Indeed, possibly the tendency was towards a much higher standard.' (Walters and Iredale 1964)

The view, repeatedly put forward by building system designers, that one of the basic purposes of an industrialized system is to decrease the demand for skilled labour which may be in great demand is again corrected. Certainly the experience of those involved in the school consortia and other systems is that a much greater attention to precision is required. This argument is then extended to the belief that the building industry must be capable of great precision. This is a further example of the traps that await those who see the building as a machine, or use the car assembly line as a suitable analogy for the built form. Certainly the building industry has been slow to change in Britain, in common with many other aspects of life, but many of the attempts to encourage it towards greater rectitude and precision seem to have been in the face of quite basic facts concerning materials and their assembly on an open site by men who have a tough independent tradition, and work in all weathers.

An enormous development effort went into NENK, with some 700 drawings being produced. A series of buildings were built (**8.3.7**) but Ministry inertia soon told and as enthusiasts departed (Iredale to work for Ehrenkrantz in the United States) it was gradually phased out. The large, captive market offered by Ministry work plus considerable development resources failed to make the lasting impact initially expected of it. What is left of such an effort, such enthusiasm and such expenditure? A residue of experience within the Ministry perhaps, personal experiences developed elsewhere and a few buildings. Some members of the NENK group (Jobson, Morgan, Pain) joined the Component Development team under John Redpath and made significant contributions to the development of performance standards and the Ministry's Method of Building. The lesson appears, as before: major changes in the whole building system cannot be made via technological means, however powerfully supported. Even the recognition by the NENK group that a building system has to have more openness than was hitherto available was insufficient to overcome the basic difficulties posed by such a totally new series of methods.

8.3.7 Invicta Park, Maidstone, Kent, England. MPBW, 1965. Military residential accommodation using the NENK system

Because architects have placed such a high premium on one or two aspects of the relation of the machine to architecture, and because they saw these as especially relevant, they have elevated them to a position which often does not seem to allow them to see the whole problem any more. This, of course, is one result of accepting the 'systems engineering' view of the making of a building: the view which is basically a mechanistic one concerned with input, processing and output. Its acceptance by architects has given rise to the constant interest in output and productivity, particularly during this period of the systems explosion. Squeeze more houses or schools from the same or fewer resources with a heavy emphasis on those aspects of design that are easily measurable. These concerns, followed more or less exclusively, have given rise to all sorts of curiosities and excesses, some of them with extreme results such as building failures, others merely demonstrating architectural naiveté.

The field of housing has been particularly prone to these pressures, and never more so than during the early 1960s. With government support for industrialized building, ministries being encouraged to establish development groups on the Ministry of Education pattern to carry out the much needed research, and the proliferation of housing systems, many felt the millenium imminent. However, most of the housing systems fell by the wayside, the development groups had their attention turned to more general issues and government support was gradually withdrawn as problems emerged — particularly those concerned with collapse or fire.

At least the development of the 5M system seems free of the latter problem, although its birth and short life have a number of points of interest to students of the systems field. It was an attempt to design a system at Ministry level which was in itself bold and controversial with its hints of a national building system. Whilst demonstrating that it was building on experience (CLASP), it was again the 'architect's system' and took little note of the implications of setting up yet another kit of parts, another closed system.

5M was the MHLG's third project, started in 1961. The decision to develop a system for local authority housing lay in a number of factors:

1. The politically stated objective of raising the national housing programme to 400,000 annually.
2. If this was to be achieved how could local authority housing output be increased?
3. The view that the shortage of site labour was increasing.

Such statements have a familiar and rather hollow ring about them as arguments for the use of a building system. The Ministry book on 5M makes the incredible claim that 'current housing targets output had to be increased by 50% without increasing either the labour force or the cost of the houses' (MHLG 1970). Some of the additional objectives aimed at by the MHLG development group were:

1. To understand the problems of using industrialized building for local authority housing.
2. To assess the reactions of committees, architects and tenants.
3. To control house building costs.
4. To 'have a finger on the pulse of the technical development' (Tindale and O'Toole 1969).

The development groups also felt that:

1. The kit of parts approach of the school building systems offered the right attitude to flexibility.
2. The slight cost increases of this method would be more than offset by savings in overall design and production costs.

Again familiar arguments, one might say, lead to familiar solutions. In this case it led the team to develop a series of components using the 1 ft 8 in (508 mm) CLASP planning grid — hence the name 5M. The dimensional discipline again is the prime mover in promoting the systems argument. However, the dimensional aspects have received so much attention that they have been bedevilled by change. Indeed one might say that it is only in this area that architects have had very much impact at all.

Although design work on the 5M project started early in 1962, with the first four prototype houses completed at the end of that year all using the 1 ft 8 in (508 mm) grid, the Ministry of Public Buildings and Works published their recommendations for a 1 ft 0 in (305 mm) planning grid as a 'key dimension' in February 1963 in a document called Dimensional Co-ordination for Industrialised Building — or DCI (MPBW 1963a). It is clear that preparation work on the crucial proposal for a 1 ft 0 in (305 mm) planning grid was going on during 1962 at the same time that the MHLG was developing the 5M system on a 1 ft 8 in (508 mm) grid.

Defending the use of this grid for 5M, Patricia Tindale at the RIBA in 1968 (Tindale and O'Toole 1969), stated that had DC1 been published a 1 ft 0 in grid would have been used. This was questioned by the Principal Development Architect to SEAC Jack Platt (the South Eastern Architects' Collaboration, a group based on Hertfordshire's work), who said he 'was surprised to find the result of the 5M development a relatively closed scheme with an old dimension as its base...' (Tindale and O'Toole 1969). The other aspects of the system which are of interest lay in the use of a steel frame with timber beams, the unusual party wall construction, and in the drawing techniques which, Tindale states, were not successful. Development costs were stated (MHLG 1970) as more than offset by savings on the first programme of 800 houses. For these 2-4 architects were employed over a period of 2½ years.

The use of a frame of course, conformed to all the myths surrounding industrialization in technological terms as well as to many previous experiments including the Dennis-Wild house of 1928 and the Key House Unibuilt of 1945. There was certainly little or no evidence to support the general viability of a steel frame for mass

8.3.8 5M housing at Gloucester Street, Sheffield, England. Ministry of Housing and Local Government, project architect P. Tindale, 1964

housing, constructed in relatively small batches. When the solution was not a steel frame but a combination of steel columns, laminated timber beams, and flitched beams it is hard to see any real support for its use. Initially the group had tested the hypothetical application of 5M CLASP components to their scheme at West Ham; it was quickly demonstrated that the system was 13% too expensive (Tindale and O'Toole 1969) for housing, having been designed for heavier loads and longer spans. With this amazing discovery behind them, the group were able to make some adjustments to the design which, according to the reports, brought 5M within the appropriate cost limits. A scheme of four prototype houses was erected at Sheffield in 1962 and others then put up at York University (16), Hull (100), Leeds (150), Catterick (370), and at Gloucester Street, Sheffield (**8.3.8**) (Crawford 1975).

Not only was there a steel frame, but a flat roof 'to give flexibility in the shape of the houses', although this was not born out by the limited shapes actually used. The external walls and the party walls were non-load bearing: the former's use of 'a variety of claddings was a matter of expediency rather than choice' (Tindale and O'Toole 1969). Apparently variety of external appearance was not one of the desired aims (unlike NENK) and this of course mirrors the whole way in which most architects seem to have viewed systems — rather as mechanisms for keeping other architects (whom they consider bad designers), and the users, in order. The use of the phrase 'in order' here is not casual, since it can be seen to relate pattern and visual tidyness, or as stated in the OED: 'regular array, condition in which every part or unit is in its right place, tidiness, normal or healthy or efficient state'. The visual hygiene of few materials, put together in clean rectangular forms, was an attempt to limit possibilities which were thought to be undesirable and in the Bauhaus tradition, to inculcate 'good design'.

The attempts to develop a lightweight, dry party wall construction were also bedevilled by difficulties, since a light framed structure conflicted with the requirements for fire resistance, incombustibility and sound reduction. After initial experiments with a lead curtain, construction (**8.3.9**) was adopted using a 1 inch fibreglass quilt between separated layers of laminated plasterboard. It may be useful to compare the complexity and ingenuity of this with a traditional crosswall and perhaps query whether the right question was being asked. Such difficulties, and a variety of other reasons related to labour and methods in the industry, led the Sheffield public works department themselves to abandon it in favour of a concrete block party wall on their development scheme at Gloucester Street, Sheffield, completed in November 1964. For the timber party walls used at Sheffield an exception (waiver) to the building byelaws was obtained, since they called for 2 hour fire resistance.

As with the NENK system, 5M was a result of the momentum and enthusiasm engendered by the confluence of the events of the period. However, it was also quite clear at this time, from the experience of Hertfordshire, CLASP and other work, that such a proliferation of effort to produce a series of closed systems was inappropriate. We have the paradox that the commitment and energy that had gone into the argument was not to be deterred, even by the weight of emerging evidence. Many paid lipservice to the necessity for an open component system whilst developing their own private building system. The choice also had a human element in that this very commitment had put many architects in influential and well paid positions within the political and administrative structure. Even those who could see the extreme dangers of continuing to develop a series of closed systems were not prepared to take the necessary personal action to halt this. Such actions would put them at odds with a system and with people with whom they felt a sympathy. The beliefs of a lifetime, it seems, were not easily discarded. Marginal changes were and could be made from within to try to encourage more flexibility and openness and some attempted this: others (like Iredale) withdrew into positions where their professional judgement could once again be independently utilized.

One of the avowed intentions of the systems designers was to look at building cost rationally. Whilst in the end, this intention rebounded on them in that the public felt that the results showed that cost was all they were interested in, anyone with experience of the British building industry will know how difficult and chaotic is costing. For the architect, cost estimates are usually arrived at by a circuitous route: a quantity surveyor works out costs based upon rates for previous similar jobs adding sums to cover increases; drawings and a Bill of Quantities are prepared (hopefully in that order, but frequently the drawings follow the Bill) and tenders called; the Bill is a schedule of quantities and bears little relation to the form and organization of the drawings, and in addition many builders remake the bill to suit their own ordering methods; at the end of sections of the work (foundations for example) and at the end of a job, the work is often remeasured by the quantity surveyor again; finally agreements are hammered out with the general contractor. The job then forms the basis for future

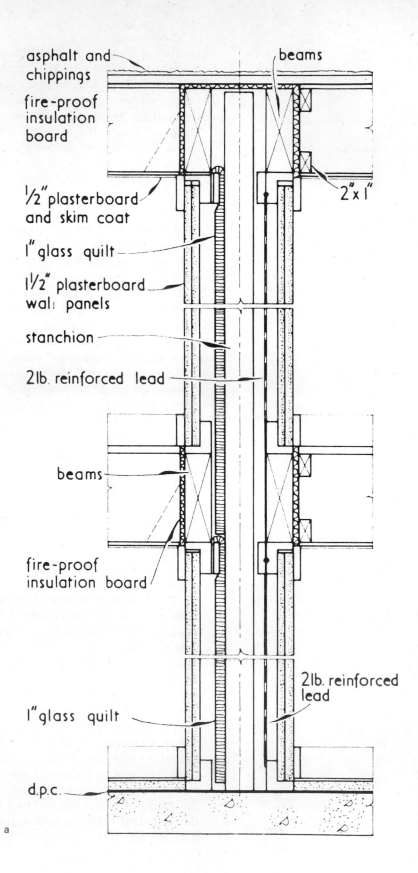

asphalt and chippings

beams

fire-proof insulation board

1/2" plasterboard and skim coat

1" glass quilt

1 1/2" plasterboard wall panels

stanchion

2 lb. reinforced lead

2"x1"

beams

fire-proof insulation board

1" glass quilt

2 lb. reinforced lead

d.p.c.

8.3.9 (a) 5M system. Original proposal for party wall using lead 'curtain' to assist sound reduction. The steel stanchion is flanked with lead, a glass quilt and two layers of plasterboard. (b) *Opposite*. 5M system. External and party wall sections. The latter using glass fibre insulating quilt, but omitting the lead 'curtain' originally proposed

a

Constructional Details: Section—External Wall

Section — Party Wall

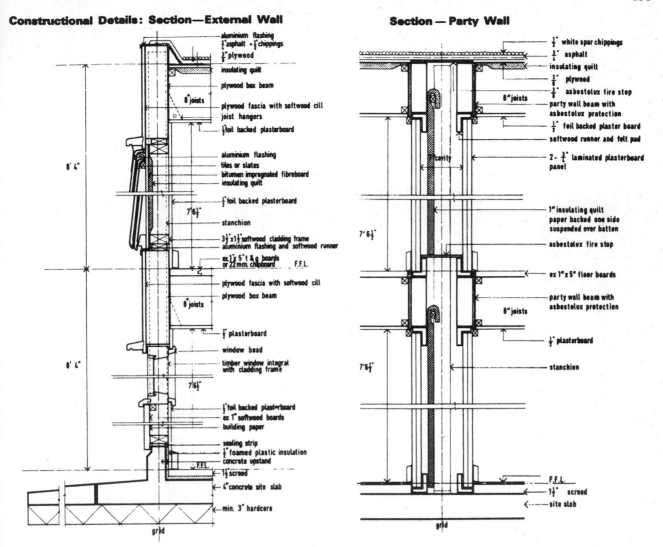

External Wall labels:
aluminium flashing
$\frac{3}{4}$" asphalt + $\frac{1}{2}$" chippings
$\frac{1}{2}$" plywood
insulating quilt
plywood box beam
plywood fascia with softwood cill
joist hangers
$\frac{1}{2}$" foil backed plasterboard
aluminium flashing
tiles or slates
bitumen impregnated fibreboard
insulating quilt
$\frac{1}{2}$" foil backed plasterboard
stanchion
3$\frac{1}{2}$" x 1$\frac{1}{2}$" softwood cladding frame
aluminium flashing and softwood runner
ex 1" x 5" t & g boards
or 22mm chipboard   F.F.L.
plywood fascia with softwood cill
plywood box beam
$\frac{1}{2}$" plasterboard
window bead
timber window integral with cladding frame
$\frac{1}{2}$" foil backed plasterboard
ex 1" softwood boards
building paper
sealing strip
$\frac{1}{2}$" foamed plastic insulation
concrete upstand
1$\frac{1}{4}$" screed
4" concrete site slab
min. 3" hardcore

8' 4"
7' 6$\frac{1}{2}$"
8' joists
8' 4"
8' joists
7' 6$\frac{1}{2}$"
grid

Party Wall labels:
$\frac{1}{2}$" white spar chippings
$\frac{3}{4}$" asphalt
insulating quilt
$\frac{1}{2}$" plywood
$\frac{1}{4}$" asbestolux fire stop
party wall beam with asbestolux protection
$\frac{1}{2}$" foil backed plaster board
softwood runner and felt pad
2 - $\frac{1}{2}$" laminated plasterboard panel
1" insulating quilt paper backed one side suspended over batten
asbestolux fire stop
ex 1" x 5" floor boards
party wall beam with asbestolux protection
$\frac{1}{2}$" plasterboard
stanchion
F.F.L.
1$\frac{1}{4}$" screed
site slab

8" joists
7' 6$\frac{1}{2}$"
2" cavity
8" joists
7' 6$\frac{1}{2}$"
grid

b

estimates. Not surprisingly, after such a procedure, during which items have to be 'rounded up', or agreed on a 'swings and roundabouts' basis, the actual items for each solution arrived at are some distance removed from the real cost. To attempt therefore to predict what a particular type of design solution will cost is a central problem for designers, and no easy one.

Many systems designers, very aware that such costing procedures were less than adequate, sought clearer ways of guiding design to make it cost effective, and to enable them to predict the likely cost effects of given design decisions. Since much of the systems building idea is concerned with rationalizations of process, it followed that it should be possible to offer useful design information in this field. The *5-Minute Guide to Economic Design in 5M System Housing* (MHLG 1966) is a clear and concise example of such an approach. On the basis of simple commonsense logic it is clear that a squarish house plan has less

434

THESE HOUSES

ARE CHEAPER **THAN**

THESE HOUSES

ARE CHEAPER **THAN**

THESE HOUSES

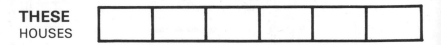

8.3.10 5M system. 'Shape of house: wide fronted houses are dearer than narrow fronted houses. Very narrow fronted houses are dearer than less narrow fronted houses'. From '5 Minute Guide to Economic Design in 5M System Housing MHLG July 1966

external wall than a long rectangle as in **8.3.10**. Equally that the simple cube of house A (**8.3.11**) should cost less than B or C with their projections containing more external wall area, more corners, and a block junction to be made. Similarly the information of length of block is eminently logical, with houses in a block of twelve shown to cost £133 more (in 1966) than when in a block of two (**8.3.12**), and the stepping and staggering of blocks also costing more. A number of points need to be made about such information.

First, although the costings may be affected by the use of the 5M system itself, it is clear that on a logical basis such statements are valid. However, it also presupposes that a general contractor's mode of costing such items is of the same order, or should be. However, costing by contractors is subject to many other influences, and these often outweigh such design rationalizations. For a simple example take in the porch projections on a house: it may be that a specific contractor has a good source of supply close at hand, whilst others do not — he would therefore be able to make such a solution quite economical. Other factors intrude: the size of the discounts contractors can obtain with suppliers and merchants (and in a system some of these may be affected), how far they are from the job itself and what their relationships are with needed subcontractors. The fact is that costing, like many other aspects of life, comes down to people and how much confidence they have in each other. Tools such as those put forward

SHAPE OF HOUSE — PROJECTING BAYS

Small projecting bays are relatively expensive

A, B & C are all equal in area —

BUT

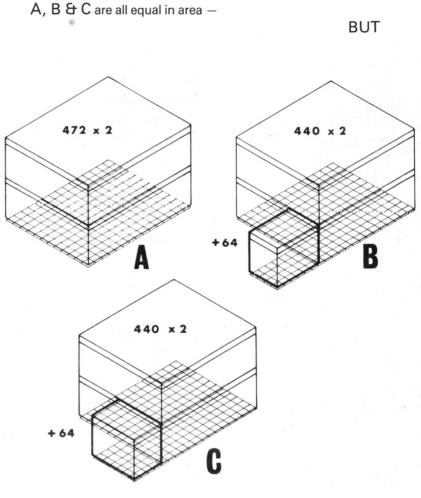

8.3.11 5M system. 'Shape of house-projecting bays: small bays are relatively expensive'. MHLG 1966)

A is £124 cheaper than B and

£90 cheaper than C

by the 5M team look eminently rational and can be argued as such: however, they make the assumption once again that technology is all. Further than this the mere existence of such information to many architects is a prescription to observe it, whereas it should be seen only as one aspect of the decision making process inherent in producing a good human design. In addition, for those architects working for central government or local government, the pressures to observe information of this sort in the interests of economy are considerable. It is interesting to note that as the concerns of energy and conservation

436

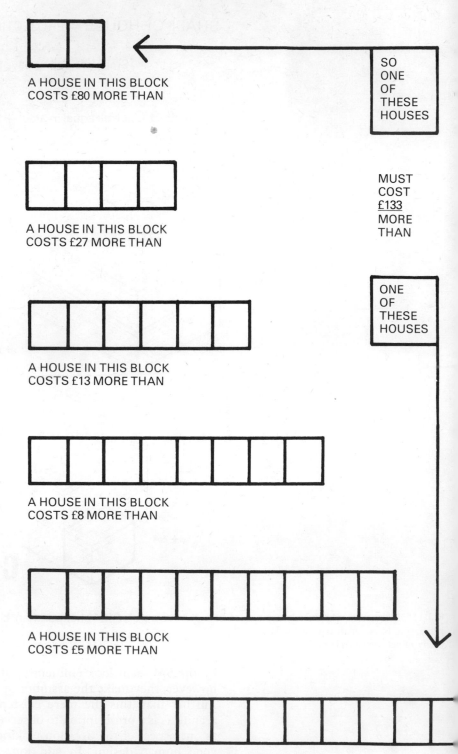

A HOUSE IN THIS BLOCK
COSTS £80 MORE THAN

SO
ONE
OF
THESE
HOUSES

A HOUSE IN THIS BLOCK
COSTS £27 MORE THAN

MUST
COST
£133
MORE
THAN

A HOUSE IN THIS BLOCK
COSTS £13 MORE THAN

ONE
OF
THESE
HOUSES

A HOUSE IN THIS BLOCK
COSTS £8 MORE THAN

A HOUSE IN THIS BLOCK
COSTS £5 MORE THAN

8.3.12 5M system. 'Length of block: houses get cheaper as the block gets longer but in ever decreasing sums, because of the extra cost of an end wall'. (MHLG 1966)

A HOUSE IN THIS BLOCK
AND they are all the same inside

generally became more central to building design, much housing again acquired a style involving projections, steps, staggers and pitched roofs of all sorts. One set of rationalizations replaced another, and a different range of expressive forms has begun to emerge. This shows the dangers of assuming that humane environments arise merely from satisfying a narrow range of criteria. The acceptance of such functional norms as primary ends reduces the rich possibilities which architecture can offer.

The 5M system was officially buried at a lecture given by Tindale and O'Toole at the RIBA in Autumn 1968. The difficulties of a Ministry promoted system together with the solution chosen clearly made this necessary. More importantly:

> 'The development of 5M assumed that a large programme would be required to bring the prices of the factory-produced components within the cost targets. Such a programme would produce repetition and continuous manufacturing conditions. From BRS studies, it is clear that these did not always occur, and there will have to be new ways or organising the design and supply of components.' (Tindale and O'Toole 1969)

Even before the 5M work, the Hertfordshire experience was questioning the view that only with a large programme run of components would costs be reduced. Unfortunately, even the 5M experience was not heeded, with the proponents of school and housing systems continuing to equate large programmes with significant cost savings. After all, even 5M demonstrated that although construction was quicker, the costs were merely 'no higher' than those of traditional construction.

More important, however, is the cost in use of such a system and here some alarming evidence is beginning to emerge. It is reported (BD 1979a, b) that authorities with 5M schemes are faced with heavy repair bills, and their difficulties have resulted in high level informal discussions between the Department of Environment and two councils, Leeds and Sheffield, at the end of January 1979. A major problem concerned concrete panels infilling the steel frame which were breaking up and falling out in a number of cases. At Sheffield these are being replaced at a cost of £300 for each house with the total repair bill for the 871 5M houses amounting to over £260,000. The 1000 houses at Leeds using the system have had panels replaced with brickwork and lightweight construction with additional insulation at a cost of some three million pounds. Houses in Hull and Manchester are to be dealt with similarly.

When a large central government agency with all its technical support, designs and promotes a technique for building houses which, in a little more than a decade, require such remedial action, important questions are raised concerning the efficacy of such a housing delivery method and its administrative context.

A project described as the biggest experiment in industrialized housing on one site in Britain commenced at Thamesmead in 1966, designed by the Greater London Council Architects Department and using the French Balency system, with Cubitts as General Contractor. The project first surfaced in 1962 and, after many changes, some 1300 acres (525 ha) was made available, 600 of them for housing, with the intention of housing some 50,000 people. The site, some 12 miles from Hyde Park on the Erith Marshes, was a difficult one. It involved, for example, covering the whole of the 50 acres Stage One housing area with two feet of sand, this resulting in no housing units being permitted lower than 8 ft 6 in (2.6 m) above Newlyn Datum level in view of the possible flooding. The project has been fraught with troubles at every level, from overall siting to detail functioning and the provision of facilities. The concern here is the use of the building system, Balency, which the Greater London Council looked upon as a means of obtaining real data on the use of such methods. In view of this the design team attitude was somewhat surprising in that they

'decided that the correct approach would be to design a good project without reference to a specific Industrialised Building system and then apply the use of a system to the scheme' (Pike 1969)

8.3.13 5 storey linear housing built in *in situ* concrete crosswall and floor slab construction to simulate the Balency system. Thamesmead, London. Greater London Council, 1968. (Photos 1978) (a) View over lake from south, (b) *Opposite.* View of access side to linear housing. No dwellings were permitted at ground level in case of flooding. (c) *Opposite.* Detail showing balconies. (d) *Next page.* Detail showing precast concrete component junctions. The 'negative' corner here may be an honest reflection of the way the panels were cast, without the finish returning on the ends, but it leaves a very crude corner

a

b

c

8.3.13 d

This had the merit of testing, however, just how far apart such a design might be from the capabilities of a system; in the event this turned out to be quite far since for the first stage of the much photographed five storey linear housing:

> 'the highly modelled form of these blocks made them unsuitable for the economic application of the Balency system'. (Pike 1969)

The result was that the first of these had to be constructed using a rationalized *in situ* crosswall and slab method with non-load-bearing

8.3.14 Point blocks using the Balency system: Thamesmead, London: Greater London Council. View from north with community buildings in foreground. (Photo 1978)

cladding panels, designed to simulate the system cladding which was to follow (**8.3.13**). This approach also obviated the necessity of waiting for the completion of the Balency site factory before starting construction. The point blocks, which were built using the Balency system, are of necessity much simpler in design. The variation in the finish of the external cladding panels indicates the difficulty of achieving consistent quality (**8.3.14**).

The first major projects using prefabrication which were carried out by the French engineer M. Balency-Bearn were in 1949, and by the time the Balency and Schuhl system came to be used on Thamesmead many projects had been completed in France, Switzerland and Italy, most of them rigorous essays in the type of high rise block which has become familiar in parts of Europe. In his paper on the system Balency-Bearn puts forward the familiar arguments for industrialization and points out that:

> 'Quantity production in series is, of course, essentially at variance with flexibility of design. It will be necessary to submit to technical conditions and to the demands imposed by the equipment employed, which to some extent dictates the form and shape.' (Balency-Bearn 1964)

442

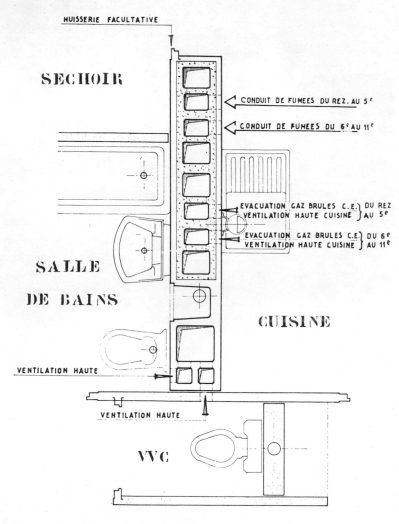

HUISSERIE FACULTATIVE

SECHOIR

← CONDUIT DE FUMEES DU REZ. AU 5ᵉ

← CONDUIT DE FUMEES DU 6ᵉ AU 11ᵉ

EVACUATION GAZ BRULES C.E.⎞ DU REZ
VENTILATION HAUTE CUISINE ⎠ AU 5ᵉ

EVACUATION GAZ BRULES C.E.⎞ DU 6ᵉ
VENTILATION HAUTE CUISINE ⎠ AU 11ᵉ

SALLE

DE BAINS

CUISINE

VENTILATION HAUTE

VENTILATION HAUTE

VVC

8.3.15 Balency system: functional unit, 1964 model (Balency-Bearn, 1964)

ENCOMBREMENT DU BLOC 46 × 2.93

However, in some ways the Balency system offered more than many others, and this is no doubt what attracted the GLC to it. The panels, for example, although of a constant height, could be extended in length and, claimed Balency-Bearn, this meant that it was not necessary for an imposed dimensional discipline to be used. The extendable steel moulds used were a key part of the system. Another was the concept of 'functional units', which were thick storey height blocks into which had been cast a series of apertures for ventilation ducts, flues, services and waste stacks (**8.3.15**). The floors, unlike many other systems, were *in situ* although a precast one was used on the three storey houses at Thamesmead. The junction of the floor and

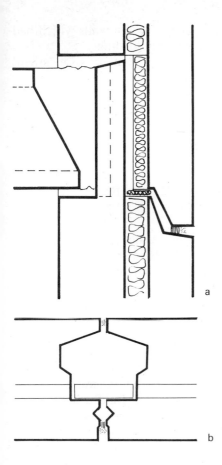

8.3.16 Balency system: Thamesmead, London. (a) Junction of floor slab and external wall. (b) Vertical joint between wall panels

wall slabs (**8.3.16**) is similar to that used by the Larsen-Nielsen system at Ronan Point and this no doubt led to the strengthening measures introduced at Thamesmead in the wake of the Ronan Point collapse (Brutton 1971).

The site factory, with an area of 114,000 sq ft (10,590 m) and a 560 ft (170.69 m) long production line, had a claimed output of 950 dwellings per year using single shift working and the rate of building (in 1969) was at £7.5 million per year which, says Pike, is 'probably exceeding twice the rate ever previously achieved in this country' (Pike 1969). However, in 1971 only a third of the expected 3000 dwellings were ready and by July 1972, 5000 residents had moved in and the delays and cost escalations on the scheme as a whole caused the GLC to instigate a major independent appraisal by outside consultants Robert Turner Associates.

The size and complexities of a project like this make simple statements on cost impossible and this is compounded by the time span involved. One report (Pike 1969) put the projected total cost at £225 million of which about £180 million would be actual building work. The first part of the contract was carried out using the Value-Cost system which although not before used for an industrialized system appeared to have advantages where prices were continually fluctuating. Simply, this method involved the council itself in financing building work whilst the contractor is paid an agreed fee to cover his management expenses. Depending upon the outcome of a comparison between the cost of work and the value against agreed rates the contractor either gains a bonus on his fee or a reduction. Whatever the control deficiencies of this method it has clearly had a marked effect on the rate of building. However, a GLC report into the cost escalation in 1971 pointed out in connection with the building system that 'It is on this part of the work that the cost-to-value relationship is most adverse' (Brutton 1971). Production line delays had occurred and the changes introduced following the Ronan Point collapse had also clearly had an effect. However, probably the most revealing figures were those given by Richard Balfe, Thamesmead Committee Chairman, in June 1973 (Feldman 1973). He revealed that the cost of a new council house at Thamesmead was about £11,530 but the funding of the capital for construction over 60 years at 9% added another £50,970 to this.

It was reported, in September 1978 (BD 1978) that £400,000 was to be spent on the scheme, following a public participation study. Proposed changes included converting the access decks in the linear blocks into individual balconies, improving security in tower blocks by the installation of entry-phones, improving the cavernous ground floor garage areas to improve surveillance, and upgrading courtyards and public areas. The Value-Cost method of contract was renegotiated in January 1972 on a conventional basis. Thamesmead has suffered more than its share of technical difficulties, for example in August 1971 it was reported that the GLC was setting up an enquiry to examine the extent of problems with leaking metal pivot hung windows and with roofs (AJ 1971): in August, 1976 it was reported that £45,000 was to be spent on reerecting partition walls which had been erected in too heavy a material for the floor loadings.

In an article where he dubbed Thamesmead 'Bennett's Leviathan' Reyner Banham said that Thamesmead was:

'In the biggest of big leagues, comparable to Hausmann's Paris or the Los Angeles freeway system, it offered to combine the economies of scale with the resources of big management and the techniques of industrialised construction, in order to create what would, for a decade or more, be a virtually self-manufacturing city, erecting itself panel by room sized panel out of a factory in its own entrails.' (Banham 1971a)

The expected economies resulting from scale and management expertise have not materialized — indeed the reverse was demonstrated by the closure of the Balency factory in 1975 and a return to an extensive use of brick in the new, smaller scale housing of the mid-seventies. The cost information that has yet to be thoroughly reported upon publicly concerns those intentions with which the architects at the Greater London Council set out: to establish viable feedback on the use of industrialized methods on this scale. The experience they have gained on this project must be immense and the dissemination of this to building research, the architectural profession and the world at large would be of great value. Indeed it could be argued that in view of the public money involved it is essential. The results of such experience, coming from one of the world's most highly industrialized nations, are of crucial importance to those countries who see in such large scale industrialized techniques the salvation of their building problems. Even a Leviathan has its uses.

A Surfeit of Systems

The success of CLASP at the 1960 Triennale di Milano brought to government attention the uses to which the systems idea could be put. Work had begun on both SCOLA and NENK devleopment by 1961 and one of the more successful client sponsored housing groups, the Midlands Housing Consortium was formed in February 1963. By 1963 Donald Gibson had formed the Directorate of Research and Development at the Ministry of Public Buildings and Works, thus bringing the systems philosophy to bear on a large area of public building programmes. Geoffrey Rippon, the then Minister for Public Buildings and Works gave support to the work of the Directorate in introducing their first statement on dimensional co-ordination for industrialized housing in February 1963 and published in the *Modular Quarterly* for Winter 1962/63 — although the Society pointed out that they had 'long opposed the basing of a national Standard upon a system of preferred sizes, maintaining that every modular interval must be available generally' (MQ 1963/64). With this sort of government support building systems multiplied until in the housing field there were an alarming number on offer, few of which could be expected to be successful. Companies outside the building industry or on its periphery invented systems. The British Steel Corporation, for example, employed a team of consultants to find a market for steel in general and the new precoated steel strip to be produced in 1965. This resulted in IBIS whose objective was the 'design of a vocabulary of modular building components for flowline production' (IBIS n.d.). These were to be used as individual units in industrialized or conventional situations or as 'a versatile kit-of-parts forming an integrated building system'. It is perhaps a measure of the euphoria produced by central government commitment which encouraged this and other ventures to ignore both the technological lessons of earlier attempts and, more importantly, the marketing problems. For in the end, the flowline production philosophy requires a market that will use its products on a large scale. The marketing problem was the one which most of these would-be building systems producers had not researched. Against this overall background the National Building Agency was established as a semi-independent government body, with a government grant and the aim of building up a consultancy earning capacity, its initial task being to examine the fast growing world of housing systems. An NBA director from 1966-79, Sir Kenneth Wood, (Chairman of Concrete Ltd, later the Bison Group, 1958-79) became Industrial Adviser on Home-building to the Minister during 1966-67. By giving encouragement, professional advice, by testing procedures and by the linking of producers with users it could attempt to bring some order into a situation where all sorts of firms were trying to establish themselves in the wake of the political industrialized building initiatives. The market was flooded with possible (and impossible) systems for housing, most of which in the event foundered for various reasons.

Cleeve Barr, once at Herefordshire County Council, later of the London County Council (GLC) Housing Division development group, LCC Deputy Housing Architect, then at the Minstry of Education, and finally Chief Architect to the Ministry of Housing and Local Government, became the National Building Agency's Deputy Chairman and Chief Architect. The appraisal of the many systems for housing on the market was intended to put them on a relative basis thus providing useful information for the many small authorities with little professional expertise but a desire to extract political paydirt from industrialized building in terms of more houses, cheaper and faster.

To do this the Agency established small groups with an architect, a structural engineer, building technologist and a quantity surveyor or estimater. They visited factories, sites, firms, offices and examined the requirements of clients. 120 confidential data files were built up for housing, 60 for educational buildings, 60 for industrial building, 50 for health and 50 for office buildings. They were also involved in the establishment of an interdepartmental committee (SCASH) whose function was to advise on the criteria for appraising housing systems. The files were used as confidential source material to advise clients or their professional advisers concerning the building systems field. This developed another important aspect of the Agency's work which was the provision of managerial, operations and technical advice to many organizations but particularly small local authorities, especially in the north where professional advice was scarce (Barr and Carter 1965). Clearly the Agency was initially very influential in encouraging rationalization and the systems drive, and this fitted well with the then government's attempt to increase the quantity of houses built. As the shortcomings of the approach began to emerge the Agency lost some of its initial purpose and turned to all aspects of rationalization and management in attempts to maintain its viability. In 1970 it was proposed to withdraw government financial support but this was subsequently reversed and it managed to stay independent and to produce further influential work such as the metric house shells — it was their work with the Housing Corporation that enabled the Agency to survive after severe grant cuts of 1972. Nevertheless in summing up the achievements of the National Building Agency's Managing Director, Cleeve Barr, upon his retirement in 1977, David Pearce had this to say:

> 'At first the NBA seemed a marvellous opportunity to lead the national housing effort. . . . Yet what is the outcome of it all? At the MHLG he (Barr) created the research and development group; instigated projects at Stevenage, West Ham and Oldham; he started the 5M and 12M housing systems; he initiated the housing cost yardsticks — all more or less failures. . . . The talent in Cleeve Barr and a few of his staff would have been wealth enough to run the construction industry of many lesser countries. Instead it had been concentrated on maintaining an organisation of which few have heard and none comprehend.' (Pearce 1977)

The National Building Agency had first encouraged and validated industrialized systems, such as those used at Ronan Point and Aylesbury, and subsequently survived by assisting in dealing with the faults arising from that very building systems drive. Some of the effects of that encouragement will be examined in the next section.

The Rise and Fall of High Rise

The progressive collapse of one corner of 22 storey Ronan Point, in the London Borough of Newham, on the morning of May 16th 1968, marked the moment of truth for those who had spent some years endeavouring to force the house-building industry to industrialize (**8.3.17**).

Ronan Point is a 210 foot tall block of 110 flats built in the Danish Larsen-Nielsen system, under licence to Taylor Woodrow Anglian, in which a gas explosion caused by a loose gas cooker nut on the 18th floor brought about an explosion which killed four people. Although by this time the tide was already turning against high rise local authority housing, the tragedy of Ronan Point caused a reappraisal of both tall blocks and of industrialized systems. Nevertheless this reappraisal seems to have only been undertaken with reluctance by many authorities, who continued to build in this way, or who were less than quick in their response to the measures indicated by the disaster.

a

8.3.17 Ronan Point. Larsen-Nielsen and Taylor Woodrow. Anglian: London Borough of Newham, Thomas North, Borough Architect. Progressive collapse after gas explosion, 16 May 1968. (a) View of block after collapse. (b) *Next page.* View of top corner of block after collapse and before upper panels had been brought down as in previous photograph

b

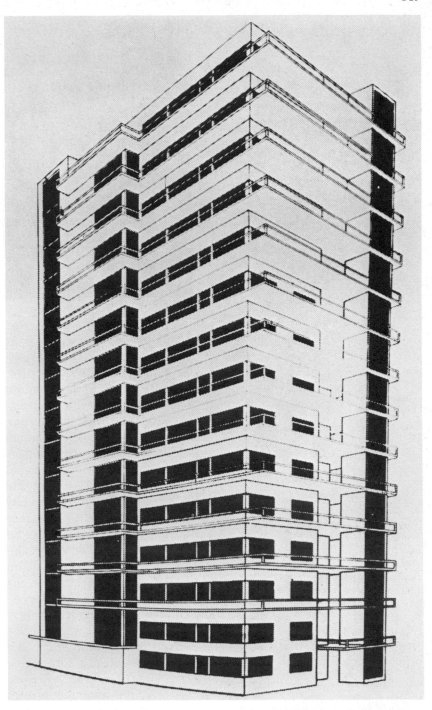

8.3.18 Apartment building project at Bauhaus. Georg Muche, 1924

The drives which led to the use of the Larsen-Nielsen system were the outcome of a long chain of events, some of which have been followed in previous chapters. Indeed Ronan Point had an uncanny visual resemblance to the apartment house designed in 1924 by Georg Muche (**8.3.18**) at the Bauhaus — both used prefabricated concrete cladding slabs, although Muche's scheme had a steel frame.

450

8.3.19 Schindler-Chase house, Hollywood, USA. Rudolph Schindler, 1921-2. Storey height, precast concrete panels with narrow gaps at joint positions occupied by windows. (Photo 1974)

8.3.20 *Opposite*. Housing at Watergraafsmeer, Amsterdam, Holland. D. Greiner, 1922-4. (a) This construction photograph shows a remarkable degree of on site mechanization for the time. The precast slabs being erected are storey height. (b) View of completed scheme in the 'Betondorp' (concrete village). (Photos 1920s?) (c) *Next page*. 1980 view of one of the perimeter streets in Betondorp. (d) Problems with roofs and walls have resulted in a variety of remedial measures over the years, including timber applied cladding (Photo 1980). (e) Part of an elegant terrace with a controlled De Stijl elevation facing the central square (Photo 1980). (f) *p. 454*. The scheme layout has an intimate nature and the rigour of the repetitive units is punctuated at corners with exuberant De Stijl compositions (Photo 1980)

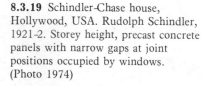

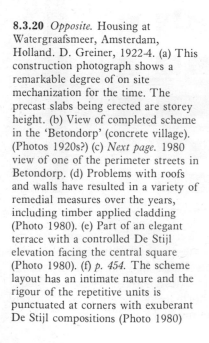

There had been a number of attempts over the years to use storey height precast concrete panels. J.A. Brodie, City Engineer of Liverpool had built a three storey block of flats as early as 1905 using the principle of the 'dove-tailed box' with panels cast off site including apertures for doors and windows (Benson *et al.* 1963).

A precast concrete system employing storey height panels was in use in 1910 by Grosvenor Atterbury who used the system on three-storey houses at Forest Hills Gardens, California, in 1916 (Handlin 1979) whilst Schindler experimented with concrete storey height panels as early as 1921/22 in his Chase-Schindler house (**8.3.19**). However, this was a single storey building. More significant is the municipal housing at Watergraafsmeer 'garden city', Amsterdam designed by D. Greiner and built during 1922-4 as part of an experiment to encourage innovative methods in the face of rapidly increasing brick prices (**8.3.20**). These two storey houses had prefabricated storey height panels which 'consisted of an outer layer of concrete, a layer of insulation, an interior layer of lightweight concrete' (Grinberg 1977). Perret introduced large precast concrete block panels into his otherwise *in situ* structures in 1929 in the apartment block in the Rue Raynouard. These were developed from his experiments at the church of Notre-Dame du Raincy where he used precast concrete bricks, and had the advantage of forming thin concrete walls. Of this Collins says:

'. . . it is important to appreciate that until this date, it is doubtful if any architect had seriously considered combining in situ and precast concrete systematically in the same design, except to make the latter constitute permanent formwork for the former.' (Collins 1959)

Grinberg's work shows that Greiner, in Amsterdam, had not only considered such an approach but had built an example.

a

b

c

d

8.3.20 f

Storey height, room sized precast concrete panels were used on a block of flats built by Glasgow Corporation in 1945. The units were 6 in (152 mm) thick and as much as 10 ft 0 in (3.05 m) wide, of foamed-slag concrete. There was considerable moisture penetration and the approach was not pursued (White 1965).

In the rush to promote and develop housing systems which followed the political pressures of 1961, many continental systems were examined with a view to use in Britain. The Larsen-Nielsen system proved attractive for a number of reasons: it had been used extensively in Denmark (although rarely more than six storeys) and it was simple to build. There is an instructive parallel between this situation and that at Quarry Hill, Leeds, in the late thirties, where a French System, the Mopin, was used, as earlier described. Had all this experience been heeded perhaps Ronan Point and its counterparts would never have been built. Of Quarry Hill Ravetz had this to say:

'building innovation, particularly when the importation of foreign systems is involved, (and) had lasting effects, material and non-material, on the estate and the life within it.' (Ravetz 1974)

The Larsen-Nielsen system started with an *in situ* concrete podium upon which were erected 8 ft 0 in (2.42 m) square, 6 in (152 mm) thick, 4 ton (4.064 tonnes) wall panels, each one containing only nominal steel mesh since the panels were said to be taking only compression loads. Outside these units was a 1 inch (25 mm) layer of polystyrene insulation, with an external skin of 4 in (102 mm) thick precast slab which 'provides a purely decorative finish and an unwanted eccentric load to the base of each wall unit above the dry packs' (Webb 1969a). A crucial detail was that of floors to external walls (**8.3.21**) where the floor slabs bear on the top of the cut back wall panel by means of only 1¼ in (31.8 mm) concrete teeth halfcone shaped and at 9 in (229 mm) centres. Glued to the underside of some of the teeth (at 4 ft 6 in (1.37 m) centres) are ⅛ in (3.2 mm) hardboard packs. Bernard Clark, a structural engineer and a witness for North Thames Gas at the subsequent inquiry said this of the joint:

> 'What holds these panels one above the other? They are literally like tall blocks of bricks 8′ 0″ (2.44 m) high, 6″ (1.52 m) wide and going up to 200′ 0″ (60.96 m). The only thing that stops these buckling is some sort of bond here (at the joint). . . . It has to be a jolly good mortar to start with to stand up to it.' (in Webb 1969a)

Subsequent tests on the joint type at the Building Research Station confirmed that the joint failed at a load of 2 lb/sq in (13.78 kN/m²), thus relying on the entire load of the block to pass through the 2 in (51 mm) 'middle third' of the inner leaf for a height of 210 ft 0 in (64 m). It was suggested by Sir Alfred Pugsley that the load would deflect through the hard pack between one unit and another thus causing rotation at the bottom of each unit and an overturning moment. Clark further pointed out that the expansion due to underfloor heating could push out the walls by as much as ⅛ in (3.2 mm) each heating season, and the nature of the joint suggested they would be unlikely to recover. This latter point, although made at the inquiry was not contained in the final report.

The report pointed out that the level of workmanship and supervision was generally satisfactory and 'it must be emphatically stated that no deficiency in either workmanship or supervision contributed to, or was in any way responsible for this disaster' (Griffiths *et al.* 1968). However, the next clause referred to two cases where 'the workmanship fell below the desired standard', and which were 'related in some respects to the design of the building' at the flank wall joints. The first of these concerned the 1½ in (38 mm) dry mortar packing under the panels at floor level which was difficult to ram home: some of this was loose and not of the specified mix. Second, some of the tie plates (which had been introduced on an earlier job at the request of a London District Surveyor) had been carelessly fitted. It is hard to see how this constitutes 'no deficiency in either workmanship or supervision'.

Behind all this is the question of structural continuity — that the forces shall be transferred through a structure by elements designed to take it. According to a report in Building Design (Brutton 1973), 700 high rise blocks using large panel prefabricates of a variety of systems relied totally on friction and deadload for this continuity. Thus any force other than that of those minimally designed for could cause a collapse.

The Inquiry which followed the explosion at Ronan Point was a curious affair: considerable time was devoted to the cause of the explosion, whereas much other evidence of a more serious nature to the building industry was not sought. According to Webb (1969a):

'it omitted to ask the originator of the system to give evidence, or to say why it rarely went above six storeys in Denmark. It omitted to say why in America this type of construction had been banned above six storeys.... It omitted to cross-examine the engineer who designed Ronan Point.'

The Inquiry report, however, did show surprise that, although the block had been designed for wind loading in accordance with the relevant Code of Practice, this only corresponded to a wind speed of

8.3.21 (a) Larsen-Nielsen/Taylor Woodrow Anglian: joint between vertical panels and floor slabs as used at Ronan Point. Redrawn from Webb (1969b) (b) *Opposite.* Larsen-Nielsen corner joint. Dodson Point, a block identical to Ronan Point and adjacent to it. London Borough of Newham (1968). Note variable finish of panels, corners damaged in erection, joint baffle failure (Photo 1977)

a

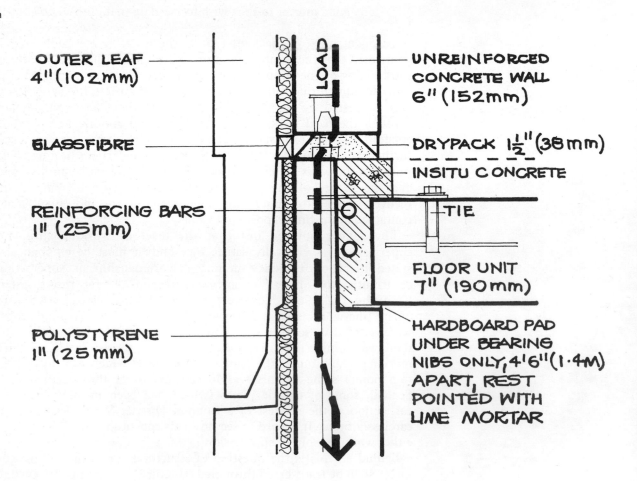

OUTER LEAF 4" (102mm)

LOAD

UNREINFORCED CONCRETE WALL 6" (152mm)

GLASSFIBRE

DRYPACK 1½" (38mm)

INSITU CONCRETE

REINFORCING BARS 1" (25mm)

TIE

FLOOR UNIT 7" (190mm)

POLYSTYRENE 1" (25mm)

HARDBOARD PAD UNDER BEARING NIBS ONLY, 4'6" (1·4M) APART, REST POINTED WITH LIME MORTAR

b

60-70 mph with 'little or no margin of strength if a speed of 105 mph is reached' (Griffiths *et al.* 1968). Clearly the implications, both for the government and for local authorities, were considerable, since many such blocks had been erected, were in the process of construction, or were planned. Ministry advice to councils was being prepared with all speed, but the preparation of such advice posed serious problems. If all blocks similarly constructed were dangerous and needed strengthening, who would pay? The local authorities would resist having to provide finance since most of them had been strongly

8.3.22 Ronan Point (on left), Newham, London. After restoration of the corner which suffered progressive collapse. At the time of this photograph, 1977, several flats in the restored left-hand side still remained unoccupied. Surprisingly, no attempt has been made to conceal the 'join' between the repaired corner and the undamaged section as can be seen at the top fascia

influenced, if not pressurized, to use industrialized systems of this sort by government itself, and this had been backed up by advice and by the additional subsidy for tall blocks. After some pressure, and in an all night sitting in the House of Commons, a clause was inserted into the new instructions to local authorities which stated that if gas was removed from such blocks there was no need to strengthen them. After the subsequent Inquiry report recommended stronger continuous jointing in new blocks using such systems, Taylor Woodrow Anglian used steel hoops cast into the wall and floor slabs with horizontal bars in all concrete cast *in situ*. In existing structures angle metal sections were bolted on to make good the lack of continuity at a cost of £400 per dwelling. Bernard Clark, who appeared at the Inquiry, believed this to be unsatisfactory, proposing a system of post-stressing — this, however, would have cost some £5000 per dwelling.

The portion of Ronan Point that collapsed was replaced in reinforced concrete and matching precast cladding panels (**8.3.22**). All the Larsen-Nielson Anglian blocks in Newham were strengthened to withstand a blast of 2.5 lb/sq in (17.23 kN/m^2). The Greater London Council was reported to have to spend £3,000,000 on strengthening the 26 Larsen-Nielsen blocks within their jurisdiction, using a system of steel angles. However, as Webb points out:

'There is no fire protection whatsoever provided to these angles, unless timber skirtings and ceiling coves constitute this.' (Webb 1969b)

Witnesses at the Inquiry such as Clark, Fairweather and Bowen all seriously questioned the safety of such blocks and condemned the system — all expert witnesses 'were under great pressure at the inquiry to agree to the re-occupancy' (Webb 1969b). Bowen was put in the position of clearly stating his opposition:

'You are really asking me, as I see it, as an engineer to say I am prepared to recommend that a considerable number of people should be put into a building, which in my view at this moment has a lower margin of safety than I as an engineer am accustomed to aim for in exactly similar usage buildings. Therefore I cannot possibly advocate occupancy.' (Webb 1969b)

Desmond Wright, the Queens Council acting for Newham, made a long statement during which he said that the Architect for Newham, Taylor Woodrow Anglian, and the consultants all considered that the 'block is perfectly safe and well designed' and he hoped that the press would give publicity to the views of the council when reporting Mr. Bowen (Webb 1969b). The purpose of all this was to permit construction to continue on similar blocks in the borough (and elsewhere) and of course for occupation to proceed in existing blocks and those to be constructed. Soon after Wright's statement permission was given for Newham to continue work on the partially completed block at Clever Road and for another block on the Mortlake Road site: these were still empty in August 1969; however, the other three blocks on the Ronan Point Clever Road site were, of course, still occupied.

In view of the limited amount of published material on the true cost of system building the exchange that occurred at the Inquiry between Thomas North, Newham Borough Architect and Mr. May, QC for North Thames Gas, is of interest. North pointed out that his department had priced Ronan Point against similar jobs of similar height in traditional construction:

'We found — I will not be popular with Taylor Woodrow in saying this — far from the claim that this is cheaper, this is not so.' (Webb 1969b)

Asked whether it was the speed of the construction that had attracted him to the system built block:

'That again, I must say is a fallacy.... The overall time taken from our first design and discussions with Taylor Woodrow until completion of the first block was about the same as the traditional built block.' (Webb 1969b)

The Ronan Point incident demonstrated that for housing the system built block was neither cheaper nor quicker and furthermore, may have deficiencies of the sort that led to the collapse. However, the manner in which many such systems had come about, with heavy government pressure and involvement, meant that the inducements to use such means were considerable: further than this, it meant that when evidence of failure arose, as at Ronan Point, the implications of effecting the necessary changes were broad. Initially neither central nor local government wished to commit themselves to the costs involved, and the measures that were utilized appear to have been governed by this factor. Important issues are raised when the level and mode of state involvement in housing and the means used reaches this position: public money can be used to encourage certain lines of development which may, curiously, receive less scrutiny than a privately financed building. It has been common practice, for example, for many local authorities to make only courtesy applications for Planning and Building Byelaw (now Regulation) approval for their own projects, whereas private applicants have to abide by all the formal procedures, observing bureaucratic protocol to the letter and often beyond.

This is particularly well illustrated in the case of Ronan Point which was certified by the Borough Architect, in the application to the Minister for subsidy, on 15 December 1965, with acceptance of the Taylor Woodrow Anglian tender authorized on 22 December 1965. However, the application for building byelaw approval was only made on 11 December 1965 and was not given until 6 January 1966 — a strange state of affairs when a tender acceptance could be authorized before byelaw approval had been given. Even stranger than this, though, was the fact that the structural engineer in the borough responsible for checking the calculations pointed out that this could not be done without further details. These additional details, consisting of some 100 sheets of calculations, were checked by him in three days — after the byelaw approval had been given. The calculations dealt with overall stability, and although five typical joint details were submitted they were without calculations and were therefore not checked. All this led the Inquiry report to observe:

> 'This is all too casual an approach, appearing to treat compliance with the byelaws as a tiresome formality rather than as an important safeguard.'

This drew attention to the practice which had become widespread in the local authority world of buildings, and was encouraged by situations where all the professionals involved were employed by one authority. It is easy to see that all sorts of pressures can be brought to bear when large and politically important building programmes have to be met. In addition lines of professional responsibility are often blurred by the bureaucratic hierarchy in which an architect or engineer finds himself. In such situations the costs of development work, construction, remedial work and the final sorting out of problems on projects become obscured, with the time consumed becoming disproportionate to the housing or other provision. This process, and its wastefulness is described by Malpass (1975) in his excellent, detailed study of local authority housing, 'Professionalism and the role of architects in local authority housing'.

An indication of the complexity of the issues involved can be seen in the period of the litigation subsequent to the Ronan Point collapse. The initial stage of this, an action for liability and negligence brought by the London Borough of Newham against Taylor Woodrow Anglian, took eleven years before a judgement was made in one of the most expensive High Court actions which, at an estimated £500,000 exceeded the contract price for Ronan Point by 10% (Webb 1980). In a judgement given on 21 December 1979, Mr. Justice O'Connor ruled that Taylor Woodrow Anglian were liable to pay for the repair and strengthening of Ronan Point and the eight similar blocks since they had failed to design and erect a building in which gas could be used with safety. The company was not found guilty of negligence. Mr. Justice O'Connor pointed out that:

> 'Unlike the Inquiry, I have come to the conclusion that recommendations in CPIII and CPII4 (and indeed CPII6) were not complied with, and contrary to the contention of TWA, I hold that they were relevant to the design of Ronan Point... and (having) reached a finding that the H2 joint was inadequate,... I hold that there was a failure to comply with byelaw 21.' (Webb 1980)

Mr. Justice O'Connor said that since Taylor Woodrow Anglian had guaranteed the building, and that it was an 'express term of the contract that the block of flats should be piped for gas', the failure of the block upon a gas explosion showed that it was one in which gas could not be safely used. Taylor Woodrow Anglian were therefore in 'breach of this warranty' (Webb 1980). In view of this breach of contract in failing to comply with the relevant byelaws Newham could therefore charge the company for the cost of repairing the blocks. At time of writing it is unclear what these costs will amount to, and in any case there is likely to be an appeal which could prolong the settlement for a further lengthy period. The Inquiry of 1968 had ruled that, broadly, Ronan Point complied with byelaws and directly applicable codes whereas Mr. Justice O'Connor has ruled, in 1979, that the block did not so comply. As Webb (1980) astutely points out this calls into question the 1970 regulations on progressive collapse which were brought in following the initial Inquiry, on the grounds that if the design failed and conformed to the regulations, it is the latter that must be strengthened. If it is now held that the regulations were adequate, and the design failed to meet them, then it may be that many subsequent measures to avoid a future similar disaster may have been directed at the wrong target.

Strangely enough, immediately prior to the Ronan Point collapse, a searching inquiry into the way in which the public building programme in Britain was handled was being conducted by the

Estimates Committee. Amongst other things the committee examined relations between client departments, the Ministry of Public Building and Works and the Treasury who provided the finance. They looked at methods of contracting, controlling the use of public money, and they questioned the way in which research was co-ordinated and the use of industrialized methods. One of the rather startling items elicited was that 'there was no point at which the final cost of projects, together with the reasons for variations over the original estimate, is presented to Parliament' (Estimates Committee 1968) and this led to the recommendation that such a report be instituted. Clearly the committee were very concerned about the modes of accounting employed up until this time since they made a number of recommendations, one of which certainly would be thought basic to any organization:

> 'the provision of information which would enable a regular comparison to be made of all elements of cost against a budget plan with an appropriate analysis of variations.' (Estimates Committee 1968)

In the questions asked by the subcommittee, the importance that the Ministry attached to industrialization and standardization continually recurs, and the Estimates Committee itself looked with favour upon their efforts:

> 'The Ministry feel that if the public sector can achieve their target of standardization by 1972 "it will not be long before the private sector swings into line". Your Committee consider that this will be immensely beneficial to the whole industry, and they wish the Ministry success in their efforts.' (Estimates Committee 1968)

Clearly then, in spite of the questionable accounting procedures, the Ministry of Public Building and Works saw itself as masterminding the final stages in that drive to industrialization dreamed of in the early years of the modern movement, and it saw the change to metric now in progress as 'a heaven-sent opportunity to cut down the proliferation of components of even the systems which were more systematic than normal building' (Redpath, in Estimates Committee 1968). However John Redpath, then Director General of Research and Development at the Ministry, was questioned closely about the way this was being handled:

> 'when you were talking about systems of industrialised building it did not seem to me from your reply that, in fact, anyone in the industry is accepting responsibility for research into this very important part of the future. . . .' (Estimates Committee 1968)

It was explained that the public sector was the only area where there was such co-ordination and that this part of the building industry's

MINISTRY PUBLICATIONS ON DIMENSIONAL CO-ORDINATION

1963 (February)	Dimensional Co-ordination for Industrialised Building	(DC1)	(Imperial)	MPBW
1963 (September)	Dimensional Co-ordination for Industrialised Building: Preferred Dimensions for Housing (DC2)		(Imperial)	MPBW
1963	Space in the Home (Imperial Edition)			MHLG
1963	Dimensions and Components for Housing: with special reference to industrialised building (Imperial) Design Bulletin 8			MHLG
1964 (July)	Dimensional Co-ordination for Industrialised Building: Preferred Dimensions for Educational, Health and Crown Office Buildings (DC3)		(Imperial)	MPBW
1964	Controlling Dimensions for Educational Building		(Imperial)	DES
1964	Dimensional Co-ordination for Crown Office Buildings		(Imperial)	MPBW
1964 (February)	Dimensional Co-ordination and Industrialised Building: Hospital Design. Note 1		(Imperial)	MOH
1967 (January)	Dimensional Co-ordination for Building: Recommended vertical dimensions for Educational Health, Housing, Office and single storey general purpose Industrial Buildings (DC4)		(Metric)	MPBW
1967 (May)	Dimensional Co-ordination for Building: Recommended horizontal dimensions for Educational, Health, Housing, Office and Single storey general purpose Industrial Buildings (DC5)		(Metric)	MPBW
1967 (August)	Dimensional Co-ordination for Building: Guidance on the application of recommended vertical and horizontal dimensions for Educational, Health, Housing, Office and single storey general purposes Industrial Buildings (DC6)		(Metric)	MPBW
1967	Recommended Intermediate vertical controlling dimensions for educational, health, housing and office buildings and guidance on their application			
1968	Space in the Home (Metric Edition)			MHLG
1968	Co-ordination of Components in Housing: Metric Dimensional Framework. Design Bulletin 16		(Metric)	MHLG

8.3.23 Relevant Ministry publications on dimensional co-ordination immediately before and after the metrication programme in the building industry commenced. Although the start of a changeover to metric, SI, units was planned to start in 1965 (and planned to be completed by 1975), a series of guidance notes were published all using imperial up until 1964

work amounted to 50% of total construction. Hence the hopes that initiative from central government would bring the private sector along with it. The concern of the Estimates Committee at the apparent lack of co-ordination in research behind the drive to industrialize building in the first half of the sixties is supported by an examination of some of the publications (**8.3.23**).

In 1963 *DCI, Dimensional Co-ordination for Industrialised Building* (MPBW 1963a) had been published. A single A3 sheet, it set out in imperial dimensions the idea of Recommended Preferred Increments: these were 1 in, 4 in and 1 ft 0 in, with 3 in referred to as suitable for 'specific purposes'. How the introduction of *DCI* caused the School Building Systems to move from their already 'metric modular' 3 ft 4 in (almost 1 metre) grids to the one foot recommended in this Ministry of Public Building and Works document is described elsewhere. *DCI* was the first of a virtual shower of documents, applying the concepts to all building types. *DC* documents 1 to 3 were in imperial terms and those commencing with *DC4* (MPBW 1967a) were in metric. DC4 itself attempted to summarize in metric terms the three previous documents — unfortunately many architects had already acted upon the imperial documents and had just completed costly changes to procedures and systems. This document deleted reference to 'industrialised' in its title and was clearly directed at all building.

The Parker Morris Report, *Homes for Today and Tomorrow* had been published in 1961. This is normally seen as an attempt to raise standards across the country, but many local authorities saw it for what it ultimately became — a set of minimum standards which, in many cases, caused them to reduce space standards thought by central government as too lavish. *Space in the Home* quickly followed this in 1963 (imperial version (MHLG 1963b), metric version (MHLG 1968a)). This attempted to set down activities in the home, to set out 'suggested space and furniture requirements' and to show specimen solutions based on the Parker Morris Report. In the same year (1963a) the Ministry of Housing and Local Government produced *Dimensions and Components for Housing* (Design Bulletin No. 8) whose chief claim to the confusion was the recommendation that the one foot 'Preferred Increments' on plan should be taken to the CENTRE LINES of structural walls. Since they suggested elsewhere that walls could be in 1 inch increments of thickness, it is clear that face to face component dimensions could often be non-modular. The proposal did, however, favour framed construction, which is not surprising since the majority of systems experience to this date had been with school building systems where this was the norm. However, it was necessary to wait until 1968 for this to be resolved in Bulletin No. 16 *Co-ordination of Components in Housing: Metric Dimensional Framework* where 'the conclusion is that in the long run the use of the face grid for all forms of construction will be most advantageous' (MHLG 1968b). By this time a key British Standard had been published, BS 4011 *Recommendations for the Co-ordination of Dimensions in Building* (1966) which had stated:

'The first selection of basic areas for the co-ordinating dimensions of components and assemblies should be, in descending order of preference, as follows (where 'n' is any natural number including unity)

First	n × 300 mm	
Second	n × 100 mm	
Third	n ×	50 mm up to 300 mm
Fourth	n ×	25 mm up to 300 mm.' (BSI 1966)

It was recommended that the third and fourth preferences should not be employed for sizes over 300 mm without good reason, and that the fourth preference was put forward provisionally. Bulletin No. 16 carried with it the weight of statute since a Ministry Circular of 1968 required that all housing schemes that were designed in metric were to follow the recommendations. This was a powerful influence on local authorities to employ such methods. *DC6,* published in 1967, has an interesting footnote for the sociologists of housing:

'In the case of housing, only one floor to ceiling height is needed to suit user's requirements.' (MPBW 1967c)

Such recommendations became part of the evaluative method used when applying the housing cost yardstick. In the meantime, other building types were also being encouraged along similar paths — although the Department of Education and Science efforts on school building had set the pattern for all. For example, 1964 saw the publication of their *Controlling Dimensions for Educational Building,* whilst the Ministry of Public Building and Works produced *Dimensional Co-ordination for Crown Office Buildings* (MPBW 1964b) and the Ministry of Health produced a Design Note entitled *Dimensional Co-ordination and Industrialised Building* (1964). Curiously enough the latter three publications, all produced in the year before the commencement of the metric programme in 1965, were all in imperial terms.

Just a few weeks before the Ronan Point collapse then, we have the Estimates Committee conducting an enquiry into the philosophy and procedures which had partially given rise to systems of the sort being employed, and giving some support to the approach. There was already concern about research and the way in which it was co-ordinated and there was further concern about the financing of certain aspects of the programme — which included the efforts to industrialize building processes. The changeover to metric, commenced in 1965, was being used by the Ministry to encourage this approach, and many of the senior staff there were imbued with this philosophy: John Redpath himself had worked on the early experiments at Hertfordshire in 1948. So what emerged at the Estimates Committee was that same important philosophy writ large. Considerable attention had been given to industrialized methods ever since Donald Gibson had reorganized the Directorate of Public Building and Works in

a

b

8.3.24 Housing for armed services, 12M Jespersen system: Gosport, England. Architects, Farmer and Dark, *c.*1969. (a) First stage: 2 and 4 storey, largely timber clad. (b) Detail of cladding junction. (c) Second stage: 3, 5, and 8 storey blocks around south facing court. Precast concrete cladding. (d) Detail of court. (Photos 1971)

1958. By the time of Ronan Point many attempts had been made to pursue this policy: direct action such as the development of the 5M system and indirect collaboration such as the work on 12M Jespersen and encouraging housing consortia on the lines of the schools consortia, such as The Midlands Housing Consortium.

Jespersen is a Danish system using precast concrete cross wall floor and roof panels. Its planning module was 1 ft 0 in (305 mm) in width and 4 ft 0 in (1.22m) in depth with maximum spans of 18 ft 0 in (5.49 m). The fact that any sort of cladding could be used and a certain flexibility in plan arrangements caused it to be considered as a relatively open system. The schemes where the system was used however, all developed standardized cladding panels and deviated little from the basic philosophy applied to the concrete sections. The Ministry Research and Development group spent considerable time adapting Jespersen to the British market with John Laing — dubbing it 12M Jespersen (**8.3.24**). The design manual, prepared by the group, is an object lesson of clarity and explains in simple terms how best to employ the system. It is also direct about the cost implications of its use, and is a powerful example of the effects of such systems on built form:

'Schemes consisting wholly or mainly of houses should be avoided since "houses constructed in any concrete system, quite apart from 12M Jespersen, are at present (January 1967) more costly than those constructed by traditional methods".' (MHLG 1967)

c

d

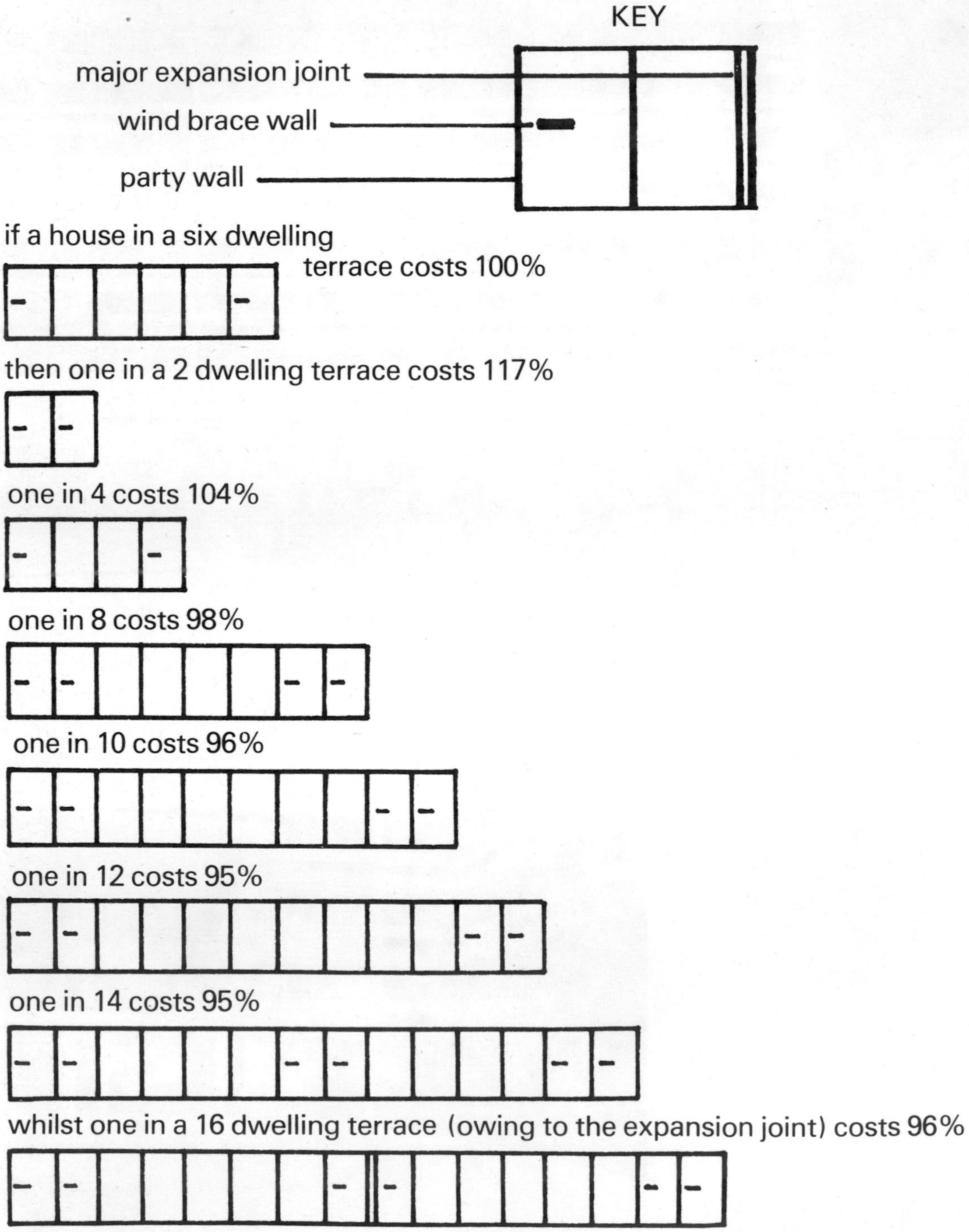

468

KEY

major expansion joint

wind brace wall

party wall

if a house in a six dwelling terrace costs 100%

then one in a 2 dwelling terrace costs 117%

one in 4 costs 104%

one in 8 costs 98%

one in 10 costs 96%

one in 12 costs 95%

one in 14 costs 95%

whilst one in a 16 dwelling terrace (owing to the expansion joint) costs 96%

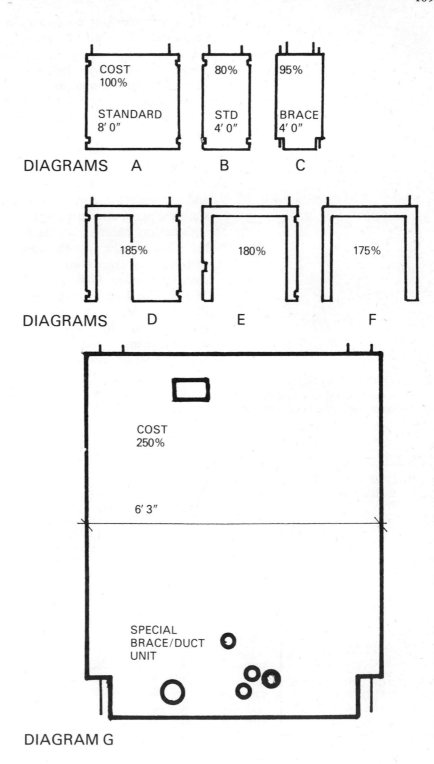

COST
100%

STANDARD
8′ 0″

80%

STD
4′ 0″

95%

BRACE
4′ 0″

DIAGRAMS   A   B   C

185%   180%   175%

DIAGRAMS   D   E   F

COST
250%

6′ 3″

SPECIAL
BRACE/DUCT
UNIT

DIAGRAM G

8.3.25 *Left.* 12M Jespersen: length of blocks. 'Flats naturally tend to be designed into long blocks — their mass making them acceptable from both an aesthetic and social point of view' (Design Guide, MHLG *c.*1966)

8.3.26 *Right.* 12M Jespersen: cost comparisons between standard, standard variant, and special wall components: Elevations of panels. The increase in cost for those panels using more complex formwork is clearly shown. (Design Guide, MHLG *c.*1966)

470

The reasons for this were twofold: 'that concrete floors and roofs are too good, in the context of housing, for the job they are used for' and the high cost of the gable wall, some 150% more than traditional construction. Clearly then it followed that any designer using the system should eliminate as many gable walls as possible, thus resulting in very long blocks (**8.3.25**). A further important limitation was the use of the necessary cranage and this also has severe implications on block layout.

A central issue however, is the question of component variety and Jespersen offers an illustration of how this works in a concrete system. The panels can be seen as of three sorts: standard, standard variants and non-standard. As can be seen (**8.3.26**) the difference between the standard panel at a cost level of 100% and the 'Deck goalpost' panel with far less material, is plus 80%, whereas a standard size panel with holes for services puts the cost up to 250%. The manual points out that in all Jespersen contracts up to May 1966, standard variant panels cost, on average, 90% more than standards, and special panels cost 380%. The length of floor components can also be used as an illustration of the effects of cost, with a sharp increase as the span decreases: another result of the standardization of sizes and of manufacturing technique. Here we can see the detrimental effects of such standardization very clearly, since the use of smaller spans can offer more flexibility, more choices. Such choices would be perfectly feasible with timber but become totally uneconomic in mass produced concrete (**8.3.27**). A further complication over floor spans was that the

8.3.27 12M Jespersen: cost of spanning units related to length of span (Design Guide, MHLG, c.1966)

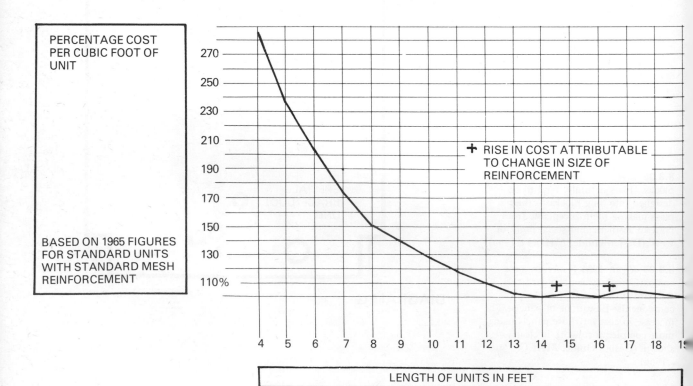

PERCENTAGE COST PER CUBIC FOOT OF UNIT

BASED ON 1965 FIGURES FOR STANDARD UNITS WITH STANDARD MESH REINFORCEMENT

+ RISE IN COST ATTRIBUTABLE TO CHANGE IN SIZE OF REINFORCEMENT

LENGTH OF UNITS IN FEET

original Danish version only spanned 16 ft 0 in (4.9 m) whereas the British requirements for a possible two bedroom frontage caused the Ministry to stretch this maximum span to 18 ft 0 in (5.49 m) with resultant possibilities of deflection up of 0.55 in (15 mm) in the long term, with possible effects on partitions and external cladding.

The combined result of all these factors was that building houses was difficult, and the emphasis was on high and long continuous blocks — as at the Aylesbury Estate in the London Borough of Southwark (**8.3.28**). However, by 1971 Laing's Jespersen factories had closed in Scotland and Lancashire; and the later stages at Aylesbury, in 1977, had modified windows and other details — most significantly to include pitched roofs (**8.3.29**). This estate contains the longest system built block in Europe, and according to the *Architects' Journal,* is the largest single scheme to use the 12M Jespersen system. The total scheme covers a site of 64 acres (25.9 hectares) and will house some 8000 people. Construction commenced towards the end of 1967 and, in addition to the use of this system, the scheme employed an extensive deck system along which it was planned to run specially designed electric trolleys to carry out the servicing (**8.3.30**). At Aylesbury we see two forces at work: the concentration on one large site of wholly council tenants, and the use of a heavy concrete building system. Both of these forces have a logic which when working together give highly questionable results.

The economics of cranage (**8.3.31**) and the production line added to the single ownership and management, creates an environment of unremitting rigour and one which, according to some commentators has created considerable social problems. Even accepting the policy of housing council tenants on large parcels of land, which many councils favour, the use of more traditional techniques permits a variety to be introduced at least into the visible environment: if different architects were employed this could be increased and solutions varied. Such a use of system building as at Aylesbury comes close to the 'inevitable logic' propounded by Gropius (1965, 1935) in the thirties and is probably only mirrored by large scale system built housing in the Soviet Union and in Eastern Europe. According to Sebestyén in a book published in English in Budapest in 1965 and distributed in Britain:

'The housing problem, in general, can be completely solved in the social and economic environment of the Socialist countries. The Soviet Union has developed into the most advanced house-building country in the world by virtue of its building methods, based on the use of large units.' (Sebestyén 1965)

The book, *Large Panel Buildings,* is clearly aimed to relate to the interest in Britain and is a technical primer on methods, upon which no critique is offered. The philosophy inherent in industrialization is accepted and put forward icily and without question:

'Buildings constructed with large precast elements should be designed with the simplest possible elevation. This will lead to

a

b

8.3.28 Aylesbury Estate, Southwark, London. 12M Jespersen system. London Borough of Southwark architects department. Commenced building *c.*1965, estate still under construction 1977. (a) Garden side of the long deck access block, shown on the left. (b) Detail above walkway, showing difficulty in achieving precision between industrialized components intended to line up on site. (c) *Opposite.* Detail at gable end with horizontal 'flash gap' joints, but with verticals sealed from the outside. (Photos 1977)

simple structural and functional elevations, and the amount of plumbing and sheet metal work required will be reduced.' (Sebestyén 1965)

It is also pointed out that the inclusion of shops, nursery schools and kindergartens in the blocks should be avoided since it impeded the 'unity of concept' involved with large panel buildings — another reduction. A similar volume from Poland, *Building with large*

8.3.28 c

8.3.29 Aylesbury Estate. 12M Jespersen system. View of blocks completed 1977 showing modifications to the system, pitched roofs, upgraded windows and trim. (Photo 1977)

Prefabricates (Lewicki 1966) covers the same field and, bearing in mind what happened at Ronan Point in 1968 and the reports on the Bison System in 1978 (Pearman 1978) makes interesting comment on jointing problems in such buildings:

> 'The designer should aim at spreading the load from an upper component over the whole area of the joint with the lower prefabricate. It is not advisable to provide joints in which the load is transferred at two or more levels. . . .' (Lewicki 1966)

The author is careful to emphasize the attention to detail necessary with such construction pointing out that with load-bearing panels

a

8.3.30 Aylesbury Estate. 12M
Jespersen system. (a) Walkway on long
central block. (b) *Opposite*. Notice on
walkway indicating the difficulties of
defining which vehicles shall use the
'street in the air'. (Photos 1977)

'careless grouting of the joint may result in a loss of 80% of the load
carrying capacity of the wall' (Lewicki 1966). But it is the rate of
production claimed that is central to this work in Eastern Europe and
the Soviet Union. The rigorous standardization and large output have,
says Diamant (1965), 'enabled far more automated methods to be used
than would pay in the West.... Five floor blocks of flats can be
erected in about 30 days by the Lugatenko method'.

The problems raised by the jointing of precast concrete panels had
received some attention in Britain, although weathering seems to have
been the major preoccupation during the 1960s. A paper entitled
'Methods of Construction — Concrete', presented by Fraser (1967) to
the RIBA/NBA conference, *Industrialised Housing and the Architect* in
January 1967, had little to say about structural matters, being mostly
concerned with variety reduction, the weathering of joints and
production problems. In 1960 the Building Research Station had set
up an experimental rig to monitor the performance of joints in natural
exposure conditions. This, together with later Scandinavian and
British experience, led to their *Digest 85*: 'Joints between concrete wall
panels: open drained joints' (BRS 1967) which set down four criteria
important to successful performance of such joints (**8.3.32**):

b

1. To prevent air filtration an air barrier is necessary at the back of horizontal and vertical joints.
2. Almost all water entering a vertical joint drains within a 2 inch (50 mm) zone from the front of the joint.
3. The inner edge of the drainage zone in a vertical joint should have a baffle to stop wind driven rain.
4. The protection of horizontal joints should be protected by a 2 inch (50 mm) upstand on sheltered sites and 4 inches (100 mm) on exposed sites with efficient protection at the junction with vertical joints.

Flirtations with a variety of new materials and techniques is at the core of the industrialized system approach: and the Greater London Council has been in the forefront in this respect. SF1, used on the Elgin Estate, Paddington, was an attempt to take advantage of the speed and precision of steel frame high rise construction together with storey height glass reinforced plastic (GRP) panels of a high strength to weight ratio, easily transportable and maintenance free. Drawing on the experience of using GRP in GLC ambulances the panel was developed with Indulex Overseas Ltd and used (**8.3.33**) on the two 21 storey blocks. The panels, to facilitate erection, were bolted six at a time onto a preassembly of steel channels and lifted into place. The GRP panels embody an aesthetic long sought by some architects, and can be seen in Le Corbusier's Loucheur house (see **4.4.9**) of 1929: pressure moulded in one piece they are identical and each has one bus type window with round corners jointed with gaskets. Since the necessary technology turned out not to be available in Britain the units were imported from the United States and jointed by means of polysulphide seals and gaskets. In some contradiction of the prefabrication argument the panels had to be backfilled with lightweight concrete to provide the appropriate fire resistance and insulation levels. Construction commenced towards the end of 1966: by mid-1975 it was reported:

> 'Water has been entering their flats for years now, and previous attempts at waterproofing have not been wholly successful'. (Scott 1976)

Soon after erection the panels began to fracture and crack and the replacement of the leaking windows was to cost £400,000.

8.3.31 12M Jespersen system. Effect of crane types on construction and construction sequence. (a) The rationale of using a crane with precast components. (b) Diagrams showing crane types and the effects of the jib length on construction and layout. (c) *Opposite.* The result of crane economics on block spacing and length of block

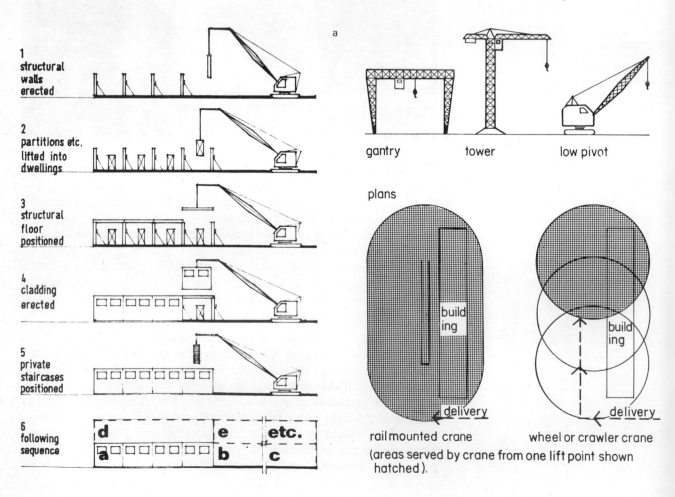

1 structural walls erected

2 partitions etc. lifted into dwellings

3 structural floor positioned

4 cladding erected

5 private staircases positioned

6 following sequence

a

gantry   tower   low pivot

plans

rail mounted crane

wheel or crawler crane

(areas served by crane from one lift point shown hatched).

8.3.31 c

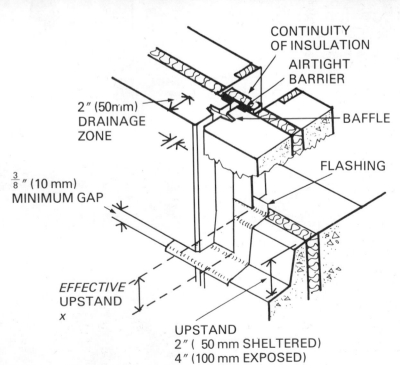

CONTINUITY
OF INSULATION

AIRTIGHT
BARRIER

2" (50mm)
DRAINAGE
ZONE

BAFFLE

FLASHING

$\frac{3}{8}$" (10 mm)
MINIMUM GAP

EFFECTIVE
UPSTAND
x

UPSTAND
2" (50 mm SHELTERED)
4" (100 mm EXPOSED)

8.3.32 Dry joint between concrete panels: criteria as set out in Digest 85 by Building Research Station (now BRE), August 1967

478

8.3.33 Elgin Estate, Paddington, London. Greater London Council, 1968. (a) View of point block and adjacent low rise. (b) *Opposite.* Corner detail showing GRP cladding panels and 'bus' type windows

a

b

More recently the Greater London Council embarked on another adventure into the industrialized world — this time to pursue another cherished architectural ideal, that of adapting containers for use in housing. The social argument was summoned up this time by Gladys Dimson, Chairman of the GLC Housing Development Committee in April 1974:

> 'As we need more and more homes desperately and today, not tomorrow, we are very interested in this pilot scheme of such unusual construction.' (Murray 1977)

What Building Design called 'an amazing saga of technical adventure, financial manoeuvres, contractual nit-picking leading to bankruptcy and police investigations' (Murray 1977), followed this claim, a further dramatic example to place alongside the many with which industrialized housing is littered. The panacea which proved so attractive this time, developed by Parking Systems Ltd, was based on International Container sizes (for production reasons no doubt) and was of steel frame construction with a cladding of plastic faced corrugated steel sheeting. The ten old persons' flats in Hackney took some three years on site and was 55% over the initial cost. The report on the enormous difficulties which beset the system supplier and the GLC caused Building Design (Murray 1977) to report that it was unlikely that they would go for such a scheme again. However, the history of industrialization does not lead one to be so confident that this will be the case. Only ten years previously there had been the SFI system, and before that the postwar prefabs where the production run was halted because of escalating costs and other problems. The total industrialized housing package is one that seems irresistible to both politicians and architects alike.

As is evidenced by the questions of the 1968 Estimates Committee there was clearly unease at the preceding events, which had started with a worthwhile attempt to introduce the rather simple basic idea of modular co-ordination into the practice of the building industry and had arrived at the point where a complex set of dimensional rules had become embodied in law. The Ministry of Public Building and Works saw, as Redpath pointed out, the metric change as 'a heaven sent opportunity' to press industrialization — and industrialization in the context of housing at that time meant largely the heavy systems. Although Reema, Bison and Wates were British, the central government pressure encouraged other contractors to bring in systems quickly from elsewhere. The 12M version of the Danish Jespersen was imported by Laings, at Thamesmead Cubitts were to use Balency, at Aldershot there was Tracoba and at the ill-fated Clever Road site which included Ronan Point, Taylor Woodrow Anglian were using Larsen-Nielsen. For such heavy systems to be cost effective, they needed large sites with a continuous throughput of housing units and it was soon clear in the early sixties that the many proffered systems, almost 300 at the prototype stage by 1963, could not all survive. It is odd that the business side of the construction industry put so much faith in government intention without taking into account both the normal stop-go policies of financing building and the fact that only very few heavy investment systems could operate the available market.

The ministries involved with housing had attempted to emulate the work carried out on schools, that is to press the market into a number of building systems. The Department of Education and Science had first encouraged the commercial systems and then killed them by the introduction of the client sponsored system such as CLASP, SCOLA, SEAC and others. In the best British tradition a form of informal centralized control of school building developed, together with many unwanted side effects such as poor ability of most of these systems to fit into landscape and local communities, maintenance problems and poor environmental control. Whilst housing tried to emulate the client sponsored system idea by means of housing consortia such as the Yorkshire Development Group, or the London Housing Consortia, political presures were such that development such as had occurred with regard to schools was out of the question. Even the Ministry's own 5M system was seen by industry as a first attempt to create a centralized government sponsored system.

In the event the enthusiastic support given to government efforts by the large contractors in making such housing systems commercially available is not so remarkable. It is a powerful combination which often worked against the user and certainly worked against the average builder in Britain who tended to be small to medium sized. During this period many an architect found himself 'working for the contractor's computer', and able to give less and less time to basic design. So not only was the encouraging of the building of large council estates socially unwise, but it increasingly eliminated the small local builder from such work. One of the results of this course of action was that more and more resources were needed to continue the public building programme, since investments had been made in plant and procedures which needed sustaining. It should not be forgotten that this was indeed one of the purposes behind the systems drive in Britain. The argument ran like this: the Treasury controls the financing of all public building; the Treasury is prone to use the building industry as one of the key capital expenditure regulators; public building is a social service; if, like other industries, building could show that there were clearly defined disadvantages to cutting the investment programme, there would be less possibility of this occurring; therefore use building systems to plan carefully ahead and tie in the manufacturing industry with the process.

By this line of reasoning, never very explicit perhaps, many system builders hoped for two results; one was to ensure an even flow of money into schools, housing and other public building programmes and the other was to have more control of the building industry itself. What happened was neither of these. The Treasury has in recent years more than once dramatically cut expenditure and far from the system designers controlling the industry, often industry (through the systems) gained control of the designers. This might be thought to be an advantage or a disadvantage, depending upon one's standpoint; what is not in doubt is that the policy of the institutionalized centralized system is a far cry from the open industrialized vernacular of building components envisaged so many years earlier.

In the broader perspective, these attempts to rationalize housing and other building programmes encouraged and strengthened centralized control. It often led to stereotyped solutions of such crudity that only a few short years later they are seen as failures, socially and technically. The solutions which developed from these approaches tended to be those used in other parts of Europe and the Soviet Union with their centrally controlled production of vast estates using heavy concrete prefabrication. During this period, Britain had managed to produce the centralized public control of these countries, but have it implemented by private enterprise.

AND THEY WENT FORTH...

'The realisation of our perceptions of the world in the forms of space and time is the only aim of our pictorial and plastic art. In them we do not measure our works with the yardstick of beauty, we do not weigh them with pounds of tenderness and sentiments. The plumb-line in our hand, eyes as precise as a ruler, in a spirit as taut as a compass.... We construct our work as the universe constructs its own, as the engineer constructs his bridges, as the mathematician his formula of the orbits. We know that everything has its own essential image; chair, table, lamp, telephone, book, house, man... they are all entire worlds with their own rhythms, their own orbits.

'That is why we in creating things take away from them the labels of their owners... all accidental and local, leaving only the reality of the constant rhythm of the forces in them.'

Naum Gabo (with Antoine Pevsner)
The Realistic Manifesto 1920; in Chipp 1968

The School Building Consortia

Simultaneous with the rise, and subsequent fall, of the total building systems idea in the world of housing there was a parallel and more successful proliferation of school building systems. However, the seeds of this were laid long before, with much of the work in housing influenced by the success of the early school systems: these ultimately developed to the point where they covered a large proportion of the school building programme in Britain. It must be a matter for concern that what developed was a number of consortia, roughly on the CLASP pattern, each with their own 'system'. A series of such consortia, with their own development groups and organizational structure produced a situation where each closed system was similar to the other, but not quite similar enough to encourage real interchangeability: enough alike to inhibit sufficient regional tailoring; insufficiently alike to posit a real change in approach and open the way to greater choice and an industrialized vernacular. Impetus was given by the Report of the Estimates Committee in 1961, which dealt with school building, and which made as its first recommendation 'greater co-operation between local authorities, particularly in the extension of the Consortia principal' (AJ 1961c).

A government circular of 1964 (MOE 1/64) then encouraged extension of the consortia idea in the sixties. Each of the consortia have their own particular stance: SCOLA, unlike CLASP, was spread out all over England and Wales; Method Building opted for a more basic modular open approach; and MACE (the last of the schools consortia) proffered a set of propositions about flexibility and environmental control partly learned from Ehrenkrantz. This section deals with this expansion of the idea in the school building field, although MACE itself is dealt with in the next chapter, a place more appropriate to its chronological position in this account. The establishment of CLASP as the first client sponsored school building system and its political credibility as demonstrated at the Triennale di Milano of 1960 was an important link in the chain of events which resulted ten years later in almost 50% of school building in Britain being constructed in one or other of the client sponsored systems (**8.4.1**) (DES 1976). Indeed it may be more accurate to describe them as architect sponsored systems since in practice much of the pressure came from architects who could in this way see hopes of establishing their belief in the production line philosophy and regaining control of a difficult situation. It also offered both architects and the Ministry of Education a way out of the dilemma posed by their own enthusiasms with regard to the commercial systems. As described earlier, the Ministry's development group had been particularly interested in encouraging and developing prototype building systems with various manufacturers during the 1950s. This arose quite naturally since many of those involved in the earlier work in Hertfordshire had by this time moved to the newly established Architects and Buildings Branch Development Group at the Ministry of Education and were keen to develop further many of the ideas generated at Hertfordshire.

8.4.1 Client sponsored school building consortia share of the total annual school building programme, 1976 (After DES, 1976)

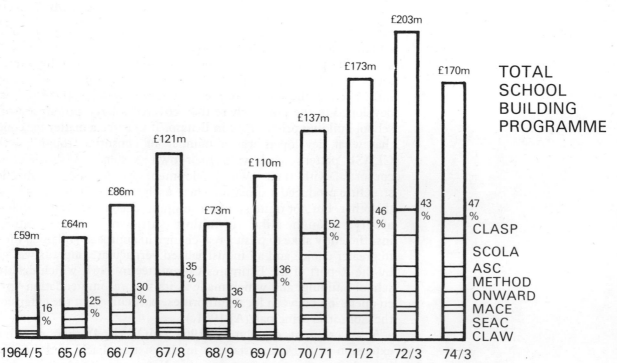

This they did by means of the 'development project', a school designed by them for one of the local education authorities. A number of these projects significantly assisted in the development and acceptance of several commercial systems for school building then on the market. The fact was that a great many school buildings during the 1950s were being built in one or other of the commercially available systems, BAC, Hills, Derwent, Medway and so on. Many examples of these can be seen dotted about England, the manufacturers having been encouraged to invest money and ideas in this direction.

Such commercial systems had been welcomed by many architects for two reasons. First, they often involved only minimal drawing office work in that the structure, fabric and services were all predetailed, and second, by using them the architect was taking advantage of that mass production idea with which he felt so morally associated. Having welcomed this utopia it quickly became apparent that it also brought a loss of control for the architect and, it was argued, for the client and user. Many architects in local authorities were heard to complain that it left them little to do and in some cases this was all too true. Perhaps the possibility was foreseen that such rationalizations might in fact bring about (to paraphrase Marx) 'the withering away of the architect'. Of course it was possible to make out a strong case, if somewhat covert, that the commercial systems were sacrificing user standards in the interests of filthy lucre. However, we shall be able to see with what standards they were replaced, because replaced they were, once the lesson of CLASP had politically struck home.

The Ministry of Education had, for some while, endeavoured to encourage Hertfordshire to become the nucleus for a consortium so that their ideas could proliferate but this had always been resisted on the very practical grounds that their studies had led them to the view that with most building components being produced on a batch production basis there were few further gains to be made by growth and quite clearly there were very many administrative and organizational disadvantages. Only after several other consortia had been established did they ultimately agree to such an arrangement, and then much more on the basis of a collaboration of designers and an exchange of information. This gave rise to the name of the group formed around the Hertfordshire work, the South Eastern Architects Collaboration. But before Hertfordshire could be pressed into the institutional mould a variety of authorities grouped together to form in quick succession SCOLA, CMB, CLAW, ONWARD. This profusion of initials and acronyms gave a clue to the agglomeration of local authorities and the almost total institutionalization of the component building idea.

Second Consortium of Local Authorities

One of the central concerns of CLASP had been to produce a building system which would deal with the problems created by mining subsidence and which would offer more choices in design than the currently available commercial systems. Initially Nottinghamshire was joined by Derbyshire and the City of Coventry who were also facing mining subsidence difficulties, thus forming a compact group of authorities. The first four members of SCOLA, however, were spread more widely over England — Gloucestershire, Hampshire, Shropshire and West Sussex: these were then joined by Dorset and Cheshire County Councils, with the Ministry of Education (now DES). SCOLA members initially considered joining CLASP but the then programme size of CLASP was thought to be the limit if suppliers were not to be overextended and if good communications between authorities were to be maintained. As it happened the CLASP programme size continued to increase anyway.

The first SCOLA publication, *Second Consortium of Local Authorities* (SCOLA 1962?), described the group as 'a going concern' in October 1961 and development work was well under way towards the end of 1960 on the basis of a set of drawings prepared for Shropshire County Council. The aims set out in the brochures were familiar ones:

'1. A variety of individually designed buildings for differing requirements (including non-educational building) together with a kit of components for different situations.
2. A high degree of standardisation to gain the benefits of quantity production.
3. The bringing together of the members' building programmes to facilitate bulk purchase and better value for money.
4. Fast construction times with minimal labour.' (SCOLA 1962?)

As with the work at Hertfordshire and within CLASP, the SCOLA development group straightaway set about designing its own system, a kit of parts which were inter-related. The great myths of mass production were once again about to be applied: dry construction and standardized factory made components, rationalized to reduce construction times. This was to be embodied in the avowed policy that no matter where a project was in the country, component prices must be the same: what one member gained by being close to a supplier would be offset by his distance from others. In view of the spread out nature of SCOLA this was both a practical as well as a geographical necessity if the consortium was to function. There seemed no doubt about the technical solution that was to be adopted. It was, of course, a light steel frame surrounded by a light, dry skin and a flat roof. The Mark 1 system, used in the 1962/63/64 building programmes (which spilled over into 1965 and 1966) employed a fixed end steel frame with a 3 ft 4 in (1.016 m) structural grid. This meant that, in theory, columns could be placed at any multiple of 3 ft 4 in (1.016 m). In practice columns occurred on the external wall at either 6 ft 8 in (2.032 m) or 10 ft 0 in (3.048 m) spacing. This in itself was dictated by the decision to employ an external skin construction of horizontal timber

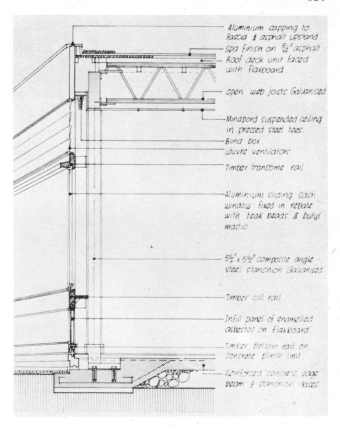

Aluminium capping to
fascia & asphalt upstand
Spa finish on ¾" asphalt
Roof deck unit faced
with flaxboard.

open web joists Galvanised

Minabord suspended ceiling
in pressed steel tees.
Blind box
Louvre ventilators.

Timber transome rail

Aluminium sliding sash
window fixed in rebate
with teak beads & butyl
mastic

5½" x 5½" composite angle
steel stanchion Galvanised

Timber cill rail

Infill panel of enamelled
asbestos on flaxboard

Timber bottom rail on
concrete plinth unit

Reinforced concrete edge
beam & stanchion bases

8.4.2 Second Consortium of Local Authorities (SCOLA), 1961 onwards. Section showing construction of Mark I with timber rails, plywood faced fascia bolted to steel stanchions. School in photograph: Wrockwardine Wood, Shropshire. County Architect, Ralph Crowe. From First SCOLA Report (1962?)

rails which, because of the spanning properties of timber, limited the maximum column spacing to 10 ft 0 in (3.048 m).

The constructional design was most notable, perhaps, for its omission of any services integration apart from the existence of a roof void (**8.4.2**). The use of the 3 ft 4 in (1.016 m) structural grid was seen as an important act of faith in the continuing modular metric argument although its coarseness for planning purposes was felt by some to be inhibiting and a planning grid of 1 ft 8 in (508 mm) was established, if not encouraged. Whilst the steel frame design was changed to a pinjointed design in projects started after 1964 the rest of the system remained superficially the same until the changes of the Mark 2 development programme of 1965/66. Like the Volkswagen perhaps, the Mark 1 and 1A versions (**8.4.3**) adhered to the original stylistic image but this concealed constant changes to the 'works', and a constant flow of revised drawings and instructions.

The floors consisted of 3 ft 4 in (1.016 m) × 1 ft 8 in (508 mm) precast concrete slabs which sat on cork strips on the steel beams. Floor finishes were laid on a ⅜ inch (9.5 mm) levelling screed which in practice, and just as it had in Hertfordshire, caused considerable problems since the cumulative tolerances of frame and slabs made such precision a rarely attainable ideal. The roof was of prefabricated timber panels with three layer felt, or occasionally mastic asphalt, laid

488

a

b

8.4.3 SCOLA Mark I, 1964. Details of steel frame. Supplier, Sanders and Forster. (a) Stanchion base. Note stanchions of 4 angles with welded battens at intervals. (b) Stanchion head at building perimeter showing standard main beam and perimeter beams. (c) *Opposite.* Stanchion at internal position showing main beams and wind bracing angles. (d) Primary and secondary beams, one double in foreground. (Photos 1964)

c

d

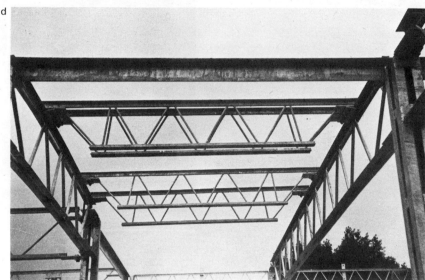

flat. An early experiment with a plastic roof finish was not a success. Staircases were of prafabricated steel units with precast concrete treads; partitions of honeycomb core plasterboard; a suspended ceiling grid with acoustic tile infill. Sanitary fittings, ironmongery and internal fittings were included in the design and bulk purchase arrangements. The external skin consisted of the timber horizontal rails and plywood fascia bolted onto the steel columns by means of steel brackets (**8.4.4**): the rails in turn could receive aluminium sliding windows, glazed asbestos fixed panels, louvres and various claddings of the traditional type: clay tiles, slates, timber boarding or brickwork, the latter being rather frowned upon by the purists. It is readily apparent that SCOLA was redeploying the same elements as others

8.4.4 School for handicapped children, Crawley, England. County Architect F.R. Steele, project architect, Barry Russell, 1965. Using Mark 1 SCOLA. (a) South elevation showing horizontal rails bolted to columns, window, and panel inserts. The design of components was such as to enable them to fall within the category of a 'two man lift'. (b) South elevation completed. The school included two covered, outdoor teaching/play areas; the edge beams fronting these being the only two non-standard components used (Photos 1970)

a

b

8.4.5 Public library, Pulborough, England. County Architect F.R. Steele, project architect M. Heffer, 1965-6. As with CLASP, SCOLA was used on a wide range of building types other than schools, with varying degrees of success. (Photo 1966)

before, with one notable change: that of the external cladding. The restyling of the building systems image in favour of the then currently acceptable stylistic norms (**8.4.5**) provoked Banham's comment 'a too clearly defined concept of architecture is being built into the system from the outset' (Banham 1962b), and created a number of problems of supply and performance that showed a disregard for previous systems experience.

The styling concept which led to this came directly out of Le Corbusier's use of deep concrete horizontal members, and the subsequent use of these formal concepts at the London County Council (now GLC) in the late fifties. The Alton West estate at Roehampton was built between 1956 and 1961 and clearly shows the influence that Le Corbusier had on architects at the time. Apart from the Ville Radieuse planning notions evident in the slab and point blocks particularly, the formal language of the elements, their proportions and materials were derived from a close study of his published buildings and L'Unité d'habitation (1952) in particular. Le Corbusier's double height living room was ingeniously translated into a 'double height' balcony framing the maisonette units — which did not have double height rooms. The use of deep concrete floor and balcony members was a part of this stylistic code and provided a powerful critique of the thin brittleness of forties and fifties styling

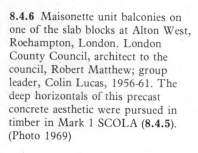

8.4.6 Maisonette unit balconies on one of the slab blocks at Alton West, Roehampton, London. London County Council, architect to the council, Robert Matthew; group leader, Colin Lucas, 1956-61. The deep horizontals of this precast concrete aesthetic were pursued in timber in Mark 1 SCOLA (**8.4.5**). (Photo 1969)

(**8.4.6**). Other LCC work of similar vintage plus the revival of the English brick and concrete tradition by Stirling and Gowan at Langham House, Ham Common (**8.4.7**) (1958) gave authority to a gathering movement away from the earlier shibboleths of the machine age and towards a new set of devices with a new rationale. It is not surprising perhaps that the key people in producing the original SCOLA concept came directly from the London County Council Architects' Department to Shropshire County Council. There is no question that SCOLA buildings of the early sixties carry an unmistakable stamp given them by the determination to emphasize the horizontal; but the effort to emulate in timber a set of formal devices developed in connection with another material — concrete in this case — shows both a lack of awareness of the relation of formal solutions to technology and a blind disregard for previous experience in the system field. There was a constant problem with the warp and wind of the timber rails and the thin plywood fascias were an immediate source of trouble in the British climate. There appeared little interest in other system work by the originators of SCOLA, and certainly much initial development work was carried out without reference to existing data from elsewhere in the field.

The problems of supply and performance arose quite clearly from the way in which development of the system was set up. The

8.4.7 Langham House Close, Ham Common, London. James Stirling and James Gowan, 1958. (Photo 1969)

establishing of another closed system on the lines of CLASP was in itself questionable and the attempt to create a specific sort of stylistic animal inappropriate. There was of course a healthy sense of identity around the new project but no-one with direct experience from either the Hertfordshire or CLASP system. Apparently, yet again, a new order was to be established and it was not to be tainted by mere history. If policy making for the new system held such an unhealthy disregard for experience, it was not improved on at developmental level. In CLASP Nottinghamshire led the first development work, with Derbyshire and Coventry allocating staff to assist, but this soon gave way to a central development group. Similarly the initial development work for SCOLA was distributed among the widely spaced member authorities — each one putting one member of staff in his office on to this work until the central group was formed in 1967. During this early period communication was by means of periodic meetings, telephone and letter, and it is perhaps an example of the strength of the systems idea as it was then seen that this could work at all.

In many ways the problems were compounded since most of those preparing the standard drawings for each element fell into two basic types. The first was the younger architect keen to embark upon 'new methods and materials' but with limited experience of the actual behaviour of either; the second type constituted older, more experienced architects who could find few ways to use this experience

8.4.8 SCOLA Mark 1, 1964. External wall detail. Fascia plywood flush jointed; transoms, horizontal rails and junction component in timber; louvres and windows aluminium. The detail shows brickwork, infrequently used during the early days of SCOLA. Note that all elements, except transom are detailed to be in one plane. (Photo 1964)

in this new situation where, not a building was being designed, but a kit of parts. The policy decision to accept, for example, an external wall design in which a minimum of seven separate suppliers were involved was one problem; to then use a development team who appeared to have little experience of the fixing, tolerance and jointing problems of such dry construction was another (**8.4.8**). After all, none of the technical problems was new; both Hertfordshire and CLASP had experience in the schools field, the ARCON group and others involved in the postwar prefabricated houses had gained knowledge of dry construction techniques, and furthermore experience and documentation existed on curtain wall techniques both in the United

States and in Britain. This had already been excellently covered by Michael Rostron in his series of articles in the *Architects' Journal* during 1960, subsequently published in book form as *Light Cladding of Buildings* (Rostron 1964). Between the baffled men brought up on more conventional construction and the young zealots whose architectural illiteracy encouraged not the new but the same old mistakes, there existed an uneasy truce from which the first version of the system emerged.

In all this could be discerned the seeds of some important straws in the wind. They were not, however, very much to do with the technology of the system but concerned with changes that could be brought about in relationships within and between local authorities by the advent of consortium working. The forces for change in this direction had a valuable new ally in SCOLA: it carried moral weight (albeit weakly founded), it had Ministry support, and it gave the impression of an interest in new techniques. The diplomatic manipulations of one or all of these were valuable opportunities in the normal local authority situation, largely hitherto characterized by inertia.

The SCOLA machinery, bringing together as it did Chief Architects and others, energized the situation in the local authority building field, where, although staff frequently moved about in search of better positions, there was surprisingly little interchange of hard information between authorities engaged on very similar work programmes. Naturally territory was jealously guarded although consortia working helped gradually to bring about more mutual confidence. Nevertheless the extra local authority decision making bodies created by a consortium such as SCOLA tended to assume a distant status, their decisions seeming to emanate from on high to the less implicated authorities.

The committee structure (**8.4.9**) seemed logical enough but was remote from the site implications of those decisions. Further, it may well be fundamentally questioned that such a structure is necessary to deal with a building programme that was annually (as one humorist put it) only equal in size to the building of one London office block. Some felt that almost any changes that could be wrought within the curious local authority situation must be welcome, and the externalized influence of consortium decision making and methods, however creaky, offered possibilities to produce buildings more quickly and thus fully utilize Treasury allocations via the Ministry of Education. Looked at coldly the occurrence of several well paid Chief Architects meeting quarterly, working parties of some 30 members meeting monthly for two days at a time, plus the frequent development and other meetings would in themselves produce an annual salary and expenses bill sufficient to pay for a good sized school.

In one sense it might be said that the general pattern of local authority working had brought systems upon its own head. Quality, speed of construction and costing procedures (apart from a few notable exceptions) were all in doubt, and until architects themselves could be

SCOLA ADMINISTRATIVE STRUCTURE: 1962 onwards

ANNUAL
ELECTED MEMBERS MEETING

> Councillors, Officers and Chief Architects
> Policy and approval of building programme

QUARTERLY
BOARD OF CHIEF ARCHITECTS OF MEMBER AUTHORITIES

> Chief Architects, Clerk and Treasurer to Consortium
> Co-ordination of programme and development work

MONTHLY
WORKING PARTY

> Architects and quantity surveyor
> Representatives from each authority
> Representatives from development group
> Day to day running of Consortia including detail development policy

MONTHLY AND AS REQUIRED
DEVELOPMENT GROUP

> Architects nominated by authorities
> Technical development work, programming and quality control

8.4.9 Second Consortium of Local Authorities (SCOLA). Committee structure used during its early years to administer the consortium

more convincingly predictable to their clients, the users, the Ministry and Treasury, they stood in a weak position to cry hands off. The postwar move in Britain towards the insulated paternalism enjoyed by most local authorities has its implications in terms of professional responsibilities. This situation provided fertile ground for the systems proponents and ultimately paved the way for the institutionalized building system situation that quickly developed.

Local authorities took up different positions in the face of these pressures. One member county of SCOLA, Hampshire, with a very large building programme, became an enthusiastic user of the system. However this enthusiasm also involved a fundamental misuse of the idea since, having acquired the use of a bank of small scale standardized components permitting varying plan configurations and designs, it was in addition decided to standardize whole plan forms for schools. In the case of primary schools some 20 buildings were erected employing the same plan type on a variety of different sites. This was something to which Hertfordshire, with its huge building programme never succumbed, and which created unfortunate compromises with regard to orientation, access and site relationship.

The flush facades of the SCOLA Mark 1 and Mark 1A versions continued until the 1966/67 building programme, although as is usual in such building programmes, many authorities continued to build in these superseded versions for some time after. This was due to a combination of factors such as familiarity and attachment to the system (or plain reluctance to change even from this brief length of 'tradition'), late starts on site where buildings projected initially to be in one year's building programme slip into the next and sometimes the one following that. However, beneath this continuing facade treatment of Marks 1 and 1A there had been some important changes: from the rigid to the pinjointed frame, a change in foundations and the discarding of the timber roof deck in favour of a steel deck of the diaphragm type (Bryan 1973). However the period 1962 to 1966 saw a mounting series of difficulties with components, particularly those of the external wall. Delivery, quality control, assembly, subsequent performance and maintenance were areas for complaint. Many of the timber components would arrive on site only to be rejected as not meeting the specifications. Far flung manufacturers naturally resisted replacing rejected components and here the enormous problems involved by having both the consortium members and the consortium suppliers spread out all over England became only too apparent. Development group architects and sometimes the working party were drawn into this since both manufacturers and contractors tended to take the view that the consortium was responsible.

Such pressures for change may have been resisted for much longer had there not been an additional factor. This was the drive from Ministry level for all the school building systems to change their dimensional basis to a common one — in the case of educational buildings to a 1 ft 0 in (305 mm) planning grid and 2 ft 0 in (610 mm) structural grid. Ironically, this followed closely on the heels of Ministry pressure to encourage all systems or intending systems to use the 3 ft 4 in (1.016 m) grid as used by SCOLA, and thus created its own backlash. In principle the decision to encourage interchange-ability by putting all the systems on to the same grid was perfectly sensible and it was paralleled by the acceptance of parts of the modular doctrine by the various government bodies and at the British Standards Institution. The first public proclamation of this was a document on dimensional co-ordination (MPBW 1963a) which basically called for the co-ordination of dimensions on the basis of a preferred size of 1 ft 0 in (305 mm). The effects on housing of the series of documents which followed have been examined earlier. Their relation to programmes in the educational field can be summarized as follows.

DC2 dealt with *Preferred Dimensions for Housing* (MPBW 1963b) and recommended the 1 ft planning grid followed by *DC3, Preferred*

498

Dimensions for Educational, Health and Crown Office Buildings
(MPBW July 1964b). This latter was a summary of a more extensive
document, published by the Department of Education and Science,
Controlling Dimensions for Educational Building (DES 1964). This
document, in accepting the move to the 1 ft 0 in (305 mm) planning
grid, became of great import to the building system designers since it

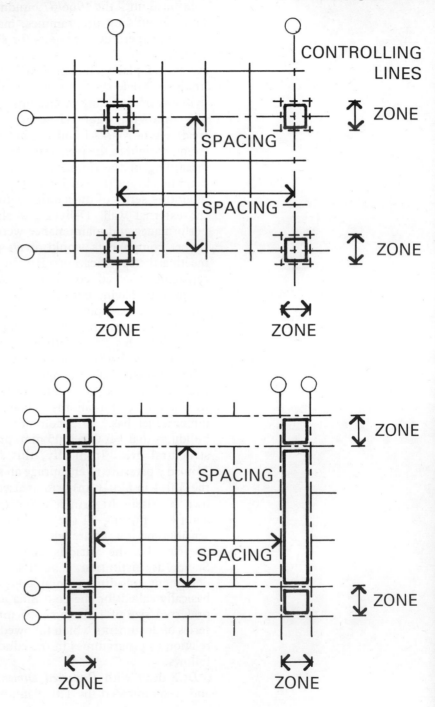

8.4.10 The conflict of centre line and
face grids. The two alternative
methods as shown in *Dimensional Co-
ordination for Building* (DC6)
published by the Ministry of Public
Building and Works, 1967

formed the new basis for co-ordinating the dimensions of all the building systems and their components. The document introduced itself as follows:

'Once dimensional decisions are made on a common basis, it should prove increasingly possible for educational buildings in all forms of construction to utilize industrially manufactured components whenever it is practicable, economic and desirable for them to do so.' (DES 1964)

Having established such a basis, two subsequent DC documents, *DC4* and *DC5*, embodied recommendations for a wide range of building types, proposing a limited selection of preferred dimensions for each type. In *DC6*, specific recommendations were made for the positioning of components which appeared, in part, to contradict the earlier recommendations in *Controlling Dimensions for Educational Building*. In particular, in attempting to satisfy both the Ministry of Housing and Local Government and the Department of Education and Science it pointed out that in planning structural elements it was permissable to measure either to the boundaries of the structural zone or to its centre line (**8.4.10**). Clearly, where the zone containing a structural element was an odd multiple of the basic module (4 inches or 100 mm) a centre line grid would cut straight through such a module giving a 2 inch (50 mm) split either side. The resolution of the component problems around a column that was, say, 12 inches (or 300 mm) were considerably complicated (**8.4.11**).

This dilemma pointed up a basic rift between those designing schools and those designing housing and various strategems were employed in an endeavour to resolve it. However, the school building systems had always been well ahead of the work done in housing and there seemed little hope of them altering one of their great traditions — that of the centre line steel column. However, all the time columns

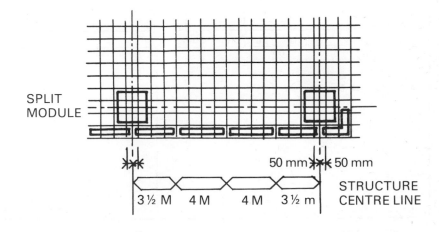

BASIC MODULE GRID (M : 100 mm)

SPLIT MODULE

50 mm 50 mm

STRUCTURE CENTRE LINE

3½ M   4 M   4 M   3½ m

8.4.11 The conflict of grids. An example of the type of component problem which can arise where the structure grid is not coincident with the component grid. Using standard components the split module results in components not of modular size

500

8.4.12 SCOLA Mark 2. SCOLA development group (1965-6). A design guide sheet showing how typical elevation drawings were prepared. Such a drawing contained all the information required by the window wall supplier. From this schedules of quantities were prepared and punch cards produced for setting up the machines. The areas of painting involved were also provided. The horizontal and vertical dimensions are all in modular units of 4 inches (101.6 mm): a figure 12, for example, means 12 times M or 4 inches giving 48 inches. Junction components and partition closers are coded and glass thicknesses shown. Note how the key plan reminds the architect to show all faces of his building, including returns, something that is often ignored in traditional elevations.

could be held at 8 inches (or 200 mm) or any even multiple of the module, the problem did not arise.

Mark 2 SCOLA, in divorcing the external skin more forcibly from the structure, attempted to provide an opportunity for further work to make the interchangeability of components a reality. The sequence of Ministry documents attempting to be more and more precise concerning the co-ordination of dimensions merely demonstrates the law of diminishing returns. The basic point was a simple one — that of the agreement to build and design components around a multiple of a basic dimension. When an attempt is made to apply this as a complete grammar, produced overnight, it is bound to be self-defeating. The various groups set up to work on the problem, as with most government endeavour, tended to proliferate the work merely by their existence. Parkinson said it more elegantly with his 'work expands to fill the time available for its completion' (Parkinson 1958). Documents dealing with dimensional co-ordination multiplied left and right, and in the end undoubtedly pressed the idea beyond the point of no return.

It is within this context, and against the background of the first DC documents, *DC1, DC2, DC3,* and *Controlling Dimensions for*

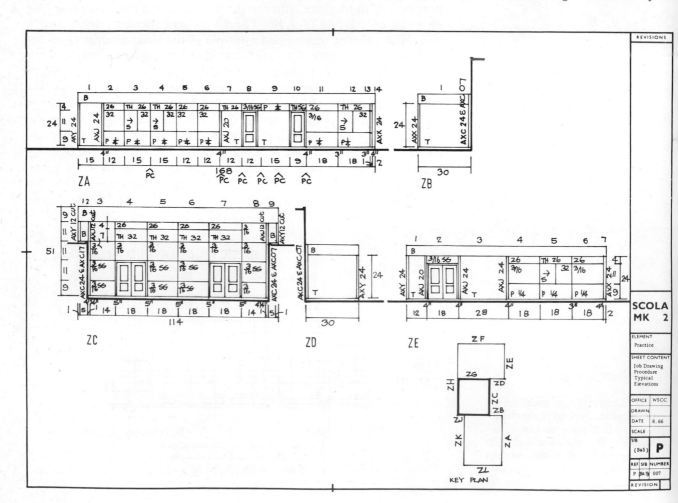

Educational Building (DES 1964), that development group work started early in 1964 on the Mark 2 version of the SCOLA system. The brief was to bring the system into line with government recommendations on dimensions, to simplify foundations, roofing and external cladding. Together with this there was to be an examination of the documentation, about which there had been continuing complaints. By mid-1966 the system had been revised and a prototype building, a school in Bognor Regis, Sussex, was under construction. The re-examination had led to some changes that marked a change of emphasis for the system.

The study of the dimensional structure and the performance of the external wall led to a search for a solution more in keeping with the realities of supply and manufacture and indeed more in keeping with the philosophic basis of the idea of component catalogue building (**8.4.12**). The desirability of inventing yet a further architect designed external walling system was questioned by the development group with the result that a survey was made of possible window wall systems already on the market, particularly those used by similar organizations. It was ultimately decided to award the contract to the Crittall Company, suppliers to Hertfordshire (and SEAC), on the basis that the architects and the suppliers had been working with the company for some years and had therefore a mutual expertise upon which it would be sensible to draw. Furthermore it was felt that if the concept of interchangeability and component building meant anything the attempt to use the external window wall from one building system on the frame of another would constitute an important step forward. However, the political importance and separate identity of the consortium could be detected in the remark of one County Architect to the effect that SCOLA did not have much of a development group if they could not design their own window system.

The manner in which the growth of the interest in documentation has paralleled, and been dependent on the growth of the systems movement in Britain is important and significant and the re-examination of the SCOLA documentation generated some interesting results. The drawings for Mark 1 and Mark 1A SCOLA were an amalgam of design, assembly and component information that had been rationalized into SfB groupings. Further the division between the drawings and the specification was still very confused as indeed is the tradition in Britain. However, the way in which such specifications and price lists are used with a system is rather particular and requires a different approach. Similarly the very nature of the system of component manufacturers and suppliers means that designers must be given very quickly the ground rules against which they can work, especially where these differ from standard practice as they do in most systems. The question of efficiency and costs was also examined and although the nature of local authority working makes printing and distribution costs difficult to ascertain it was felt that the cost of the frequent revision and circulation of standard drawings amongst widely spread members and throughout the jobs (their consultants, suppliers and contractors) had become an exceedingly complex and expensive business. The proposals for the revision of the documentation were twofold: first that information should be tailored to the recipient and second that a very high standard of production be aimed at with its control and distribution identified as the major task it had become. It was also considered that a well ordered approach to the preparation of documentation, its correction, revision and issue, would have significant effects on the rather poor standard of drawing organization evident in many local authority departments. The documentation was produced in the following sections:

1. Design data: largely intended for architects and others involved with the design stage.
2. Procedure data: these described the processes and procedures that had been developed with suppliers and manufacturers and also guides to the clear preparation of drawings that would be commonly understood. These were particularly important since they involved all those at the design stage in the implications of decisions in relation to order and delivery times, and the economics of alternative solutions.
3. Assembly data: these carried site assembly drawings for the variety of constructional situations likely to be met using the standard kit of parts.
4. Component data: these contained information concerned with the manufacture of each of the components. Sometimes prepared by the architect but sometimes prepared by the supplier. They related to shop drawings in traditional practice.
5. Specification data: these contained the written specification and schedules of rates (cost) related to each component or group of components

In addition a comprehensive index was provided. Apart from the attempts to tailor each section of information to those persons most likely to use it, the division had the added advantage that a change in one part did not have to effect every other part. For example the specification of a subsystem (say roofing) may change but the drawings remain the same. Alternatively modifications could be made in the details of components without affecting the assembly drawings or specification. This sort of flexibility is particularly important in the administration and use of a system. Each section of the SCOLA Mark 2 manual was printed by offset lithography on to coloured A3 size sheets to denote sections. This method allowed selections of sheets to be made and built up into specific books for specific jobs and purposes. It was a bank of information looked after by a full time administrative officer, serviced by the development group.

In summary then, the two main thrusts evident in the SCOLA Mark 2 system were in the direction of:

1. endeavouring to provide a more open framework for component and subsystem interchange through
   (a) using the same dimensional basis as other systems.
   (b) so designing the components, particularly the external wall,

that they were not locked into each other on the closed system principle.

2. recognizing that the actual component parts of a building system were, although important, enmeshed in a whole system of communication with people and that this aspect should receive proper attention, through documentation and other means.

The first, the drive to re-establish the original purpose of component building as a bank of industrialized parts, was given some sense at least by the early DC1 documents which attempted to co-ordinate the, by this time, disparate activities of the school building consortia. The traumas of bringing all the systems on to a 1 ft 0 in (305 mm) planning grid were in this sense a prerequisite. Ironically at the very time it was taking place it was clear that metrication was at hand and a further change would be necessary. A proposal was made within SCOLA to design the Mark 2 system in metric since it would arrive on the market early in the metrication programme starting in 1965. This, however, was rejected, at policy level. What was pursued in the redesign of the external envelope were the technical difficulties: first a reduction of the technical problems encountered with the many components and an attempt to 'open' the design more so that individual local authorities and individual project architects could introduce a wider range of components. This meant dimensionally disentangling the external wall from the structure (**8.4.13**) and detailing the parts so that real interchangeability could become a reality. In an endeavour to encourage this an innovation in jointing was introduced. This was in the form of a series of four junction component types: one for straight run conditions, one for internal corners and two alternatives for external corners (**8.4.14**). These, of standard profile extruded aluminium, came in a range of heights and were fixed top and bottom: with the addition of suitably designed preformed flashings any material could be junctioned to the units. This approach gave precision whilst at the same time dealing with the problem of tolerance usually associated with the junctioning of any two materials of varying thickness. This promising development, however, did not survive long and soon became a casualty of cost pressures. Nevertheless, having separated the external wall from the structure, it became possible for the first time to introduce non-system components and subsystems from other consortia — the well tried Hertfordshire/SEAC window assembly was in this way applied to Mark 2 SCOLA with only minor modifications. Further, it also meant that brickwork and other components could more easily be introduced. It is of interest that this immediately gave rise to a large number of brick clad SCOLA buildings on the basis that brick was the most economical finish. Hitherto almost all authorities had paid lip service to the view that the light claddings used by SCOLA were the most economical. This, in turn emphasized the illogicality of using two load-bearing systems (frame and brick) in one building.

A further aspect was the attention given to the flexibility of site activity. In a steel frame system the frame is erected rapidly and ideally

504

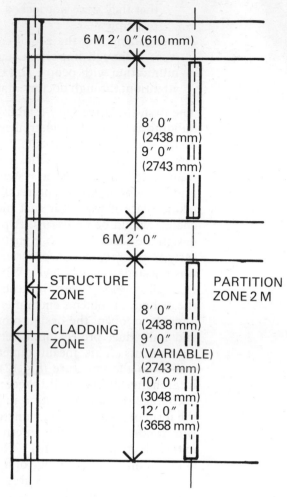

6 M 2' 0" (610 mm)

8' 0"
(2438 mm)
9' 0"
(2743 mm)

6 M 2' 0"

STRUCTURE
ZONE

PARTITION
ZONE 2 M

CLADDING
ZONE

8' 0"
(2438 mm)
9' 0"
(VARIABLE)
(2743 mm)
10' 0"
(3048 mm)
12' 0"
(3658 mm)

a

SECTION

8.4.13 SCOLA Mark 2: SCOLA development group (1965-6). (a) Plan and section showing structure, cladding and partition relationships. (b) *Opposite*. Plan showing separation of structure and external wall zones. (c) Plan showing how the standard straight run junction component allows disparate materials to come together with a common joint detail (flashings not shown). This also permitted any thickness cladding to be used, something not possible in SCOLA Mark 1 and 1A with its adherence to the flush facade

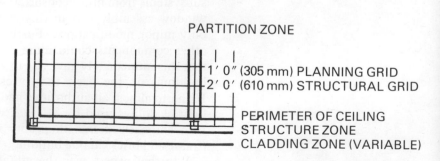

PARTITION ZONE

1' 0" (305 mm) PLANNING GRID
2' 0' (610 mm) STRUCTURAL GRID

PERIMETER OF CEILING
STRUCTURE ZONE
CLADDING ZONE (VARIABLE)

PLAN

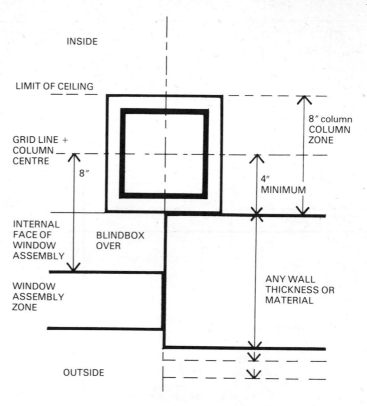

INSIDE

LIMIT OF CEILING

GRID LINE +
COLUMN
CENTRE

8″

8″ column
COLUMN
ZONE

4″
MINIMUM

INTERNAL
FACE OF
WINDOW
ASSEMBLY

BLINDBOX
OVER

WINDOW
ASSEMBLY
ZONE

ANY WALL
THICKNESS OR
MATERIAL

OUTSIDE

b

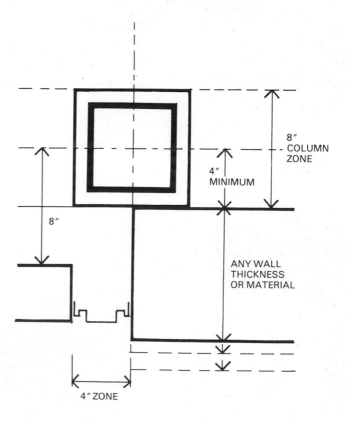

8″
COLUMN
ZONE

4″
MINIMUM

8″

ANY WALL
THICKNESS
OR MATERIAL

c

4″ ZONE

506

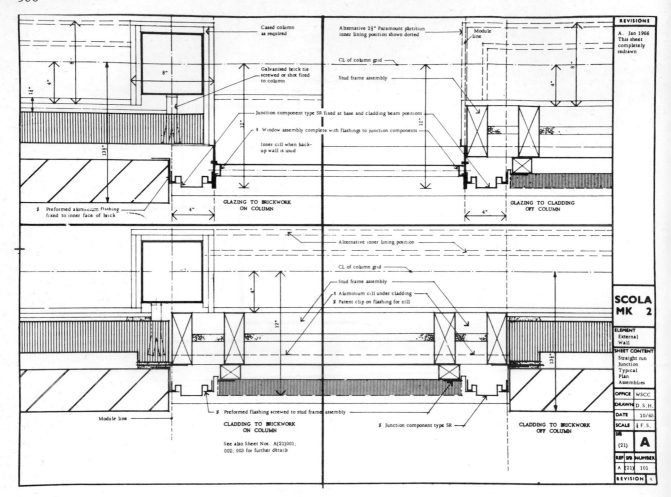

a

8.4.14 SCOLA Mark 2. SCOLA development group, (1965-6). (a) Straight run junction assembly plans. A sheet from the assembly section of the manual showing standard details for some of the possibilities. (b) *Opposite*. External corner assembly plans, showing the two types of junction component available. (c) The aluminium straight run junction component at sill. (Photo 1966). (d) Aluminium external corner junction component at eaves. (Photo 1966). (e) Aluminium external corner junction component at sill. (Photo 1966)

the roof should then go on so that all work can take place under cover. However, suppliers appointed by the consortia, even with the programming available, were frequently unable to arrange for components to arrive on site at the right time. A build up of jobs, slow contractors or architects, produced inevitable bottlenecks. The response, in SCOLA Mark 2 was to design the junction between the roof and external wall so that each could be erected independently of the other (**8.4.15**). Traditionally of course it is necessary for the roofer to await the completion of the wall to the eaves level before he could complete. It now became possible for whoever arrived on site first to proceed — either the wall or the roof. This offered considerable advantages on site and was a useful innovation. The second aspect, that of the organization that was now necessary to administer such a system, also received attention. The many authorities, their professional staffs, outside suppliers, contractors and others all needed to understand the system and the somewhat amateurish methods so far in use left a great deal to be desired.

Decision making, always slow and complex in such an organization, was at its most crucial when on-site problems arose with components

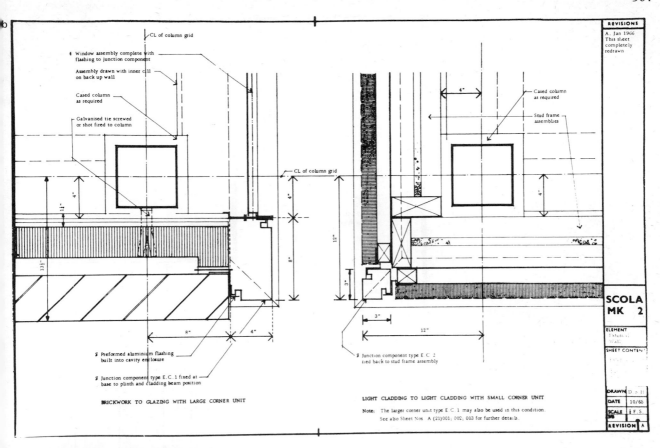

REVISIONS

A. Jan 1966
This sheet
completely
redrawn

CL of column grid

Window assembly complete with
flashing to junction component

Assembly drawn with inner cill
on back up wall

Cased column
as required

Galvanised tie screwed
or shot fired to column

CL of column grid

Cased column
as required

Stud frame
assemblies

§ Preformed aluminium flashing
built into cavity enclosure

§ Junction component type E.C.1 fixed at
base to plinth and cladding beam position

§ Junction component type E.C.2
tied back to stud frame assembly

BRICKWORK TO GLAZING WITH LARGE CORNER UNIT

LIGHT CLADDING TO LIGHT CLADDING WITH SMALL CORNER UNIT

Note: The larger corner unit type E.C.1 may also be used in this condition.
See also Sheet Nos A (21)001; 002; 003 for further details.

SCOLA MK 2

ELEMENT	External Wall
SHEET CONTENT	
DRAWN	D ? H
DATE	10/65
SCALE	½ F.S.
REVISION	A

e

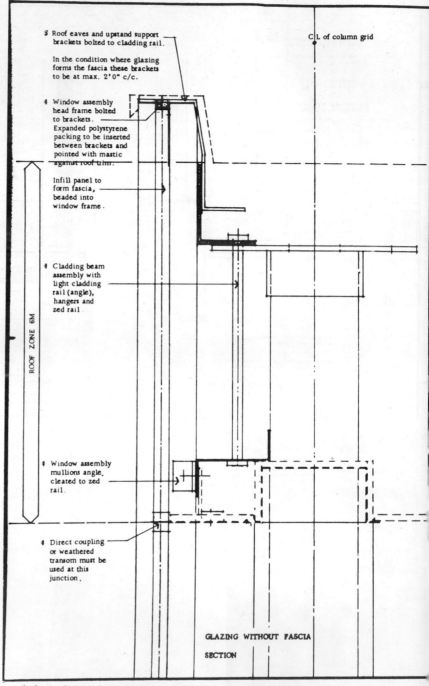

a

Roof eaves and upstand support
brackets bolted to cladding rail.

In the condition where glazing
forms the fascia these brackets
to be at max. 2'0" c/c.

C L of column grid

Window assembly
head frame bolted
to brackets.
Expanded polystyrene
packing to be inserted
between brackets and
pointed with mastic
against roof trim.

Infill panel to
form fascia,
beaded into
window frame.

Cladding beam
assembly with
light cladding
rail (angle),
hangers and
zed rail.

ROOF ZONE 6M

Window assembly
mullions angle,
cleated to zed
rail.

Direct coupling
or weathered
transom must be
used at this
junction.

GLAZING WITHOUT FASCIA

SECTION

8.4.15 SCOLA Mark 2. SCOLA development group, (1965-6). Eaves detail options without fascias. To reduce on-site delays the design was developed to allow either the roof and finish to go on before external cladding erection, or the cladding to be erected before the roof. (a) Section showing window walling. (b) *Opposite*. Section showing lightweight cladding, timber. (c) *p. 510*. Section showing masonry cladding, brickwork

and the related documentation. Changes in procedure, agreement with manufacturers, programme start dates, and detail design are frequent in a building system and need careful monitoring. Here the documentation is crucial. The rationalization of the SCOLA Mark 2 documentation attempted to move the system from hurriedly altered sets of dyeline prints to that of a well organized, controlled feedback and properly distributed manual. Manual sheets, as earlier described,

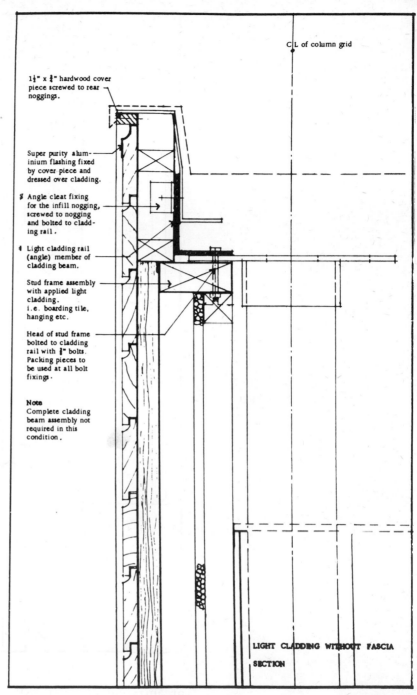

b

1½" x ½" hardwood cover piece screwed to rear noggings.

Super purity aluminium flashing fixed by cover piece and dressed over cladding.

⅜ Angle cleat fixing for the infill nogging, screwed to nogging and bolted to cladding rail.

¼ Light cladding rail (angle) member of cladding beam.

Stud frame assembly with applied light cladding. i.e. boarding tile, hanging etc.

Head of stud frame bolted to cladding rail with ⅜" bolts. Packing pieces to be used at all bolt fixings.

Note
Complete cladding beam assembly not required in this condition.

C L of column grid

LIGHT CLADDING WITHOUT FASCIA

SECTION

would be issued systematically from a centralized source in binders to be stored in an orderly fashion by each party concerned. The manual included clear guidance as to its use and organization and through this it set something of an example, being subsequently used by King and Everett (1971), in their standard work, as the example of system documentation.

This recognition that a building system of the complexity of

8.4.15 c

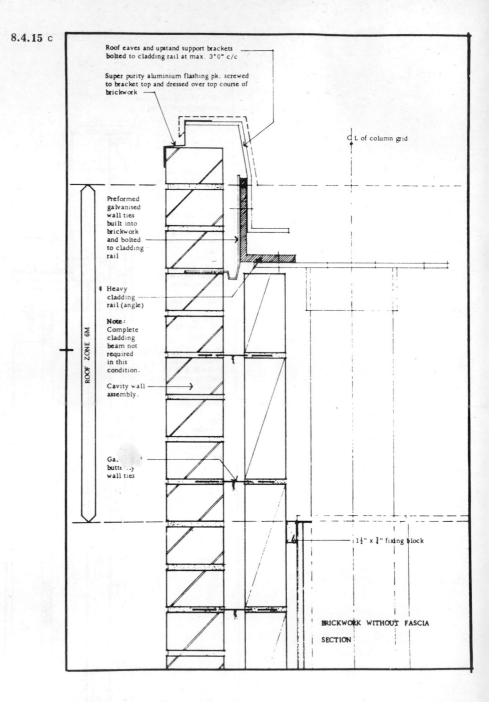

Roof eaves and upstand support brackets
bolted to cladding rail at max. 3'0" c/c

Super purity aluminium flashing pk. screwed
to bracket top and dressed over top course of
brickwork

C L of column grid

Preformed
galvanised
wall ties
built into
brickwork
and bolted
to cladding
rail

ROOF ZONE 6M

Heavy
cladding
rail (angle)

Note:
Complete
cladding
beam not
required
in this
condition.

Cavity wall
assembly.

Ga.
butt....y
wall ties

1¼" x ½" fixing block

BRICKWORK WITHOUT FASCIA

SECTION

organization of a CLASP or SCOLA needed different sets of operational procedures than would a single building appears to be a short advance. However, even such a modest attempt at organization was something that many local authorities appeared to have insufficient expertise to assimilate. After all, many had never seen or read the literature upon which much systems activity was based, and cared little for its dictates and much less for the complexities inherent in component interchangeability or the industrialized vernacular.

However, within a short time yet a further major system change was undertaken by SCOLA and, with the development group now centralized on Gloucester (1967), work was again commenced to revise many components. More important, however, this group were able to undertake a number of studies that, in the long term, were probably far more important than the constant redesign of the parts that seems to be the stock in trade of all the systems. SCOLA, like other school building systems, settled into its own routine and style, although its use by member authorities diminished greatly in recent years. This is partly due to the overall reduced school building programme and to the shift in emphasis from new schools to the adaptation and improvement of existing stock. It also reflects a disenchantment with the systems idea in general and a somewhat overdue recognition that the particular model adopted offered rather poor performance characteristics in the crucial areas of maintenance and energy. The external timber rails and the flat roofs require constant remedial work: rotting transoms are being replaced and, in 1980, tiled pitched roofs are being added to existing flat roofed schools in an endeavour to solve the more serious problems of roof failure. Hampshire County Council under the direction of County Architect Colin Stansfield-Smith had, by the late seventies, found it necessary to positively reassess the performance characteristics of its large stock of some 500 SCOLA buildings. Energy studies provided the impetus for wide ranging reappraisal and some interesting transformations of the SCOLA system. Frogmore Comprehensive School at Yately, Hampshire, the last large scale use of SCOLA in the county, had its first phase built using SCOLA Mark 3 to a 'standard plan form' (Perkins 1980). Phase 2 of this building involved a reassessment of the range of SCOLA subsystems available and resulted in a 'kit of parts in a "local" economy rather than a consortium economy' (Galloway 1980). Universal beams were used instead of web joists, permitting the height (and therefore the cost) of the building to be reduced. The storey height double skin wall had an outer glazed skin and an inner skin of lightweight panels which slide to give variable window area equal to 20-40% of the south facing elevation. Air from outside introduced into the space is preheated by solar gain and discharged into the building by means of fan convectors. It is expected that the aluminium brise-soleil will eliminate unwanted solar gains. It is interesting to recall that those early examples of lightweight construction schools by C.G. Stillman in the thirties, and already discussed, had recognized the necessity for such solar shading even in Britain's uncertain climate. Whilst this is an interesting attempt to modify the SCOLA approach Nelson (1980) points out that 'all the environmental control "systems", which have been incorporated into the new building, must be considered as being an attempt to compensate for the faults of the original method design (sic)' and that the success of the measures used in this school rely heavily on their being understood and operated efficiently by the staff involved. Further, if this approach works it could be used to improve the many system schools in the United Kingdom.

Just as individual buildings reflect the changing concerns of designers and the influences upon them, so one can see in the establishment of buildings systems a similar process at work. A building, to create a major impact, in some way demonstrates the designer's ability to reorder the problem which confronts him. This can be seen all too clearly in historical terms: Wright and Le Corbusier were able to do this just as have, more recently (and closer to the systems argument), Ehrenkrantz and Foster. In this sense a significant building offers a critique of what has gone before, at its best a crystallization of a set of ideas into a whole with characteristics which clearly distinguish it from its predecessors. As discussed later, Foster's building for IBM at Cosham takes a group of ideas that had been developing around building systems, environmental control and flexibility and orders these in such a way that their expression is made absolutely explicit, made public. To the many designers to whom these ideas were before implicit or unrecognized such an architectural statement has the power of innovation and acts as a critique on the situation leading up to it.

In the same way the school building systems, for example, can be seen as exhibiting the same process. The establishment of each new organization from CLASP (or even Hertfordshire) onwards demonstrates the changing forces and interests in the systems idea. Each new system constitutes some sort of critique of the previous ones in the way in which some intentions fresh to the situation are incorporated. With Hertfordshire it was an open ended exploration of the component building philosophy, with CLASP the idea of the grouping of authorities and bulk buying and with SCOLA the consolidation of the CLASP idea over a wider area — although demonstrating less new thinking. The creation of SCOLA seemed to many interested in the idea of an 'industrialized vernacular' a step in the wrong direction. Forming another new entity to emulate the CLASP idea, commencing again and producing yet another closed system, constituted a number of misunderstandings. Unfortunately these misunderstandings encouraged school building in a direction which has made subsequent innovation very much more difficult. The critique of CLASP and SCOLA as closed systems was to be found first in Method Building, a grouping of authorities based on Somerset County Council.

The need felt for a new consortium offering a change of emphasis can be seen quite clearly in that formal discussions between Chief Architects to set up some new type of grouping commenced early in 1962, a year after development work had commenced on SCOLA. This in itself reflects the discussions that had previously gone on to endeavour to create a situation which would somehow mitigate the dangerous trend by then already evidenced by the closed systems, and the resultant dismay amongst those interested in a genuinely open component situation in Britain.

The six initiating authorities, Berkshire, Bristol, Cornwall, Devon, Somerset and Wiltshire ultimately inaugurated the Consortium for

Method Building (CMB) in July 1963 although it was not until 1965 that the first bulk purchase contracts were let. The length of time that it took to become established does perhaps reflect the difficulties of establishing a fundamental idea as opposed to designing another one-off kit of parts with all its instant architectural appeal. Alan Diprose summed up very well their intentions in an article in the *Architects' Journal*:

> 'The initiative for establishing the consortium and the broad philosophy on which it is founded stem from Somerset, where a strong conviction was held that it was not only feasible but urgently necessary to break open the closed systems if the advantages of industrialised building were ever to be obtained without the disadvantages.' (Diprose 1966)

Whilst CMB were clearly interested in standardization, factory production and the bulk purchase of components they were also concerned to provide a much broader range of structural possibilities and considerably more design freedom than was possible with the then existing systems. There were two key factors amongst their ideas: one was the need for some sort of dimensional framework, the other was a large continuing programme of building work. This had the result of Method Building deciding first to put together the market in the form of the consortium rather than to design initially a system or kit of parts which then had to be sold to all concerned. The inevitable backlash that there would be to such an artefact could in this way be avoided. In Method Building can be seen many of the postwar arguments reduced to their most direct form: the belief that the component vernacular could best be established through a set of dimensional relationships, and that the more you buy of building components the cheaper they become. Whilst there is clearly an element of truth in both of these, and indeed Method's attempt is particularly laudable, the omissions from the argument have, in the ultimate, probably had more effect upon the development of the idea than its commissions.

A dimensional framework was one thing but its relation to the real world of materials and components production was another. True, at this time the modular argument was gaining weight, and government weight at that. Nevertheless the vast majority of components in buildings related more to the dimensional framework of the process by which they were produced than to some abstract industry-wide formula. More importantly there was the jointing problem. Louis Kahn may have elegantly pointed out that 'the architecture lies in the joints' but J.F. Eden's important, but largely overlooked, paper probably put the problem more directly to the dimensional co-ordinators when he cited the development of the engineering industry whose success in mass production architects were trying to emulate. He pointed out (Eden 1967) that the accurateness and interchangeableness of parts in engineering relied upon the definition, early in its development, of sets of standards for joints — that the way in which the parts fitted together was very much more crucial than that the parts be themselves dimensionally co-ordinated. This had resulted in the development of standards for tolerance and fit which were accepted industry wide, and allowed the parts themselves to change provided that the interfacing requirements were met. Parallels to this can be seen in the development of the electronics industry, one exposed to particularly rapid change. Here the connections between parts can remain the same whilst the components used can differ. Of course as with the car analogy for building, it is unwise to draw too close a parallel between a field such as electronics which is changing so rapidly that even agreed standards for 'jointing' change almost overnight, and the building industry where change, whatever many architects might like, is relatively slow.

The Consortium for Method Building was concerned to develop an approach which would not only apply to school buildings but which would embrace the whole of local authority building: health, welfare, libraries, housing, fire and ambulance stations. It was indeed to be a method rather than a system. Their application of the dimensional ideas is particularly interesting in view of the way in which the use of grids had developed. All the systems up to Method had repeatedly compromised the theoretical basis of the modular idea and the flexibility inherent in Albert Farwell Bemis's (1936) original concept of a 4 inch (102 mm) module in three directions by settling for a two way planning grid which was then used to size components. In the case of the Hertfordshire 8 ft 3 in (2.5 m) grid this then resulted in components of 8 ft 3 in (2.5 m) in length, in the case of the systems on the 3 ft 4 in (1.02 m) grid this resulted in component sizes of 3 ft 4 in (1.02 m) or multiples of it. With the introduction of the 1 ft 0 in (305 mm) planning grid consequent upon the publication of *DC1* (MPBW 1963a) the systems resized their components generally to 1 ft 0 in (305 mm) multiples. This latter occurrence was in spite of the fact that the *DC1* document itself made no recommendation that components should be sized to the planning or structural grid but expressly drew attention to the fact that a 4 inch (102 mm) increment was likely to be

8.4.16 Consortium for Method Building (1963 onwards). The CMB handbook pointed out the effects of differing sized grids when it was required to increase a given building by a small amount. Using a 1ft 0in (305 mm) grid the minimum increase in area possible is 8% of the original, with a 3ft 0in (914 mm) grid it is 26% and with a 6ft 0in (1.83 m) grid it is 56%

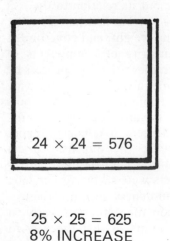

24 × 24 = 576

25 × 25 = 625
8% INCREASE

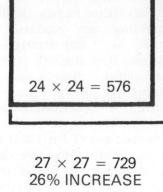

24 × 24 = 576

27 × 27 = 729
26% INCREASE

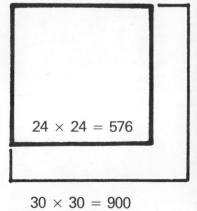

24 × 24 = 576

30 × 30 = 900
56% INCREASE

most suitable for the sizing of non-structural components such as windows, external cladding, door frames, internal fittings and so on. The opportunity to gain more flexibility and move towards interchangeability offered by *DC1* was partially lost by the concern of the systems to keep the variety of components in a given range as few as possible. Using 3 ft 4 in (1.02 m) they had components available at 3 ft 4 in (1.02 m), 6 ft 8 in (2.03 m) and 10 ft 0 in (3.05 m) whilst using the 1 ft 0 in (305 mm) preferred size they might have almost any sizes in multiples of 1 ft 0 in (305 mm) and end up with many more components. In not taking up the possibilities offered by the use of the 4 inch (102 mm) increment as set out in *DC1* and as developed by modular theory there had, unwittingly perhaps, become firmly fixed in many an architect's mind a totally wrong idea concerning the relation of grids and components. It did not seem to be thoroughly understood that a grid used for planning must relate to component sizing but does not need to be bound by that sizing.

It is against this context, then, that the approach adopted by Method Building must be viewed. They recognized very clearly that a grid and a series of restricted component ranges would create many severe limitations and lead them into the same impasse that other systems had reached. They were able to make out a strong case that the coarser the grid the more uneconomic it became the more generally it was applied. A coarse planning grid could give oversized or undersized spaces and be very uneconomic on other than the most straightforward sites. It is no accident that most system buildings sit on flat open sites although there are some notable exceptions to this. **8.4.16** illustrates very simply the effects of varying sized grids. The argument was carried further in that the consortium held that although wider ranges of components may cost more than smaller ranges (although this had never been conclusively and publicly proven with regard to the building industry), this was offset by the savings that could be made in the planning of individual buildings. The result of this was that the dimensional framework became based upon modular reference grid lines at 4 inches (102 mm) and 1 ft 0 in (305 mm) intervals in all three dimensions, with the large size being used for the location of components. However the sizing of most components was to increments of 1 ft 0 in (305 mm). One of the results of this dimensional approach was that it was possible to use a number of different structural types and to use them separately or together: steel frame, concrete frame (**8.4.17**), load-bearing brickwork, cross wall masonry construction, or mixed constructions.

Diprose makes the point that in spite of their intentions CMB were still unable to avoid a number of quite serious shortcomings. He cites the limited ranges of components as giving little design flexibility in practice and restrictions on the way the components are used in relation to the structure which was the result of attempting to simplify the detailing. One example of this is that Method have produced exactly the same difficulties that have been produced by other consortia with respect to the relation of the frame and the external skin (**8.4.18**). Columns freestanding within the envelope present planning

516

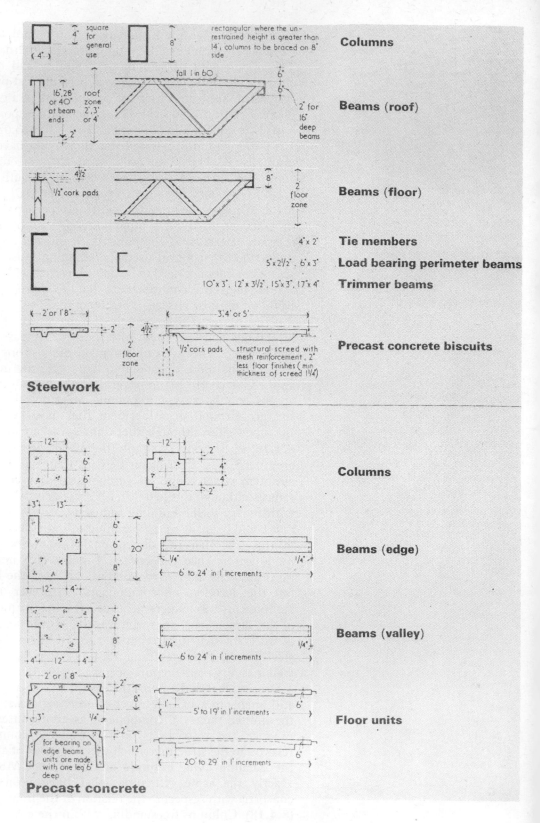

8.4.17 Consortium for Method Building (CMB), 1966. Basic ranges of structural components: steel and precast concrete

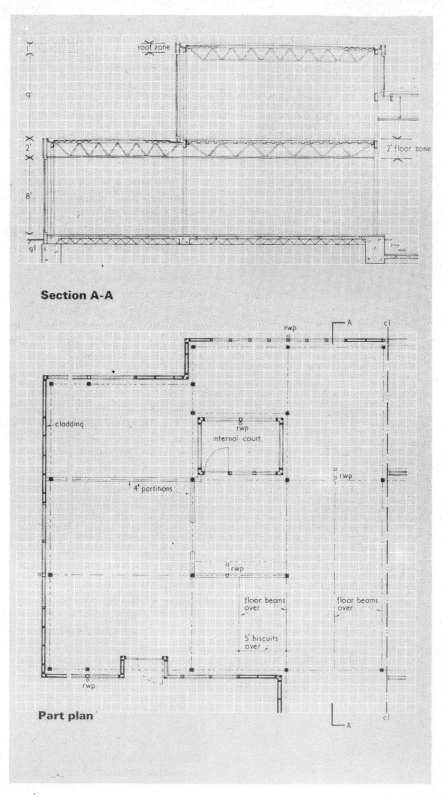

Section A-A

roof zone

2' floor zone

9'

2'

8'

gl

Part plan

rwp

cladding

internal court

rwp

4' partitions

rwp

rwp

floor beams over

floor beams over

5' biscuits over

rwp

A

cl

A

cl

8.4.18 CMB 1966. Typical plan and section

8.4.19 Bauhaus building, Dessau, Germany. Walter Gropius 1925-6. Northwest corner of the workshop building and entrance

difficulties internally with regard to furniture fittings and services, and with regard to the perennial problem of closing the gap between column and external skin where a partition junctions. In the demonstration yet again of this problem it is interesting to see how one of the most powerful concepts of the modern movement of the twenties and thirties has created a difficult planning and technical problem for countless architects and building users. In *Space, Time and Architecture* Giedion's caption to a photograph of Gropius' Dessau Bauhaus of 1926 (**8.4.19**) encapsulates the case:

'in this case it is the interior and exterior of a building which are presented simultaneously. The extensive transparent areas, by dematerialising the corners, permit the hovering relations of planes and the kind of "overlapping" which appears in contemporary painting.' (Giedion 1967, 1941)

When related to the many other ideas upon which this view was dependent, such as planning flexibility and the use of the lightweight skin, this can make sense: but it must be reinterpreted in total, and afresh, as it has been by many architects. As a concept used as a technical panacea, out of context, with little or no architectural literacy to back it up, it can produce little else but trouble. Many of the school building systems show the idea in an alien context, for the subtle interlacing of a set of ideas such as these cannot easily be systematized. Further, heavy reliance is placed upon those involved in the execution of such ideas. This calls upon a wide range of architects of differing experience, age and cultural experience to instantly 'know' how to use such a language, and moreover to get it right. It is not surprising that many system built buildings (whatever the promise of the system) lack coherence. It would perhaps be more fruitful to draw more on the real context of design experience available. Here, however, it is often also painfully true, that the level of such experience and awareness is dismally low.

In the case of Method Building, an attempt was made to match the real world of local authority design and construction with the hopes for improvement offered by the modular concept. In the event, no clearly definable 'system' emerged and this must count in its favour. Built solutions drew upon components as and when appropriate and this has produced a wide range of solutions. Most of all, perhaps, the authorities concerned are to be congratulated for defending the fundamentals of the idea against the obvious pressures and attractions of another 'closed' kit of parts.

More Acronyms

The growth of the local authority client sponsored consortia was considerable during the 1960s in Britain. By 1966 work built by these consortia amounted to over one-third of the whole school building programme and by 1970 it accounted for over half (DES 1976). The map (**8.4.20**) shows how almost all of England and Wales was covered

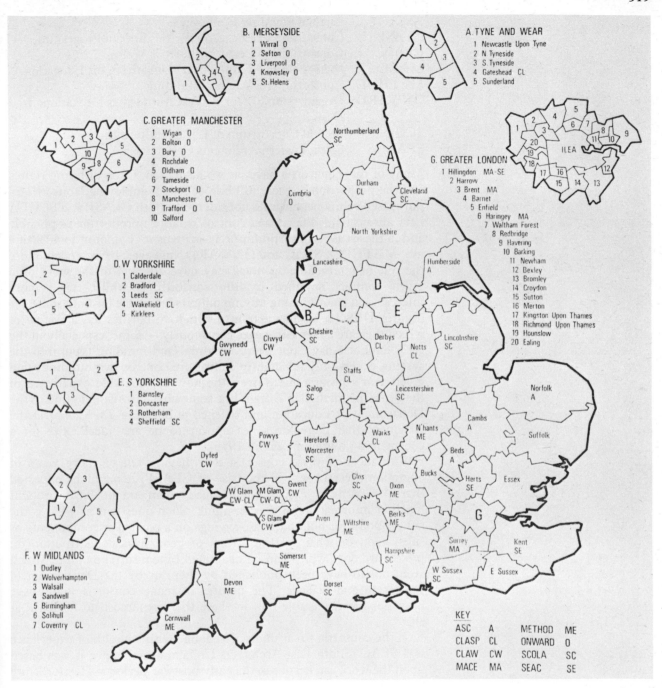

B. MERSEYSIDE
1 Wirral O
2 Sefton O
3 Liverpool O
4 Knowsley O
5 St.Helens

A. TYNE AND WEAR
1 Newcastle Upon Tyne
2 N.Tyneside
3 S.Tyneside
4 Gateshead CL
5 Sunderland

C. GREATER MANCHESTER
1 Wigan O
2 Bolton O
3 Bury O
4 Rochdale
5 Oldham O
6 Tameside
7 Stockport O
8 Manchester CL
9 Trafford O
10 Salford

G. GREATER LONDON
1 Hillingdon MA-SE
2 Harrow
3 Brent MA
4 Barnet
5 Enfield
6 Haringey MA
7 Waltham Forest
8 Redbridge
9 Havering
10 Barking
11 Newham
12 Bexley
13 Bromley
14 Croydon
15 Sutton
16 Merton
17 Kingston Upon Thames
18 Richmond Upon Thames
19 Hounslow
20 Ealing

D. W YORKSHIRE
1 Calderdale
2 Bradford
3 Leeds SC
4 Wakefield CL
5 Kirklees

E. S YORKSHIRE
1 Barnsley
2 Doncaster
3 Rotherham
4 Sheffield SC

F. W MIDLANDS
1 Dudley
2 Wolverhampton
3 Walsall
4 Sandwell
5 Birmingham
6 Solihull
7 Coventry CL

KEY

ASC	A	METHOD	ME
CLASP	CL	ONWARD	O
CLAW	CW	SCOLA	SC
MACE	MA	SEAC	SE

8.4.20 Membership of local education authorities (England and Wales) in client sponsored consortia: DES 1976)

by one or other of these consortia, although each had a different approach. To encourage this development a Ministry of Education circular, a powerful means of persuasion on local authorities, was used in 1964 (MOE Circular 1/64). By May 1976, of the 106 Local Education Authorities, 71 were either full or associate members of one of these consortia, named as follows:

ASC: Anglian Standing Conference
CLASP: Consortium of Local Authorities Special Programme
CLAW: Consortium — Local Authorities Wales
MACE: Metropolitan Architectural Consortium for Education
METHOD: Consortium for Method Building
ONWARD: Organization of North West Authorities for Rationalized Design
SCOLA: Second Consortium of Local Authorities
SEAC: South Eastern Architects Collaboration

Many of these consortia have, as we have seen, taken differing routes and the proportion of each building contract employing 'consortium' purchased components varies considerably. Both CLASP and SCOLA have always tended to try and control as large a proportion as possible and in these cases consortium purchase items may amount to 40% of a job. METHOD, SEAC and ONWARD, however, having taken more the role of intervening agencies may have 20–30% of the work on a given contract supplied by the consortium, whilst ASC, more interested in co-ordinating buying policies, accounts for only some 6% (figures DES 1976). The extent to which authorities use a system of which they are a member varies enormously — some, especially in the early euphoric days, put every building (and building type) into the system, others only reluctantly contributed one or two buildings to a given year's programme. Given the now considerable experience of the consortia method in Britain it is somewhat surprising to find in the Department of Education and Science publication *The Consortia* the statement that there 'does not appear to be an "ideal" size for a consortium programme' (DES 1976).

With most of these consortia now having ten or more years of operation (over twenty in the case of CLASP) it might be expected that the Ministry which encouraged them might have brought together information on their running. Such questions as what is the cut off point for bulk purchase savings on a particular components or subsystem, or what are the effects of geographical distribution on prices, are ones which can be examined by consortia data but which seem not to have been published by them or by the Department of Education and Science. The failures of many consortia items have been not inconsiderable but on these there is almost no feedback at all.

Of the consortia so far undiscussed, one with a wealth of experience was SEAC (South East Architects Collaboration), since it was based upon the work carried out in the early postwar period at Hertfordshire County Council first under Charles Aslin and then under Geoffrey Fardell. So powerful and convincing was the Hertfordshire experience that many of those who worked there went on to involve themselves in other system ventures — Henry Swain at Nottingham with CLASP and Cleeve Barr with the GLC, 5M, 12M and at the National Building Agency. However, with all their experience, Hertfordshire resisted for some years suggestions from the Ministry that they should form a consortium, and many of their arguments for such a view were well founded on experience. First, they pointed out that their success relied upon proximity — a geographical closeness that assisted design team, contractors and suppliers to build up real long term confidence; second, and perhaps more significantly, they had discovered that there were limits to the reductions in the costs of components that could be made by bulk purchase. For some component groups the upper limit occurred quite quickly with the result that there was little or no advantage beyond that point. Hertfordshire therefore could show that there was little to be gained from increasing the size of their orders by combining with others. Even as late as 1961, the *Architects' Journal* in publishing a Hertfordshire brick crosswall school proposal at Cheshunt, pointed out that they 'have a big enough programme to pursue experimental and other work without a direct need to set up a consortium to put it into effect' (AJ 1961a). Ultimately in 1963 SEAC was formed (**8.4.21**) and included, in addition to Hertfordshire, Essex and Kent County Councils (**8.4.22**), the Department of the Environment and the London Borough of Hillingdon. A pilot project went on site in July 1965. However, by 1974 some of the predictions of those at Hertfordshire seemed justified since SEAC had to be reorganized with the following aims:

1. To establish annual building programmes of sufficient size to achieve economic production of components.

8.4.21 Southeastern Architects' Collaborative (SEAC) Mark 3 system: 1972. Axonometric showing range of external cladding alternatives

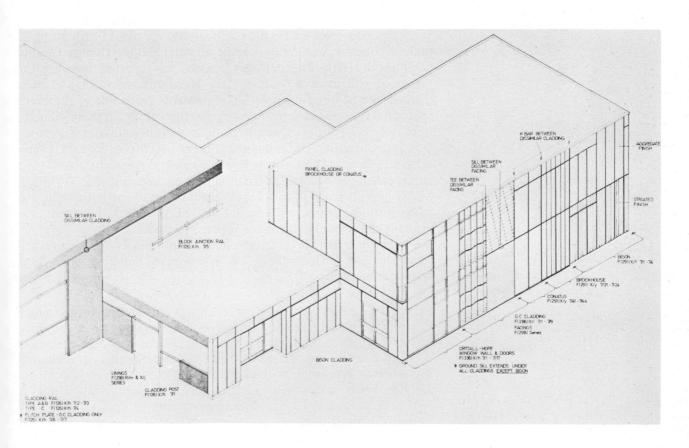

8.4.22 South Avenue County Junior School, Sittingbourne, England. SEAC and Rice, Roberts and partners; project architect, David Aplin, 1975. (Photo 1978)

2. Continuous research and development work.
3. Joint programme quotations for standard components and sub-contracting work.
4. Integration of activities amongst all those participating.

At this time SEAC also set up an arrangement to sell their system commercially as had CLASP and SCOLA, and as with those consortia this was done through the steel frame supplier. Finally, in December 1976, *Building Design* (Stevens 1976) reported the disbanding of the South East Architects Collaboration, effective from March 1977. Workload had dropped from an annual average of £15 million to £2 million. Projects running were to be serviced and the Deputy County Architect to Hertfordshire, Roger Watson, was reported as saying 'the system is still available'. The early Hertfordshire work had been marked by a sense of purpose and a panache which few of the client sponsored systems managed to catch. Further, although these pioneers were convinced of the value of rationalization they still did it within a strictly architectural and humanist context — evidence for this is their considerable use of non-system methods where sensible, and their development of several alternative component systems. Although SEAC took on much of this, inevitably it began to show the signs of a more institutionalized approach. The SEAC Mark 1 system was virtually the whole Hertfordshire system revised.

By the time the SEAC group formed in 1963 Hertfordshire had established its own tradition of component building with three systems available:

1. Steel frame on a planning grid of 2 ft 8 in (813 mm) used up to four storeys. This consisted of light steel double channel box columns, carrying lattice beams and a timber joist roof. Column centres at 21 ft 4 in (6.5 m) and 32 ft 0 in (9.75 m). Significantly internal partitions were brick or block plastered.
2. DISC brick method: a load-bearing brick crosswall method at 24 ft 0 in (7.32 m) and 12 ft 0 in (3.66 m) centres using a 1 ft 0 in (305 mm) crosswall zone.

3. Concrete frame system: a modular concrete frame system developed originally for multi-storey colleges of further education and not used very frequently.

All the documentation was on colour coded sheets: blue for the steel frame, pink for the brick, green for the concrete. In addition all those drawings which contained approaches common to the three systems were referred to as 'basics' and produced on white paper.

Other groupings of local authorities, such as CLAW, ASC and ONWARD mainly used the consortia idea as a method of bulk purchasing components and for the exchange of ideas, and little has been published on their mode of operation, difficulties and results. Each of the consortia has a Management Board consisting of elected members and Chief Officers, a Co-ordinating Committee or Working Party of Officers and Development Group members and a Technical Development Group. In addition specialized working groups are established from time to time to deal with specific issues. There are regular meetings between the Chief Architects of the various consortia and a Technical Co-ordination Working Party (TCWP) both meeting under the chairmanship of the Department of Education and Science.

The proliferation of the consortia brought its own problems, not the least of which was co-ordination, for although they had staked out their own territories clearly much work was duplicated, especially at a technical level. It is true that there emerged a common dimensional framework for the consortia but this came about by government direction. Collaboration has taken place on structural design (stressed skin roof decks for SEAC and CLASP building systems — Bryan 1973), on component jointing, the economics of component production and procurement, the measurement of on-site labour and fire protection. Little of the results of this, if any, have been published and so its effects must be a matter of conjecture. However with the pressure on resources, and the emerging problems of the systems, all found their programmes diminishing. Many authorities became increasingly resourceful at avoiding taking up consortia components and, usually in response to energy and conservation pressures, to use those parts of hitherto closed systems that did make sense — usually the steel frame and steel deck roof at a minimum. In this way they began to move gently towards a more open system approach although this has had its own repercussions on system component procurement policies where planned runs have not been taken up and suppliers have therefore made counter charges on the consortia for the lost business.

Through all this almost no component interchangeability took place between consortia, each having its own unique approach. Public building programmes diminished as problems of maintenance, poor environmental control and aesthetic limitations became, somewhat tardily, recognized. Most consortia attempted to modify their approach, but a dramatic drop occurred in the number of system-built schools under the pressures outlined above and the emphasis on the adaptive re-use of existing buildings. Fortunately, the rumoured creation of a single centralized system did not materialize, no doubt in part due to the avowed aim of the Department of Education and Science which was:

'... not to turn all building systems into a single system, but to expand collective variety by harnessing the experience and resources of all consortia.' (DES 1976)

What has the harnessing of experience left us with? In the event very little systematic documentation of the advantages and disadvantages of such approaches and very little in the way of an overall picture of the performance of such buildings. The promotion of the closed systems may well have mid-directed efforts from more open attempts to introduce co-ordinated components.

9

'THUNDERBIRD AND MODEL T'

A NEW
FORCE

'It is as if, after years with a Ford Model T, one is given a Ford Thunderbird.'

Robert Fawcett, 1968

In the last section the enormous effort put into the development of a multiplicity of building systems in Britain was examined. The political pressures of the early sixties created a climate in which system after system was generated. Having created the momentum it was very unlikely that, whatever the shortcomings of the solutions proposed, the various institutional bodies involved could avoid their attractions and, in the public eye, such systems could be made to appear efficient, cost effective and fast. However, few of those outside the bodies concerned appreciate just how little real cost knowledge many of them possess about their buildings, or realize the hand to mouth basis by which much work is carried out — it was not until the introduction of 'cash limits' to local authority spending in 1975/76 that the overall implications began to be thoroughly looked at. Very few educationalists or housing experts actually liked the building systems, but very quickly they could be heard using the architectural reasoning of the thirties in their defences of systems in which they were now enmeshed. The institutional habit of defending, regardless, whatever it is that you happen to be doing at the time, saw its full flower in the case of the building systems. Equally, however, with the first signs of cracks in the argument and the literal collapse of part of Ronan Point, the depth of this understanding was revealed for what it was — political convenience. The housing systems began to disappear almost overnight: if they did not disappear they continued on sufferance, having become part of the local government administrative machine.

One of the more interesting aspects of this series of events is how each of the client sponsored school building systems was created in the same basic image with few of the major changes of emphasis that were quite clearly necessary. The limited nature of CLASP was emulated almost word for word by SCOLA albeit in another style, and often enormous design changes were made from mark to mark in all the systems, with little real benefit. The fact that some co-ordinating device was necessary if component interchangeability was to ensue was ignored except in the expensive and time consuming pressure for all systems to be put on the same modular basis; and this shortly before another major change — that of the move to metric. Even more crucial was that vital changes in user requirements, technology (particularly environmentally), and conservation could not be assimilated. SCOLA, for example, was not able to introduce pitched roofs until 1976. The lessons to be learned from the changes in housing standards or in educational methods, from European and North American developments in systems were openly treated as unimportant by British building systems proponents. The result is a complex impasse which left both professionals and public almost no room for manoeuvre.

It is against such a background that the work of Ezra Ehrenkrantz must be seen. Commencing in the mid-sixties with his Schools Construction System Development (SCSD) he has offered a number of radical strategies for looking at building problems. His work on SCSD, whilst drawing on British experience, succeeded in throwing a critical light on the way in which the closed systems had developed in Britain and although the Metropolitan Architectural Consortium for Education (MACE) endeavoured to implement some of the SCSD experience it generally went unheeded by the existing building systems. At a more general level the work of Ehrenkrantz on SCSD and subsequently URBS and other approaches, offers another fascinating illustration of the interplay of system ideas at an international level.

THE EHRENKRANTZ ICON

'The systems approach is the only new concept of significance in educational planning in the last 100 years. Teacher education still lags; we still teach teachers to teach in a conventional classroom. They are competent in their mastery of their subject matter, but generally unresourceful in adapting new audio-visual techniques to instruction.'

Frank Fiscalini in C.W. Griffin, *Systems: an Approach to School Construction*, 1971

At a conference in London in 1968 organized by the Ministry of Public Building and Works (now the Department of the Environment) a series of turgid papers were read on the great value of the industrialized building going on in Britain. The last main speaker was Ezra Ehrenkrantz, who described the Californian SCSD (School Construction Systems Development) and his subsequent work (**9.2.1**). After he had finished a performance which demonstrated a mastery of the implications of design, organization, technology and cost rarely publicly attempted by any of those involved in building systems in Britain, Robert Fawcett rose and crystallized the feelings of many present when he suggested that for all the effort in Britain our successes in solving the problem were trivial compared with what we had just heard. 'It is as if', said Fawcett, 'after years with a Ford Model T one is given a Ford Thunderbird.' At this, there was considerable applause and, with others rising to their feet with questions, Sir Donald Gibson in the chair quickly drew the proceedings to a close.

The Thunderbird analogy elegantly portrayed the views of a small minority of those architects interested in building systems in Britain. For in spite of the public acclaim accorded CLASP and its progeny, many architects with first hand experience were aware that school building systems in Britain conformed to a series of technological and stylistic norms that were inhibiting real development. News of the SCSD project began to filter through to interested architects in the early sixties. The first SCSD report, *SCSD: an Interim Report* was published in 1965 and the more elaborate *SCSD: The Project and the*

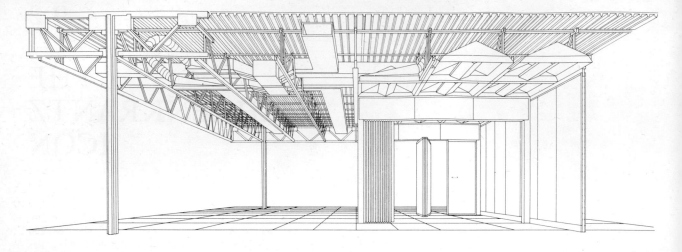

9.2.1 School Construction Systems Development (SCSD). Ezra D. Ehrenkrantz, project co-ordinator. Project commenced 1961; first school using the system, Fountain valley, opened 1966. The cross-sectional perspective showing the various sub-systems which became a powerful model for many other architects

Schools in 1967, both published by the Ford Foundation's Educational Facilities Laboratory in New York (EFL 1965; 1967). Although it received a number of mentions in the non-architectural press it is perhaps significant that it did not achieve major coverage in Britain in the architectural press until the features in *Architectural Design* in July 1965 and November 1967, and in the RIBA journal in August 1965 (Cartmell 1965).

Whilst freely drawing on the mass production argument the SCSD approach focused the shortcomings of the British school building systems sharply: where were the user studies upon which the arguments of the latter were based? Why had they concentrated almost totally on architects constantly revising solutions to the same technical problems with no great advantage? Why was the question of environmental control and services barely studied? Why were maintenance costs so high? Ehrenkrantz also showed that the mass production argument does not mean vast closed systems with guaranteed markets: indeed, the indications were that, in many ways, this was a disadvantage to development. A leader in the *Architects' Journal* in November 1967 vividly drew attention to the questions raised by the success of SCSD whose clients were 'very small indeed and could offer no promise of continuity', and caused the author to wonder 'whether these vast continuing programmes (in Britain) and their administrative embarassments are really necessary' (AJ 1967).

Ezra Ehrenkrantz came to England in 1954 as a Fulbright fellow working at the Building Research Station. During this time he became aware of the work of Hertfordshire County Council in school building, and this ultimately became transmuted into SCSD. His doctoral thesis was an extensive study of number theory and was ultimately published as *The Modular Number Pattern* (Ehrenkrantz 1956). This work reflects the importance that systems builders have attached to the dimensional co-ordination aspects of the work and relates to the interest shown at Hertford in grid sizes, component sizes and number systems during this period. Together with the Ford Foundation's Educational Facilities Laboratory, Ehrenkrantz and a group of school districts in California established the SCSD project in 1961. In *SCSD: The Project and the Schools* the argument is boldly stated:

> 'What is this project that has called forth such praise? Basically it is a means of using the efficiency of modern industrial production to construct schools, while still avoiding standardised plans or monotonous repetition of either rooms or general appearance. It is also a way of introducing specific educational requirements into the manufacturer's part in the building process at an earlier stage than usual.' (EFL 1967).

In the development of the systems argument the SCSD project has brought into focus a number of issues:

1. The study of user requirements and the way this was related to the building requirements.
2. The way in which criteria were defined, particularly the performance criteria.
3. The attitude to first cost and maintenance cost; that these be examined together if value for money is to be achieved.
4. The attitude to bulk purchase and the market.
5. The attitude to structure/fabric and services.

However, before examining the implications of SCSD it is important to know something of the context of education and indeed of architectural design in the United States. With the advent of the Russsian Sputnik on 4 October 1957, a great trauma swept the land. This soul searching, which culminated in United States landings on the moon, gave rise to many new lines of enquiry but in particular provoked a reappraisal of education. The intentions of such a reappraisal were to encourage a more creative approach to the teaching of science and technology, and by this means to unleash new ideas. This, it was hoped, would help to close the 'ideas' gap and the technology/manpower gap. One of the results was a report that has become known as the 'Trump Report', properly titled *Images of the Future* by J. Lloyd Trump, published in 1959. This is a remarkable document in many ways, not least for the concise way in which the case is set out. It is hard to avoid the conclusion that a direct clear statement such as the Trump report has encouraged more innovation and rethinking than the innumerable vast wordy reports on education produced in Britain over the years. The intentions of reports like Newsom and Plowden quickly become diffused by the very means they used to communicate. The Trump report has a number of sections which are closely connected with the way in which EFL and SCSD rethought the school building problem. First was a deceptively

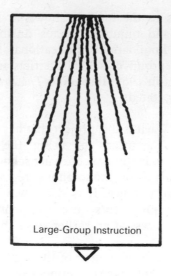

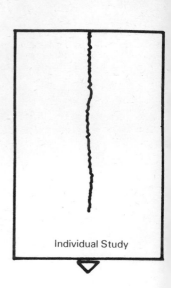

Large-Group Instruction

Small-Group Discussion

Individual Study

Introduction	Group exami- nation of terms and concepts and solution of problems	Read
Motivation		Listen to records and tapes
Explanation		View, Question, Analyze, Think
Planning	Reach areas of agreement and disagreement	Experiment, Examine, Investigate, Consider Evidence
Group Study		
Enrichment		
Generalization	Improve inter- personal relations	Write, Create, Memorize, Record, Make
Evaluation		Visit
		Self-appraise

PLACE
Auditorium, little theater, cafeteria, study hall, classrooms joined via television or remodeling, other large room

about 40 per cent

PLACE
Conference room, classroom

about 20 per cent

PLACE
Library, laboratories, workshops, project and materials centers, museums — inside or outside the school plant

about 40 per cent

9.2.2 'Organization of Instruction' from the Trump report of 1959, showing links between teaching/ learning experiences space requirements, and percentage curriculum time allocation. Notice the emphasis placed on individual study as opposed to the more traditional large group instruction. Such an emphasis gave support to more open, flexible plan forms and encouraged experiments with individual work stations and mobile furniture

simple division into three types of school activity: large group instruction, individual study and small group discussion. The first would take 40% of the students' time, the second, 40% of the students' time, and the last about 20% (**9.2.2**). This concept is then expanded (**9.2.3**), showing visually how the normal 40 or 45 minute time blocked day could be reconstructed more usefully.

The section on the usage of staff is particularly relevant since it is concerned with rearranging the school organization and its financing in a manner which matches the types of activity undertaken, or which could be undertaken. The report particularly shows that in appointing administrative staff to deal with the enormous amount of such work attendant upon such an organization and also using more positively the various levels of teacher activity (aides, assistants), available money and time could be used more efficiently. The chart from the report demonstrates the effects this reassessment would have, and was backed

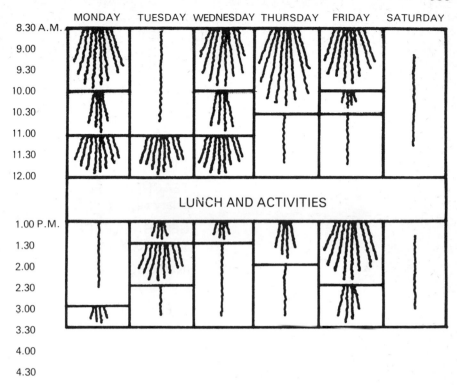

MONDAY TUESDAY WEDNESDAY THURSDAY FRIDAY SATURDAY

LUNCH AND ACTIVITIES

 Large-Group Instruction: various subjects, activities, and lengths of periods — 12 hours

 Small-Group Discussion: various subjects, activities, and length of periods — 6 hours

Individual Study: various subjects, activities, and places — 12 hours, not including time on Saturday and after 3.00 P.M.

9.2.3 'How a student might spend time in the secondary school of the future' (Trump 1959)

up by another potent example in the form of a redistribution of expenditure on staff salaries (**9.2.4**) to gain greater flexibility in time and in space usage for students and teachers. In the absence of such a well integrated argument, attempts in British schools to relieve teaching staff of duties which could easily be undertaken by others have been half-hearted and inadequately supported. Such aides or assistants could assist with teaching material organization and distribution, setting up experiments, audio-visual aids, form filling, typing, general administration and the inevitable communication problems of a large organization. With schools ranging in size from 50 pupils to 2000 there are inevitable management problems. Yet almost no Head teachers have the background to run such an organization nor are they very often supported by anyone who does. Hence the extraordinary inability possessed by most schools when it comes to the distribution of resources within and, even more critically, the

NOW (1959)

16	TEACHERS — average salary $5,500	$88,000
	Some clerical help available but now charged to the principal's budget	
	TOTAL	$88,000

FUTURE

5	TEACHER SPECIALISTS — average salary $8,000	$40,000
5	GENERAL TEACHERS — average salary $5,000 same as at present	27,500
	INSTRUCTION ASSISTANTS, 200 hours per week, $1.80 per hour	12,960
	CLERKS, 100 hours per week, $1.40 per hour	5,040
	GENERAL AIDES, 50 hours per week $1.30 per hour	2,340
	TOTAL	$87,840

9.2.4 The Trump report example of the way in which the normal salary structure could be changed to introduce more administrative and teaching help within the same overall budget. (Trump, 1959)

obtaining of resources from without. The teaching profession has itself contributed to its own difficulties with arguments about when is a non-teacher teaching, and whether such posts are acceptable. In Britain some schools have created the post of Bursar, but this is a minimal response to the considerable managerial implications of the modern school organization. More enlightened counties, such as Leicestershire, have two deputy heads in a large comprehensive school: one of these will be virtually a managerial post. Having examined types of learning situations and types of organization to facilitate possible changes, Trump dealt with educational facilities under three heads:

1. 'Educational facilities will no longer be merely a school building and its grounds.'
2. 'Space within the building will be planned for what will be taught and how it will be taught.'
3. 'Installations for effective use of electronic and mechanical aids will be provided.' (Trump 1959)

These seemingly innocuous statements brought in the change that is still taking place in teaching/learning situations under the pressures of sources other than the 'teacher speaks' variety (**9.2.5**). The report also has comments to offer on community and school, team teaching, in-service training, the use of clerks and aides, the use of university and other consultants from other areas of education. Those interested in

9.2.5 'Avenues to learning', Trump report (1959)

BEFORE WRITING
BEFORE PRINTING
UNTIL ABOUT 1900
THE 20TH CENTURY

TEACHER WRITING PRINTING FILM RADIO TV RECORDINGS MACHINES

the idea of the diffusion of the school will also see its emergence in this remarkable document. Like many documents from North America its simple directness is almost too much for the European: or too little. The European requires his arguments more densely formulated in keeping with his literary traditions. The potential of such simplicity may be dismissed as naïvity but such cultural chauvinism can blind us to its real virtues in solving problems. The climate of thought evident in Trump's slender book permeates the work of SCSD — first in its study of the requirements and then in its argument for the approach to building adopted. This marriage of American concepts about education with a British concept about how buildings can be made has produced results of considerable potency. Before examining the SCSD system itself two other contextual aspects are relevant: the traditional development of school buildings and the position of the building industry in North America.

The one teacher to 30 children concept, which grew from the 'little red schoolhouse' has resulted in the one teacher to 30 children multiplied by 4 or 6 or 8 concept. The built result was a string of boxes, often alongside a double-banked corridor with mechanical extracting along the inner walls. The development of school building forms is a separate study, but such work as that of TAC (The Architects' Collaborative) at Wayland in Massachusetts in 1960, John Lyon Reid with Hillsdale High at San Mateo, California in 1955 (**9.2.6**) and Caudill Rowlett and Scott in Texas in the fifties constitute

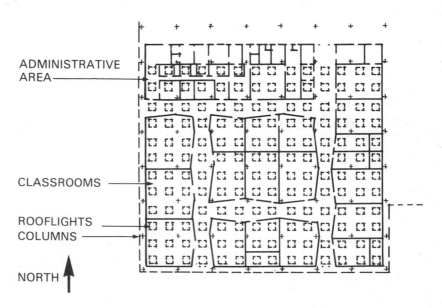

ADMINISTRATIVE AREA

CLASSROOMS

ROOFLIGHTS
COLUMNS

NORTH

9.2.6 Hillsdale High School, San Mateo, Cal, USA. John Lyon Reid, 1955. Part plan. An early example of the 'loft' plan school using a deep plan, rooflights, and moveable office partitioning to offer a flexible interior

such a fundamental rethinking of the problem that one can see clear connections with Trump, and later with SCSD. At Hillsdale High the so-called 'loft' plan was used: a large hangar like space with steel frame and flat roof. Internally the subdivision was by means of standard office partitioning to provide a totally flexible space which could be changed at very short notice, by then already a necessity in Californian schools. This project also had internal classrooms. At Moore Junior High School in Tyler, Texas, 1955, Caudill Rowlett and Scott used a light steel frame with partly glass partitions between classrooms, which was a step towards the open plan schools to follow. At Peter Pan School at Andrews, Texas, 1955, the same architects produced the first carpeted classroom in the United States, with a ceiling lighting grid which bore great similarities to the SCSD solution of ten years later. This is perhaps not surprising since, as Caudill points out, Charles Lawrence served on the SCSD advisory board and 'we "loaned" SCSD one of our people, Bert Ray, for a two year period to help develop the ceiling system' (Caudill 1971).

The questions raised by the new types of internal environments being created by these schools, particularly those that employed a 'loft' plan and had classrooms without windows, were to have resounding effects on architectural design — and not only in the field of school building. The deep plan, with its savings in external wall and need for good internal environmental control systems has created bitter arguments amongst architects. Behind such arguments are more fundamental feelings about just how far we should go in isolating those using buildings from the outside world. One of the most important pieces of research into the question of windowless environment was carried out by Theodore Larsen (1965) and his team at the University of Michigan in the early sixties. In this study Larsen took two Unistrut system built schools of similar form and made a wide range of studies covering three basic stages: the first with the classrooms as normal, with windows; the second with the windows removed and the addition of good artificial lighting and environmental control; the third stage with the windows replaced. The study, as Larsen readily admits, produced very little that was conclusive either way — at least in measurable terms. One unexpected effect was an unexplained higher absence rate for the kindergarten children in one of the schools with windows removed. This Larsen suggests should be investigated further. 'Some concern for an outside view' on the part of pupils was also evident but the report points out that 'the test school children have shown very little personal interest in whether their classrooms had windows or not' (Larsen 1965).

The most positive statements concern the responses of teachers about whom the report states:

> 'There is no question as to their preference for windowless classrooms, once they have had the experience of teaching in such an environment, and they are unanimous in their reasons for not wanting the windows.' (Larsen 1965)

These reasons included the elimination of outside distraction for the children and the additional extra wall space. However, as the report also points out, professional educators were concerned at the educational implications of the stimulus that is often provided by a view of the outside. Such a discussion perhaps says more about the place of education in an age of production than anything else. Indeed in many senses the child in the windowless clasroom perhaps with a not so good teacher, is not far removed from the worker on the production line. However it remained for later commentators to raise such issues more forcibly. It is easy to see, then, that within the American school design field the pressures for change associated with the lessons of the space race crystallized into the proposals of Trump and SCSD.

It is also relevant to establish the position with regard to the North American construction industry since it is sometimes assumed that SCSD is a major revolution that merely appeared from thin air. Compared to the British construction industry, and that in a number of European countries, the US industry is very efficient, although much of this is subtle and only evident with experience. The American industry is very pragmatic, competitive and responsive, and tends not to have its operational modes neatly conceptualized. However, there is a hard-headed openness to new possibilities at many levels which make the campaigns in Britain to introduce, for example, tower cranes or critical path programming, seem somewhat comic. It is to the credit of SCSD that in promulgating their ideas they were able to use the existing concepts of competitiveness and responsiveness to great effect rather than attempt to impose some total version of order, social and architectural, on the building situation.

This natural responsiveness can be seen at work when the performance standard concept is examined. It is sometimes ignored that, albeit in a rather crude form, specification by result (or performance), rather than by the means for achieving it, has always been a feature of architectural design and bid (tender) practice in the United States and Canada. To describe the performance you require in a window, for example, the specifier will list each item and often will give a preferred brand name but follow this with the statement 'or equal'. Deceptively simple. This leaves it to the General Contractor to provide the required performance in the most economic way and the fact is that he usually has far more reliable information on which to do this than does the architect. Further he has a more direct relationship with those subcontractors who will work with him. Added to this, and developed to a high degree of sophistication, the defence department have used tender by performance extensively, for many years. One does not design a space vehicle and put it out to tender: one calls for answers to the performance requirements stated. This approach is obviously of great importance when it is necessary or desirable to take advantage of developing technologies which a designer may know little about in detail. It will also allow cross-industry solutions to be offered. For example, although certain solutions in building may be thought of by architects around a concrete frame, a steel frame of a new type may solve the problem equally well or better. This approach requires the problem to be set out in a particular way, by defining the performance required. The basic procedure adopted by SCSD was as follows:

1. Identify a number of school districts and gain their commitment to the use of components and subsystems dependent upon process and design being satisfactory.
2. Examine the way schools are used in the area, organizational and other structures and assess likely future trends.
3. Convert the requirements into performance criteria and performance specification of subsystems in relevant categories.
4. Survey relevant areas of industry who would be interested in this market.
5. Take in competitive bids (tenders) and designs and select best value.
6. Integrate the various subsystems.
7. Prepare documentation on the systems including unit prices to individual architects appointed by each school district.
8. Each architect then designs each school using the preselected subsystems and components.
9. Each individual school is bid (tendered) locally and the appointed general contractor will assemble the components and carry out all other work as normal.

This, however, must be placed in its context since that context may yield information as to how transferable are the ideas to other countries, other situations. Further, the specific solutions, the way the system developed technically, is of equal importance, and will also be examined.

The school district in the United States has many significant differences from local education authorities in Britain. The most important is in its autonomy as opposed to the centralized co-ordination and control function practised by the Department of Education and Science in Britain. The latter, whilst superficially open and democratic, has a number of highly questionable features, the most crucial of which concerns the way in which innovation can occur. Because of the financial and administrative chain of control, school heads and teaching staff recognize implicitly that change or innovation has to be argued through a series of administrative levels which by their nature will be unlikely to react favourably. Innovators do not have to get the support of their community, or of the public, or even of publicly elected figures, but often of a series of educationalists and administrators. At the top end of the scale, Ministry level, both in education terms and building terms, innovation and change is constantly promulgated. However, at this level, a paternalistic hostility to ideas emanating from a grass roots level is often detectable.

American school districts (run by school boards) are in a totally different position since each is administered by a board of Trustees who direct policy: these boards are composed of volunteers elected at local level. Each School Board appoints a full-time director and maybe a staff who carry out the day to day running of the school system. The amount of finance available is dependent largely upon local agreement to necessary expenditures with contribution from federal funds of only some 5%. The local community itself votes to increase its taxes to finance each school or group of schools proposed. This can give rise to serious anomalies between affluent and poor areas but also has many advantages. The most important of these is that there are much more direct links between the community, the resources and the result. Policies can be seen to be operating or not operating and innovation can occur much more directly. It also places considerable responsibility on those operating the particular school system and this can encourage caution or timidity, but can also act as a healthy brake on change for its own sake. However, the amount of innovation that does go on seems to suggest that giving individuals responsibility in this way is more just and more likely to encourage full participation.

It will be seen from this that the sorts of pressures that created groupings such as CLASP, SCOLA, SEAC and other British consortia cannot easily exist. The creation of a market was the first prerequisite for the SCSD group although their attitudes to this were more open than their British counterparts. This involved gaining the co-operation of a number of school boards, each one autonomous. In the event 13 came together and the work involved repesented some $25,000,000 in school building work, over a two year period. This should be related to the fact that both CLASP and SCOLA carried much less annual building work in their early years: the CLASP programme at inception in 1957 was £1,000,000 whereas in 1961 (the year SCSD commenced) it was £6.75 million.

An examination of the needs of each school district was first carried out and this culminated in the production of a document entitled *General Educational Specifications* published in 1962. These outlined user requirements and gave design aims, including illustration by means of a hypothetical school for 1800 pupils called 'Hypo High'. This concept was subsequently used in design terms to gain a knowledge of quantities involved and other such problems. The main points which came out of this study were:

1. The need for varying size group spaces down to 450 square feet (41.81 sq m), each with their own environmental control and servicing.
2. Subdivision of these spaces for individual work spaces. A flexibility requirement on the basis that a flexible school can be used for conventional teaching patterns but a non-flexible school cannot be used for new methods requiring different sorts of spaces.
3. Related to this was a requirement for long spans together with flexibility in control systems and internal space division.

These general aims were then expressed in this way:

1. Long span structures.
2. Varied mobility in the partitions.

3. Full thermal environmental control with the ability to adapt to changing plan configurations.

4. An efficient and attractive low brightness lighting system which is adaptable to new plan configurations.

The performance standards derived to answer these requirements were seen as a means of gaining a creative response from industry to the specific problems posed by schools and in this way avoiding the mere adaption of components (for example, office partitions) developed for use in other building types. A further important implication is that in the United States it is unlikely that an architect appointed to design a school would accept components designed by another group of architects and would be more sympathetic if the components arose from industry in response to established standards. The following section of the performance specifications, concerning the heating, ventilating and cooling requirements (category 7) shows the approach:

> 'The structural systems shall allow the various District Architects freedom to plan the structure of the individual schools in a 5 ft × 5 ft (1.52 × 1.52 m) module or multiples of this module. The structural systems in all areas of the school except physical education shall be designed to meet the requirements of an integrated structural, mechanical, and lighting-ceiling sandwich (hereinafter referred to as the integrated sandwich). Bids will be evaluated on the basis of consistency, compatibility and a composite total price of the integrated structural, mechanical and lighting-ceiling solutions.

> 'Mechanical equipment, or components, which are exposed on building interiors and exteriors shall be well organised and detailed. Component contractors are encouraged to think of the final appearance of their equipment as contributing to the character and interest of the general architecture of the building. Controlled expression of function is suggested as a guide to the desired design approach in exposed equipment (a ship's ventilator is suggested as a good example of this approach)'. (AD 1965)

This approach involved not only trying to interest manufacturers but integrating the work of different firms. A number of conferences were held for interested manufacturers where the concept was explained. This had the added advantage of bringing manufacturers from different fields together, something that in normal circumstances rarely occurred. For example, where the performance specification called for an interface or integration such as occurs with roof, ceiling and internal partitions, it was the responsibility of the subtrades to make this integration — with advice and help from the SCSD group. This is a long way from the development group approach used in Britain where not only are components designed within the orbit of

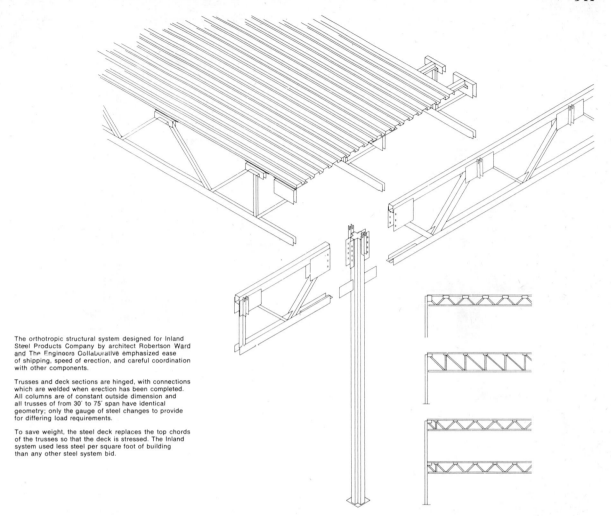

The orthotropic structural system designed for Inland
Steel Products Company by architect Robertson Ward
and The Engineers Collaborative emphasized ease
of shipping, speed of erection, and careful coordination
with other components.

Trusses and deck sections are hinged, with connections
which are welded when erection has been completed.
All columns are of constant outside dimension and
all trusses of from 30′ to 75′ span have identical
geometry; only the gauge of steel changes to provide
for differing load requirements.

To save weight, the steel deck replaces the top chords
of the trusses so that the deck is stressed. The Inland
system used less steel per square foot of building
than any other steel system bid.

9.2.7 School Construction Systems
Development (SCSD). The structural
system. Robertson Ward, architect
and The Engineers Collaborative for
Inland Steel Products Co. Trusses and
deck hinged, welded when completed.
Steel deck replaces top chords of
trusses

the group, but also all interfacing is done in the same way, albeit with
advice from manufacturers. Ehrenkrantz's approach demonstrates
again, and with great clarity, one of the traditional differences between
British and American practice: the balance of responsibility between
architect and contractor or subcontractor. In the United States the
contractor takes more responsibility when the work is in contract, as
indeed he should since he is financially in charge at this stage. Whilst
the architect makes periodic checks it is not his job to do the day to day
work of the contractor, and to check or measure his every move. The
specification-by-result approach is clearly bound up with this
tradition.

After the first SCSD conference manufacturers submitted an initial
submission called the 'Evaluation Submission'. No prices were
requested. Since no schools had at that time been designed bids were
requested on the basis of a quantity of material and equipment to build
1,400,000 sq ft (130,060 m) of floor space, a guaranteed minimum,
and subsequently bidders were asked to provide unit prices. Fifty

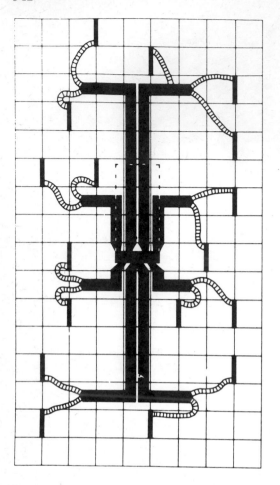

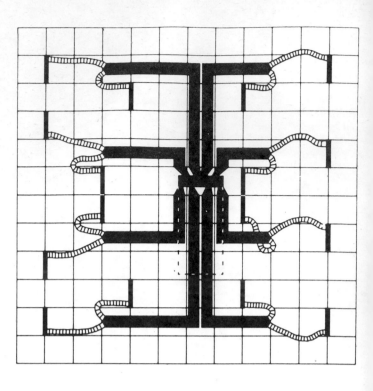

9.2.8 SCSD: ceiling duct layouts.
Lennox Industries. Heating and
cooling module serving 2500-5000 sq
ft (232.25-464.5 sq m) of space with
fixed and flexible ducts

firms submitted Evaluation Submissions in September 1963 and this
was followed by a period of development work with final bids
submitted by 26 firms in December 1963. After evaluation six firms
were nominated on January 7 1964, as follows:

Structural: Inland Steel Products with an ingenious folding roof at
$1.81 sq ft (conventional construction $3.24) **(9.2.7)**

Heating, ventilating and cooling: Lennox Industries with
composite roof mounted units and distribution *including* maintenance
contract for 5 years, at $2.24 per ft (target had been $1.90 with
minimal air conditioning and no maintenance) **(9.2.8)**.

Lighting-ceiling: Inland Steel products at $1.31 per ft (target $1.58)
(9.2.9).

Demountable partitions: Hauserman Company **(9.2.10)**.

Operable partitions: Hough Manufacturing and Western Sky
Products.

The aggregate partitions bid was $1.52 as compared to the target cost
of $1.67. The aggregate of all component bids was $6.88 per ft against
an assessed cost of $8.39 using conventional components, representing
a cost saving overall of about 18%. The Inland Steel roof, designed by

Robertson Ward and the Engineers Collaborative of Chicago, had been developed from a feasibility study by Ward 'after pressure from one of their (Inland) executives and persuasion from Ehrenkrantz' (Hislop and Walker 1970). During the building of the schools due to site and factory welding problems it was agreed not to use the hinged connections and to weld one side of each deck unit to the beams at factory.

In view of the amount of attention given in Britain to dimensional co-ordination and the modular argument the requirements in this respect for SCSD are of interest. The structural subsystem was required to conform to a planning grid of 5 ft (1.52 m) × 5 ft (1.52 m)

9.2.9 SCSD: Lennox heating ventilating and air conditioning system. A self-contained heating and cooling unit feeds into mixing boxes which feed to the fixed ducts. Air is returned by the open plenum

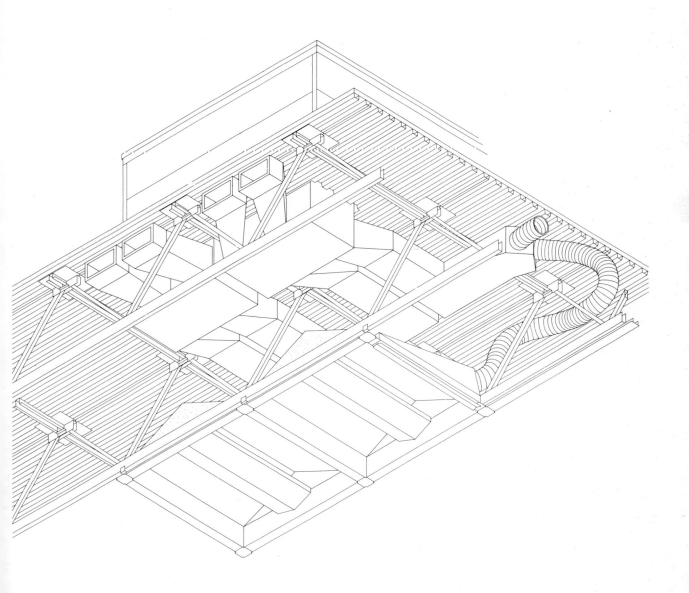

544

and to have vertical increments of 2 ft (610 mm). At the same time all subsystems had to 'acknowledge' the requirements that would allow district architects to plan interior spaces to a module of 4 in × 4 in (102 × 102 mm). This was called the partition planning module. The thermal criteria are as set out in **9.2.11.** The investigation during the educational requirements stage had shown the importance of maintenance. Indeed it was found that some school districts paid out an additional 40% of the first cost on maintenance during the first year, and that 10–20% was quite common. The inclusion of a maintenance contract with, for example, the heating, ventilating and air-conditioning installation, had the effect of discouraging

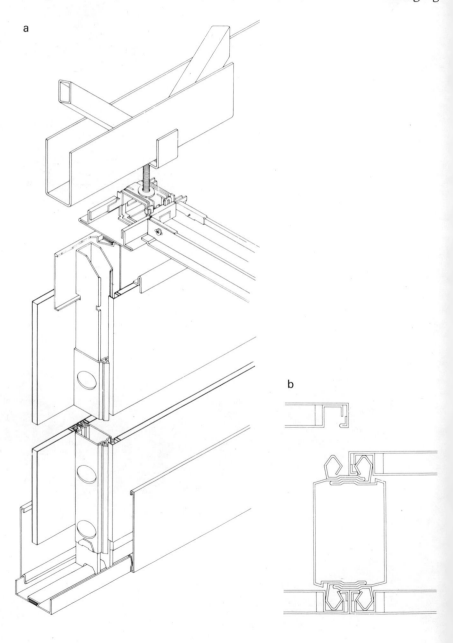

9.2.10 SCSD: fixed and demountable partition system by E.F. Hauserman Co. 'Spider' connecting ceiling runners to structure every 5ft 0in (1.5 m) by Fastex Division of Illinois Tool Works Inc, designed to take up deflection and movement. (a) Isometric of partition system and connection to structure. (b) Plan of vertical support and clip on panels

9.2.11 SCSD: Performance criteria for heating, ventilation and air-conditioning

Temperature	Plus or minus 2°F
Outside air	Minimum 8 cubic feet per minute per person
Total air supply	Minimum 30 cubic feet per minute per person
Air velocity	Between 20 and 50 feet per minute
Solar heat gain	6000 Btu per hour per 200 sq ft of exterior wall
Outside temperature range (for design)	30–100°F

manufacturers from cutting materials to the bone in order to obtain the job and indeed manufacturers expressed enthusiasm for the orderly way that SCSD offered to deal with the long term effects of the practices forced on them by the market. Ehrenkrantz quotes the later bids for carpeting as a further example of the operation of the laws of maintenance. In bidding for floor covering one manufacturer offered to install carpet free provided he was awarded the maintenance contract. Calculations demonstrated that this would be cheaper than the conventional method of differing floor surfaces everywhere and considerable annual expenditure on cleaning labour, materials and equipment. This again throws an interesting light on the way maintenance has been handled (or not handled) in most British examples of systems. There is no evidence that a proper account has been taken of the maintenance costs of the materials and methods used in British school building systems. This results from a number of factors:

1. That accounting procedures on the part of central and local government have constructed a situation where the money for capital investment and the money for maintenance and upkeep come from two different pockets, and these are normally administered by different departments.
2. An interest on the part of many architects in what they call short life building, lightweight structures and obsolescence.
3. The training of the architect encourages innovation for its obvious personal values and in turn encourages a transference of this directly to the world of building materials: the philosophy that in changing the technology, behaviour and society can be changed, is still strong amongst architects.

Evidence of all these trends can be seen in school building throughout Britain where cheapness has been substituted for value for money under the pressure of forces such as those outlined.

The way in which SCSD dealt with external walls offers a further example of a recognition that some of the shibboleths of the British system builders could be cast off. CLASP, SCOLA and other systems have spent enormous time and effort designing and redesigning the external skins of their system. Great care and attention has been lavished upon the details and junctions, although it would be difficult to determine this by inspection, or by an assessment of performance, since so little information has been published. Many of the systems have had constant water leakage problems, together with expansion, contraction and tolerance problems, often because the architect designers have had to learn or relearn the languages relating to the materials and construction being used.

Ehrenkrantz and the SCSD group, when studying the cost problem found that only 7% of the cost was in the external skin and that whatever saving could be made there, would be marginal when set against savings made elsewhere. Additionally the level of performance in the industry in the United States suggested that, whether using concrete block or curtain wall, good performance standards could be acheived. Clearly the political problem of gaining the co-operation of district architects was important and by excluding the external wall from the system advantage could be taken of local conditions and materials together with the design aims of individual architects and their clients.

This is particularly instructive in the face of the attitude of most British system builders who seem to have been concerned to control the whole building in the interests of controlling the architecture itself. It also illustrates the open and pragmatic nature of the SCSD approach. This, which caused Fawcett to describe it as a Ford Thunderbird rather than a Model T, embodied a really hard look at what was actually happening in schools and education and a very thoroughly worked out method of work. The development of the performance criteria and performance standards idea is of particular import as is the related attempt to make use of the dynamics of the industry. The innovations that this produced are significant in themselves and have had far-reaching repercussions throughout school building internationally and in architecture generally. Many of the subsystems were marketed commercially outside the SCSD programme, with even some of the unsuccessful tenderers like Macomber and Butler offering their structural systems generally. Inland Steel were least successful here and their components were not continued. Lennox and Hauserman were very successful in marketing the units they had developed for the SCSD programme. This floating of subsystems on the open market tests their performance and cost effectiveness more clearly than if they were permanently protected in a system environment.

The awarding of subsystem contracts for the whole group of SCSD schools to single subcontractors created delivery problems similar to those experienced by the schools consortia in Britain. An early design and construction schedule for SCSD indicated that there was a bunching of start and completion dates in the first year and this posed problems for component suppliers who 'could not be expected to work on all of the projects at one time' (Boice n.d. c 1970?). In addition the school districts could give little firm information until state funding intentions were clear — this was not until March/April 1965 and yet the majority of start dates called for were during the last two weeks of August in the same year.

Enormous difficulties arose with Inland Steel, who had promised that no other work would be permitted to interrupt the SCSD commitment. However, it was revealed during a painful process of negotiation, that Inland had on 20 August 1965 accepted a large contract from the Lockheed Georgia Co. This resulted in a reduction of the original production schedule from 12,000 to 8000 sq ft per day (1105 to 783 sq m). After a series of meetings Inland Steel were served with a Temporary Restraining Order to stop them supplying Lockheed which, according to Boice, was not observed. Inland had also taken out an injunction restraining SCSD and the contractors from interfering with its Lockheed contract.

After further discussions over scheduling, a compromise settlement was reached. Inland went on a three shift, six day week to increase production, which was already behind schedule. However at no time did actual production equal expected production (Boice n.d.). Further concern arose when Inland requested price increases on all 'optional', non-system items supplied with their components. This, a traditional response of contractors in difficulty with fixed prices, was partially solved by compromise agreements. The summary of reasons, put forward by Boice, for the production and construction problems are familiar to all involved in building systems work:

1. School districts could not offer or guarantee firm dates.
2. Development work completed late by Inland Steel.
3. Production capacity of plant overestimated by Inland, and their inability to meet required schedules.
4. The difficulty project architects had in accurately estimating building costs, causing delays.

Such difficulties should be carefully considered by those holding ideal views of the production capabilities of factory and site. It is to the credit of SCSD that, unlike any other system, they have published a thorough assessment of the work by John R. Boice, *A History and Evaluation of the Schools Construction Systems Development Project 1961-67* and this deals with design, construction and user feedback.

It is clear that there have been a number of problems with SCSD, the biggest of which concerns acoustic control. In one survey, at Harbor High School in Santa Cruz, 'noise isolation between rooms' was rated 'poor' by 13 of the 23 teachers and by three-quarters of the 156 students. The Educational Facilities Laboratory summary of SCSD failures says:

> 'Acoustical design is, beyond all doubt, the weakest aspect of the SCSD schools. It is far less a failure of hardware than a failure to use it properly.' (Griffin 1971)

The report points out that acoustic control is a major problem in open plan schools and many of the architects using the SCSD system failed to recognize this and also ignored the guidance prepared for them by an acoustic consultant. This draws attention to a common 'system' difficulty — that when almost the complete kit of parts is changed from more or less traditional and understood techniques to a system, the architect on the receiving end may suffer an overload of change. Often, it seems, even his normal design skills desert him and an implicit judgement is made that the system can deal with everything.

The built-in flexibility of the system presented problems of a different nature. In spite of the enormous efforts put into this aspect by the Ehrenkrantz team there was a communication breakdown between them, the school principals and the teachers (Griffin 1971). In the survey carried out by Building Systems Information Clearing House (EFL 1972) it was clear that a majority of school staff members were unaware of the facilities for change available to them **9.2.12.** Somewhat surprisingly, the survey found that the longer a teacher had been at the school the more likely he was to suggest changes. Whilst 73% of the staff rated the overall appearance of the schools 'good' only 36% of the pupils did so, and the lack of colour was a major student complaint.

The individual school boards and their architects failed to take advantage of the system in other ways. The offered maintenance contract on the heating, ventilating and air-conditioning (HVAC) subsystem is an example, since it was not taken in one case, and thus the school lost out in two ways: the HVAC subsystem had been designed on a more durable than normal basis (with the prospect of having to maintain the equipment) and they then had to pay for another firm to maintain the equipment in the usual way.

The surveys carried out by EFL are commendable and clearly point to the close intertwining of solution and problems. It is pointed out that in some cases schools are not being used as the designers anticipated. However, it is clear that the heavy emphasis on open planning built into the concept was inhibiting:

> 'Large open spaces, which were designed for team teaching, are being used as self contained classrooms with disastrous results acoustically.' (EFL 1972)

Although individual architects often failed to use the available solutions (sound absorbent wall coverings) the ceiling plenum in itself was a major source of sound transfer.

Clearly then the system itself created quite new problems, both in terms of communication between the many parties and the precise nature of the technical solution evolved. Unfortunately much of SCSD appears to have been misunderstood and its specific solution given more importance than they would claim. The very drawings produced have created a whole architectural style equally as powerful as that produced by Le Corbusier's Dom-ino drawing. The 'Ehrenkrantz Icon', the precision drawn cross-sectional perspective which shows structure, construction, servicing and interior is a useful and powerful communications vehicle (see **9.2.1**). The approach and the drawing method recurs in Foster's work on Newport and IBM, Wright at Redbridge, and in the MACE system. Performance

9.2.12 Faculty knowledge of the flexibility of various SCSD subsystems. From BSIC Research Report No. 2 (EFL 1972)

Subsystem	Response (%)		
	Can be changed	Cannot be changed	Don't know
Interior walls	47	34	19
Accordian partitions	48	28	25
Operable panel partitions	33	28	39
Lighting fixtures	18	51	31
Ceiling panels	28	37	35
HVAC outlets	15	43	42
Interior of cabinets	67	18	14

standards have become a Ministry panacea and are being built into some British Standards. The attempt however to establish sets of performance criteria for buildings, building systems or components in Britain has met considerable difficulty since there is little traditional basis for it. Fifty years ago adequate performance was achieved by employing the right people, the right craftsmen, and although this can no longer operate, little in the way of a new tradition has been firmly established to take its place. The lessons of SCSD then are considerable but the dangers inherent in assuming its procedures are easily transplantable are equal to those involved in assuming that the 'English' spoken on the North American continent has the same meaning as that spoken in Britain.

TORONTO'S SEF

'...a system is nothing more than the subordination of all aspects of the universe to any one such aspect.'

Jorge Luis Borges
Tlon, Uqbar, Orbis Tertius: Labyrinths,
1971, 1964

It is in the nature of building systems that they seem to burst upon the public scene soon after their initiation and then fade into oblivion as far as further public information is concerned. After the initial wave of publicity, partly necessary of course to gain credibility for the system, the buildings get built and sometimes reported upon. Indeed some of the systems herein described have been followed up. By and large, however, remarkably few really comprehensive follow up studies have been published. Of course such studies would have many negative qualities which would stand poorly alongside the positive crusading view that the launching of a system can command. It would be unfair to say that the Toronto School Board Study of Educational Facilities (SEF) project falls into this category since subsequent studies have been carried out; nevertheless, these have hardly been able to match the publicity achieved by SEF, in the attempts to get the approach originally well established. The project's technical director, Roderick Robbie, originally from Britain, settled in Canada in 1956. Educated at Portsmouth and Regent Street (now Central London Polytechnic) Schools of Architecture he had worked under Roger Walters at the Eastern Region of British Rail, developing aspects of prefabrication. Robbie's best known project is probably the Canadian Pavilion at the 1967 Expo in Montreal upon which he worked as one of a team of eight architects, and where he developed his ideas of an overall environment and the implications of the co-ordination of many subcontractors. It was one of the few pavilions to be finished for the opening day.

SEF started a comprehensive study in 1966 with Robbie as technical director and Hugh Vallery as academic director. The problems faced by the Metropolitan Toronto School Board were familiar ones: very fast population growth, capital shortage, vast numbers of portable classrooms which were seen as undesirable, and obsolete buildings. It should be pointed out that the School Board was classing schools of 35 years old or older as obsolete and demolishing them; further there was talk of doing the same for anything over 15 years old. This will be recognized as a familiar attitude in North America summed up by Sir Hugh Casson in his observation: 'in New York they tear all the permanent buildings down and leave the temporary ones up'. These pressures, in addition to those for educational change in a rapidly growing city where half the population are first generation immigrants, are clearly immense. Nevertheless one cannot help but look carefully at such arguments when they form the basis for the establishing of systems the world over, whether it be in North America, Eastern Europe or in Britain — where change can hardly be said to be rapid when set against Canada or the United States.

Having established the problem criteria, the SEF group set about their task with three study areas. These fell into the following categories, (E) Educational, (T) Technical, (A) Administrative.

E: a series of educational studies dealing with educational specifications and user requirements.
T: dealing with the development of a building system, high rise and mixed used structures (i.e. including educational and non-educational aspects), and the study of temporary building provision.
A: dealing with the administrative and execution aspects.

The stated purpose of the SEF system was 'to build schools in reduced time, at no extra cost over existing traditional methods, with total internal flexibility and to a higher standard' (Robbie, 1970a). The initial programme was to apply to some 2 million square feet of construction (185,800 sq m) on a budget of 41.7 million dollars. The building system developed contained ten subsystems. The work draws heavily on Ehrenkrantz's SCSD project although it embodies a number of interesting extensions of the approach. As with SCSD there was a performance specification for each of the subsystems with bids (or tenders) being called for from prequalified bidders, these usually being combinations of subcontractors and manufacturers. A complex system of interfacing subsystems during the bid stage was developed. For example the structural subsystem bid has to include an interface with both the vertical skin and atmosphere subsystems. Further than this the bidder had to give two alternative prices for interfacing with two subsystems in each case, in this way offering a wider range of choices through the large number of possible subsystem combinations. Bidders were then asked to attach penalties against those nominees who, for them, would have poor technical interface or poor business interface (i.e. poor co-operation from a business point of view). This complex procedure involved the consideration of over 1 million subsystem combinations, producing, says Robbie, '13,040 complete building systems' (1970). Variety with a vengeance. It is particularly interesting in its attempt to encourage co-operation between firms and to place upon them a responsibility for making explicit both the technical and managerial implications of their decision making.

Robbie's avowed intent was to create an open system which no doubt led to this sophisticated method of encouraging responsive co-operation from suppliers and subcontractors and to the possibility of a series of cost performance options for the client body. This demonstrates a confidence in the systems approach never accorded by, say, any of the building systems in Britain where architect control over decision making has always been firmly held onto at all levels. Robbie says this about his approach:

> 'In defining the methodology of the programme, I was not interested in technical innovation, but rather in rationalisation of the human relationships and skills of the industry.' (Robbie 1970a)

He goes on to describe how, at the many pre-bid meetings, he emphasized that industry was not to get involved in technical innovation and heavy retooling to execute the programme. This was another interesting departure from previous approaches where innovation had always been stressed. The cost targets were set at 20.85 dollars (Canadian) per square foot compared to 26 dollars for the same buildings to SEF standards if built by traditional procedures. The construction time reduction sought was 30%, which, in view of the speed of construciton already in North America was a considerable request. The low bid combination of subsystems ultimately approved gave a building cost of 18 dollars which was increased to 19.10 dollars per square foot, 'to allow for reasonable freedom of architectural design' (Robbie 1970a). It is reported that of two identically programmed (briefed) schools, the SEF Roden school was built for 18.52 dollars (Canadian) per square foot whilst one designed by the Toronto School Board cost in excess of 26 dollars per square foot. However, Oddie (1975) reports that:

> 'The first series of schools built in this system proved in the event to be more expensive than others being built at the same time and exceeded the standard limit of expenditure then in force. The advocates of the SEF system believe, although others do not agree, that this was the worthwhile price for the extra adaptability achieved.'

The building system which eventually emerged was based on a 5 feet square (1.53 m) planning module with available floor to ceiling heights of 10 ft (3.1 m), 14 ft (4.3 m), 18 ft (5.5 m), and 24 ft (7.3 m) with roof and floor zones both at 4 ft (1.2 m) thick. The subsystems involved (**9.3.1**) were structure, atmosphere, lighting-ceiling, interior space division, vertical skin, plumbing, electric-electronic, furniture (casework, seating, standard furniture), roofing and flooring (generally carpet). The spans available ran from 10 ft (3.1 m) to 30 ft (9.1 m) in primary spans and 5 ft (1.53 m) to 65 ft (19.8 m) secondary spans all in increments of 5 ft (1.53 m), with buildings up to three storeys or five storeys with an additional range of columns. The atmosphere control involves full air-conditioning in ten individually controlled zones for each 4000 sq ft (371.6 sq m) of floor space.

Almost all internal partitions were designed to be relocatable, a particularly important feature of the system since, as with Ehrenkrantz, Robbie's belief was that the users can by this means be given more and more control over what they do with their own

554

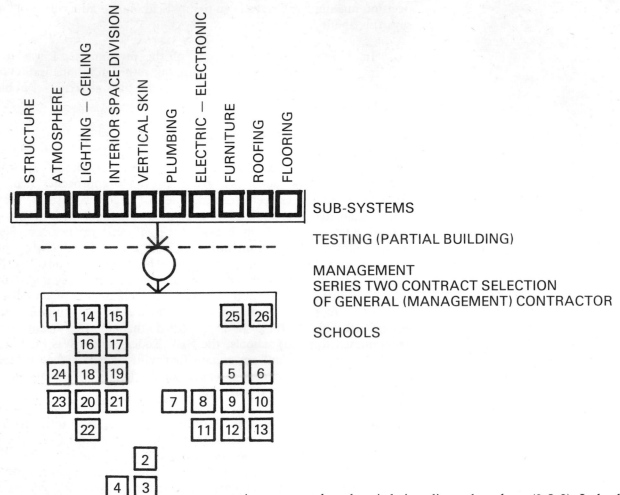

STRUCTURE
ATMOSPHERE
LIGHTING — CEILING
INTERIOR SPACE DIVISION
VERTICAL SKIN
PLUMBING
ELECTRIC — ELECTRONIC
FURNITURE
ROOFING
FLOORING

SUB-SYSTEMS

TESTING (PARTIAL BUILDING)

MANAGEMENT
SERIES TWO CONTRACT SELECTION
OF GENERAL (MANAGEMENT) CONTRACTOR

SCHOOLS

9.3.1 Study of Educational Facilities (SEF): Toronto, Canada. Roderick Robbie and Hugh Vallery, 1966 onwards. Range of subsystems and two stage contract procedure. (Robbie (1970a))

environment rather than it being dictated to them (**9.3.2**). Indeed he even extended this to the exterior of the buildings, where, unlike SCSD, the external wall system was part of the system and upon whose concrete panels he hoped the teachers and children would get to work and redecorate every so often in accordance with their wishes (**9.3.3**). In all this one can see many strains of the then current concerns: it had become quite popular for groups in many large North American cities, e.g. New York, Los Angeles, San Francisco, to decorate buildings with painted murals and there was a strong move for more public participation in the internal and external environment. All this led to a structure which had long clear spans internally and little avowed attempt at formal architectural elaboration externally. Here, as with so many attempts at systems building, such rationalizations caused great argument, not least because, in the end, it always seems to transpire that the particular architectural solution chosen is one that precludes rather than encourages user interaction. The somewhat rigorous attitude to the external skin was embedded in the philosophy of the whole programme which was an attempt to make explicit the performance and cost options in a very clear manner

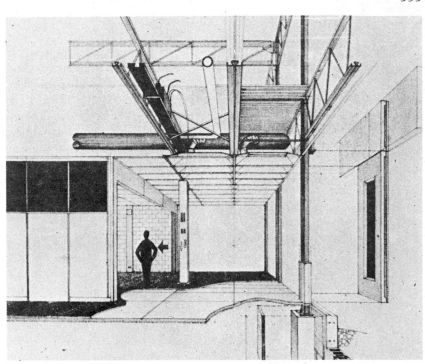

9.3.2 SEF. Cross-section showing subsystems

9.3.3 SEF. View of exterior of school using concrete cladding

to the sponsoring body. In this way they would, it was hoped, be able to make better informed decisions concerning how they wanted to spend their money rather than that these matters should become so fused into specific architectural and educational problems that such clarity was lost. One of the intentions was to encourage project architects, for educational and heat conservation purposes, to reduce the large areas of glass then popular. This was done by establishing criteria for heat loss and gain and resulted in glass areas of the order of 12% of the overall external wall areas. Such an approach has, because of the energy crisis, subsequently become commonplace.

The performance specification approach set out to compare the basic options available for the external walls — in concrete, brick and metal. The basic option in concrete was a smooth Portland cement finish; for brick, the local common red clay brick; for metal, anodized aluminium or porcelain finished steel. If these materials and their prices were basic then alternatives, be they enhancing or otherwise, could be compared against them. Indeed, Robbie's view on this was very clear:

> 'In making these decisions, we sought to present the owner with a clear set of choices, based on a realistic breakdown of the budget between the 14 subsystems and non-systems work'

and his view of the position of the project architect for each job was that 'a clear responsibility was forced on the architects: to spend more on the exterior walls, a decision had to be made which other subsystem to rob' and in forthright manner he then admits that:

> 'Architectural critics have described the exteriors of the SEF schools as criminal, stark, grim, a disgrace and an affront. I think they are severe, and as much of a shock as was board marked concrete and most other attempts at aesthetic honesty when they were first introduced.' (Robbie 1970a)

In this statement we can see the difficulties faced by many systems designers. There is a fundamental view that there is some sort of baseline which is morally correct. For example, Robbie's reference to 'aesthetic honesty' and its implication that anything over and above this is decoration, architectural whimsy or possibly even unnecessary, reflects this attitude. In equating his concrete finish with that of Le Corbusier's introduction of board marked concrete at Marseilles there is perhaps a claim to architectural innovation of even greater proportions than the costly architectural game playing he seeks to bring out into the open. In addition board marked concrete is, in itself, a deception in that by using rough sawn boards as formwork it can be given a romanticized patina akin to timber itself. In addition to this the public reaction to concrete buildings in many town centres has been severe, no matter how it has been finished.

By taking such a 'cool', utilitarian view of the external skin, SEF sought to demonstrate clearly to both client and public the cost implications of an elaborate exterior. The unadorned expression that resulted brought forth the strong reactions referred to above and showed again that it is not possible to stand aside from aesthetic judgements. The Ontario Association of Architects set up a committee, under the Chairmanship of Howard Walker, to examine SEF: their report emphasized the reaction to the external walls but had little to say about the other subsystems which went to make up the larger part of the SEF approach. It must not be overlooked that SEF, as with SCSD, laid bare much of the mechanics of decision making in design and in this way encouraged such a discussion by all parties. It was a useful instrument for making comparisons between architectural design skills and it is, perhaps, inevitable that any weaknesses would be seized upon by more traditionally minded architects.

> '. . .it is our opportunity to do this, to divert the human species from its present path, which leads to almost inevitable robotisation, and towards the explosive and generalised development of human creativity. It is our responsibility to do this as architects and builders through a radical and accelerated change in the means by which the built environment is produced, and the procedures through which it is designed and evolved.' (Robbie 1970a)

The approach developed for SEF was not confined to Toronto, and Robbie set up a programme and designed two schools in Boston, one in Ballston Spa and another in Albany — the latter with Richard Jacques. This project consisted of a High school for 3000 students, the design and construction of which was completed in two and a half years employing construction and project management techniques developed for SEF. This formal development of construction management as a skill in its own right was one of the central innovations of SEF, and this has frequently been overlooked — largely no doubt because it is a non-physical system and leaves only the building in its wake as evidence. The importance of the techniques can be judged from the fact that the Ontario Government utilized the SEF construction management form of contract, general contractors and individuals established themselves as construction and project managers, and most projects over 10 million dollars in Canada are now built using such project and construction management techniques.

In SEF we see how the emphasis has shifted, and the means have become more sophisticated; nevertheless the argument again concerns the right use of the machine, although Robbie offers a better informed and more complete view in that he calls into question the 'overly scientific and technical attitude towards people' (Robbie 1970a). The important innovations with regard to user study and construction management have, however, other implications in addition to those intended in that they allow the architect to extend his rationalizations into other areas. Indeed, those other areas are in many ways even more controlling and all powerful for the user than were the architect's previous concerns with the mere fabric of the building. Design now permeates right back through the client organization and forward to involve the users of the building. A comment on this aspect of the SEF work is offered by J. G. Wanzel who prepared one of its studies, 'The Role of the School in the Community'. This, he states:

'brought me directly into contact with the conflict between the interests of the "client" and industry for efficiency through normalisation, and the demand of "users" for specificity. It was in attempting to clearly define "user" requirements with respect to the community use of school facilities that I found considerable "client" resistance and hostility to canvassing the prime users of such facilities — the community. The School Board's planners implied that such a project would not be feasible due to the ignorance of the public in respect to its environmental needs. In this and in other situations, this particular corporate client insisted, through its "experts' on arrogating to itself the right to make decisions for the "user".' (Wanzel 1969)

The strength of the technological imperative, deeply embedded in society, is an argument which Wanzel develops in connection with his experience of SEF. He is very aware that much work of a so called systems nature in building has been concerned with efficiency and normalization — further that the systems view, properly understood, takes account of more than this:

'Fundamental to the systems approach is the notion that each problem must be considered with its social, economic and political context. However, careful observation of "systems" in operation, suggests that the more ethereal considerations are consistently glossed over in the face of the practical realities of economy and efficiency.' (Wanzel 1969)

The protection of the building user against corporately efficient solutions becomes the role of those professionals involved. Clearly such a role can be in opposition to the architectural and technological commitments held by many architects. Such commitments are, in the light of the history presented earlier, more likely to have grown up within a climate of a fundamental acceptance of the normalizations associated with the mass production ethic. In seeking systems solutions, then, architects have been bound by their own history. They often see users through this particular window of efficiency and consequently any request by users that is in any way at odds with this (from more space to more meaning) is seen as an attack on their total view. The architect, by his commitment to the very philosophy espoused by the machine apologists, may have implicitly relinquished his claim to be an impartial mediator between the user and the technology of building. In name he may still be an independent professional but in placing production efficiency above user enhancement rather than as equal to it he will inevitably be seen in this light. In this the architect has failed to recognize one of the most significant factors in the true systems approach. This as Rapoport indicates is that:

'The "psychology" of the system may be entirely independent of the psychology of its human components. If this conclusion is correct we need not search the human psyche for attributes that explain the murderous tendencies of certain forms of human organisation.' (Rapoport 1974)

METROPOLITAN ARCHITECTURAL CONSORTIUM FOR EDUCATION

'MACE-oschism'

ACID **8** (undated)
Greater London Council,
Architecture Club Magazine

The impact of Ehrenkrantz's SCSD on architecture in Britain has been considerable, although the many transformations which the ideas have undergone make the sources almost unrecognizable. Ehrenkrantz's impressive lectures, a combination of a strong philosophy and command of the facts of life of the building industry, were influential in spreading word of his approach. Curiously enough his audiences often consisted of those architects in Britain who were already involved in systems and, as the early part of this chapter indicated, were largely unable or unwilling to take action along the lines suggested by the success of SCSD. There was scepticism because of its North American source and because of the clear way in which it cut across many of the sacred cows of building systems as they had developed in Britain. Nevertheless the ideas which SCSD offered gradually permeated the architectural scene in Britain. At a blow Ehrenkrantz, with his development of the performance standard idea, had offered a way of giving controlled responsibility back to the industry and encouraging them to be innovative within sets of distinct criteria. Further, he focused attention onto the real study of user requirements and environmental criteria in distinct contrast to the British obsession with the mechanics of components. From the mid-sixties the performance standards concept was repeatedly discussed and elaborated until, in a total different form, it became incorporated in the metrication programme for components. Towards the end of 1966 the Department of Education and Science announced that a performance standards exercise for different elements was being set up, in conjunction with the schools consortia. This ultimately resulted in their document *Common Performance Standards for Building Components* (DES 1974?b and LEAC).

Ehrenkrantz's postulations on flexibility were also misinterpreted. This was largely on the grounds that the finished buildings (in California) with their carpeted floors, uniform ceiling and partitions, were bland and educationally inappropriate to Britain. A summary of the prevalent British attitudes to these issues is contained in the paper by David Medd (of the DES Architects and Buildings Branch), 'Respnding to Change' (Medd 1972). That Ehrenkrantz had actually made a close study of user requirements, building usage and environmental criteria, unlike most British building systems, largely went unnoticed. The only school building system to be developed in Britain after the advent of SCSD was the Metropolitan Architectural Consortium for Education (MACE) which attempted to introduce some of these ideas: unfortunately it also clung to the closed kit of parts concept, and this was probably its single biggest error, especially since the other consortia were already offering the feedback that the actual hardware of the systems were merely part of a much bigger system, the problems of which needed urgent attention. A clearer and more direct application of the SCSD approach had to await Wright's Ilford Jewish Primary School, Foster Associates' Newport Comprehensive School Project, and their IBM building.

In this context the Metropolitan Architectural Consortium for Education is especially interesting as an example in the chronology of building systems development in Britain. On paper, and in practice, it embodies most of the mainstream arguments concerning the use and misuse of the industrialization argument. Although established under the pressures emanating from the DES during the euphoric post-CLASP success of 1960 (for example MOE circular 1/64), the inaugural meeting of interested bodies did not take place until April 1966, which locates it in the tail end of the great school building system drive. The mock up building to test components was not finished until 1969 and the pilot scheme, Poyle Infants School at Colnbrook in Surrey, was not completed until early 1970. A significant aspect of this timing is that the main building programme did not get going until after this, at a time when many of the disadvantages of earlier established closed systems were beginning to be publicly and seriously talked about. By 1974 some 90 projects had been completed using the system. Since it had been DES policy to cajole, encourage and coerce local authorities into either joining one of the existing systems or banding together to form their own consortium, the grouping of a number of the London boroughs with the Inner London Education Authority, Surrey CC, and the more distant East Sussex, Brighton and Eastbourne authorities was of some importance (MACE 1968). By 1976, prior to its virtual disappearance, MACE had only three full members with Surrey as the major user (DES 1976).

The four basic resolves of the consortium inaugural meeting in April 1966 were sensible and in themselves predicted no rush to judgement:

'1. Exchange of ideas and experience in the field of educational building.
2. Undertaking new development work in building techniques for educational purposes.

3. Promoting the use, where appropriate, of available systems of industrialised building.
4. Rationalised organisation of purchasing procedure for standard components to take advantage of bulk buying.' (Killeen 1968)

A number of the local authorities had been reluctant to involve themselves in what they saw as the restrictiveness of building systems and this cautiousness can be seen in these resolves. No one would quarrel with item 1: there certainly was, and is, a great need for local authorities to exchange such experience. Item 2 suggested a slight reflection of the Ehrenkrantz philosophy in that it was felt that there were far too few building components designed specifically for the educational building market. Items 3 and 4 are particularly interesting in that they suggest, but only suggest, that the meeting was aware that the many already available school building systems should be looked at together with their component ranges.

The involvement of the Inner London Education Authority in one of the school building system programmes was considered of major import in view of their previous implacable questioning of the implications of the school building systems approach. This is particularly interesting in view of subsequent events where we shall see a powerful critique of MACE emanating from the Greater London Council Schools Division as architects to the Inner London Education Authority.

The MACE committee structure differed slightly from that of other consortia in being divided into an Education Advisory Committee, a General Purposes Committee and a Technical Steering Committee. Other consortia like CLASP and SCOLA tend to co-ordinate these through the consortium Working Party which is thus occupied with a great deal of detail across a very wide range. The MACE development group, set up in September 1966, was a multidisciplinary team working under the direction of the technical steering committee initially under the Chairmanship of the Architect to Surrey Council Council, Raymond Ash. The group was located in 'neutral' premises at Tolworth in Surrey, rather than in the offices of any one authority. Rather surprisingly, and in spite of the four aims of the consortium referred to above the brief given to the development group was described as follows:

'To design an industrialised system of component building, integrating the architectural and educational performance standards required in the schools to be built by the members of the consortium.

Architectural performance standards

The system was to be:

1. an industrialised system of component building
2. in metric dimensions

3. capable of building up to four storeys
4. designed to accept components from other systems

Educational performance standards

The system was to have the potential to give:

1. planning solutions for primary, secondary and special schools
2. maximum flexibility in use
3. protection from external noise sources
4. correct levels of natural daylight.' (Killeen 1968)

It can be seen immediately that the four aims of the inaugural meeting and the development group brief do not relate. Further, the aims quoted side by side with the brief on the same page in the initial article on MACE (Killeen 1968) call for 'the use, where APPROPRIATE, of AVAILABLE systems of industrialised building' (my capitals) with the development group apparently then being asked to 'design an industrialised system of component building'. The group opted to do what almost all similar groups had done before which was to themselves design another kit of parts. This decision to design yet another system is particularly surprising for the time (1966) since it was already common knowledge in the systems world that what was needed was not to design more closed systems, but an approach that would not lead to the major effort going into further technological development. The problems which had already emerged from the systems experience were of quite a different order: that components were important but that the administration, organization, marketing and programming aspects of the systems were by far the more significant ones requiring attention. The particular solution to specific building problems was, of course, crucially important to the user and to architecture but within the closed system world of the consortia the non-hardware implications of the total system became the determining ones.

Many of the existing building systems were at this time under pressure to take account of the demand for flexibility and change inherent in new educational approaches. This had particular implications for the structure, the interior subsystems, and especially for the servicing. The impact of Ehrenkrantz on the school building situation in Britain was significant but like so many influences on the British way of life it became absorbed and dissipated by a situation that has a high degree of closure to new ideas. MACE attempted to introduce some of these ideas at a general level and claimed to give 'the project architect a design tool which would give him freedom in overall planning' (Killeen 1968). A series of five flexibility ratings (**9.4.1**) were set down which were an attempt to be explicit about this problem. This then affected the structure, its maximum economic spanning properties, and indeed the very way components were jointed together.

The decision to design a metric system in 1967 was certainly correct in the context of the proposed change to metric which was planned to

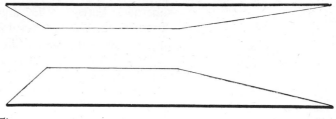

The space

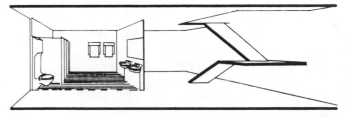

Static zones: staircases, lavatories, kitchens

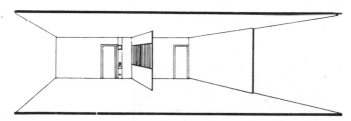

Annual flexibility: partitions, doors, glazed screens

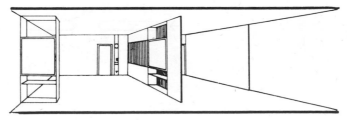

Termly flexibility: lightweight partitions, full height storage units, glazed screens

9.4.1 Metropolitan Consortium for Education (MACE). John Killeen and development group, 1966 onwards; pilot project completed 1970. Flexibility ratings. The connections between the Trump report proposals, the SCSD work and MACE are here made very clear

Weekly/daily flexibility: sliding or folding doors and screens, lightweight furniture

take place between 1965 and 1975. It would have been folly to have commenced a long development and production programme in imperial measure when the industry was being encouraged to 'go metric'. Here the politics of building systems intrude again. The MACE decision to design in metric and their decision to use a 100 mm basic module for planning with a square tartan grid of 900 × 900 mm with 100 × 100 mm bands was, however, bitterly opposed by the DES. One of the other school building systems, SCOLA had undergone a major dimensional and component change from one set of imperial dimensions to another in 1965 and 1966, this, in spite of the metric programme which was already under way and knowledge that there would need to be a further redesign to 'go metric', when the

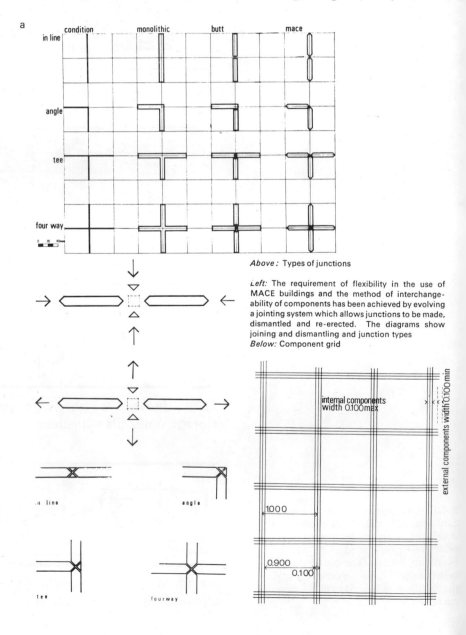

Above: Types of junctions

Left: The requirement of flexibility in the use of MACE buildings and the method of interchange-ability of components has been achieved by evolving a jointing system which allows junctions to be made, dismantled and re-erected. The diagrams show joining and dismantling and junction types

Below: Component grid

9.4.2 MACE: jointing methods. (a) The principles of the MACE jointing proposals. (b) *Opposite*. The original joint proposal (Killeen, 1968). Not used but revised prior to construction of first MACE school. All components brought to a standard joint condition by means of the extruded rigid PVC jointing components

call came from central government. The use of the tartan grid in MACE also went seriously against the other consortia traditions in school building systems which all had their structures on centre line grids which in turn meant that the skin always had to be outside the structure if component integrity was to be maintained. A further important theoretical recognition by the MACE team concerned the importance of the joint. Killeen states:

> 'Achieving a proper dimensional basis for components in a system should not overshadow the problems of joining components together.' (Killeen 1968)

One of the main interests, technically, in MACE lay in this approach to the jointing of components. Although SCOLA Mark 2 had attempted to bring the problem into the open by its introduction

b

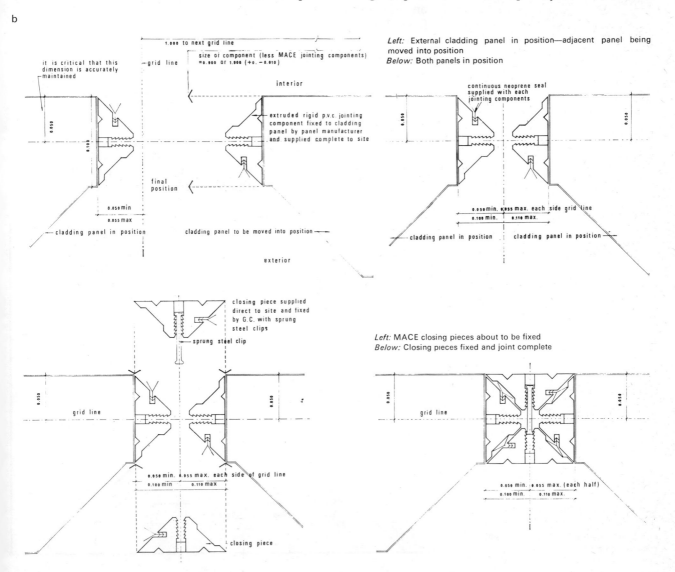

of special component junction pieces in 1966, almost all other systems momentum was concerned with purely dimensional development. Indeed the many organizations and ministries producing different prescriptions during this period demonstrates how a good and simple idea can be made thoroughly unusable by the vicissitudes of professional power politics. The basic idea in MACE was that all vertical planes should come together at a point on the grid, thus giving all components a standard end profile which allowed them to meet regardless of component thickness. Flexibility had been placed above dimensional rigour. Such a standard end profile considerably reduced the ranges of components required since each would not have to be then individually designed to suit each specific junction situation. The standard triangular junction profile was applied to each component (**9.4.2**) as a 'third' member depending upon the type of junction. This principle was proposed for use internally, on partitions, and externally on the exterior skin. It had a direct connection with the view held on flexibility: that if a building was to be responsive to change over short and long terms the components should be capable of being individually dismantled and repositioned with ease. By stripping the jointing sections and keeping the integrity of each component this would be achieved: however, the sophistication, complexity and cost inherent in this undoubtedly caused its own problems. Certainly in the short term it appeared to many using the system as unnecessary, and Padovan (1974) suggests that 'a more truly flexible solution would have been to accept non-standard junctions as a general rule, as Hausermann did in the SCSD system'. Here again we see the conflict between the 'needs' and the arguments surrounding the development

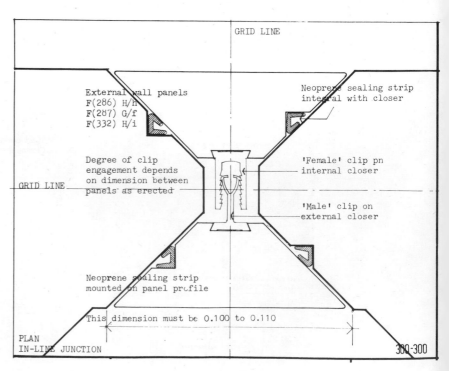

9.4.3 MACE: revised joint 1970. This involved a profiled end to components and a rigid PVC cover piece to joints

group approach with its attempt to take in some overall view of the building process, and the requirements of the specific building to a specific time and cost and in a very specific place.

Such pressures can be seen in the changes which the joint proposals underwent in practice. The original four part external wall joint (**9.4.2**) was revised prior to the construction of the first MACE school. Instead the ends of external wall panels were profiled to a near triangular form and, in straight run conditions, an extruded rigid PVC closer was applied outside and inside. These incorporated neoprene sealing strips and male/female fixing clips (**9.4.3**). In 1973 a further modification was introduced when timber (Siberian redwood) closer pieces were substituted for the PVC. These were screw fixed against foam sealing strips (**9.4.4**).

Such a series of changes are a good example of what has happened with most building systems. A solution, often innovative, is proposed and as feedback from use is accumulated this solution becomes repeatedly modified. At some point the solution settles into a stable state, although in some cases there has been almost continual change. Often as in this example, the initially sophisticated technology has given way to simpler solutions: here the aluminium sections were replaced with the more adaptable timber cover pieces. It is unfortunate that, in most systems, the publicity surrounding the initial proposals has not been matched by that for subsequent development. The broader questions raised by such a compressed development process in building are discussed elsewhere.

The attempt to introduce more flexibility into the British school building systems can also be seen in MACE at the structure level. The

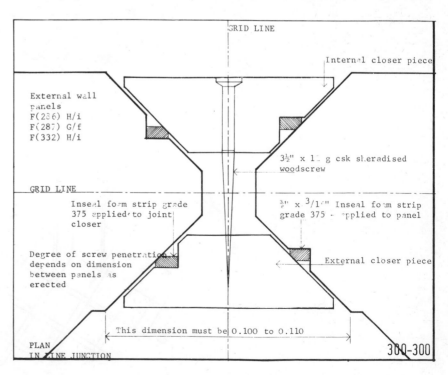

9.4.4 MACE: detail showing further joint revision, 1973. Detail showing profiled components and timber (Siberian redwood) closer pieces with foam strip seals. This progression, from the ingenious technology of the first suggestion to the simple application of a piece of wood, is a good example of the sort of changes that system components undergo

a

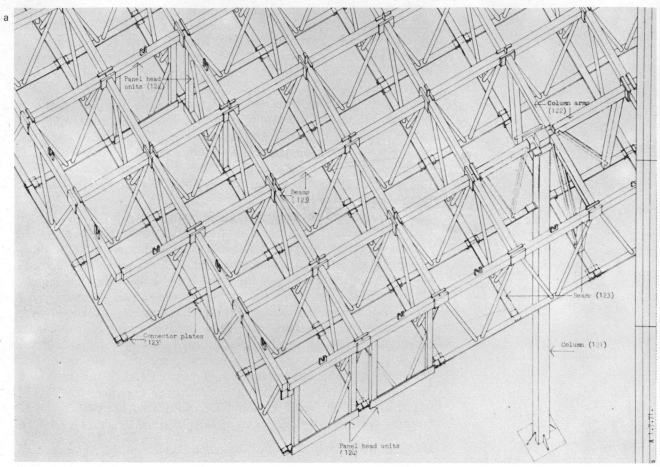

Panel head
units (122)

Column arms
(122)

Beams
(123)

Beams (123)

Connector plates
(123)

Column (121)

Panel head units
(124)

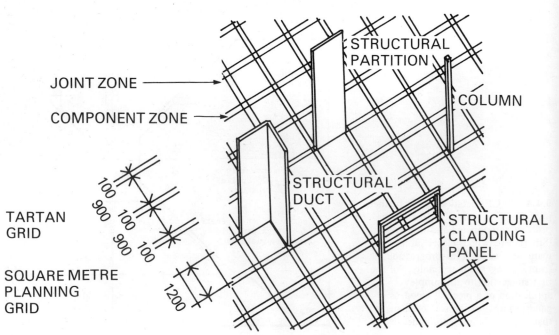

JOINT ZONE

COMPONENT ZONE

TARTAN
GRID

SQUARE METRE
PLANNING
GRID

STRUCTURAL
PARTITION

COLUMN

STRUCTURAL
DUCT

STRUCTURAL
CLADDING
PANEL

100

900

100

900

100

1200

b

9.4.5 MACE. (a) Axonometric of A-deck roof (1969). (b) Modes of vertical support. (c) *Opposite*. The A-deck roof mock-up. (d) A 4 storey structure using A-deck

c

d

self-validating economic logic of the steel frames used in all the other systems meant that it was difficult to justify longer spans than those required in classroom based teaching methods. From Stillman in the thirties (Stillman and Cleary 1949), through the Hertfordshire schools to CLASP and SCOLA the maximum spans required were centred around the 30–40 child standard class unit. The amount of steelwork required, plus the historical momentum acquired, had established the economic viability of spans in the region of 30 ft (9.14 m). Gibson and Iredale, at the Ministry of Building and Public Works (now DOE), had, with the development of the NENK system (1961), endeavoured to escape from the limitations imposed by this post and beam situation, where they used a version of Denings Spacedeck which allowed columns to be placed anywhere. Similarly, MACE were concerned to build in more freedom structurally and were also interested in the Spacedeck idea but ultimately opted to develop their own system. This, named A-deck, was a two way lattice girder spanning system made up of repetitive steel units (2 metres long) nominally 900 mm deep overall which could give floor spans up to 10 metres and roof spans up to 16 metres. However, a number of alternative support methods were introduced as part of the system (**9.4.5**).

The three methods of supporting the A-deck floors or roof (which were identical structurally and thus led to extreme cost difficulties) were:

1. Box columns in steel either 100 mm or 150 mm square.
2. Load-bearing floor to ceiling panels 900 mm or 1900 mm wide in lightweight concrete or braced steel panels.
3. Load-bearing floor to ceiling duct columns occupying 1100 × 1100 mm square and made up of load-bearing panels.

Project architects thus had a choice. They could use steel columns but also they could eliminate some or all of the columns by the use of

a

b

9.4.6 (a) Thamesmead No. 1 Primary School Thamesmead, London. MACE: GLC architects department for ILEA (Inner London Education Authority). (b) Holland County Middle School, Oxted, Surrey. County Architect, Raymond Ash (1973-4). South west elevation

the panels. This was an attempt to answer the many critics of the frame and envelope school, who were constantly drawing attention to the limitations posed by having freestanding columns just inside the external wall line and standing about awkwardly elsewhere. Equally however MACE project architects found that the limitations imposed by the use of panels were severe (**9.4.6**). Further they could not easily be subsequently moved. Ironically a number of MACE users seemed keen to press for the system to become a frame system like SCOLA and CLASP. Nevertheless MACE did offer a limited choice to the project architect and, if he was prepared to understand the rules of the MACE 'game', he could use this choice to some architectural and educational purpose. Surrey County Council's Godstone County First School, completed early in 1973, demonstrates that, even within the extreme limitations of MACE, an architect can produce a reasonable interior within the grim exterior offered by the system (**9.4.7**). However for the large central space in this school the project architect (John Wright who earlier designed Ilford Jewish Primary School) replaced the expensive A-deck with a castellated beam structure.

The failure of the standardization and mass production argument is clearly present in the standard MACE roof structure as Padovan succinctly indicates, having first drawn attention to its theoretical attractions:

> 'But, in practice, MACE steelwork is not so simple as the publicity handout suggests. It is claimed that "three units make up 75 per cent of total production". However there are 175 different column permutations plus six types of "braced column", seven column head elements, nine different beam units, 10 panel head units, seven types of diagonal bracing and ten "secondary components": a total of 217 different bits of steelwork, some of which are of considerable complexity. Is this what is meant by "simplicity achieved by minimising the variety of jig built/mass produced components"?' (Padovan 1974)

One of the most important design criteria which MACE had to endeavour to satisfy was that pertaining to external noise. A large proportion of the proposed building programme was in the London area and many of the projects were near or beneath the flight paths to and from London Airport. The options available within the roof and external wall elements were intended to offer the protection required. As with the other school building systems the external cladding panels were designed to take full advantage of the factory made concept, and arrive on site requiring only to be installed with the windows and panels preglazed by the use of neoprene gaskets. Special 'services drops' were available and these were unikely to be moved, and a demountable partition system in 100 mm thick components of 900 mm and 1900 mm was prefinished to take chalkboards, pinup boards or other fittings. As with the other school systems lightweight concrete floor 'biscuits', an acoustic ceiling and staircase were included. Again the wind from California can be seen in the attitude to environmental

9.4.7 Godstone County First School, Surrey. County Architect, Raymond Ash; project architect, John Wright, 1972. (a) Exterior showing covered outdoor play area. (b) Interior showing SMP mobile units to right of picture. (c) *Opposite*. Interior showing variety of uses for SMP mobile units

a

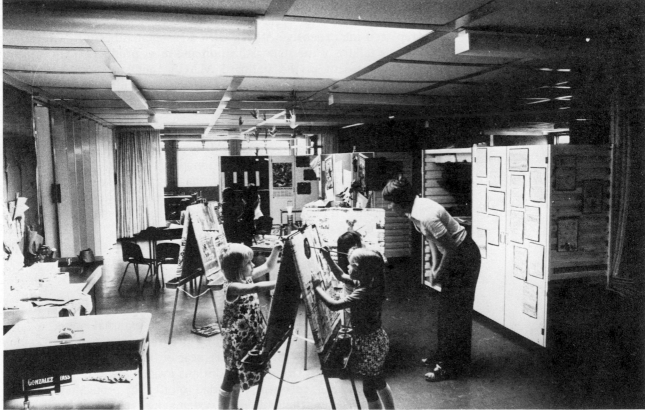

b

control and services: a welcome wind indeed since the other school building systems pay little more than lip service to these problems. The MACE drawings (**9.4.8**) indicate that servicing can take place vertically in controlled areas and horizontally through the latticed floors and roofs. A major decision was to make these vertical and horizontal zones pressurized areas for the conveying of warm air, which would be aspirated through the suspended ceiling membrane to the areas below. This was certainly the first time that one of the

9.4.7 c

9.4.8 MACE, 1972. Cross-section showing general construction and services integration. The coding refers to the location of drawings for the subsystems and components

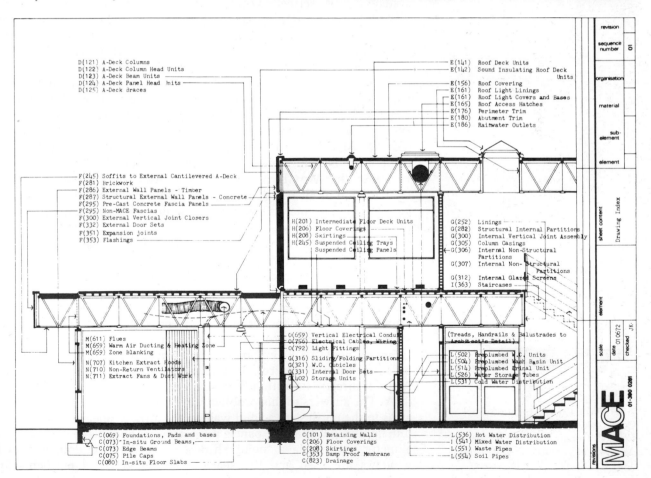

D(121) A-Deck Columns
D(122) A-Deck Column Head Units
D(123) A-Deck Beam Units
D(124) A-Deck Panel Head Units
D(125) A-Deck Braces

E(141) Roof Deck Units
E(142) Sound Insulating Roof Deck Units
E(156) Roof Covering
E(161) Roof Light Linings
E(161) Roof Light Covers and Bases
E(165) Roof Access Hatches
E(176) Perimeter Trim
E(180) Abutment Trim
E(186) Rainwater Outlets

F(245) Soffits to External Cantilevered A-Deck
F(281) Brickwork
F(286) External Wall Panels - Timber
F(287) Structural External Wall Panels - Concrete
F(295) Pre-Cast Concrete Fascia Panels
F(295) Non-MACE Fascias
F(300) External Vertical Joint Closers
F(332) External Door Sets
F(351) Expansion joints
F(353) Flashings

H(201) Intermediate Floor Deck Units
H(206) Floor Coverings
H(208) Skirtings
H(245) Suspended Ceiling Trays
Suspended Ceiling Panels

G(252) Linings
G(282) Structural Internal Partitions
G(300) Internal Vertical Joint Assembly
G(305) Column Casings
G(306) Internal Non-Structural Partitions
G(307) Internal Non-Structural Partitions
G(312) Internal Glazed Screens
I(363) Staircases

M(611) Flues
M(659) Warm Air Ducting & Heating Zone
M(659) Zone Blanking

N(707) Kitchen Extract Hoods
N(710) Non-Return Ventilators
N(711) Extract Fans & Duct Work

O(659) Vertical Electrical Conduit
O(756) Electrical Cables, Wiring
O(792) Light Fittings

G(316) Sliding/Folding Partitions
G(321) W.C. Cubicles
G(331) Internal Door Sets
G(402) Storage Units

(Treads, Handrails & Balustrades to Architect's Detail)

L(502) Preplumbed W.C. Units
L(504) Preplumbed Wash Basin Unit
L(514) Preplumbed Urinal Unit
L(526) Water Storage Tubes
L(531) Cold Water Distribution

C(069) Foundations, Pads and bases
C(073) In-situ Ground Beams
C(073) Edge Beams
C(075) Pile Caps
C(080) In-situ Floor Slabs

C(101) Retaining Walls
C(206) Floor Coverings
C(208) Skirtings
C(353) Damp Proof Membrane
C(823) Drainage

L(536) Hot Water Distribution
I(541) Mixed Water Distribution
L(551) Waste Pipes
L(554) Soil Pipes

revision
sequence number   01
organisation
material
sub-element
element
sheet content   Drawing Index
element
scale
date 01 06/72   checked J.K.
revisions   MACE   01-360 0281

British school building systems had attempted such an approach, and historically the lightweight, 'kit of parts' school of opinion placed a very low emphasis on a close control of environmental standards. Most British systems prior to MACE had paid some lip service to adequate control by means of servicing but generally this merely involved designing the structure and the fabric and then fitting the services in wherever possible, and to minimum standards.

By accepting the floors and roofs as pressurized distribution zones MACE were in the position of having to provide a sealed building, something that all the systems so far had found extremely difficult. Not only did system jobs often let the water in but they also let the heat out. Indeed the whole concept of light, dry, clip-together components has had great difficulty in coming to terms with this problem. The types and multiplicity of joints demanded by the component ethic created enormous problems regarding the penetration of wind, water, sound, cold and heat, and the loss of heat from the interior: similarly condensation presented problems. One reaction to this was to cite the car industry and their success in dealing with similar problems in a motor vehicle. The introduction of neoprene gaskets and window weather stripping in SCOLA and other systems can be ascribed to this analogy. Nevertheless there are distinct differences between a car and a building, not least in the cost per square foot of floor area. Thus, sealing a building of this sort, made up of light, dry factory made components put together by a still largely traditional construction orientated industry can pose considerable problems in addition to those normally encountered concerning manufacturing and erection tolerances, fixings and waterproofings. Until such practices become the norm additional expense is bound to ensue. Thus such innovations, however worthwhile, contributed to the cost of MACE since contractors were grappling with both a component building and a new attitude to environmental control. However the MACE plenum system made no provision for the recirculation of air nor does it have a cooling system: the latter is of particular importance as cost and educational pressures created deeper plans.

In MACE, albeit somewhat disguised, we see the clash of a number of forces: the light dry mass production idea; flexibility of plan, structure and envelope; and the environmental control high servicing concept. It is no surprise that MACE was not really able to resolve these nor indeed that it has come in for a good deal more overt criticism than the other systems, many of whom tried to do far less.

The MACE pilot project number 1 at Colnbrook, Poyle County Infants School (**9.4.9**), embodies a number of the more critical aspects of environmental control. It is sited to the northwest of London Airport, directly beneath a flight path which gave noise levels of 108 dBA. The answers to this problem were several: the facing of teaching areas to the north with a solid wall on the airport side; the roof of 140 mm lightweight concrete panels projects over the walls to the extent of a metre to provide a 'noise shadow'; non-standard double glazed windows are used and the sealed building allows a mechanically

9.4.9 Poyle County Infants School, Colnbrook, Surrey. Surrey County Council, County Architect Raymond Ash; project architect T. Dolan, 1970. This was the MACE pilot project

controlled environment to be provided to suit changing educational needs. Flexible ducts take the warmed air from a central plant and it is delivered via the pressurized roof zones. The building is air-conditioned thus allowing such spaces as the assembly hall to be windowless and classrooms have double glazed sealed windows. Even so, conversation was impossible in the classroom when an aircraft was passing, even on a high flight path (Diprose 1970). The contract period was 50 weeks, but the building was completed 3 months inside this, by Wates construction.

Padovan points out that the programming of jobs and components, a crucial aspect of running a system, has 'worked smoothly and this seems to be confirmed by many contracting firms and their operatives with continuous experience of the system' (Padovan 1974). This is largely due to the early MACE recognition of the importance of appointing at its inception an administrator and programmer with experience in this field from another consortium. The co-ordinating of component delivery dates with manufacturers and suppliers is particularly critical in a consortium since the general contractor on a project can more easily blame delay on the project on such consortium nominated firms and, further, can claim a time extension under the RIBA form of contract. This in itself puts the consortium and its programmer in the difficult position of having to reconcile pressures from a number of local authorities for delivery to sites at specific times. The programmer then has to ensure that such demands are made 'in good faith' and that the general contractor will indeed be ready for the components. The other schools consortia had found that it is not unusual to find such demands being made only to discover that the site in question is nowhere near ready for them. Conversely any overt contradiction of an authority's claim by the consortium development group, or a redistribution of delivery dates may be seen as interference between the local authority concerned and its general contractor who still, under the terms of the contract, is responsible for the job.

Much of the criticism of MACE followed the pattern familiar to all the systems but a great deal more had itself been generated by the particular approach which they had adopted. For example since MACE, in common with the other school building systems, claimed to hold costs and improve performance, this came in for severe comment. Internally, within the consortium, reports circulated which referred to the extreme difficulty in meeting cost limits and pointing out that comparisons with traditional construction were not favourable to MACE. Thus again the classic systems argument reappears. The ideal of cheapness through mass production here cannot easily be proved or disproved and for a number of reasons:

1. Building costs are constantly rising, for all building types and components.
2. Comparisons between bulk quotations for system components compared to those available on the open market are rarely useful since the former are usually purpose designed.

3. Comparisons between a bulk quotation for a given component between one year and the last give no indication of the amount of redesign that may have taken place, or whether standards were reduced. Certainly there have been many cases of cost reductions being achieved by a lowering of standards. Many would say that this is the reason that, in spite of the use of the 'development group' method, the environmental, constructional and detailing standards of system built schools still seem surprisingly similar to Stillman's experiments of the mid-thirties.

4. Whilst it may be possible to demonstrate savings on components it may well be that installation costs have been affected adversely in the process. Of course, this will reflect in the costs of jobs but such costs are much more difficult to disentangle than straightforward component quotes. Indeed, more and more sophisticated components can often mean higher costs on site.

In MACE this latter can clearly be seen in the difficulties which contractors experienced during the earlier projects when erecting the concrete panels and their ancillary components to the tolerances necessary to produce weathertight and draughtproof joints. As in SCOLA and CLASP contractors developed all sorts of methods for dealing with this problem. In the case of setting out they made all the dimensions slightly longer than shown on the drawings to give themselves the working tolerances they required. Clearly the attention to precision necessary (however laudable in the eyes of the machine purists) and the additional materials are then included in future tender prices.

Prices are further complicated by the fact that contractors soon become aware, as they have on all system jobs, that there is a division of responsibility between the project architect and the components designed by the system development team. Thus the former often feels no obligation to reject claims incurred by errors which, he justifiably feels, are not of his making. Contractors soon learn that although some areas of the Bill of Quantities are heavily controlled, a building system offers considerable scope for the exercise of their rights under the contract documents, and there is very little anybody can do about it. In this way the systems have actually assisted in escalating costs by their implicit encouragement of a method of design which splits control and responsibility.

A group of architects at the GLC, who worked on MACE 'found the MACE system to be expensive, involving on-costs of between 5% and 15% when compared with non-MACE schemes' (Wilson *et al.* 1974). That MACE has been the subject of more outspoken attacks than any other of the schools consortia is, however, only partly due to its own specific shortcomings. Had it been launched in the early sixties in the wake of CLASP and during a generally favourable climate to building systems, it may have received less scrutiny. Certainly, SCOLA, CMB and SEAC had their skeletons in the cupboard: however, like CLASP they have taken a somewhat cautionary attitude to publicity. By the time the MACE pilot scheme

579

was completed in 1970 the limitations of the myriad of closed systems were well known and the creation of yet another technical solution was bound to draw fire, even from seasoned systems men. Add to this its espousing of metric before being given the official go ahead, and of the 'alien' Ehrenkrantz attitudes to flexibility and environmental control and it is clear that their lines of support were tenuous. A number of critics have directed attention to the shortcomings of MACE:

> 'The MACE system unfortunately achieves hardly any of the claimed advantages of system building. It is not cheaper to construct, it does not lower production costs and it does not even reach technical competence let alone excellence.' (Wilson *et al.* 1974).

The writers, architects in the GLC Schools Division, suggest that the published evidence available from CLASP and other consortia does demonstrate these claimed advantages, although their very use of the phrase 'claimed advantages' indicates doubt. The fact that these consortia have well established public images and have had time to produce a few reasonable buildings appears to have deflected such critics. Indeed the fact that MACE was introduced to the GLC (with its history of autonomy, particularly in school design), focused critical comment in a way not possible in the other consortia, with their members spread through the provincial county offices. Where critical discussion did take place however, it has been contained at county level and has not gained the support that the Wilson report was able to gain in the GLC. However, to examine this more closely it is necessary to go into the connection between the GLC, MACE and the consortium idea. In this relationship the appointment of Gordon Wigglesworth, as Principal School Architect to the GLC in 1972 plays an important part. Wigglesworth, for some years with the Department of Education and Science Architects and Buildings Branch and later Deputy to the Chief Architect, was considerably involved in the promotion of the building systems idea. The GLC Schools Division, who carry out all the work for the Inner London Education Authority, had under Michael Powell produced many very interesting schools. The division had, however, seemed to be resistant to what they saw as compromises made necessary in the use of one or other of the closed systems. Although they had experimented many times with similar ideas they had, until the advent of MACE, kept outside the building systems and consortia orbit. In the event, the MACE development team was to be headed by John Killeen from the GLC Schools Division.

It should not, therefore, be a surprise that with this incestuous history MACE should very quickly find itself under extreme attack from all sides, but most powerfully from the GLC Schools Division architects themselves. For unlike most of the architects employed by the counties involved in building systems, the GLC had always been large enough to spawn powerful minority interests of its own. Indeed in many ways, since (later Sir) Robert Matthew's appointment in 1949, this had been one of its strengths. Hence the introduction of

MACE to the division created the first possibility for a unified response to the ideas proposed. A report in the *Architect's Journal* of a meeting between Principal Schools Architects Wigglesworth and Pace and Project Architects indicates the strength of the feelings about MACE at the GLC:

> 'A resolution was then proposed to the effect that MACE was unacceptable to the meeting in its present form and it was carried by an overwhelming majority.' (AJ 1972)

This followed a discussion of the shortcomings of the system in which project architects described MACE as being, in its present form, unable to give the best service to the client compared to more conventional building. Many said it to be costly, inflexible, technically crude, unsympathetic to existing old building, and aesthetically ugly. In a subsequent letter Wigglesworth indicated quite clearly that as a member of MACE, the Inner London Education Authority was committed to it fully and in this was supported by the GLC Architects' Department. The ILEA withdrew from MACE in April 1974. It is perhaps interesting to note that the newly appointed Architect to the GLC, Sir Roger Walters, headed Donald Gibson's development section (at the War Office Directorate) at the time of their introduction of systems ideas with NENK, the use of the Coignet system at Aldershot by Building Design Partnership, and the subsequent work on the 5M System discontinued in 1968. Clearly there were architectural pressures on the GLC to commit its school building programme to the system philosophy.

The magazine of the GLC's Architecture Club, *ACID*, had been particularly outspoken on the shortcomings of MACE. Under the title 'MACE-oschism', *ACID* 8 (GLC Architecture Club, undated) reports on project architects' criticism of MACE, summarized as follows; the area per place possible for primary schools using MACE is 3 metres compared to 3.5 metres in traditional construction in the London area. They point out that the first group of MACE schools received an extra 5% to cover development costs and claim that the DES provides more money for abnormal site conditions. This latter is a sum agreed job by job, and originally intended to deal with special site or access difficulties. They also criticize the attempt to provide flexibility on the gounds that even the use of the frame requires the use of load-bearing concrete panels. This seems only partially accurate for, although in MACE Mark 1 it is not possible to use a totally framed structure it is possible to obtain long clear spans by using the load-bearing properties of the panels on the perimeter. Nevertheless, the subsequent report by Wilson *et al.* (1974) indicates that MACE was successful on speed of erection, reducing 15/16 month contracts to 12 months, also that the discrete trade sequence, eliminating the overlapping of traditional work, worked well. It is hard to see in this criticism that there is very much difference structurally between what MACE can offer and what can be done traditionally. They gave a list of technical failures:

1. The external walls leak.
2. The thermal insulation in walls and roof are below standard: here it is not only the components but the problem of the joints between components.
3. Internal sound reduction is poor. Here again the judicious use of non-MACE materials can show that project architects need not be entirely cowed by the faults of the system.
4. The 1 metre by 1 metre planning grid is criticized as producing useless sizes for toilets, corridors and the smaller rooms, being too coarse a planning grid.
5. The prefabricated plumbing units are inadequate and expensive: these have been omitted on many jobs.
6. Difficulty with manufacture and erection of concrete panels to the precision necessary. (This was largely overcome by the use of the same erection subcontractor on all projects who evolved a particular technique.)

Their catalogue of errors is concluded by describing it as a 'short life, high energy, tight fit' system which, because it is machine intensive, is contributing to ecological imbalance at a number of levels.

> 'An Architect', says ACID, 'is trained to solve a problem using the solution he considers best for his client.' (The authors presumably include the user in that.) 'MACE is one of the many solutions available to schools architects . . . is it the best?' (*ACID* 8, undated)

In this last lament rests one of the most important issues evident in the use of government run systems. When an architect, who has been educated to exercise his judgement as to available options for the user, is employed in central or local government, which is itself committed to offering basically one choice, is he any more that same architect? Certainly the difference between what he is doing and what an architect in private practice is doing is enormous. Certainly the responsibiity which the latter personally takes means that, had he invented MACE (or one of the other systems) he would probably be sued into the ground.

From MACE, then, a great deal can be learned of the difficulties confronting not only architects but architecture in the period since 1945. For architects the attractive possibilities of a public service architecture, untrammelled by commercial concern has inexorably led to good ideas being institutionalized and bureaucratized. This is pointed up by the report on MACE prepared by the GLC Participation Movement Working Party where a section is devoted to professional responsibility:

> 'The architect (says the report) directed to use MACE, faces a cruel dilemma. He may either compromise his professional responsibility to give the client (in this case both the users of the building and the community financing it) the best possible result

from the resources available, or he may defy the management, and risk damaging his career prospects within the organisation.' (Wilson *et al.* 1974)

The author's complaint, a good one, is, however, not only the result of MACE, but the result of the changed status of the architect within a large bureaucratic organization. The political exploitation of the naïvety inherent in the utilitarian argument has become all too apparent: for the institutional socialism which fired many architects and planners between the wars has, in one sense, been achieved. The system built schools look uncannily like those drawings from the Bauhaus and elsewhere. The image has been changed to conform to a view of the world long discarded by avant garde political theorists, who are now much more concerned with the individual. As a result the building user and the building designer seem somehow to have been left with fewer options and choices, rather than more. The richness of possibilities inherent in the making of environment has often, with the connivance of architects and their systems, been considerably limited. The troubles that surrounded MACE caused a continual reduction of programme after the withdrawal of the GLC/ILEA. The DES publication *The Consortia* of 1976 includes MACE as existing but production of buildings using the system appears to have ceased about that time.

SYSTEMS AT LARGE

'Good or bad, there is one thing certain. In 14 years we will either be working for General Electric, General Dynamics, General Motors or General Cities — or they will be working for us. Which do you prefer?'

Karl Koch
quoted in Pike (1968)

Performance Standards in High Places

Some of the reasons for the failure of MACE lay in the way in which it had reordered the priorities accepted by other systems. Although in the event there was just another set of components, a great deal more emphasis had been placed upon the infrastructure of the consortium programming, co-ordination, strategies for procedure in specifying, testing and purchasing. Attempts were made to define the quality of performance required both from the buildings and the subsystems, and much of the impetus for this had come from the work of Ehrenkrantz. The notion of writing performance standards which subsystems and components would have to meet, as developed by SCSD had a profound effect upon those engaged in similar areas in Britain. This approach had been introduced, some would say confused, by the many efforts to describe it in Britain. Early papers were those by Harrison (1969) and in the *Architects' Journal* of the same year, the latter pointing out that:

'A performance specification is a description, preferably in quantitative terms, of the performance or function required of a building or its component parts, including statements as to how these will be maintained during the specified life of the building or component.' (AJ 1969b)

Such an approach serves to delineate precisely the qualities sought and at the same time provide those tendering with the opportunity for innovation and for using their own methods in answering these requirements. Unfortunately, many British companies seemed unable to operate in this climate, declining to think afresh about their approach for a given product.

A determined effort to establish the viability of the performance standard approach allied to some sort of open system of components was made by a group set up in 1968 at the Ministry of Public Buildings and Works. This group comprised architects who were working in the NENK system and Component Development teams and their brief was 'to consider the development of a method of building capable of meeting the majority of requirements encountered within the Ministry's building programmes within the cost limits laid down...' (Rabeneck 1973). The Ministry's building programme is an extremely large one and embraces a range of some 150 different building types. The group found that 75% of the portion of the market available to such an approach would consist of nine types of building:

1. Single living accommodation
2. Social (clubs and messes)
3. Offices
4. Education
5. Medium workshops and laboratories
6. Telephone engineering centres and vehicle workshops
7. Single and two storey automatic telephone exchanges
8. Multi-storey automatic telephone exchanges
9. Postal sorting offices

The group was started by Colin Pain, who went to see Ralph Iredale in the United States during 1968. Iredale, it will be recalled, had worked for CLASP, been a prime mover in the development of NENK and had then emigrated to California to work with Ezra Ehrenkrantz. The new group set out to avoid creating the sort of closed building system that already existed elsewhere and to create an open compendium of components based upon performance levels which could be tested and agreed. This resulted in the Method of Building (MOB) which is

> 'primarily a way of thinking and working to achieve goals that are the common concern of all building professionals. These goals are:
>
> 1. Better standards — to improve the value obtained for initial capital outlay on PSA (Property Services Agency) buildings.
> 2. Saving of time — to shorten the average period spent in design and construction of new works projects.
> 3. Design efficiency — to obtain a better return, both in quantity and quality, from the professional effort available, to the department.
> 4. Financial control — to improve building performance while keeping within Treasury cost limits.' (Rabeneck 1973)

From the beginning a wide range of constructional choices was to be made available with five basic categories of structure: lightweight steel, composite/steel concrete, precast and *in situ* concrete, load-bearing brick and blockwork. Allied with this was the concept that most of the components would be capable of being used with any of the types. Such an approach on the part of the Ministry sought to bring together a number of important threads. The first was to show that, unlike the closed system consortia, an open component based approach was possible and desirable. The second was to make further progress towards dimensional co-ordination, the third, to relate the work to the change to metric which had been programmed to take place between 1965 and 1975 and also to the work currently going on at the British Standards Institution. In this way, through the concept of agreed performance standards, a link could be made between the considerable building programme of the Ministry and the BSI revisions of standards to create a bank of components to agreed quality levels and which had a ready market. This latter was of course of vital importance to those manufacturers involved in discussions on changing to metric and the possibility of such a market would encourage agreement.

The Method of Building work has a number of detail aspects which are worth examining. The dimensional framework proposed was based on BS 4330 but the MOB proposal gave a preference for floor to ceiling dimensions rather than floor to floor dimensions on the basis that it was this dimension that affected the largest number of components (**9.5.1**).

Another deviation from the strict rules of modular theory concerned the question of components 'keeping station' within their modular

9.5.1 Method of Building (MOB). Property Services Agency, London, 1968 onwards. Diagram of vertical controlling dimensions

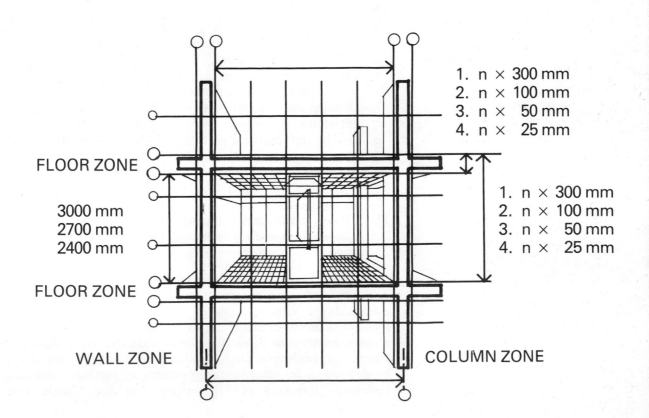

1. n × 300 mm
2. n × 100 mm
3. n × 50 mm
4. n × 25 mm

FLOOR ZONE

3000 mm
2700 mm
2400 mm

1. n × 300 mm
2. n × 100 mm
3. n × 50 mm
4. n × 25 mm

FLOOR ZONE

WALL ZONE   COLUMN ZONE

space, which the group thought an unnecessary restriction. One example shows a vertical cladding component penetrating the floor zone 'in order to lap the adjacent component for jointing purposes' (Rabeneck 1973). Although small in themselves, such moves from the hitherto rigorous tenets of modular theory indicated a recognition of the practical implications of building to such rules. Jointing also received more comprehensive attention than had been the case in many of the closed systems. The positions for fixing holes dimensioned from reference grid lines rather than from the edges of components was clearly set down so that compatibility at this level could be achieved (**9.5.2**). The basic agreement was that for all components the primary fixing convention should be offset from the reference grid by 50 mm, at 100 mm centres. Although ARCON during the 1940s had a clearly defined policy on fixing and SCOLA Mark 2 had briefly addressed itself to the question of defining fixing conventions in 1965/65, this is an issue that has received far too little attention: the Method of Building proposals recognized that it was a matter of some importance if progress was to be made with interchangeability. Windows, partitions and doorsets were the first subsystems to be put out to tender on the basis of performance standards, in April 1970. The procedure followed was:

1. Advertisements in national and technical press.
2. Each applicant for documents sent brief summary of tender documents and a questionnaire.
3. Appraisal of questionnaires.
4. Full documentation sent to selected applicants.
5. Applicant after three months provides a product concept and outline cost information.
6. Confidential reports on each of the proposals prepared and firms advised of suitability or otherwise.
7. Detailed designs and costs were called for from the remaining applicants.
8. Design appraisal and preliminary cost assessment by the MOB team.
9. Invitation of test samples.
10. Successful firms asked to make any minor changes, compatibility arrangements.
11. Those components offering value for money included in compendium and agreements for 3 years with government signed.

Testing was an important part of the process and indeed one of the difficulties of using the performance standard approach is that of establishing a suitable range of tests. Often the tests available are inadequate or non-existent. In this case it was agreed that MOB would arrange for tests to be carried out on manufacturers' designs, using British Standards tests where applicable. A number of new tests had to be developed, for example the resistance of partitions to changes in humidity. Rabeneck (1973) reports that the 'failure rate was

IT IS PROPOSED THAT FOR ALL
COMPONENTS THE PRIMARY
FIXING CONVENTION SHALL
BE AT 100mm CENTRES
DISPLACED FROM THE REFERENCE
GRID BY 50mm

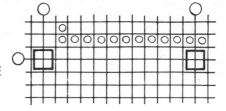

IN THIS WAY A 3 DIMENSIONAL
FIXING POTENTIAL CAN BE
ESTABLISHED WITHIN THE
BOUNDING FRAMEWORK AS
$50 + (n \times 100)$ FROM THE
REFERENCE GRID

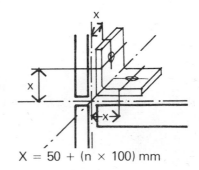

$X = 50 + (n \times 100) \text{ mm}$

THE COMPONENTS FITTING TO
THIS FRAME REQUIRE A
FIXING PROVISION WITHIN
THE SAME DIMENSIONAL
SYSTEM. FOR EXAMPLE
A COMPONENT WITH 3 FIXING
PROVISIONS AT $n \times 100$ CENTRES
CAN BE FIXED TO A BASE
HAVING A FIXING POTENTIAL
OF 100 min CENTRES

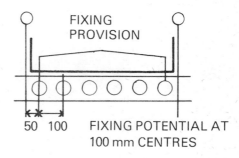

FIXING
PROVISION

50 100   FIXING POTENTIAL AT
100 mm CENTRES

THE LOCATION HOLES OR
SOCKETS TO RECEIVE
FIXINGS MUST BE TAKEN
FROM THE REFERENCE
GRID TO THE CENTRE OF
THE HOLE OR SOCKET AND
NOT FROM THE EDGE OF
THE COMPONENT. ANY
MEASUREMENT THAT IS
MADE FROM THE EDGE MUST
TAKE ACCOUNT OF ANY
DEDUCTION MADE FOR INACCURACIES

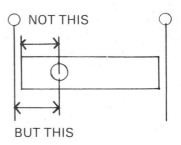

NOT THIS

BUT THIS

9.5.2 MOB. Dimensional
conventions for fixing

SIMILARLY AN ALLOWANCE
MUST BE MADE IN THE
TYPE AND SIZE OF THE BRACKETS
TO ALLOW FOR SITE INACCURACIES
AND TO PROVIDE ADJUSTMENT ON
SITE IN THREE DIMENSIONS

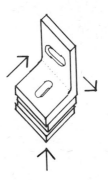

considerable, and in particular some products claiming British Standard levels did not pass the test'.

As with the schools consortia bulk purchase arrangements none of this procedure implied any contractual arrangement between the proposed suppliers and government. Only when an order was placed was a contract made, in the normal way. Clearly the amount of work necessary in preparing and submitting proposals limited the type and number of firms who could enter the compendium. This had the advantage of some notional degree of quality control and advanced the idea of an instantly available bank of components but at the same time it must in many cases clearly be more economic to go to those firms who were either too small to be in the compendium or not interested in the complexities of a centralized performance standard tendering set up. It would appear that Method of Building, as with many of the building systems in schools and housing could claim value for money with components to a specific level of tested performance. Equally with these systems their costs were almost always higher than products obtained outside the arrangement. Again as with the client sponsored school systems one of the attractions of the MOB procurement arrangements lay in the institutions which developed them. Central government, even more than many local authorities, has created such a complex series of checks and balances that as Rabeneck points out, 'within the context of present PSA contracting methods for nominated products, the forms of agreement for compendium components represent a major economy of procedure'.

So once again we find ingenious system designers employing their skills to balance the lugubrious complexity with which such an institution surrounds its work. The documentation initially produced consisted of five handbooks:

1. Design discipline and design requirements;
2. Design guidance including early cost advice;
3. Technical data;
4a and 4b. Product data.

By February 1971 the dimensional framework had become mandatory and by September 1972 the use of compendium components had similar status. Links were being created with GENESYS, the PSA's suite of computer programmes and with CEDAR (computer aided environmental design analysis and realisation) in an attempt to automate much of the tedious process of evaluation and data processing that an approach such as Method of Building needs. Clearly the whole procedure set up is a sophisticated one and required a great deal of effort to keep it up to date and relevant. Indeed the requirements with Method of Building in this respect are even more demanding than most of the closed systems, and Rabeneck indicates that its very wide approach would raise considerable problems. The claims being made are, after all, familiar ones — although more ambitious — than earlier attempts:

'The essence of the approach is to make the best use of professional skills within the design offices by setting down as many procedures as possible from drawing office practice to project management. As these become second nature to the design teams, more time can be devoted to the actual task of design.' (Rabeneck 1973)

Once again the promise of some golden future is held out: a time when designers will be freed from the day to day trivia of procedures and quality control. However it is a strange view of design that divorces the inter-related aspects of design such as these from the broader questions inherent in any architectural design. As we have seen there is a break point in the commitment felt by professional designers. If large areas of decision are removed from them the total process, or whole system, begins to suffer.

The Department of Education and Science had also been impressed by the SCSD approach and had continued to press the idea of the use of performance standards with the result that after several years of consultations and work, they published in 1974, for Local Education Authority Consortia, *Common Performance Standards for Building Components: Structure, External Envelope, Internal Subdivision* (DES 1974?b). The criteria were for use in Consortia building programmes and emphasized prefabrication. The Consortia had joined forces to agree common standards, a significant move: 'it is hoped', said the introduction to the document, 'that this alignment of standards will reduce unnecessary duplication of effort by industry when tendering for Consortia components and some degree of interchangeability of components will result' (DES 1974?b). Unfortunately it had taken from the early sixties to this publication in 1974 for any positive move to even 'some degree' of component interchangeability between these client sponsored consortia.

The question of jointing had also simultaneously been under consideration at the Department of Education and Science, with a series of proposals being considered in the early seventies. The Chief Architects of Consortia (CAOC) had discussed, in June 1969, the Demonstration Rig to examine component fixing conventions set up by the Building Research Establishment (DES 1969). They had then received a theoretical study on jointing in November 1970 and another paper on component compatibility in 1972. A concluding paper was then considered by the Consortia Interdepartmental Working Party in June 1974. There was considerable liaison between the DES group and the work at the MPBW (later DOE) with the Component Co-ordination Group (CCG) attempting to integrate the work so that a useful advance could be made.

Whilst these 'systems at large' can be seen as an attempt to guide the industry in the direction of an industrialized componentized vernacular it must be recalled that this, after all, was the intention of those pioneers of the 1940s. The creation of the mini-empires of the client sponsored systems had in many ways delayed progress along these lines, since there was little internal incentive for them to interchange components among themselves much less permit alien components to be used on a large scale. It took the impact of the Ehrenkrantz work to cause a reappraisal of what had happened in Britain. The performance standard idea is an elegant and persuasive one. However, it should be remembered that it grew naturally from the technological approaches of industry in the United States and, in the British cultural context, creates quite new conditions both for the suppliers and for the designers.

Rethinking the Housing Industry, Again

An attempt to develop industrialized building in the United States, somewhat on European lines, was made in the late sixties when the Johnson administration focused attention on the housing problem. 'Operation Breakthrough' was set up by the Federal Agency, Housing and Urban Development (HUD) to invite proposals from consortia of designers and developers. New and radical solutions were encouraged and many interesting ideas were put forward, several drawing on new technologies. The programme claimed to encourage ways of increasing production so that substandard accommodation could be cleared away and the cities revitalized. The programme showed some understanding of the problems in that it attempted to tackle the housing delivery system, but the suggested claim that it could put together adequate new markets for the new technologies it encouraged proved to be unfounded. The belief that new technologies, particularly those of the automobile industry, could answer housing problems was fired on this occasion, in 1969, by George Romney, Secretary to the United States Department of Housing and Urban Development and one time head of General Motors.

Clearly the 'Operation Breakthrough' approach had been influenced by the adoption of industrialized building by government in Britain and other European countries during the sixties. Another factor had been the success of the Ehrenkrantz work with SCSD and URBS, and its implied suggestion that such an approach could be applied in other areas. Habitat, at the Montreal Expo 67, had also renewed popular interest in the possibilities of such an approach. Here Moshe Safdie, with a forceful and entrepreneurial enthusiasm encouraged the Canadian government to finance his vision of the possibilities offered by a full use of industrialized methods. Habitat was composed of a series of 17 ft 6 in (5.3 m) wide load-bearing reinforced concrete box modules, pedestrian streets acting as beams and stair and lift towers, designed together as an indeterminate structure by engineer August Komendant. Safdie put forward his proposal, early in 1964, based upon the box modules stacked in a variety of ways so that the top of one could form an extensive outdoor space for the one above. His account (Safdie 1970) of how the idea developed and was built is a rare and revealing example of the interaction of the various forces in such a situation. It shows very clearly how most of the ideas associated with industrialized building could not survive the impact of the real world of building and political forces.

The scheme was initially based on a design for 2000 families, with all ancillary facilities. This was ultimately reduced to 150 units under financial pressure from the Canadian government, and of these it was only possible to finish internally two-thirds in time for the Expo opening. Safdie points out that this tripled the unit cost of the units and thus was an unfair and inaccurate reflection of the possible cost reductions that could be made using such methods (Safdie 1970). His book shows how a number of the ideas prominent in architectural thinking fused together: the additive cell approach of van Eyck the Dutch architect; Le Corbusier's notion of dwelling units as a series of drawers in a cabinet; the ideas of Buckminster Fuller; d'Arcy Thompson's book *Growth and Form* (1961, 1917); and vernacular forms such as Pueblo villages. It also demonstrates with the utmost clarity that a proposal which sees a solution to housing problems in the ideal rational terms of some industrialized nirvana is doomed to failure. Safdie describes how he became:

'aware of the basic shortcoming of the building industry. Its whole tradition is to build with what materials happen to be available. Every other industry defines its requirements and then develops the material best suited to the problem.'

This approach may well be the industry's strength and we must surely recognize that any innovation must take account of such human factors and their history. The constant technological, political and financial compromises recounted by Safdie testify not only to his enthusiasm and sense of purpose but also to the inherent unreasonableness of many of the underlying assumptions. He saw the building as a prototype, as a total piece of building research, and draws a parallel with the development of the Concorde supersonic airliner. Seen in this light, the 20 million dollars spent on the design and construction of Habitat (2 million of which was on design) can be put down to research to develop new solutions:

'We were stretching the existing state of the building art far beyond its accepted capabilities. For these lessons the Canadian taxpayers paid twenty million dollars.' (Safdie 1970)

Once again the solution to housing problems was seen to lay in a technological revolution, but at Habitat this was associated with a romanticized imagery derived from a selective view of vernacular forms. The conflict inherent in this relationship is apparent not only in the technical difficulties encountered but also in the final form of the building. Operation Breakthrough, however, emphasized performance and the process of housing provision.

The Breakthrough approach had two parts, developed together: from the public sector, state and local governments were to identify

a

9.5.3 Acorn House, USA. Carl Koch, 1947. (a) House package being erected. (b) *Opposite*. Plan. (c) *p. 594.* Section showing core and foldouts and view of roof foldout. (d) House erected

and assemble continuous assured markets. The private sector was to develop high volume production and delivery systems. The Housing Act of 1968 had called for five large scale experimental housing schemes. HUD had enlarged this to 250,000 housing units per annum. The procedures adopted called for two types of proposal — type A, which was for the design, testing, prototype construction and evaluation of complete housing systems which could lead to large scale production; type B which was for advanced research and development of ideas not yet ready for prototype construction or which might form part of other total systems. The claims made by Breakthrough were considerable:

> 'The long range results of building twenty six million new dwelling units in the next decade ultimately rest on the ability to change the market structure and the popular concept of what a home is.' (Nomitch 1970)

621 proposals were submitted on September 19 1969, of which 236 offered full building systems ready to go into production. Not surprisingly, some of the most viable systems to emerge from 'Operation Breakthrough' were based upon the already highly

b

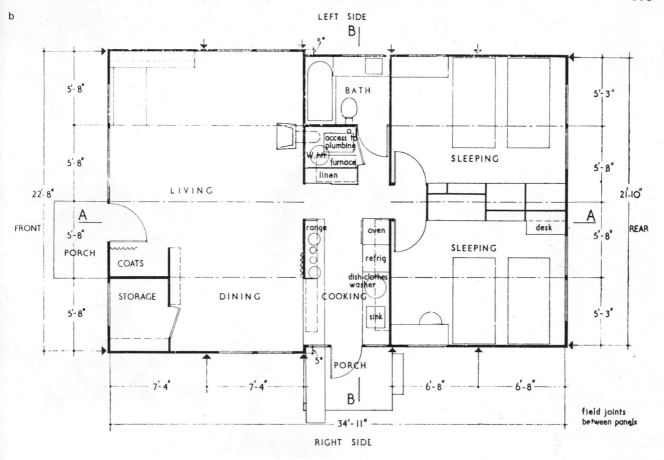

LEFT SIDE

BATH

access to plumbing

W. htr.   furnace

linen

LIVING

22'-8"

FRONT

A

PORCH

COATS

STORAGE

DINING

COOKING

range

oven

refrig

dish- clothes washer

sink

SLEEPING

desk

SLEEPING

A

REAR

21-10"

5'-3"

5'-8"

5'-8"

5'-3"

5'-8"

5'-8"

5'-8"

5'-8"

5'-8"

PORCH

B

7'-4"   7'-4"   6'-8"   6'-8"

34'-11"

RIGHT SIDE

field joints between panels

rationalized timber frame house. These included Levitt Technology's Georgian style box system which showed an undoubtedly accurate response to 'the market' and Boise Cascade, who were already producing 1000 units per week. One of the more innovative proposals was that from Ehrenkrantz's Building Systems Development, Fibershell. This system:

> 'uses non-critical materials of glass fibre reinforced plastic and paper honeycomb to manufacture shells and shapes by wrapping the materials on a revolving mandrel.' (Rabeneck 1970)

Karl Koch made a proposal in the Breakthrough programme based upon his Techcrete system which had been developed in the early sixties and had a completed scheme in Boston by 1965. In the work leading up to Techcrete Koch also had some highly pertinent comments to make about the available systems in Europe since he was far from a newcomer to prefabricated methods in building, having designed the Acorn house in 1947 and the Techbuilt house in 1953. The former, of timber and steel construction, arrived on site with floor, roof and walls folded against a central utilities core — this was placed on a prepared foundation and unfolded (**9.5.3**). The Techbuilt

9.5.3 c

9.5.3 d

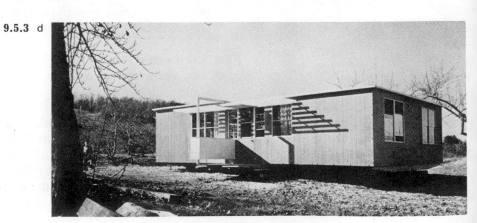

house was of modular component design and produced by Techbuilt Inc. (**9.5.4**). The company provides a house package consisting of timber structural posts and beams, exterior wall panels, roof and floor and this is erected by local builders with interior partitions and finishes provided by him to the client's requirements. A precursor of Techbuilt can be seen at Concord, Massachusetts where Koch again produced a development of 100 homes on a wooded site, each house built from standard elements. Koch's move into concrete prefabrication seems to have been occasioned by his work for the Boston Redevelopment Authority who commissioned a housing study which resulted in his Academy homes project in Roxbury, a part of Boston with extreme housing problems. The Academy homes project used precast prestressed concrete panels which spanned 32 ft (9.75 m) between walls, with apartments stacked one on top of the other. This scheme was completed in 1965 and was the close precursor of a larger scale proposal across the road which developed the system and became published under the Techcrete name (**9.5.5**).

Koch's commitment to some form of industrialization is clear from his several attempts to produce rationalized solutions. Nevertheless, he does not seem to have fallen into the European trap that has awaited so many system builders — that of designing an 'ideal' system abstracted from market realities. Koch's proposals have always been firmly founded in an existing developing technology. His timber houses posit the idea of a standard kit of parts in the existing timber vernacular context and around specific schemes. The Techcrete proposal was developed as a response to a particular, urban, set of problems and after a perceptive review of European building systems had demonstrated their unsuitability for use in the United States. To European eyes Koch's systems may seem not to be systems at all since they offer few radical new visual innovations of the type sought, it seems, by most system designers. Nevertheless, whilst committed to a set of architectural ideas he was also concerned that these be put to use. He points out that

'Pure design in the grand manner was, is and always will be part of good building. Functionally adequate buildings can be built without an architect. Our goal as architects should be to strengthen our function, that of a co-ordinator of the building process, into a meaningful accomplishing position, where we can apply our design talent effectively.' (Pike 1968)

Koch received a research contract from the US Department of Defense in 1967 to study the feasibility of obtaining significant unit cost savings for military family housing by the use of advanced technologies and its possible applications in the urban realm. He was joined in this study by the Batelle Institute, Kaiser Industries Corporation and Sepp Firnkas Engineering. Koch was already obviously very aware of many of the implications of using innovative industrialized methods — the most crucial being the existence of a guaranteed market so that development costs could be recovered and

a

9.5.4 Techbuilt House. Carl Koch, 1953. (a) Exterior view. (b) *Opposite*. Exploded perspective

capital costs written off over time. In the past the markets have never really existed in a large enough form to justify complex tooling up to give the required flexibility. In his study Koch asked four basic questions:

1. What is an industrialized building system?
2. What are the shortcomings of the European systems?
3. What is the current status of building systems in the US?
4. Are the European systems applicable to the USA?

The research team examined many of the building systems then available in Europe and came to the basic conclusion that the technical aspects were far and away less important than the managerial, organizational and marketing aspects of a system. The findings of the research team were significant in view of the subsequent development of systems in Europe since they identified several problems which until that time had remained unstated. These led the team not to recommend the use of these systems in the United States and, although clearly Koch had an interest in promoting his own 'system', the team's findings have proven more than accurate — even to the extent of the subsequent progressive collapse of Ronan Point in 1968 as implied by the first of their findings on European systems:

1. Their 'stacked' method of construction would not meet most US building codes because of lack of structural continuity.
2. The overall dimensions of the rooms were small and below the standards expected by the potential occupant in the US. This

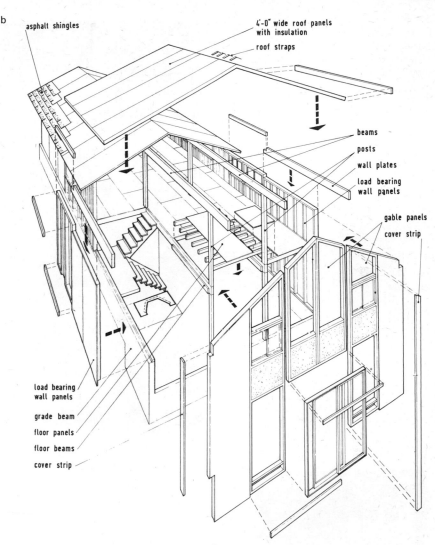

b

asphalt shingles

4'-0" wide roof panels with insulation

roof straps

beams

posts

wall plates

load bearing wall panels

gable panels

cover strip

load bearing wall panels

grade beam

floor panels

floor beams

cover strip

would be a difficult problem to overcome because most of the systems employed cross wall construction, and the majority appeared to produce panels with a maximum span of about 24 ft (7.315 m).

3. The adoption of a European system by an American Company would entail payment of a royalty fee to the European sponsor which would undoubtedly make it difficult for the system to compete economically with conventional methods of construction.

4. To utilize a European system a large capital investment would be necessary, which would be excessive for most of the organizations currently active in the US building industry, which has always been oriented towards low capital investment. (Pike 1968)

In brief, their doubtful methods of construction, poor space standards, high development and capital investment costs clearly did not recommend them to the North American market. Nevertheless Koch's attitude to the machine is a familiar one:

9.5.5 Academy Homes, Roxbury, Boston, USA. Carl Koch and associates, architects for Techcrete system and scheme, 1965. (a) View showing facade modelling achieved by projecting balcony boxes and fins. (b) View of interior court. (c) *Opposite.* View of 4 storey units. (Photos 1969)

c

'that the machine, with its possibilities and limitations, is a better tool for architecture than any ever available before; and that the machine, creatively used, is the essential means of achieving economy, quality and splendour.' (Pike 1968)

The result was, not surprisingly perhaps, that Koch set about evolving a more suitable approach for the context in which he was working. The basic design decision involved the use of concrete precast wall panels supporting precast, pretensioned, extruded, sawn off floor slabs of 32 ft (9.75 m) span up to a height of twelve storeys (**9.5.6**). The floor panels are in 1, 2, 3 and 4 ft widths (305, 610, 914, and 1220 mm) and the wall slabs can be any lengths. Prestressing rods run continuously through the walls from roof to foundations for structures up to five storeys — above this additional shear walls are introduced. End panels can obviously be of many sorts — on the Roxbury project a range of infill panels, windows, heating and air conditioning units was designed. Internal fittings can be designed to the standards of each project, according to Koch's office. The standard of components such as windows were to suit North American

600

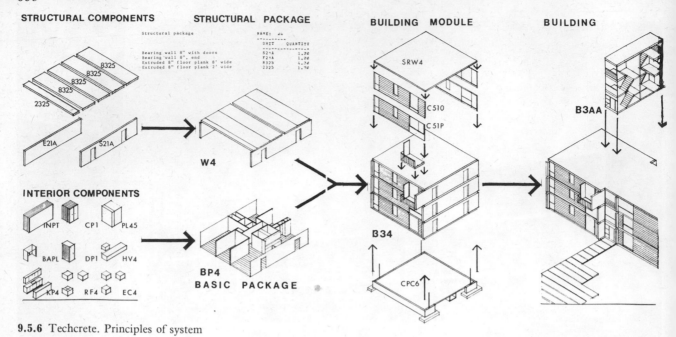

STRUCTURAL COMPONENTS   STRUCTURAL PACKAGE   BUILDING MODULE   BUILDING

INTERIOR COMPONENTS

9.5.6 Techcrete. Principles of system

requirements — double glazed, well insulated and draught sealed. Nevertheless the scheme appears to suffer from many of the problems of similar attempts at apartment building in Britain — such as poor floor and wall sound insulation and low standards of finish.

The author's visit to the completed scheme at Roxbury, Boston in 1969 showed that it had many interesting design features, not least the three-dimensional handling of the scheme and the use of playdecks. These latter, however, seemed to be barely used whilst the planting and landscaping on the site had been virtually destroyed demonstrating again the social difficulties surrounding the provision of public housing. The published acount of Koch's Techcrete (Pike 1968) embodies a whole series of proposals for possible developments but the fate of these has received less publicity. There are the inevitable architects' drawings of a multilevel good life with the obligatory helicopter dropping a component into place.

Techcrete was put forward in a massive document as part of Koch's application to the Operation Breakthrough programme in 1969, but with the passing of the Johnson administration into that of Nixon on 20th January 1969, this attempt to revitalize housing was gradually shelved. Nevertheless the American response to the Johnson administration's call for a rethink of housing problems is of interest since it embodied many of the myths familiar to Europeans. As in the British public housing drive of postwar years, and again during the politically inspired impetus of the early sixties, vast systems of building, complicated procedures, and technical proposals proliferated around Operation Breakthrough. In general, such methods quickly proved their limitations and their demise was only matched in rapidity by the speed with which political support was withdrawn. In the United States as in Britain, the formulation of housing problems into a 'crisis' of political dimensions created a surge of enthusiasm around government funds and the belief in industrialized building systems as the solution. In Operation Breakthrough many ingenious methods were presented — those that made any headway, however, were those firmly based in developing existing technologies.

The lesson was again clear: that technological ingenuity can assist in a solution to housing problems, but when it becomes the central, driving force attempting to override its social context, fundamental issues are raised. The development, marketing and support required for most building system approaches need a degree of centralized state direction which, in a democratic society, few would wish to encourage.

10

INTEGRATED SYSTEMS

THE ENERGY IMPACT

'The conservation of our natural resources is only preliminary to the larger question of national efficiency.'

Theodore Roosevelt
quoted in
Frederick Winslow Taylor 1911

'The fact remains', says Reyner Banham, in his introduction to *The Architecture of the Well-Tempered Environment* (1969), 'that the history of architecture found in the books currently available still deals almost exclusively with the external forms of habitable volumes as revealed by the structures that enclose them'. Anyone who has attempted to refer to the history of building systems will know that, as far as it exists, it is also concerned almost exclusively with structure. Indeed what has happened in the development of building systems is an even further reduction than this. For, as Banham makes evident in his study, the history of architecture may have ignored the impact of environmental engineering but buildings themselves have not. Larger and larger parts of building budgets have been devoted to ever wider ranges of servicing, and building structure and form have changed to accommodate it. Nevertheless the fact that it has been barely recognized at the theoretical and educational level has meant that these practices have not been crystallized into accessible knowledge — except that gained by the hard practicality of building.

As has been seen in earlier chapters the history of building systems has been particularly bedevilled by this problem. Questions of structure and fabric (lightweight fabric at that) have been the major obsessions of those concerned with the mass production argument. They have been concerned with the technology of construction. This is not to say that no-one has concerned himself with a more total view of technology or architecture: only that the major models that have provided the source material for the systems designers have been those of a very particular and limited sort. From the powerful effect of the Crystal Palace as a model for lightweight mass produced buildings, through Le Corbusier and the Bauhaus the argument is clear, and it is clear in visual terms. The models are buildings and projects which architects can understand: they are still concerned with the traditions of form, space, colour, light and shade. It must be no surprise, then, that many of the building systems available have, as their central concern, those kits of parts which make up the fabric and structure of the building. Architecture is not, however, without its examples of a more holistic approach although these generally have tended to fall outside that boundary which the building systems have managed to draw around their field of available knowledge.

Many of Frank Lloyd Wright's buildings are holistic, more systemic, than most of those produced by a building system. And this is not, as the systematizers would have it, merely because of the 'prima donna' qualities of the architect, unattainable by lesser mortals. It is because, generally, the concepts pertaining to building systems have subsequently been described in such a way as to exclude a range of concerns which would normally be the province of the architect. Such matters as the symbiotic relationship between the inside and the outside, the manmade object and its site and location, the construction and the people using it — all these have been subjugated to the imperatives of continual reassessment of constructional method — Constructivism with a vengeance.

Banham points out that, although most of Wright's buildings have been carefully considered as integrated wholes, only those that do this in a formally obvious manner have caught the attention of architects and their historians. A prime example is the Larkin Administration Building of 1906, in Buffalo, New York. Here the external service towers have been lauded for their purely external formal qualities, whilst the servicing functions which they so well integrate have been largely ignored. Similarly with Buckminster Fuller: his Dymaxion house, Dymaxion bathroom and other work certainly indicate that he was not unaware of the proper balance between the environmental control and all the methods of achieving this. His attack on the International Style 'simplication' is particularly cogent in this respect: 'It (the International Style) peeled off yesterday's exterior embellishment and put on instead formalized novelties of quasi-simplicity, permitted by the same hidden structural elements of modern alloys that had permitted the discarded Beaux-Arts garmentation. It was still a European garmentation' (quoted in Banham 1960).

Equally, whilst the avowed concerns of the Bauhaus were with the right use of the machine and with the industrialization process, these do not appear in practice to have embraced very thoroughly the available machine technologies. Further, the superficialities of styling referred to by Fuller seemed dominant. It remained to those involved in work on the theatre (e.g. Schlemmer) and sculpture to draw more heavily on these machine technologies. As with Meyerhold in the twenties, and others, the dramaturgical model whilst the most immediately powerful is also unfortunately the least durable. Thus the physical object, and its photographs subsequently gain more and more credibility.

It is with this background in mind that we must view the failure of the building systems to deal adequately with aspects of environmental comfort and control: reaction against this failure was a factor in the subsequent emergence of the 'environmentalists'. This term, an imprecise and unwieldy one, is useful in describing a group of designers, together with their projects and buildings, who have tried to redefine the systems concept to take account of such deficiencies. Although many of them are well conversant with the sources of systems ideas these have often been retranslated into buildings of a sort very different from those which we have come to expect from the self-styled system builders.

Fuller's reference to the European approach as 'garmentation' is, I think, not a casual one. The way in which technology has been integrated into North American society is very different to the manner in which it developed in Europe, particularly in Germany and Britain. An important systems manifestation of this is the SCSD system of Ezra Ehrenkrantz and his team, discussed earlier. The influence of this work has been a strong but subtle one. For those with eyes to see it SCSD offered a powerful critique of most of the British system work, even though somewhat reliant on it as source material. It was all the more easily officially buried, coming from America — and California at that. The initial British suspicions, and then gradual absorption, of ideas American was here as much in evidence as in other areas. With Ehrenkrantz being invited to talk to conferences, to system designers and to the DES development group the scene was set for the gradual discrediting of the SCSD ideas when set against the more 'humane' British approach to building schools (Medd 1972). Nevertheless SCSD attitudes to flexibility, environmental servicing, the role of the external wall, specification by performance rather than solution, the definition of environmental criteria around thorough user studies — all these have found their way gradually into the vocabulary of a powerful group of designers, both inside and outside the 'official' systems movement in Britain.

As already described MACE was the only one of the latter to attempt to redefine the British client sponsored systems argument by taking this work into account and this explains, in part, the opposition which this system created. Much of the criticism subsequently heaped upon MACE could equally be applied to the other 'client sponsored' systems. Both CLASP and SCOLA have their share of drab concrete factory-like schools and both have severely limited choices in components. Their attention to environmental comfort and servicing has been confined to attempts to exhort their engineering sub-committees to reduce services cost. MACE has been the sacrificial lamb of the 'client sponsored' systems and this ritual act has deflected the attacks of the GLC schools architects, referred to earlier, from the real issues. For solutions which identified, with more clarity than MACE, the new directions the systems idea was taking, we must look to a series of buildings which in their way have acted as catalysts in a situation of some criticality for the institutional systems.

BIG BOXES, DISAPPEARING WALLS

'Clients pose problems to us that are initially expressed in building terms, but, with analysis, the critical areas of the solution frequently emerge as non-architectural. This results in a change of emphasis in our role as prime adviser to the client.

We are interested in extending our activities in the direction of a total systems approach to any building type.'

Foster Associates
Architectural Design May 1970

Wright at Redbridge

The excitement engendered by the publication of the SCSD work was considerable. In calling into question many of the fundamental notions of systems designers it served a valuable purpose although it had little immediate effect upon them. For those architects working or interested in the school building field it crystallized in architectural terms many of the then current concerns of progressive educationalists at the way in which the physical form of the building may be rethought to take account of the changes in teaching methods which many predicted. If practice was unable to draw immediately upon the possibilities suggested by Ehrenkrantz, his work was to have long term repercussions on architectural education and research.

This influence can be seen in the work of John Wright who, during 1966/67 whilst at Portsmouth School of Architecture, prepared a research thesis (Wright 1967) which first surveyed the environmental conditions in a range of schools and then went on to argue the case for an agreed and considerably upgraded set of environmental criteria for school building. This work brought out many of the major problems apparent in school building design: poor response to changes in teaching methods, lighting problems (calling in question the Department of Education and Science's adherence to the 2% daylight factor), poor standards of artificial lighting and inadequate heating. The result was that some parts of the buildings were very much overheated and some (usually those rooms where children removed their clothes) very much underheated. The overglazing to achieve the 2% daylight factor gave rise to massive heat losses and downdraughting in winter and enormous solar gain in both summer and winter. There was also inadequacy of acoustic control with very little attention to sound absorption, sound transmission problems engendered by the (generally) lightweight construction being used, and a whole range of problems related to the interior furnishings, fittings, materials and colour.

Drawing on international research and his own studies Wright proposed a set of environmental criteria in which the building and its internal environment was looked upon as a total system. With such an approach not only were the existing standards brought under critical examination but also the planning and organizational suppositions which related so closely to them. From this it was possible to suggest technological solutions that might arise from such an approach. This drew out the necessary relation between the environmental servicing and the structure and fabric of a school building. In theory Department of Education and Science cost limits allowed the designer to distribute the money within his overall budget in any way thought necessary to a particular educational solution. In practice, however, there was continual and strong pressure to cut down environmental engineering costs and passive modes of control — such as more insulation, in the interests of creating more floor area. If indeed internal environmental standards were to be improved within the existing cost limits it was clear that the building as a total system needed to be re-examined.

Wright found, not surprisingly perhaps, that if major simplifications were made in the external envelope (both the walls and the roof) considerable cost savings could be made which could then be redistributed elsewhere, incuding an upgrading of the insulant quality of the envelope itself. For example, the adoption of deep planning coupled with good artificial lighting standards not only produced simpler plan forms with fewer costly corners and changes of construction, but was of considerable importance to the use of internal space on a flexible basis. Where 30–40 child class units are no longer the only mode of learning there is a strong requirement for spaces to hold different sized groups ranging from the single child through groups of 4 and 5 to seminars of 10-14, class groups of 30-40, house groups of over 100 and, on occasion, several hundred for a film or top class speaker (Wright 1967).

In view of the difficulties being experienced by schools designers in providing adequate environmental conditions, let alone more flexibility, such an examination was long overdue. It further indicated that the regulations governing school premises initially promulgated under the Public Health Act of 1936 and separate from the Building Regulations, were an extremely inadequate mechanism. One of the few defined environmental requirements, in keeping with the health and sun obsessions of the twenties and thirties, was that which called for the 2% daylight factor to areas used for teaching purposes. Quite apart from the fact that the daylight factors are an exceedingly arbitrary device, their implementation created many problems, involving large areas of glass. Indeed one of the major sources of complaint from teachers has to do with problems resulting from overglazing. The inadequacies of venetian blinds, attempts to provide cross-ventilation, and so on, make a mockery of such regulations in practice. Wright's work then, drew explicit attention to the problems arising from not considering the environmental control, the construction, and the building users as a whole.

His work in this field was to be of considerable importance in his first major project: the Ilford Jewish Primary School in the London Borough of Redbridge completed in 1969, opening in January 1970. Designed whilst with Scott Brownrigg and Turner, this attempted to apply what amounted to a version of the SCSD system to the English situation: not surprisingly it met with considerable opposition — not from the client group or staff of the school, who were very committed to the whole idea, but from the Department of Education and Science. Wright points out that:

> 'Flexibility was the fundamental design philosophy: modern educational techniques such as team teaching often involve constant change from large group instruction to small group discussion and individual study, thus teaching spaces clearly need to match these requirements at a moment's notice with accommodation of appropriate scale without compromising the environment... The teaching space in this school can be subdivided without compromising standards of air, temperature, light, sound.' (Wright 1969)

This aim resulted in a building which embodied some major propositions for school design in Britain; inevitably it also resulted in considerable problems. That the innovations were attempted is a credit to Wright, his firm, and Michael Booth the then Borough Architect who gave the scheme his support despite reservations. That this project was not supported and monitored more positively to assess its successes and weaknesses is a further example of institutional attitudes to major innovation.

The school complex comprises four basic elements: the primary school itself, the nursery school, the youth club and the caretaker's house. Of these, it is the primary school which embodies the major innovations and which attempts to provide an internal environment more related to the concept of flexibility of internal space set out by Wright. The constructional solution follows one of the classic systems solutions, a large rectangular box, but with important modifications (**10.2.1**). Hollow box section columns support long spanning castellated primary and secondary beams with a corrugated steel deck roof. A number of solutions were proposed for the structure before this one was finally adopted. Initially it was thought that one of the consortia school building system frames would be the most economic solution but prices received from one of the major suppliers involved indicated that this was not so — similarly with regard to the external wall and roof, both of which had a high insulation requirement to avoid excessive heat loss and cooling loads from solar gain. Here, in considering carefully, and in total, the internal environment in relation to the external wall and to external environment Wright had

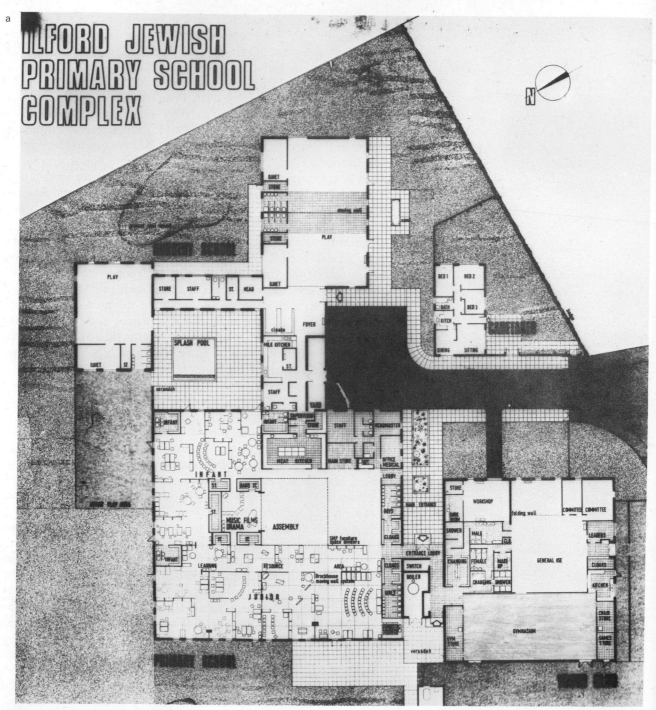

10.2.1 Ilford Jewish primary School,
Redbridge, London. Architects, Scott,
Brownrigg and Turner; project architect John Wright, 1969. (a) Plan
showing Infant and Junior area at
bottom left, kindergarten at top.
(b) *Opposite.* Axonometric

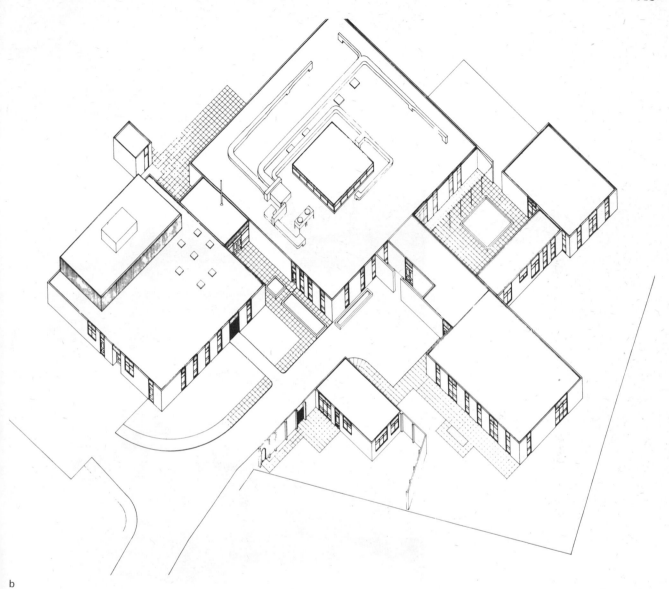

b

to discard his initial intention to use prefabricated units produced by one of the existing systems since their performance in thermal and acoustic terms was inadequate. Not only did they not meet thermal requirements in themselves but the jointing presented considerable problems — as did the excessive number of joints which all the walling systems used by the school consortia possessed. The inability to seal system built schools properly against heat transfer through the external skin is one of the most serious faults of the component philosophy as it has developed. Architects seem often not to have realized the significance of, for example, providing joints that only left one or two thicknesses of metal between the inside and the outside. A number of architects using these building systems had resisted this reduction in environmental standards in the only way left open to

10.2.2 Ilford Jewish Primary School.
(a) Exterior showing vertical strip
windows. (b) Interior. (Photos 1971)

them — by the continued use of brick, block or stone as an external
skin. Brickwork was not a 'standard' detail in the CLASP system until
1972 although some authorities using SCOLA had insisted on this and
other masonry materials from its inception in 1961, and it became
'standard' in 1965. Generally however, the espousing of the purist
philosophy of mass production meant that it was thought to be
somewhat heretical to use heavy materials in a building system.

Therefore, although initially committed to the systems technology
view Wright found that, when examined from the environmental
point of view the available solutions were largely substandard. In the
case of Ilford Jewish Primary school the external wall became, on
grounds of thermal capacity, noise transmission (an adjacent railway

10.2.3 Ilford Jewish Primary School.
(a) Interior view down 'resources
zone', showing SMP cupboard units,
Track Lockwall ceiling grid and
partitions. (Photo 1971). (b) Plan:
Ceiling grid, 6ft 0in × 6ft 0in (1.83
× 1.83 m), for moving wall system
can be seen dotted on lower half

a

b

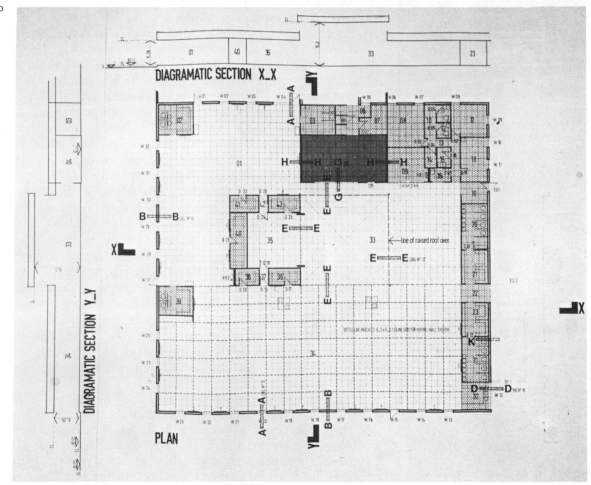

10.2.4 Furniture units (SMP), developed for Ilford Jewish Primary School (1969), here shown in use at Godstone County First School, Surrey. (a) Used as bunks in rest period. (b) Used as desk and storage

line) and cost, a brick external skin and an internal skin of blockwork with a cavity filled with foam. The square, compact plan form of the Redbridge school gave a very low wall to floor area ratio and in addition the ceiling and external wall height were kept low. In keeping with the deep plan philosophy external windows are small compared to the average system built school up to this time, with its obsessional use of glass. Windows are in the form of 900 mm vertical strips glazed from floor to ceiling (**10.2.2**), partly because in primary schools children spend a great deal of time sitting on or near the floor level. Therefore any solid panels below sill height would discourage contact with the outside. Further it is often found that furniture is pushed up against large glazed areas — sometimes improvised curtains or other protection is being used. Wright, in choosing to use such a window type, was influenced by the work of Manning (1967) and Markus (1967). The latter made an important reappraisal of the function of the window in which he pointed out that a vertical cross-section of the landscape outside, showing not only the sky, but the distance, middle distance and foreground, 'offered a great deal more to the viewer' than the continuous horizontal strip window which allowed often only a view of the sky.

In keeping with the philosophy of flexibility a moving wall was developed jointly by the architect and Carlo Testa, then Development Architect for Brockhouse structures. This employed a square ceiling grid of 1828 mm location channels (**10.2.3**). The panels, 914 mm wide and faced with PVC coated steel were pin mounted into the ceiling grid and had a skirting level jacking system, which allowed them to be lifted and moved by the teaching staff. In practice the difficulties of moving panels of this type seems to have inhibited their use to any great extent. In addition Wright developed a mobile furniture unit into which could be slotted shelves, cupboards, worktop display boards (**10.2.4**). Both the partition system (SMP's Track Lockwall), and the mobile furniture units (SMP's Educat) have, after modification, been extensively used in schools and other building types — most notably the William Morris school at Merton, and Godstone First School, in Surrey. This is an example of one of the gaps in available component systems identified by Wright and demonstrates how this 'open systems' approach can float new components on the whole market. Attention to the acoustics of such a large open space gave rise to an air-conditioning system designed to provide a background so that speech frequencies are masked. The whole floor area is carpeted and the ceiling has acoustic tile throughout. Not only does the carpet provide some absorption for sound but a soft surface for children to work on the floor and a low maintenance cost.

The plan form provides a very high percentage of teaching area (70%) compared to many schools since there are virtually no corridors. The 'resources strip' running the length of the school upon which are located the library, sink units, benches and so on, is an attempt to put these in a position through which children pass daily and which are therefore a crucial part of the school's activities. The area designated for music is contained within a band of insulating stores with a folding door at one end. This seems, with the other acoustic considerations, to give a very adequate sound cut off. The central assembly space has a higher ceiling level with a clerestory strip running round it (inserted as a concession to DES pressures) which seems rather out of place, and allows a considerable heat loss. Toilets and kitchens are grouped along the south and entrance front, thus providing a further noise and solar gain barrier to the body of the school. The infant area, on the other hand, opens out onto a southeast facing courtyard, and splash pool.

A building of this sort which relies, in more obvious terms than is normally the case, on understanding the control systems has suffered some difficulties. The air-conditioning located on the roof to save money, has caused considerable roof problems. More important perhaps are the difficulties of a teaching staff who, whilst very enthusiastic about the intent, are not used to working in a building with such close environmental tolerances. Therefore, a door will be opened to 'get ventilation' and this in turn will unbalance the air handling system, which in turn causes further malfunctioning. One solution would have been to provide airlock lobbies at all doors — this was rejected on cost grounds but their absence has proved perhaps more costly in maintenance. An air-conditioned building has to be a sealed building, a lesson hard learned in Britain, as many other examples show. Indeed cultural reasons may preclude the successful use of air-conditioning in many building types as an acceptable solution.

Although receiving support from the client and the Borough itself, the design met with considerable criticism from the Department of Education and Science and exposed clearly some of the difficulties such an innovation can encounter. In theory, the DES approval is a formal process provided that cost considerations and the School Premises Regulations are adequately met. In the case of the Jewish Primary School, a long series of queries were raised which considerably delayed the start of the project. Although a number of these were valid the basis for the opposition aroused was that the school proposed a set of design suppositions with which many at the DES profoundly disagreed. A natural suspicion of anything North American plus the long tradition of favouring the throwing open of large windows and allowing the breeze to waft in over the carefully arranged flowers on the window sill seemed to take precedence over a positive evaluative approach to this set of innovations. The design was criticized for having less than the 2% daylight factor; for being air-conditioned (it was pointed out by the DES many times during design that this could not be done at all within the cost limits) and for having an interior that was soulless and inhuman. Much of the basis for this reaction was to be sought in the early flexible schools, like Hillsdale High School and others in the United States which had used simple plan forms and such devices as office partitioning to create very flexible interiors within acceptable cost limits.

Wright's use of furniture, materials and colour save the interior from the worst characteristics of open plan schools. The exterior, however, with its vertical and repetitive slit windows proved more difficult to handle and presents a grim appearance. The roof mounted air handling units and ductwork are all exposed on the roof and whilst they are unashamedly painted in primary colours they have not been integrated into the overall external design to the extent we see in the later work of Foster and Rogers.

The Department of Education and Science did not take the view that might have produced some valuable research data. Had they been more encouraging or more open to possibilities it may have been seen as a unique opportunity to follow and properly monitor how such a school as this worked. This may well have yielded some useful information — much of it in favour of their view. However, the curious centralization of the British educational machine does not in reality make it easy for innovators from the periphery, especially as in this case the designers and clients were both in the private sector. Therefore a thorough study of such a school remains to be done in this country — although a brief study of the Redbridge school was carried out by two social scientists, Powell and Matthews (1976) and a study of the design genesis of the building by Wang (1970). Indeed the Department of Education and Science Architects and Buildings Branch have not so far shown that they have either the skills or resources for adequately researched school studies. Their building bulletins which monitor the major work done by them in designing and building 'development' schools to test ideas, whilst of value, are less interested in the appraisal of buildings in use than in positing models or patterns for local authorities to reproduce. The sad fact is that because of the form of this work the more general implications are rarely adequately substantiated or explained. This has often resulted in the slavish repetition of these bulletin proposals all over England. It remained until 1977 for the Department of Education and Science to publish more thorough studies of school building, and these confirmed many of the failures of environmental performance which Wright had drawn attention to in 1967. Clearly the energy crisis and the effects of the falling birthrate had made their impact at government level. However, a closer attention on the part of government to 'outside' research such as that done by Wright (1967), Hardy and O'Sullivan (1967a,b,c) and others may well have enabled them to take action earlier to deal with the many problems besetting school building in Britain.

Wright's work illustrates two important points. The first is one that has a respectable history in the development of architectural ideas: that of taking a set of ideas and recombining them in other situations. The second has been much rarer in architecture, although a commonplace mode of advance in other disciplines: that of research carried out under the aegis of an educational establishment and then being drawn upon by practice. In bringing together current research ideas with a strong architectural idea the Ilford Jewish Primary School of 1969, although flawed, presaged a new swing in direction towards a greater concern for environmental comfort. Within ten years of this building we find a general return to smaller glazed areas and more insulation. The oil crisis has caused a move away from fully air-conditioned solutions to ones relying merely on air circulation and the failure of many of these indicates that the question is still an open one.

Foster's Boxes

Another important project, but one which was not built, was put forward by Foster Associates in 1968 in their entry for the Newport Comprehensive School Competition (**10.2.5**). Not only was it important in translating, almost directly, the SCSD concept into a large school design proposal but it became the forerunner of Foster's later building for IBM at Cosham, which for English eyes crystallized the whole Ehrenkrantz philosophy into a strongly visual form. Published material on the school describes the design approach in true systems terms as an interdisciplinary one between architects, consultants, manufacturers and suppliers. The aim was to 'develop a system incorporating usage for the total range of educational servicing needs, from primary to university levels, and to optimise the advantages of industrially produced components to achieve flexibility for choice, change and growth' (BG 1969).

The proposed building was a large rectangular shed with a long spanning steel lattice roof structure. Extra height for gymnasia, pools and so on were achieved by excavations at appropriate points. The teaching space was shown on the plans as a large open area which could be divided at will by moveable partitions and cupboard systems. The sports halls, gym and kitchens occupied one long side of the plan and between these and the flexible teaching/circulation/administration areas was an internal circulation route or social 'mall' running the length of the building with moveable toilet blocks located on a service trench acting as a perforated buffer between it and the teaching area. The servicing occupied an important place in the concept and roof plans similar to those produced for SCSD were naturally in evidence together with cross-sectional elevations of the interior (**10.2.6**). The way in which the flexible interior would work was indicated by a series of line drawings of the type found in the Ford Foundation's Educational Facilities Laboratory books on school design. These emphasized the way in which the open area could be multi-used. The internal areas, it was claimed, would have a well controlled internal environment in a deep plan. This was to be achieved by a flexible heating and cooling system and 'partial' air-conditioning. However, the designers claimed that the required 2% daylight factor was 'inadequate at a qualitative level' (BG 1969) and they therefore aimed at 5% achieved with natural top lighting through sealed rooflights. The aim was to reduce internal heat gains in teaching and administration areas on extreme summer days to external shade temperatures:

> 'Conditions would not be those of a fully air-conditioned school but would be superior to an ordinary sideglazed school with opening windows.' (BG 1969)

620

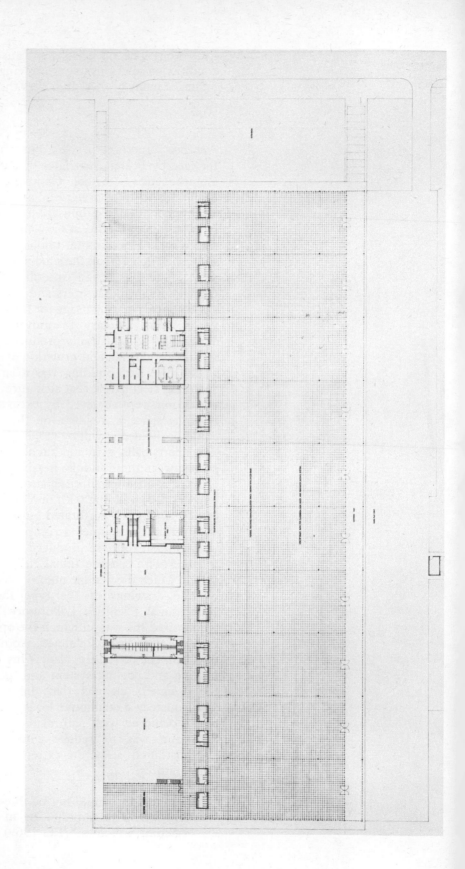

10.2.5 Newport Comprehensive
School, S. Wales. Foster Associates.
Competition entry, 1968. Plan

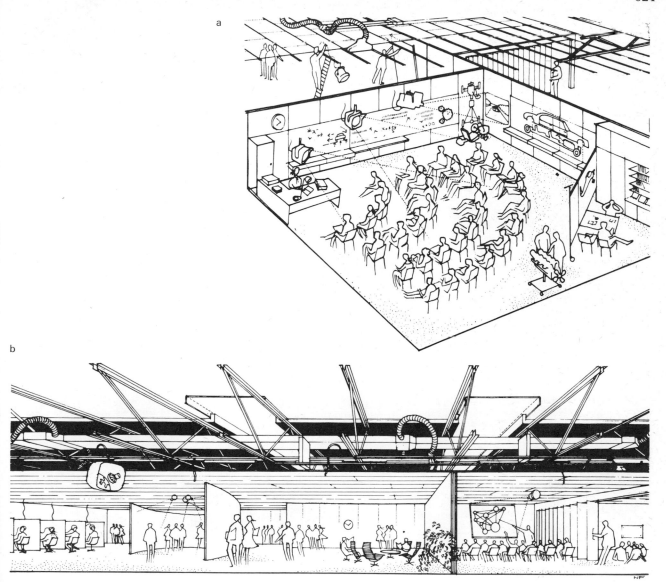

10.2.6 Newport Comprehensive School, (a) Teaching space with its view of how diverse activities could simultaneously take place, and be easily changed. (b) Cross-sectional perspective

The cooling requirement would be considerable since not only was the building deep plan, but it had a flat roof and a vast number of rooflights and there would be heat build up from occupancy, machines, electric lighting and particularly solar gain under such circumstances. The roof mounted multizone heating, ventilating and air handling units with their eight flexible outlets follow SCSD very closely. The problems posed by the rooflights and amount of servicing in the roof would in themselves have required further compromises to be made in one direction or another, and does call into question how fully examined were some of the implications of the deep plan in this project. The effects of such a servicing system require firm decisions across a whole range of subsystems. There appears to be no easy path between the environmentally machine-managed internal space, and

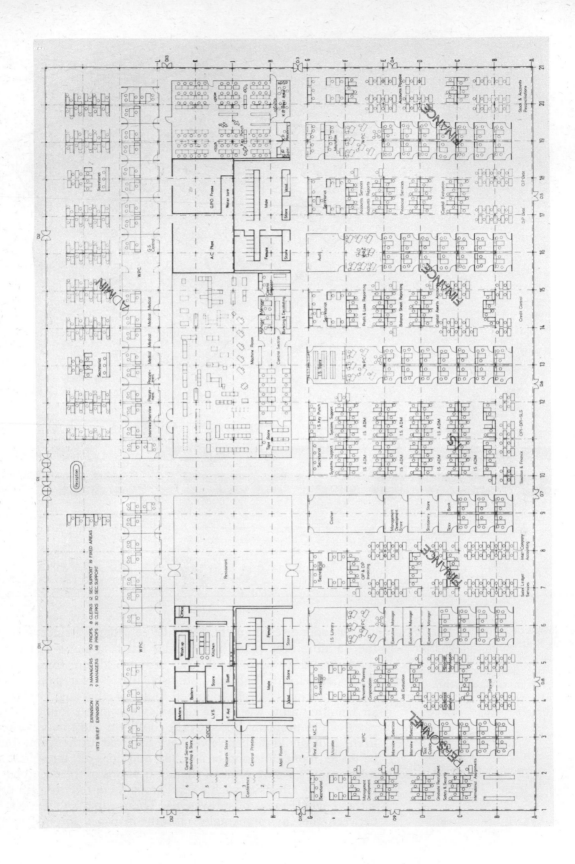

b

10.2.7 Office building for IBM, Cosham, England. Foster Associates, 1971. (a) *Left*. Plan showing unbuilt 3 bay section on north side. (b) View from south. Note roof mounted air handling units

10.2.8 Office building for IBM. Part of south elevation. Sunblinds visible below eaves. (Photo 1972)

that which relies less heavily on the technology of servicing systems. This is especially pronounced in solutions involving deep planning, and flexibility of the sort described here. In this case one can see how the problem was resolved in practice since subsequently the Foster Associates IBM building at Cosham, completed in 1971, uses almost the complete Newport concept excluding the rooflights (**10.2.7**). *Architectural Design,* in its appraisal of the IBM head office (Rabeneck 1971), showed the plan of Newport comprehensive school as the basis for the IBM proposal. Certainly there are similarities in concept but it is probably in the differences, the translation from a school to an office, and from a project to a building, that are of most interest. The plan is again a long rectangle with dining facilities, computer room, toilets and so on ranged along the north side of the plan with somewhat over half the plan given over to a large flexible space.

10.2.9 Office building for IBM. View of corner showing the carefully wrought simplicity of the detailing. (Photo 1972)

Although it is somewhat uncertain as to the full extent of air-conditioning in the Newport proposal there is no uncertainty in the IBM building. The box is a fully heated and air-conditioned environment with no rooflights, and the external wall is bronze glass in neoprene gaskets on aluminium mullions (**10.2.9**). The same principle of roof mounted multizone air handling modules is employed, and the main space conceived as a flexible interior.

However, the flexibility of the interior is conceptual rather than real in that IBM management policy insists that managers have their own private offices, and the space is therefore divided by bands of separate offices with open areas between. The fixed nature of this layout contradicts the philosophy of frequent change and evolution that lies behind the Ehrenkrantz, Wright and Newport proposals. Indeed both in plan and fact the desk layouts at IBM are rigorous in the extreme and seem to be more concerned with surveillance than anything else. Further change from the Newport proposal can be seen in the deletion of the large overhang proposed to mitigate solar gain. At IBM Foster argues a case for the 6 ft 0 in (1.8 m) internal perimeter circulation strip taking up excess solar gain, but the use of this area as office space, and the subsequent introduction of sun blinds is indicative of some of the problems being encountered here (**10.2.8**). The deletion of the overhang, the use of the air-conditioned internal environment plus the fully glazed external wall is a somewhat dubious conjunction. The use of spectra float bronze mirror glass, which has a light transmission of 51% and a claimed heat transmission of 61% (Adams 1972) (only one-third reduction on clear glass), is an architectural device here carried out with panache. However, even with such high heat transmission as that admitted by the manufacturers, the use of it around the whole building must surely be in question. A further case may be made out for it as a simple factory produced repetitive set of components — its surface merely reflecting what is outside.

Foster was initially retained by IBM to carry out a feasibility study for the location of new pilot head offices on a site at Cosham, north of Portsmouth. IBM's policy hitherto in such situations was usually to first buy an 'off the peg' temporary timber building system with medium spanning properties and to light and ventilate by means of courtyards. Since most such systems are merely structure and fabric systems with questionable environmental standards they are in distinct contrast to the prestigious permanent buildings IBM then later erect. An example of this approach could already be seen quite close to the Cosham site at Havant where Arup Associates had designed what Balfour (1970) describes as 'a veritable temple to information processing'.

Foster Associates put a range of alternative solutions to IBM for the site. The first was a 40 ft 0 in (12.2 m) × 40 ft 0 in (12.2 m) bay single storey patent timber building system using a courtyard system to produce adequate lighting and cross-ventilation. The second was a two storey version of a similar system (**10.2.10**).

The limitations of such solutions were:

1. Site coverage necessitated by the internal courtyards severely restricted available car parking.
2. Natural ventilation of each office space adjacent to a large traffic interchange would produce intolerable noise levels as well as dust and draught problems.
3. The inability to cope adequately with the services installation required — including a computer. The latter in any case needed air-conditioning.
4. The poor company image such a building produces.

In view of their previous work on the Newport school it is perhaps not surprising that the solution preferred was another version of the Ehrenkrantz SCSD approach. It is described as an 'open industrialized

626

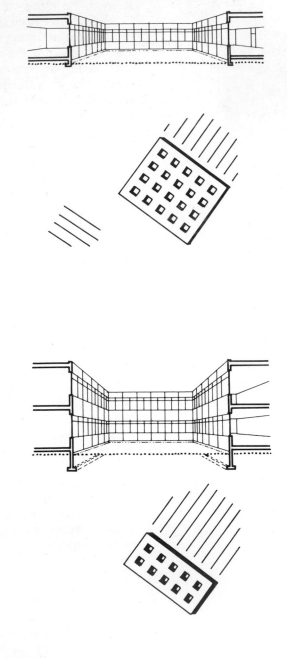

10.2.10 Range of options indicated to
IBM in the concept report prepared
by Foster Associates

component system', is supported by a 'systems approach' drawing
(10.2.11) that is similar to those produced by Ehrenkrantz and
reproduced in the Ford Foundation's Educational Facilities
Laboratory publications on SCSD.

In the IBM building the systems approach adopted relies upon a
thorough design integration of siting, structure, envelope, environ-

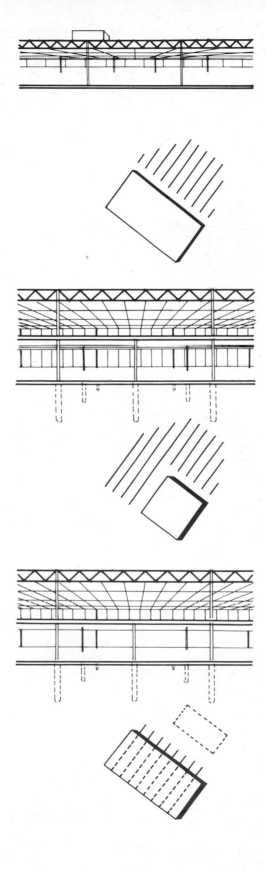

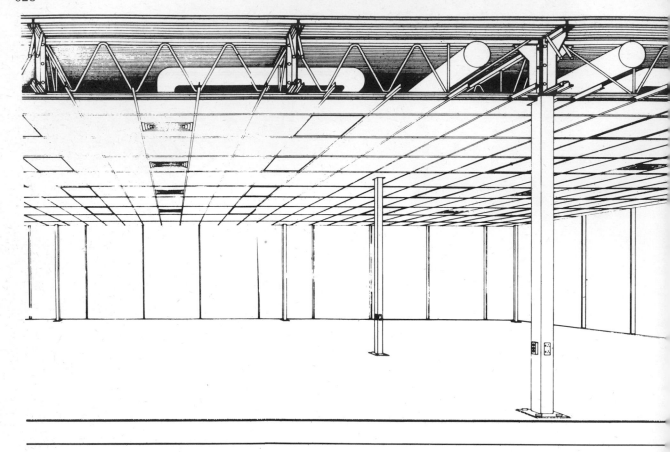

10.2.11 Office building for IBM.
Cross-sectional perspective

mental control and internal planning possibilities, including a possible expansion to twice the size. A structural and service module of 24 ft × 24 ft (7.3 m × 7.3 m) was adopted with a constant ceiling zone of 2 ft 0 in (600 mm), with the 5 in (127 mm) hollow steel columns used as the outlet points and telephone lines. The environmental criteria to be achieved were carefully set out listing winter day and night, inside and outside temperatures, and humidity levels. Solar gain, lighting and acoustics were also set down. The whole was a brilliant putting together of the then current systems arguments and clearly demonstrated that when a designer can put up such a well supported and argued case he has tools at his command which few package deal firms comprehend.

Foster Associates have subsequently developed their minimalist approach to a high degree of technical and architectural sophistication in the Willis, Faber, Dumas building at Ipswich (1974) (**10.2.12**) and the glittering silver serviced shed that is the Sainsbury Centre for the Arts (1978) at the University of East Anglia at Norwich (**10.2.13**). Their buildings do utilize rather literal applications of notions surrounding mass produced components, but they do this with a precision and

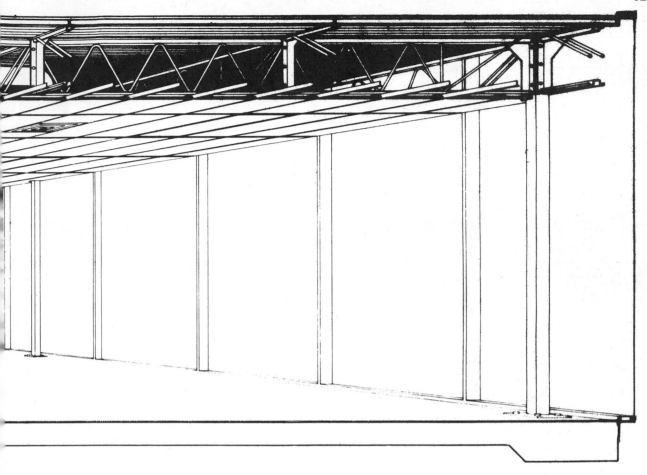

panache which add up to works of considerable architectural presence.

Both Wright's school at Redbridge and Foster's IBM building posit systems models which in their single-minded enthusiasm show that if architects are to do more than pay lipservice to such concepts as flexibility and environmental comfort they must understand in an integrated way the interactions of heating, lighting, ventilation, air handling, acoustics, colour and materials. Of course, some of the best architects have always done this and usually with less reliance on mechanical servicing systems: they have done it through a thorough and often intuitive understanding of the relations between climate, the site, the materials used and the likely internal conditions. However, skill of this order has to be built up by supplementing theory with considerable practical experience. In earlier times an architect like Frank Lloyd Wright, or Voysey built house after house and from this they learnt quickly and built what they learnt into their next project. The explicitness of the 'systems approach' is thus doubly attractive to young architects and indeed to many clients, since it exposes a range of integrated factors only available to those with experience. Further I maintain that it is no accident that such a movement gained ground in

a

b

10.2.12 Willis, Faber, Dumas
Offices, Ipswich, England. Foster
Associates, 1974. (a) The curvilinear
form simulates the rounded corners of
many existing street blocks in the
town as well as reflecting adjacent
buildings in the repetitive glass skin.
(b) Detail of exterior showing four-
way glass-to-glass junctions and fixing
cleat. (Photos 1980)

a

b

10.2.13 Sainsbury Centre for the Arts, University of East Anglia, Norwich, England: Foster Associates, 1978. (a) The building is a long aluminium clad tube with fully glazed ends, seen here with Foster's helicopter lifting off (b) Detail of interchangeable cladding units. (Photos 1980)

the Britain of 1945 and after, where over half of the qualified architects became employed by central or local government. Whatever its advantages, it must be faced that in order to gain the necessary promotions in such severely hierarchical organizations young professionals have perforce to move from job to job rapidly. Thus it is very frequently the case that jobs are not seen through by one man or team: as a result the feedback open to those designing and building one job after another, often in the same area, is gone. As discussed earlier, the systems philosophy is a convenient support, if not a cause, of that argument which grew out of the twenties and thirties to the effect that the ubiquitousness of transport and modern industry had made regional implications irrelevant.

It is also no accident then, that such architectural solutions have, quite literally, a sealed boundary condition. All interchange with the environment is controlled; air is funnelled in, cleaned, heated or cooled, passed through the interior, inhaled and exhaled by the occupants and expelled from the building. Unwanted punctures in the external skin cause the whole system to oscillate and move towards breakdown. Similarly with people: they must enter and leave via prescribed holes in the skin. When they are inside they are inside and when they are outside they are outside. There can be no intermediate position, neither can the user have any effect on this aspect of his environment. Where John Wright at Redbridge did provide doors to the outside they put both the administration and the user in an impossible position: if they open a door the system goes out of balance with the result that rules of conduct have to be imposed. The response to this approach has often been, as with the DES over Redbridge, a partial or outright rejection of such solutions. To do this however, is to miss the point and the opportunity for reinvesting architecture with that consideration for the user which has undergone a series of attacks from the powerful philosophies of the modern movement.

The 'total technology' of Redbridge, IBM, and others can, if they are to redress this balance, be seen as very useful positions indeed. This is not to say that their complete approach must be followed, since in many ways this has all the marks of the total control beloved of many architects. What it can do and has done is to make crystal clear the necessity for giving more careful and well backed up thought to such traditional concerns as warmth, lighting, ventilation, and for seeing these concerns as part of the more obvious ones of fabric, structure and services.

Integrated Design

For an example of a less profligate use of energy sources it is worth examining a small village school at Eastergate near Chichester completed in 1970. Designed on a principle described as 'integrated design', it is an attempt to shift the design approach a little towards a more holistic systems view. In this respect it is significant that the design, carried out at the West Sussex County Architects Department (B. Peters, J. Paterson, C. Isaacs), had contributions from the

Electricity Council and the Building Science Section of the Newcastle-on-Tyne School of Architecture (O'Sullivan and Cole 1974). Shepherd points out that the concept of integrated design

> 'means designing a building to a standard of physical performance related to the heat, light and sound. It recognises that the building itself, its fabric, its shape, its relation to local climatic conditions, and its internal layout are no less important than installed services in effecting a high standard of internal environment with economy and efficiency of means.' (Shepherd 1971)

In this design the whole question of building shape and fabric design was looked at in relation to internal performance criteria concerning planning and the thermal environment. Just as Wright at Redbridge had re-examined open planning, building shape and the design of building envelope, at Eastergate the designers took a similar approach, although the result has one crucial difference (**10.2.14**). This is in their attention to achieving adequate comfort conditions by rethinking the heating, ventilation and lighting subsystems in relation to the external skin rather than just providing a well insulated box and then using full air-conditioning to achieve the required effect as at Redbridge.

10.2.14 Eastergate Primary School, W. Sussex, England. County Architect, B. Peters; project architects, J. Paterson and C. Isaacs, 1970. Plan

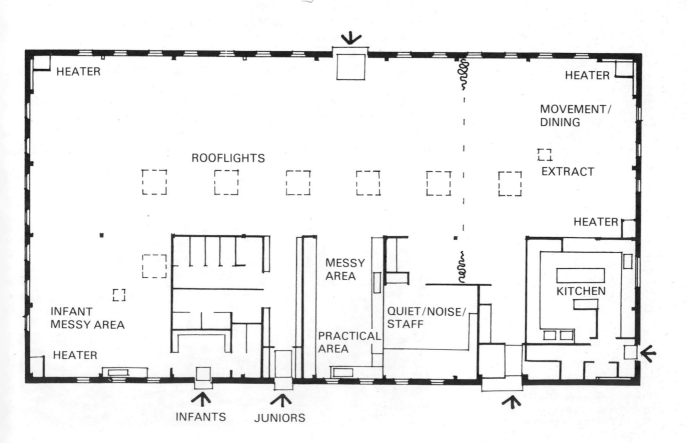

The Eastergate School has a total floor area of 520 sq m (5610 sq ft) and posits a way of using parts of the SCOLA building system which hitherto had not seemed easy since consortium members were tied by 'gentleman's agreement' to take up the agreed quantity of the component programme. The possibility of using 'non-system' components can be seen as one outcome of the considerably remodelled Mark 2 system as earlier described. It will be recalled that the total interdependence of the frame, external cladding and roof in the SCOLA Mark 1 system was here discarded so that at least a choice of cladding and other components could be made without the whole kit of parts having to be redesigned. Not only was the external wall redimensioned to allow a wider range of materials to be introduced but an eaves detail was devised to allow the external wall or the roof to be installed independently of each other. An immediate result was that a number of member authorities began to make a wide use of brick. At Eastergate it was thus possible for West Sussex to take the SCOLA frame, some system components and combine this with a foam filled brick and block external wall together with additional roof insulation. This in turn made possible a rethinking of window type and function, resulting in a tall thin proportion with splayed reveals and sealed glass, constituting only 20% of the external wall area (**10.2.15**). By very similar means to Wright at Redbridge, the designers had managed to return to a sealed building, thus overcoming one of the major problems created by the school building systems: the considerable and rapid energy transfer from inside to outside in winter and outside to inside in summer. This was a positive step in dealing with the problems of solar gain, poor control over ventilation, and heat loss in winter, which had bedevilled most of the lightweight system built schools.

That all the problems were aleady well documented long before Eastergate was completed in 1970 is appropriate enough comment on the power of the school building consortia in holding local authorities to a given set of norms. It has already been pointed out that SCOLA

10.2.15 Eastergate Primary School. View of entrance side showing the reduced window areas (Photo 1972)

(Second Consortium of Local Authorities) chose at its inception of development work in 1961 to ignore the already apparent environmental deficiencies of similar approaches and place the main weight of their argument (as with CLASP) upon the myths of mass production and the technology of structure and fabric. In thus committing themselves to consortia working local authorities would contribute to a set of norms already untenable on environmental comfort criteria. The rash of masonry cladding following the slight freeing of the dimensional rules in SCOLA Mark 2 is evidence of these pressures among others. However, it was left to MACE to carry the full brunt of the difficulties arising from the overapplication of technological virtuosity.

The Eastergate plan is largely an open area with the addition of separate spaces for noisy (or quiet) activities, and another for practical and messy activities. The overall floor area is 520 sq m (5610 sq ft) with a main teaching area of 354 sq m (3815 sq ft). Two of the published accounts (BS 1971, Shepherd 1971) give the total pupils as 100 and 120 respectively. To those versed in the art of school planning such a difference may be significant since it is important to achieve the correct balance between the floor area per child and the overall allowance, which was (under the then regulations) found by multiplying the cost per pupil place by the nominal number of pupils. Often costly elements in a school building will be 'paid for' by a reduction in floor area, or the nominal number of pupils used for cost calculation is larger than actual intake, thus permitting more overall finance. In the case of Eastergate it should be borne in mind that all circulation except the lobbies has been absorbed into the teaching area, and the external form is of the simplest. The published accounts give some attention to the environmental engineering approach and set out the way energy is used — the heat gains and losses are given as in **10.2.16.**

This in itself is a considerable advance on normal practice in published accounts of school buildings which usually only describe the planning criteria together with extensive coverage of the structure and fabric. Occasionally daylighting receives attention but in very general terms. Indeed one has only to read published accounts or visit a range of schools to realize that environmental comfort conditions are often the last item to receive positive attention. The performance,

10.2.16 Eastergate Primary School. Energy gains and losses

Losses		Gains	
Windows and doors	5.4 kW	Lighting	10.1 kW
Walls	1.8 kW	Occupance	6.0 kW
Roof	4.6 kW		
Floor	3.7 kW		
Total structural losses	15.5 kW		
Ventilation (occupied)	19.7 kW	Gross heat gain	16.1 kW
Gross loss (occupied)	35.2 kW		

however, of the Eastergate building was carefully monitored by O'Sullivan and Cole and published in 1974.

Clearly then, even on the data so far published the school at Eastergate offers an important change of emphasis. Its shortcomings are in other areas of the 'design system' — the relation of the inside to the outside, the site and the location of the building in relation to the village. The *Architects' Journal* (1970) referred to its 'fortress like appearance' and certainly it pays very little respect to the village in which it finds itself. In this it follows the unfortunate tradition of the building systems, in divorcing the design of the building from its location. After all, if all the components are predetermined and procedures institutionalized via local authorities it takes a very strong minded architect to stand above all this and create something that relates to the place upon the earth's surface upon which the building stands. As we saw earlier the whole concept that has grown up around the industrialization of building has pre-empted that possibility. In calling up the deities of mass production and the control of components the designer has very frequently totally excluded the external environment in which his object sits.

As the Eastergate building was being constructed, Gloucestershire County Council (a member of SCOLA) also set up a study group to look at 'integrated design'. In addition to the County Council representatives, the study included, P.E. O'Sullivan (Welsh School of Architecture), A.C. Hardy (Newcastle School of Architecture), W. Hillier (RIBA) and representatives of the Electricity Council and Board. The initial stage of the work, carried out between September 1969 and March 1970, clearly showed the benefits of reducing glazed areas and increasing roof and wall insulation in a SCOLA school of rectangular plan. A thorough cost study showed that:

> 'considerable improvements in thermal and acoustic properties can be provided at costs no greater than for school building of current design.' (Davison *et al.* 1970)

The feasibility study which followed this work, and which preceded the design of a primary school, was published in December 1971 (Davison *et al.* 1971) and compared the performance in summer and winter of a range of buildings of simple rectangular plan form with floor areas that varied from 460 to 5500 square metres that had been constructed using the SCOLA system. From this work a set of guidelines was proposed:

> '(i) Glazing areas should not exceed approximately 20% of the facade area in the arc south of the east-west axis. Within this limit single glazing in timber frames will generally be adequate.
> (ii) The U-vaue of the opaque part of the infill panel and the solid cladding panel should ideally be 0.40 W/m² deg. C.
> (iii) The U-value of the roof should be no greater than 0.60 W/m² deg. C. In single storey buildings, where roof areas are a greater proportion of the external surface area the U-value should be improved to 0.40 W/m² deg. C.

(iv) The building should be of a square or simple rectangular plan form. If rectangular its long axis should run in a roughly north-south direction and the ratio of length to breadth should be less than 4:1. If square, directly south facing glazing should be minimal.

(v) For floor areas of less than 930 m^2 a single storey building will have the best overall performance. If the floor areas are greater than 930 m^2 either a single or two storey building will be satisfactory with an advantage towards the two storey building as size increases. However, this decision should not take precedence over planning requirements.

(vi) The ventilation is important and must be designed to a performance standard.' (Davison *et al.* 1971)

Not surprisingly perhaps, the proposals for the design of the school which was to follow moved away from a total SCOLA package of components and back to more traditional construction — much as had happened at Eastergate:

'Alternative forms of construction which would give this standard of insulation are, for example, a traditional brick and block wall with the cavity filled with plastic foam insulating material, or 25 mm polyurethane foam board of equivalent fixed to the inside of a heavyweight panel.' (Davison *et al.* 1971)

The report also drew attention to the advantage to be gained from the 'reduction of infiltration heat loss as wider panels are used', something that almost all system designers appeared to ignore as components and joints proliferated.

Such developments as this, and the work at Eastergate show that some local authorities were beginning to question the premises of system school design that had established themselves. Further than this, the energy question gave those critics of the many unsolved problems created by the building systems a firm basis around which to argue the case for reappraisal against more traditional building techniques. By 1971, the Department of Education and Science themselves were working on a set of 'guidelines', ultimately circulated to local authorities in 1972 as a reaction to a number of attempts around the country to posit new solutions in the school design field along integrated environmental lines. Since at that time the only really 'hard' regulation concerning environmental comfort concerned the 2% daylight factor it was seen to be necessary as a matter of some urgency to set down some recommendations. These were communicated to authorities in an unusual manner, with architects in those local authorities involved first being made aware that new guidelines were in preparation when they submitted a scheme or had discussions with Ministry officials, with some being shown the drafts of the proposals. Several authorities pointed out that such an important departure as the *guidelines* contained would normally be discussed with authorities (issued for comment) prior to publication.

As a result regional meetings were ultimately set up at which a team from the DES explained the proposals but these were never formalized into a Building Bulletin. However, the blanket use of the 2% daylight factor was dropped and more vigorous attempts by some authorities resulted in improvements in thermal properties. Nevertheless most of the consortia systems were slow to account for such changes. The School Premises Regulations were revised in 1972 (Stat. Inst. 2051.1972) and came into force in February 1973.

The further mutations of the systems idea can be seen to be contributing to a more inclusive approach: those conducting research and building using 'integrated environment' can show some evidence of the recognition of the problems, with their attempts to establish a more symbiotic relationship with the environment. Alex Gordon's 'Long life, Loose fit, Low Energy' propositions (1972) were a further valid attempt to redress the balance away from the sort of use of resources that has taken place in, say, the school building systems where, reflecting government budgetary policy, heavy recurring maintenance bills have been incurred due to the pressure to cut capital costs to a minimum. The success of a new integrated approach must use subtle combinations of both traditional and new methods of mediating between the inside and the outside of a building. One effect of this has been to re-emphasize the importance of a sensitive design of the external skin of a building. The wall and roof have again come into their own.

Designers have themselves contributed to imbalance in so far as they have developed tools and arguments for reducing costs in a limited range of areas — by means of multi-use planning, the design and redesign of components and (as in the building systems work) developing far reaching purchasing and programme control mechanisms. Paradoxically, the very credibility of this work has encouraged further the gap between first cost and cost in use. In this way those areas, such as environmental comfort and maintenance, have lost ground. They have lost ground to the point where designers themselves now find it extremely difficult to argue a balanced case: especially since, on government financed work at least, the financial provisions for first, running and maintenance costs, come (on paper at least) through different administrative channels. We have had the ludicrous spectacle of cost savings being made by, say, cladding whole schools in painted timber which immediately adds a large recurring bill to the maintenance budget.

The major problem is now a political one in that Britain's highly centralized control and approval system moves only sluggishly in response to such mounting pressures. Those pressures, within a short space of years, have become very apparent: the relation of the building to its environment, the use of energy and materials — and in the end the most telling of all — rapidly escalating tax and rate increases necessary to finance such a system. There are indications that unless our institutions and procedures can modify themselves to take a more systemic view of the building process and product they may find that the public is no longer prepared to finance them.

Already a number of the consortia have faded away quietly, as did SEAC; collapsed dramatically, as did MACE; or transformed themselves, as did SCOLA in mid-1979. However, the very nature of the structure and goals of many of the systems made it almost impossible for them to respond sufficiently quickly to the pressures already outlined above. The attempts described in this chapter, whilst they had failings of their own gave clear notice to the systems that changes were necessary. They also gave clear notice that most of the systems were less than systemic. Unfortunately, the lugubrious nature of the client sponsored sytems of the local authorities meant that they first ignored such signs, and then only responded very slowly. Whilst it may be easy to brush off those examples which emanate from architects in private practice, as at IBM or Redbridge, the systems also took scant notice of those experiments occurring within their own ranks — as at Cheltenham and Eastergate. It took well over a decade, and the direct financial pressures arising from the oil crisis for the client sponsored systems to make serious responses. Hardy and O'Sullivan had published two papers in 1967 which drew attention to the issues of the building as a climate modifier, whilst Banham had published *The Architecture of the Well-Tempered Environment* in 1969. The inertia in the client sponsored systems held them to their original goals long after those goals were shown to be deficient — it was not until 1977, for example, that even general guidance was issued on energy conservation in school construction, with Building Bulletin 55. In their own way each of the examples discussed in this chapter offers a comment on that deficiency. In the concluding chapters we shall return to the building systems themselves and examine their operation and goals in more general terms.

11

ENTROPIC DRIFT

FEEDBACK, FEEDFORWARD

'The entropy of a system is a measure of its degree of disorder.'

Beishon and Peters
in *Systems Behaviour* 1976, 1972

Systems Operation

At the level of their operation the systems in Britain have developed a number of interesting approaches, many having little to do with the technologies of which they were composed. This is often a fact overlooked by most system commentators, but excusable perhaps since it requires a close knowledge of local government and other building procedures and the place of the systems within them.

Finance: government building programmes were, and are, notoriously prone to the stop-go policy and, as is shown elsewhere, the fact that the client sponsored systems had commitments to take up preagreed quantities of components from manufacturers created problems when financial cuts were discussed. Less often considered is the position which often occurs at the end of a financial year when the Ministry department discovers it has not used all of its allocation from the Treasury: rapid calls can be made to local authorities to see if they can make a 'token start' on site before March 31 — for a system with agreed component ranges and prices, this is very much easier to do.

Productivity: when the systems came into prominence, many a local authority office was less than efficient, a number having very poor records for taking up available money and for utilizing their staff properly. With the consortia programming of jobs and components, a useful check existed against which progress could be seen. There is little hard published evidence of great increases in site productivity, although fewer site operatives may be involved at a given time. The rationalizations of sequenced, dry construction can, in well programmed jobs, speed progress. One frequently overlooked side effect of the use of system components on site is the reduction of wastage by materials stolen, since these are usually less useful in other building situations.

Management: probably one of the most under-rated aspects of the client sponsored systems was that they created a reason for local authorities to meet and exchange experience. For those who have not experienced it, it may be difficult to imagine how isolated some local authorities are, how unaware of the actions and experience of others doing similar work. The variety of consortia meetings, although often tedious, created useful links and exposed unresponsive authorities to better practices being used elsewhere. In more recent times this has developed to the point where the consortia are used much more in this way.

An important side effect of the increased contact between authorities lay in the opportunities it opened for promotion. The avenues for gaining more responsibility and a higher salary are few in most local authorities, and the process is exceedingly slow. The career structure tends to reward age and long service rather than ability or effectiveness. The result of this is the movement of staff from place to place in both local and central government to achieve better paid or more responsible positions. The advantages of this are obvious: new blood and ideas can be introduced and individuals can take on new challenges to develop themselves. However, for the building process there are many disadvantages and I believe that many of the faults which have arisen in public building programmes can be traced to the excessive mobility of architects and other professionals.

The reason is that, to master the considerable complexities of the briefing, design, construction and occupation of a building, it is necessary to be responsible (at some point in one's career) for the whole process from beginning to end. The more often this is done the more the architect is able to appreciate, control and predict possibilities. Now, in Britain, even a small building can occupy a lengthy period between the first briefing meeting and occupation, whilst on very large schemes the time span may extend over several years.

The effect of the career pressures in local and central government meant that architects would start a job and then move on. Inevitably those brought in to continue such projects have less personal commitment than the designer — and this often encouraged them to move on in turn, whilst the original designer never feels the full effect of his work at a personal level. The fact that many buildings are not designed by a single designer but by a group does not detract from this argument since the effects are much the same. Gordon (1978) in a paper 'Self Motivation in Career Development' recalled John Redpath, of the Department of the Environment:

> '...telling me that in interviewing 200 architects for a job, he asked them each the same question: "Have you ever finished a job?" and in the whole 200 there wasn't a single one who had.'

Apart from the decreased opportunities to learn directly from one's own decisions, this process had a cumulative effect in that it produced a large number of quite senior architects who had rarely, if ever, seen a project through. The serious implications of such a position need not be elaborated, but must undoubtedly account for much building failure.

Programming: it was only with the advent of the systems that most local authorities took the programming of jobs seriously. From the time of the legendary: 'Programme Room' at Hertfordshire County Council, those architects concerned to try and deliver much needed buildings on time and cost, have developed techniques for work planning. The vagaries of site acquisition, government and the building industry have not made this an easy task, but with the creation of consortia suppliers with component contracts, it became a very necessary one. For all the consortia bulk purchase components of a given sort expected over the next year, or sometimes longer, suppliers gave unit prices and stated estimated fluctuations. Those who know the building industry will see the difficulties in all this: often authorities could only guess at what would be in their building programme for the coming year and it was almost impossible to estimate the numbers of each component in buildings yet to be designed.

For the subcontractors submitting the prices, the task was even more like crystal gazing. They were dependent on the 'gentleman's agreement' that there would be such a quantity of the said components and that the jobs would start in the stated financial year. At first they were prone to accept the agreements in view of the size of the orders, but they quickly found that they could be left with unwanted components on their hands unless they were careful. An increasing number of suppliers were forced into making claims from the consortia for components not so taken up: more experienced suppliers inserted no-take-up clauses right from the start and these bound the consortia. However, this introduced the very strange non-legal status of the consortia for although they had designed the subsystems and negotiated the prices they did not exist as an entity and had no money of their own. The only legal point in the whole procedure was the traditional one when the contract for each job was signed between a single local authority and a single general contractor. This placed all subsystems and component suppliers in a very difficult position when such problems arose. To endeavour to continue confidence building the consortia became heavily involved in the programming of the subsystems themselves, going into order dates (normally dealt with by the general contractor), delivery dates and completion dates. The normal position where the general contractor takes the responsibility for running the project on site began to be eroded. Let it be said that many general contractors seem totally unable to programme their own work and that of related subcontractors, and so the systems felt that such assistance would be beneficial. However, it merely introduced another link in the chain and any delay in the supply of such subsystems then immediately became the responsibility of the consortium in practice if not in law. Project architects found themselves not only having to deal with a general contractor, but also with the Consortium Development Group and the national suppliers — with the latter often under pressure from many such sources at once. As is always the case with government jobs, the bulk of them slipped to the end of the financial year thus further increasing component supply problems. In addition the bypassing of local builders merchants, a large factor to local builders who gained discounts on all building materials, created additional problems. Finally, although to the consortia the orders seemed large, the annual programmes often amounted to little more than the cost of a large office block, with many very small contracts, and these spread all over Britain. Some of the difficulties were alleviated by grouping job contracts in a 'serial tender'. Here contractors will price a hypothetical Bill of Quantities, and that tender accepted will apply to a run of jobs. Since a sizeable percentage of any contract is consumed by 'preliminaries' — administration and overheads in setting up the job — such an arrangement can offer savings.

Subsystems: those designing the subsystems and the components developed very good relations with the manufacturers and this was certainly to the advantage of everyone. For the first time the architect really did begin to collaborate with the man making the product and both learnt in the process. Quality control on site became a major problem, especially as most of the participating local authorities had no 'approved' range of samples against which to work but had to rely on the consortia development groups.

Construction time: one of the major claims of the building systems is speed of construction. However, since almost every building is a new problem on a new site it is difficult to be precise over such claims. Certainly the frame and roof, in being erected first and fairly quickly, make good one of the claims — that of working in the dry. However, the complexity of much systems detailing and the supply problems already referred to rarely make for spectacular improvements in erection time. The observation of one experienced systems development architect was that for financial reasons, general contractors like a job to go at a particular pace: a systems job will therefore take as long as a traditional job, but there will be fewer men on site.

Oddie (1975) indicated that unless two conditions were met, real cost savings could not be obtained. The first was the way in which those parts of a building which were part of the system integrated on site with those that were not, bearing in mind that even in 'system' jobs, much of the work was still of a traditional nature. The second condition concerned the familiarity necessray on the part of the construction workers — if they became used to working with a particular system, not surprisingly they became more proficient at it. 'In cases such as CLASP' where these conditions are met by careful programming and the use of serial tendering, 'the savings in construction time are considerable' (Oddie 1975).

Subcontractors developed their own responses to the systems. To the author one craftsman made the comment: 'There's no skill in it any more — I might just as well work in a factory and earn more.' Many disliked systems work for this reason, it did not encourage the continuance of any craft they had learnt, and offered little to those entering the industry. However, some subcontractors developed a more entrepreneurial response, just restyling their skills and travelling from one system job to the next using them. In recent years it is not unusual to find site operatives who have never worked on anything other than a specific system.

Drawing office: another of the claims of the building systems is that by utilizing standard drawings and procedures the design and production time in the office will be cut down. It is generally true that production times in many British offices seem rather excessive, but one of the points of using known solutions was to allow more, not less, time to do a good design. Often authorities took advantage of the situation to squeeze the time allowed for such work, with generally bad results. Many architects resisted the systems idea, either on principle or because they found it too complicated to understand (a systems manual might run to several hundred pages), but the major problem faced by most offices using a building system was the question of responsibility.

The introduction of a system often resulted in a further erosion of the already delicate balance of professionalism in local authority offices. Because most of the subsystems to be used were already worked out, many architects felt little or no responsibility to a building project. This important psychological response created something of a dilemma for, in all truth, some of these architects would not be able to create a satisfactory building with or without the building system. However, it is certainly true that far from making it easier, it became incredibly difficult to create good buildings from some of the systems, and this became more marked as issues like conservation and energy consumption came to the fore. The good architect had always considered such things, but with some of the systems there was a distinct limit to how far he could go in satisfying these requirements. On balance there is possibly some truth in the adage that whilst a mediocre architect is restrained from dropping below the minimum standard offered by a system, a good architect can, from the same system, make a good building. However, it takes a particular cast of mind, and a very tough disposition to wring the best from many systems. There was another side to the new set of work practices which architects had developed for the building systems, where the intention to predict so much of what would happen had created a complex documentation. The standardized details, elementally organized job drawings, elemental cost analyses, coded components, bulk purchasing arrangements, and programme charts, were sufficiently complex to preclude immediate penetration. This applied to the lay public, other consultants, and even to many architects not directly involved. The mystique which the systems developed from rationalization certainly served to provide, initially, great credibility: it also had a number of interesting side effects. In the rather conservative world of most local authorities, it provided many young architects with new opportunities: the non-traditional nature of the drawings also made it difficult for building inspectors and fire officers to understand fully the nature of what was to be constructed. Other involved bodies, planning committees and the like, found it difficult to comprehend the nature of the proposals unless very special efforts were made to produce suitable drawings. For the hard pressed architect these aspects proffered both advantages and disadvantages: associated bodies tended to accept the mystique and leave him much more to make decisions in his area of competence — until real user experience of the systems began to build up, when this competence began to be severely questioned. The normal insulations of local authority working creates difficult user feedback problems — with the client sponsored systems another layer was inserted between the designer and those using his buildings. Nevertheless some of the more sensible notions, such as the way sets of working drawings should be organized, began to make inroads on the attitudes of many British architects' offices. A survey in *Working Drawings in Use* (Daltry and Crawshaw 1973) under two criteria, adequacy of information and ease of search for information, found that where such ideas had penetrated normal practice they had been useful. However, the study of 15 sets of drawings by 27 architects and engineers revealed again how slow architects have been to change their own work habits whilst encouraging others to do so. One of the study's recommendations was that:

> 'the set should have a systematic structure comprising separate groups of location, schedule, assembly and component drawings,'

an idea developed many years before by building systems designers. Even more telling, perhaps, is the comment by Salisbury in his review of the Building Research Establishment's publication on the study, that:

> 'they (architects) may not have realised that cross referencing between drawings is very important' (Salisbury 1973)

It must be a matter for concern that the architectural profession in Britain has been unable to develop efficient work practices of this simple sort, especially as on this point some useful lessons can be learned from those ideas developed in building systems.

Information systems: if many of the building systems in Britain have proved less than satisfactory, a number of the lines of inquiry which they have started and sometimes nurtured, have proved of great interest. One of these concerns the large question of how information is handled for designers, constructors and users. It has become commonplace to point out that the increase in available information has overloaded designers, confused contractors and left users with buildings with no clear guidance on how to operate them. The building systems early gave attention to drawing office techniques, attempts to rationalize information on available products, and quality control. The philosophy behind their approach led to the idea of coding components similar to the codings used in the automobile industry. A system of breaking down a building into elements was sought and, arising from the work at Hertfordshire, the *Architects' Journal* fostered the use of an elemental cost breakdown in cost planning work. This work was pioneered by quantity surveyor James Nisbet who, after Hertfordshire, joined the Ministry of Education's Architects and Buildings Branch, and in this way the methods were transmitted to the world of school building and to the client sponsored systems. In their search for rational classification of product data many of the building systems ultimately employed a version of the Swedish SfB system, whilst CLASP developed its own classification system called Building Industry Code (BIC). Developed by Lars Magnus Giertz in 1949 the SfB system aimed to unite all documentation used in construction into one classificatory mode thus facilitating cross-referencing. In this way product information, research publications, specifications and drawings would all have a common basis. In Sweden the system developed rapidly, embracing the already many standardized components and cross-referencing to a nationally used standard specification. It can be argued that this rationalization was one of the influences that contributed to the great success of postwar Swedish architecture, whereas in Britain such attempts at a national system for the co-ordination of all such information was only attempted long after the systems had carved out their patches and so developments have been slow. Furthermore the United Kingdom SfB Agency quickly superseded the original English version of 1961 with a revised anglicized version in 1968, the CI/SfB *Construction Indexing Manual*. Then again in 1976 a further revision was carried out causing considerable reaction in the profession. To many, both the anglicizing and the constant revisions seemed to go against the basic purpose or at least to constitute a critique of that purpose. Clearly in such a chaotic industry as the British one, such a classificatory tool has a use. Although useful to the system builders the SfB system, in fragmenting the parts of a building in a specific way, also encouraged the viewing of buildings themselves as a series of fragments. This is another example of how a fine attention to the parts of a system may encourage designers to lose sight of the whole. Much of the interest in SfB led to the National Building Specification, formally established in 1973, and its concern to provide a range of standard specifications.

Considerable use has been made of the SfB system to classify drawings for building projects and system standard drawings. SCOLA, for example, adopted the use of SfB for this purpose right from the outset in 1961 whereas Hertfordshire and CLASP already had their own classification methods. The use of CI/SfB as a method of classifying project drawings generally has spread, and there is no question that there is need for the project drawings prepared by architects to have an organized format. The study referred to earlier, carried out by the Building Research Establishment in 1973 (Daltry and Crawshaw 1973) sharply indicates the confusion that arises from the lack of such an approach. Common but obvious faults are: no lists of drawings, so that a contractor does not know if he has them all, if they cover everything, and whether he will be receiving more; the lack of cross-referencing between drawings, and the use of inadequate explanations. Improvements in drawn communication would probably do more to improve construction efficiency than much that the systems work has acheived.

The introduction of A sized metric paper sizes paralleled this development, commencing long before the planned metric programme started in 1965. The obvious advantages commended themselves to many architects struggling with a tradition where the many drawings and brochures they had to deal with in the course of a job came in a vast range of unrelated sizes. The problems caused on site by sets of drawings some of which were of a very large size whilst others were a few inches square did not encourage confidence or a sense of order. The ratio system inherent in the A sizes offered advantages in reproduction by scaling up or down since each succeeding A size is exactly a half of the one before it: AI (594 × 841 mm) is half of AO, (841 × 1189 mm), A2 is half of AI and so on. Before long trade literature, stationery and all related documents followed this pattern. Attempts to standardize documents had a long history, having been more successful in Europe and North America (which does not use A sizes) than in Britain. Hannes Meyer was using A sizes for location and production drawings as long ago as 1916, seven years before the German DIN standard was introduced. He redrew Palladian Villas on AO sized sheets, and all this long before his tenure as head of the Bauhaus from 1928 to 1930 (Schnaidt 1965).

Although the introduction of A sizes had the advantages referred to, in many quarters it was carried to extremes as architects sought appropriate ways of co-ordinating the vast amount of information generated by the attempts of the systems to predict and control events. What should go on each size of drawing or document and how the sheets should be cross-referenced became important matters. If the fragmentation introduced in this way mirrored the approach of the component philosophy, the lessons it attempted to teach architects and the industry at large in Britain, were very relevant.

In the area of building products the growth of the classified product data catalogue systems has been one of the most influential. The United States has its *Sweets Catalogue* — a series of annually published volumes containing a wealth of such data whereas in Britain no such simple device exists. The gap has been partially filled by the commercial information services of Barbour Index and BLIS who offer sets of catalogued data which offices hire on an annual fee and which is updated monthly. Barbour also published a useful *Compendium* in which they have attempted a high standard of information and presentation. However, the aim of a comprehensive set of product information organized on a standardized basis has not yet been achieved in Britain. Nevertheless the work has improved the quality of information produced by companies, many employing professional designers to prepare their literature. The nature of the British industry has meant that it has not easily taken to the type of rationalization the Scandinavians have found possible nor has it developed a natural response as in the United States: such systems therefore, still leave the question an open one in Britain.

In a perceptive article in *Architectural Design* in 1976 entitled 'Whatever Happened to the Systems Approach', Andrew Rabeneck pointed out that:

> 'What we have now is a McLuhanesque match between new problems, which we perceive only because of our "systems approach" to problem definition, and new solutions perceived as a function of the technical change brought about also by the "systems approach".' (Rabeneck 1976c)

He goes on to describe how many architects, in becoming swept up in the technological part of the systems argument, failed to realize that it was more than that. The great strength of the systems idea is as a conceptual tool, allowing us to see things in alternative ways and to posit alternative solutions. These may or may not be novel but they will attempt to address the difficult sociotechnical problems which we now face. The major problem of this kind which emerged from the considerable systems work in Britain is the relation between the 'building systems' created and the institutions which support them. For, as has been shown, there is a common thread running through the growth of the idea: first, encouragement of both public and private enterprise to produce solutions to problems stated to be critical — such as the school, hospital or housing shortage; second, the co-ordination or rationalization of these to such an extent that a few models survive; finally these enter a stage of institutionalization where change and response is slow.

All the drives to produce systems have emerged from some sort of crisis syndrome: after World War I the 'Homes for Heroes', towards the end of World War II, a similar cry. Again in the early sixties the use of housing as a political football encouraged a further embracing of the idea by politicians and civil servants alike. In each of these cases the building industry itself has initially offered comments to the effect that total changes in approach cannot be lasting, drawing attention to the importance of skills and continuity on the construction side. Such fears have then been set aside as typical of a conservative industry. Nevertheless in the three cases referred to such fears were largely borne out in that the considerable problems created by the approaches used, surfaced only a few short years later. In the end those approaches, systems or otherwise, which have been most influential and long standing have not been those with an avowed commitment to make total changes in building methods, rather it is those that have followed the pattern of development suggested by the voices of the construction industry itself that have prevailed — that is the gradual improvement and prefabrication of those items which seem relevant to the conditions of the industry and the market at the time. All those innovations which involve a 'totally new' view of the market have encountered problems of one sort or another. Nowhere is this more true than where both the provider and the user of a system is virtually the same organization. In such a situation the normal feedback on

11.1.1 The system built legacy. Three photographs taken on the same day in September 1978, from the pedestrian deck spanning Winston Churchill Avenue, Portsmouth. These show 3 system built blocks simultaneously undergoing repairs. (a) View to the northeast: on left Wilmcote House (Reema system modified by architects Wilson and Womersley, 1968). Strengthening fins being added to gable ends. This block had already had remedial work after tenants in a similar block, Cannock Lawn, took the City Council to court in a historic action concerning condensation problems. (b) View to the south: centre, flanked by two blocks (Tipton and Edgbaston Houses) in Bison High Wall Frame system, part of Cannock Lawn 1967 (Reema system; architects, Wilson and Womersley) under repair. Repair work not yet commenced on the two Bison blocks. (c) View to the west: Solihull House 1965-6 (Bison High Wall Frame system and City Architects department). Repair work in progress on external cladding

products from the point of use is muffled, and inadequate solutions can become institutionalized.

This had been the case to some extent with the postwar prefabricated housing programme, ultimately brought to a halt by complaints concerning its enormous cost. The totally new modes of construction used created their own problems which, for those still standing some 35 years later, showed their limited life compared to more traditional construction. Both the BISF (British Industries Steel Federation) house (Stevens 1977) and the Airey house proved to be serious fire risks even though they were only of two storeys. The Airey, of prefabricated concrete slab external wall construction with no party fire walls in roofs, continuous roof timbers, and fibreboard

b

c

linings was dubbed, according to a report in *Building Design*, 'Airey by name and Airy by nature'. Following a fatal fire in an Airey House at Barnsley, it was decided to spend £500,000 on fireproofing 320 homes, and the Department of the Environment has been examining the extent of the problems left the nation by this experimental form of construction, of which 26,000 were built (McGuire 1978). In 1981 it was found that there were severe structural problems with reinforcement in the base of the columns of the houses which is causing the concrete to spall off.

As we have seen it took the drama of a Ronan Point to illustrate the problems created by the use of industrialized high rise systems for public housing in general. However, the nature of public building programmes is such that it is not easy to establish just how widespread have been the failures due to the indiscriminate use of building systems (**11.1.1**). Scott in *Building Disasters and Failures – A Practical Report* (1976) shows the possible extent of the problem, but this only deals with major failures, and of a direct technological sort. Some observers of the Ronan Point Inquiry, which followed the progressive collapse there in 1968, drew attention to the pressures to disregard the dangers inherent in much system building of the large panel sort. Central government and local authorities went to great lengths, as discussed in the chapter on high rise building, to assure tenants that there was no danger. At Birmingham 4300 flats in 86 blocks were strengthened in a programme which allowed two hours per flat. Tenants were not moved out and each was given a £70 electric cooker free (Webb 1969). The Minister of State made a personal statement to all tenants of high rise system built flats in the country on 15th August, 1968:

> 'I know that the next few days are bound to be worrying for some of you. I ask you, as far as you can, to leave the worrying to us and the local authority.' (quoted in Webb 1969a)

a

Some substance was subsequently given to the critics of the Ronan Point Inquiry and the safety measures adopted when, in 1978, it was revealed that a considerable number of blocks built in the Bison High Wall Frame system also required extensive repairs. At the introduction of the system in 1962 the *Architects' Journal* had congratulated Concrete Ltd on having produced a system which followed '...the trend of industrialised building without imposing too great rigidity on the designer's individuality' (AJ 1962). The problems reported in 1978 were on blocks in Kidderminster, Bradford, Rotherham, Slough, Ealing, Hounslow, Windsor and Maidenhead, Hillingdon and Portsmouth (**11.1.2**). An estimate of £1.5 million was quoted for restoring the two blocks undergoing repair in Portsmouth, demolishing them having been rejected on the grounds that the cost of replacing them with low rise accommodation would cost twice this sum. A number of difficulties were apparent. The external gable (or flank) end wall consisted of an inner 150 mm (approx. 6 in) load-bearing reinforced slab, polystyrene insulation and an external non-load-bearing slab of 75 mm (approx. 3 in) reinforced concrete. The panels had been cast face down and some of the reinforcement was found to be too near the surface. This resulted in the spalling and cracking of the external panels, although in some cases it had even affected the inner leaf (**11.1.3**). Another problem, similar to findings at

11.1.2 Solihull House, Portsmouth, England. Bison High Wall Frame system and City Architect. (a) *Left*. Repair work to external cladding in progress 1978. West elevation. (b) The south flank wall with the external skin completely removed, exposing the inner leaf. (c) *Next page*. A visual summary which could be applied to many British cities: the two storey house with its garden, the local pub (The Mystery Tavern), the cars fitted in with intended landscaping and the high rise system built block (Tipton House, Bison High Wall Frame). All at Somerstown, Portsmouth. (Photo 1978)

656

11.1.2 c

Ronan Point, concerned the junction between the load-bearing panels. Each of these sat on two levelling bolts which projected through an *in situ* concrete joint from the panel below. The drypacking which was supposed to have been inserted under the panels was found to be 'insufficient or completely missing' (Pearman 1978). This led to cracking of concrete surrounding some of the bolts. Other problems reported included rotting of window frames, the spalling of window cover pieces, and the breaking away of parts of panels where they had been made good (**11.1.4**). At Hillingdon, a London borough, one 12 storey and one low rise block had been evacuated and a report commissioned into housing in the borough built in the Bison High Wall Frame system. Published in 1979 the report, by engineers Campbell Reith and Partners, ran to six volumes and raised serious questions about the range of problems involved on the two to four storey low rise and the four 13 storey blocks on the six estates. The external panels in some cases suffered from inadequately compacted concrete, variable panel thickness and reinforcement cover differences. These the report put down to 'lack of control during manufacture'. Damaged corners to panels had been repaired with inadequate material which had insufficient life. Hillingdon's Housing Committee Chairman, Terry Dicks, suggested that remedial work could cost £8 million which, if carried out by the borough, would add 16 pence to the rates residents paid. The blocks were built between 1965 and 1970 at a cost of £5.5 million when the government was encouraging the use of industrialized methods for housing. The National Building Agency had issued a twelve month appraisal certificate for the system in 1966, at a time when tender negotiations were in process between Hillingdon and Bison. This was not renewed in subsequent years, although this may have been because no application was made. In 1964, the year when the Agency had been established, some 21% of public sector housing used system methods. In 1967 the proportion was 42% (Berry 1974). Until the 1968 Ronan Point collapse, local authorities had been pressed to use industrialized systems and to build high rise, a subsidy being introduced to encourage the latter. Hillingdon, not unnaturally, now looked to the government for help with the problems they were encountering but expressed concern at the response. The Chairman of their Housing Committee, when announcing the conclusions of the report on the Bison blocks to the press in July 1979, pointed out that:

> 'The responsibility lies firmly with central government, they had behaved like ostriches with their heads in the sand. When they received early reports of failures... the DOE (Department of Environment) fought to keep reports on system building secret and were very reluctant to organise an interchange of ideas about remedial work.' (BD 1979c)

This raises interesting questions concerning the uncertain relation between central and local government in such matters. Certainly the policy pressures from central government are often considerable; equally, however, the city fathers in many local authorities seemed very keen to have their own collection of tower blocks.

In the first, 1978, reporting of the problems with Bison, an engineer from one of the councils involved pointed out that it was necessary to remove the external panels to discover the extent of the problem. Internal inspection had been made more difficult by the installation of

11.1.3 Solihull House, Portsmouth. (a) The difficult process of removing external cladding panels to the flank walls begins

a

the strengthening ties used in the wake of the Ronan Point collapse. A Ministry circular (RP 68/OI) had stated that:

> 'The importance of adequate buttressing of wall panels cannot be over-emphasised... experience shows the external wall panel connections to be the weakest point of a precast building'

and Webb had pointed out in 1969 that 'strengthening angles are not connected to the external non-load-bearing walls which blew out at Ronan Point at a quarter of a pound per square inch (1.7 kN/m²)'. Apparently some 50,000 housing units were constructed using the Bison system with the last only completed in 1976.

At a more general level, *Building Design* for March 26 1976 carried an alarming account of building faults arising in, largely, buildings designed in the pubic sector. Of course not all of these problems are concerned with building systems but many were. Remedial work to some of the school system projects has been considerable, much of it remaining unpublicized. The extent and cost of such remedial work as this, as well as much on housing, normally escapes public scrutiny since it can often be absorbed in the maintenance budgets of local authorities. However, whilst there has been a great deal of publicity

11.1.3 (b) South flank wall after stripping of external cladding panels and insulation

b

for the use of a given system at its inception, the monitoring of progress is rather haphazard since the art of building appraisal is in its infancy and little co-ordinated knowledge is available. Discussions with contractors, designers and others involved in such work leads one to believe that failures have been exceedingly widespread and costly.

Difficulties of another sort emerged after the extensive fire at Fairfield Old People's Home at Edwalton in Nottinghamshire on 15 December 1974 which caused 18 deaths and a government Inquiry. This dramatically focused attention on the capability of CLASP in particular, and such system building in general, to withstand fire (**11.1.5**). The Inquiry involved CLASP officers and staff 'full-time during the hearings' (CLASP 1975) between 3rd and 20th March and

11.1.3 (c) Close up of load-bearing inner leaf of flank wall, showing condition of concrete

subsequently in the provision of additional information and examination of the system. The findings of the Inquiry were published in a government report (Jupp *et al.* 1975) which, as well as making important observations on the type of construction used, revealed the ambiguous position of such a system building within the controls and guidance then operating.

The Fairfield building opened in 1961, having been designed as a single storey structure using Mark 3 CLASP (the standard drawings for which had been prepared during 1956–7), by the architectural firm of Robert Matthew, Johnson-Marshall and Partners. The fire was

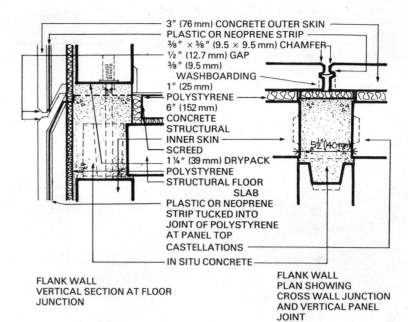

3″ (76 mm) CONCRETE OUTER SKIN
PLASTIC OR NEOPRENE STRIP
3/8″ × 3/8″ (9.5 × 9.5 mm) CHAMFER
1/2″ (12.7 mm) GAP
3/8″ (9.5 mm)
WASHBOARDING
1″ (25 mm)
POLYSTYRENE
6″ (152 mm)
CONCRETE
STRUCTURAL
INNER SKIN
SCREED
1 1/4″ (39 mm) DRYPACK
POLYSTYRENE
STRUCTURAL FLOOR
SLAB
PLASTIC OR NEOPRENE
STRIP TUCKED INTO
JOINT OF POLYSTYRENE
AT PANEL TOP
CASTELLATIONS
IN SITU CONCRETE

5 1/2″ (140 mm)

FLANK WALL
VERTICAL SECTION AT FLOOR
JUNCTION

FLANK WALL
PLAN SHOWING
CROSS WALL JUNCTION
AND VERTICAL PANEL
JOINT

11.1.3 (d) Bison High Wall Frame: vertical section through horizontal joint in flank walls and plan of vertical joint

d

thought to have been started by an elderly resident smoking in bed, and it quickly spread. Although one of the fire doors was open the report concluded that 'the main body of smoke travelled into House 2 through the roof void'. The ceiling, constructed of perforated plasterboard panels, offered almost no resistance to fire and the smoke could enter the void at junction points. A flashover occurred, set off by the build up of heat in the bedroom where the fire had started. The dry timbers in the roof — made up of prefabricated panels of ¾ inch (19 mm) boarding on timber joists — and fascia ignited together with the bitumen roof covering. The build up of pressure forced smoke into the roof void. The Inquiry report made it clear that, in view of the state of knowledge of fire experts at the time of design, the architects could not be criticized for their approach. However, the report also drew attention to the large amount of combustible material present in CLASP Mark 3, the lack of fire stopping over window heads (not in accordance with existing building bye-laws), and the lack of fire resisting compartment walls. The undivided hollow ceiling had acted as a chimney for the spread of fire and smoke.

The report also highlighted the complex and confusing nature of the regulations and the place of such a building within them. The Fairfield home was owned by Nottinghamshire County Council who were also the clients. The CLASP system had been designed and promoted by the same county. Robert Matthew, Johnson-Marshall and Partners, as the appointed outside architects for the project, designed the building in consultation with the county architects department. The CLASP system itself had been initially designed for school building in areas of mining subsidence and Fairfield was the first — although by no means the only — use of the system for other building types. School building, for which CLASP had been designed,

11.1.4 Solihull House. Bison High Wall Frame. Part east elevation showing differential weathering problems, not helped by the near flush detailing

was exempted from the building bye-laws (now Building Regulations) governing other building types, by the Public Health Act of 1936. Schools were governed by regulations made under the 1944 Education Act — the Standards for School Premises Regulations. Although there was some guidance on fire safety for schools in the form of the Ministry Building Bulletin No. 7 of 1955, there was no such guidance for the design of homes for the elderly. The only guidelines were in the appropriate bye-laws and these called for such a building to have external walls that had a fire resistance of one hour, were externally incombustible and had their cavities firestopped with incombustible material.

However, when the architects applied for bye-law approval West Bridgford Urban District Council gave no approval since they assumed such buildings to be exempt from bye-law control. The reason they gave the Inquiry for this assumption was that they considered that buildings 'belonging to statutory undertakers' — that is government, local government and other official bodies — to be exempt. They therefore only made examination of drainage and did not inspect the structure or construction on site. Whilst the report states that the UDC 'mistakenly considered the building to be exempt' (Jupp *et al.* 1975) it does not point out that, at the time, this was a

11.1.5 Fairfield Old People's Home, Edwalton, Nottinghamshire, England. CLASP system and architects, Robert Matthew, Johnson-Marshall and partners, 1961. After the fire on 15 December 1974 which caused 18 deaths

common assumption on the part of local authorities. Many of these took the view that those official bodies that had their own design departments — and this included other local authorities, gas, water, electricity, health, defence and other government buildings — did not need to gain bye-law approval. The practical politics of the situation with such building must also be considered. The long term implications of the Fairfield Inquiry in this respect are far reaching and by no means fully resolved. There are indeed more and complex regulations affecting building now than ever before, and much attention is given to attempting to satisfy their often conflicting demands. However, the question of the precise nature of the responsibility carried by those providing and designing publicly owned building within this network of regulation is still fraught with uncertainty.

The technical criticisms raised by the inquiry were carefully examined and changes put in hand. The CLASP 1975 annual report stated that 'a number of points' had already been taken into account in the design of the then current Mark 5 system — these included the provision of a one hour fire rated ceiling as standard, the elimination of flammable materials from roof and floor construction, the extension of the provision of fire stopping details, and the one hour fire resistant internal linings to external walls. The problem at Fairfield had been compounded by the fact that the ceiling void was connected to the rooms below for there were no ceilings over built-in cupboards and wardrobes.

The increasing incidence of fires in schools during the sixties gradually had its effects on both CLASP and the other school building systems. It should be recalled that school building in Britain is not generally subject to the building regulations, but is governed by the less stringent requirements of the School Premises Regulations, although in recent years there has been continuous pressure to change this situation. This led the Fire Research Station and the Building Research Establishment to carry out an investigation into fires in schools and to comment in the introduction of their report:

> 'The results of the investigation indicate that those who design, build, staff and maintain buildings do not appreciate fully the fire hazards involved in the use of non-traditional construction and materials.' (Silcock and Tucker 1976)

This only repeated the advice already given in 1975 by DES Building Bulletin No. 7 *Fire and the Design of Schools.* Similar advice had been given to designers of old persons homes (coming under the Ministry of Health) in 1966. In the 1976 study the largest single factor affecting the spread of fire was the 'undivided ceiling void', one of the key devices used by the systems. CLASP, for example, accentuated this problem by using timber roof decks until 1972, when the Mark 5 version of the system introduced a steel deck diaphragm roof such as SCOLA had already implemented some years before. Although the 1976 BRE report examines 14 school fires it makes no mention of whether these schools were system built and if they were, which system was used. Clearly many of them are so constructed judging from the described detail. It is pointed out that between the early sixties and 1972 the number of fires in schools had doubled from 900 per annum to 1889, creating a considerable problem of reinstatement. Concern over fires in CLASP structures had, however, started long before and came to a head with a severe fire in a primary school in Manchester at Easter 1972. As a result loal fire officers expressed concern to the Education Committee and a survey of 37 other CLASP schools in the city was carried out leading to the recommendation to install fire stopping and partitions at a cost of £200,000. The government refused permission for this money to be spent, pointing out that the buildings were designed so that children could leave quickly. The *Architects' Journal* reported, in 1973, that the five fires in CLASP buildings in the previous year (four in Britain and one in France) was a cause for concern with the lack of fire stopping and partition breaks as a 'recurring theme' (AJ 1973). The fire in the Paris Edouard Pailleron school building on 6 February 1973 had more serious repercussions. Twenty people were killed in a blaze which completely gutted the school in twenty minutes, having been started by two pupils setting fire to a rubbish bin. The building had been constructed by Constructions Modulaire, the French branch of Brockhouse Steel Structures, who market the British CLASP system abroad. In the subsequent trial the builder of the school, Hubert Lefevre and the British architect Michael Keyte, were found guilty of 'imprudence, negligence and non-observation of safety regulations' (Stevens 1978), Keyte was given a 15 month suspended prison sentence against which, at time of writing, he has appealed. The speed with which the fire swept through the building was said by the French fire experts called in, to have been the result of the use of unsuitable materials, the ceilings voids, and the incorporation of gas pipes within these voids. During the trial Keyte had been accused of not putting up sufficient resistance to pressures from the company to make plan modifications. This case tragically illustrates the peculiar position that architects find themselves in when using a system, particularly when the use of such a system has often been decided upon by the client body.

The nature of the relationship between a client sponsored system and the architect for each individual building is highlighted by these difficulties over fire precautions. After the Manchester fire in 1972 the Principal Architect to the CLASP Development group, Derek Lakin was quoted as saying:

> 'We provide a kind of meccano set with instructions on how to put it together. If someone does not use it sensibly, this does not reflect on the consortium.' (*Guardian* 1974)

Accurate though this may be, the fact remains that most architects when using a system, particularly a client sponsored one, feel that part of their professional responsibility has been eroded. Further, a building system with its own administrative layers forms another net of bureaucracy with which they must deal and they are often not encouraged by the fact that most such systems have a very slow response time to change. The introduction of brick into the CLASP system took until 1973/74 because, according to the 1975 CLASP report:

> 'Many buildings in the past have been isolated from neighbouring development and so the problem of blending with existing development has not been contentious; however the necessity for members to build on small infill sites requires more care to be taken with the selection of cladding material and building form.' (CLASP 1975)

The standard flat roofs were also added to with a monopitch and the

> 'Central Development Group is taking the pitched roof principle further in order to cover simple rectangular spaces and to allow the use of traditional coverings.' (CLASP 1975)

a

b

11.1.6 (a) Cannock Lawn, Portsmouth, England. Reema system modified by project architects Wilson and Womersley, 1967. Centre of picture, and Tipton House in the Bison High Wall Frame system *c.*1965. View from north. (Photo 1969). (b) Cannock Lawn, from the south, undergoing repairs, 1978, subsequent to the court action brought by tenants against the City Council over the problem of condensation. (c) *Opposite.* Wilmcote House, on an adjacent site and also using a modified version of the Reema system. Unlike much system built housing an attempt was made here to create urban spaces, and to give the system some facade modelling. All this was overshadowed by the other problems surrounding the blocks

c

Jordan (1969) had already drawn attention to the slow rate of mutation especially questioning why

> 'That in twelve years no sun and glare control methods have been developed in the vastly over-glazed British CLASP, or in any other school system for that matter, in contrast with the example from Germany.'

He further questioned the continued use of the steel frame for single storey buildings where the walls could often hold the building up (even more true when they were to be in the 'rediscovered' brick) and the continued adherence to the timber frames and surrounds. So it has been left to the twin pressures of energy conservation and the incidence of fire before any significant changes were made to the basic system. The growing concern of fire officers, insurers and those having to foot the bill caused the beginnings of a reappraisal — how far this will go has yet to be seen. On the former aspect, that of energy and the way in which such buildings relate to their place, it is perhaps significant that whereas the early CLASP reports mostly concern themselves with the organization, the meetings, the programme quotations, in the 1975 report the first section after the introduction concerns appearance. Rather than CLASP absorbing parts of the building industry it may well be that the general environment of the building industry will absorb CLASP, until it becomes just another range of available products from which to choose.

The dilemma facing the building systems in Britain in recent years is daunting, for their claim to embrace all building in a given type was made upon the necessities of mass production, on the size of the market justifying the standardization. Although as has been shown, much of this argument was spurious, in practice, the amount of time and energy that had been invested in them caused government and local government to react only slowly to the negative feedback. Once tied to a particular system many authorities have found themselves involved in a huge decision making network which cannot be unravelled overnight, and they and their professionals are often forced into actions which they would not otherwise undertake. In the field of housing this clash of the social and the technical systems can be seen at its clearest with, at one end of the spectrum, disasters like Ronan Point, and at the other continuing problems for users left with living in the results.

The problems for the latter were dramatically illustrated in an important court case involving a system built block of flats on one of two developments using a modified form of the Reema system in Portsmouth, and designed by architects Hugh Wilson and Lewis Womersley in 1967 (**11.1.6**). In the year that it was completed, Cannock Lawn had already suffered enough to warrant a report from the Building Research Establishment. Housing departments in most cities tend to view condensation fatalistically, and in this case subsequent petitions, letters, mass meetings, and an abatement notice served by the tenants on the City Council failed to have any effect. Ultimately, in September 1975, tenants took their case to the Magistrates Court where their claim that conditions were detrimental to health was upheld. A subsequent appeal by the City to the Crown Court, in spring 1976, supported the ruling in favour of the tenants. The court distinguished between traditional misuse of premises, for example by excessive condensation created by washing, and that which occurred as a result of the design. Remedial work was carried out at Cannock Lawn during 1976, with Frank Guy who had advised the tenants, acting as consultant. For the first time, a local authority became responsible for such design faults which it had created, much as it would were it a private concern. The case amply illustrates the problems created by the impact of social and technical systems. Both the provision of public housing on a vast scale and the relatively insulated position enjoyed by those providing and designing have exacerbated the situation. Whatever ones political views it is clear that any private individual or organization would be held responsible for many of the faults inherent in much public housing. Frank Guy, who gave evidence for the Cannock Lawn tenants, put it squarely:

> '...30 per cent of the population of a big city which lives in council houses can exercise only the most tenuous control over their architects while at the same time paying their fees. Here large bureaucracy insulates the architect from his client to the impoverishment of both.' (Guy 1976)

11.2

MASS PRODUCTION
AND
ITS MYTHS

'Architecture, together with all the activities of the
Werkbund, is moving towards standardisation
(Typisierung); only by means of standardisation can it
achieve that universality characteristic of ages of
harmonious culture.'

Hermann Muthesius
Werkbund Annual Meeting 1914,
in Benton, Benton and Sharp 1975

'So long as there are artists within the Werkbund, and so
long as they are able to influence its fate, they will
protest against the imposition of orders or
standardisation.'

Henry Van de Velde
Werkbund Annual Meeting 1914,
in Benton, Benton and Sharp 1975

Mass Production and its Myths

Central to the development of modern ideas on prefabrication,
industrialized building or building systems has been the notion that
mass production techniques, as used by other industries, will offer
enormous benefits. This study has followed the development of the
many facets of that argument as it emerged in the late nineteenth and
twentieth centuries, and attention has been drawn to the distinctions
between this set of attitudes and those which preceded it. In placing
production as the main plank of the argument many architects have
found themselves unable to equate this with their traditional concern
for architecture itself and with the needs of the users. J.M. Richards,
in his contribution to the influential 1937 publication *Circle*
paraphrased Mumford and at the same time characterizes the
implications of this view:

'They include an acceptance, indeed an exploitation, of mass-
production: the multiplication of standard patterns, implying the
elimination of personality from the process of manufacture.
They include therefore the disappearance of the handicraft
respect for technical virtuosity, and respect for rarity as such.
They include also the acceptance of a new formal vocabulary,
derived from the needs of machine production and influenced by
the example of machines themselves.' (Richards 1971, 1937)

The acceptance of this would, thought Richards, lead to an anonymous tradition akin to the 'standardised design vocabulary of the eighteenth century'. Although by 1972 Richards, in the RIBA Annual Discourse (1972), had repudiated much of the approach of the modern movement which emanated from such views, clearly its effects at the time were considerable.

Having been told for years that traditional methods were inadequate, the period after 1945 saw the erosion of the confidence of the building industry in Britain gradually become structural within the new social context of the period. The increasing numbers of architects being produced during the postwar years were largely educated within a framework of ideas put forward by the modern movement and this engendered a running critique of the existing construction industry and its techniques. Under such an attack much of the industry came to feel their existing expertise to be irrelevant and this in turn contributed to recruitment, training and deskilling problems. The building systems designers had been particularly vocal in this situation, with industry itself a party to its own destruction. The aftermath of the political events of the thirties and forties had found large numbers of architects who could see a very direct relationship between political theory, architectural solutions and the means of production. This had a direct effect upon their view of the role of the architect whom they now saw less as a professional adviser, more as a technician directly participating in the mechanisms of social change. Professionalism was seen as a narrow and inhibiting device for deluding the public, with the result that large numbers of architects preferred to deploy their skills in the public sector as salaried staff. The idealism of many young, and not so young, architects was closely linked with marxism: forms of socialist planning, economics and the reordering of the world were all seen as being very closely inter-related, which they are, and as being directly translatable into built form, which they are probably not. Overt political views became covert architectural ones. The basic argument was simple: the structure of society is governed by the nature of the ownership of the means of production. This proposition, which Marx was developing before 1845 and which came to fruition in *Das Kapital* published in 1867, was summarized by his friend and collaborator Frederick Engels in his Preface to the English edition of *The Communist Manifesto* in 1880:

> 'That in every historical epoch, the prevailing mode of economic production and exchange, and the social organisation necessarily following from it, form the basis upon which is built up, and from which alone can be explained, the political and intellectual history of that epoch....' (Marx and Engels 1948, 1848)

Marx had also pointed out that it was not the consciousness of human beings that determined their existence, but their social existence which determines their consciousness. Thus the widespread social, not to say socialist and communist feelings of the post-1945 period led many architects to the view that architecture must be seen as a social service, as with health and unemployment insurance. From such a vantage point they felt themselves to be personally free from the forces of the market place but, more significantly, they could effect major changes in the building industry itself by means of control and planning. Government and local government building programmes were seen as the obvious way of effecting this, and the architects' departments of such organizations offered a climate for those architects wanting to implement such ideas. Now that most of these organizations have the character of complex bureaucracies often sadly unable to deal adequately with the tasks before them, such views seem, to say the least, naive and idealistic. Jencks has described the result of all this in these terms:

> 'Finally some of the social ideals of the early modern movement were coming to fruition. But other ideals were not. For what this meant in terms of the politics of architecture was that individualism, expressionism and "Art" were to be denied in the name of the Welfare State, economy and social service.... The politics of Architecture was conducted under the basic dichotomy "Art or Social Services"... showing that it had to be one or the other.' (Jencks 1968/69; 1973)

Such a view, which assumed that in becoming a social service, architecture would be liberated from the constraints of art, is rooted in the economic determinism of Marx. His powerful analysis has itself the internal contradiction that, in proposing that the superstructure of society is determined by the infrastructure of production and exchange, this proposition itself must also be so determined. The isolation of economics and practical concern as the core of the infrastructure in the infrastructure/superstructure model is, in itself, culturally determined. In this sense, the acceptance of Marx's analysis can encourage the view that economics is all, that cost effectiveness is all. It is no surprise then, that we find such arguments playing a central role wherever the view of architecture as a social service becomes dominant. Often we find those architects most concerned to see architecture perform a useful social service are also those obsessed with cost and production methods and less concerned with the art of architecture. The infrastructure/superstructure model becomes codified in the very way the work is carried out.

The issues raised by the prominence of a model based on economic determinism are closely examined by Sahlins, who points out that although at first sight the relation of the 'cultural and material logics' appears unequal:

> 'The material forces taken by themselves are lifeless. Their specific motions and determinate consequences can be stipulated only by progressively compounding them with the co-ordinates of the cultural order.' (Sahlins 1978, 1976)

The holistic nature of the cultural order is evident in the realm of meaning and 'it is this meaningful system that defines functionality' so that functional reasoning has to be seen as part of the overall culture. There is, says Sahlins:

> '...no material logic apart from practical interest, and the practical interest of men in production is symbolically constituted'. (Sahlins 1978, 1976)

What is interesting about different cultures and subcultures is the way in which they decide to make their choices: that some animals are thought edible in some cultures but not in others, that one house type is suitable for one and not another, and that one way of making buildings is good for one and not another. That many architects have assumed that there is a single productive, and aesthetic, model appropriate to all conditions of men, in all parts of the globe is as fallacious as it is widespread. If we respect the individuality of cultures, and of individuals, an approach which constantly summons only the deities of economic determinism is seen to be totally inadequate. Nevertheless, many of the attempts to introduce industrialized methods held, and continue to hold, the economic criteria as primary. As has been shown the economic benefits that it is intended will acrue from more industrialized methods in building have been cited time and time again by architects and politicians alike, as reasons for pursuing major building system innovations.

After so many attempts to develop and apply methods in accordance with these views it might be thought that there would be clear evidence in support of its cost effectiveness. Most of those involved in building systems have, however, shown some reluctance to produce published results on this question: this is partly because of the difficulty of isolating the effect of quantity production over other factors. One example (**11.2.1**) can be found of the savings to be made on system components over traditional components in *The Story of CLASP* (MOE 1961), and there is other fragmentary evidence. A study by Oddie (1975) of eight different school building systems in six countries (including CLASP in the UK) pointed out with respect to the claims that:

> '...industrialised building is cheaper and quicker than the use of alternative methods... no generally or direct confirmation can be given.'

For example, the first series of schools in the Canadian SEF programme exceeded its cost targets, whilst CLASP demonstrated that it could stay within the government cost limits which applied to all school building whatever the constructional method. A concerted effort to provide a bank of dimensionally related, factory produced components, was made by the Department of Health and Social Security for the IHB (Industrialized Hospital Building) Programme. For this a comprehensive *Compendium* (1965/66) was produced giving

11.2.1 Consortium of Local Authorities' Special Programme (CLASP). Costs of some bulk purchased components during the first three years of CLASP, in relation to the 1957-8 pilot programme costs. (1961)

PRICES OF SOME CONSORTIUM ITEMS

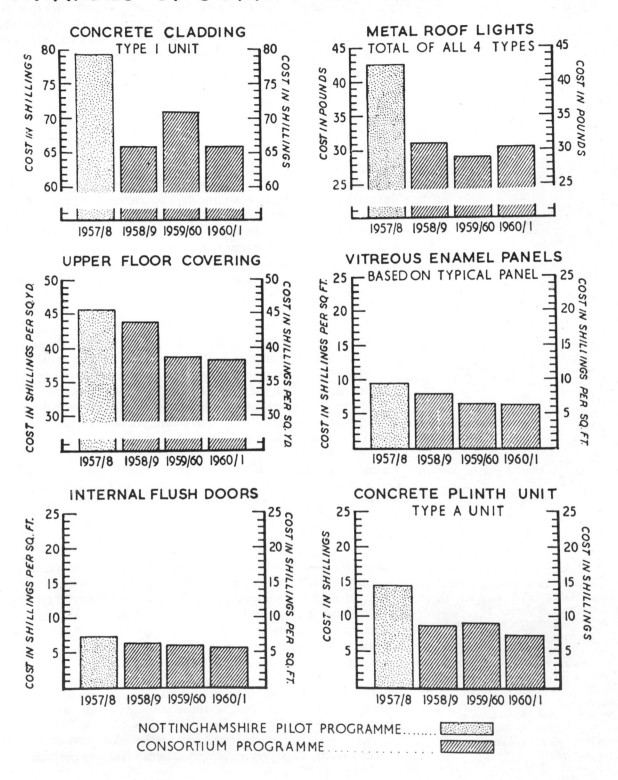

ranges of components, details, and schedules of rates for a wide range of alternatives. Some interesting information emerges from the prices quoted by manufacturers who were asked to give unit rates (per item) and an 'optimum' rate based upon quantity production. The resultant savings offered were very small. Timber windows required a manufacturing run of 500 before a 5% reduction in cost was offered, whereas steel or aluminium windows offered 2½% for over 100 units. These reductions only applied when the orders were for one type. Typical figures for softwood door assemblies were a mere 1½% reduction for 1000-2000 and 2½% for over 2000. Probably one of the largest studies of the British client sponsored consortia was undertaken by the Building Economics Research Unit at University College London under the direction of Professor Duccio Turin. One of the aims of the project was to look at the timetabling of jobs and cost fluctuations, interviewing architects, contractors, manufacturers and suppliers in four consortia — CLASP, SCOLA, SEAC, CMB. As one of the intended aims of the use of such building systems in the public sector is to reduce the unpredictability resulting from traditional methods Turin's work is of special interest. Since one of the claims of the consortia is that it is a more economic way to build, the following is a significant finding:

> 'One hypothesis we wished to test is the contention that the cost of school building (or, at least system school building) has not risen as fast as general building costs. If changes in BNCT/GA (Basic net cost on tender to Gross floor area ratio) are accepted as an indicator of changes in building costs, then inspection of diagrams suggest that system school building costs have risen in the same way as the cost of new construction, as measured by the DOE index.' (Turin 1972b)

Turin further goes on to point out that, since the school building cost limits had risen more slowly than general building costs, local education authorities were either using additional monies to deal with inflation, reducing quality, or reducing the ratio of the gross area of floor plans to cost places. In a further analysis of the frequency of use of major component types, Turin revealed that there is a very high concentration across a very small number of available types, suggesting that there is an excess of size differences. It was also noticed that of two components with the largest range of types these were the lowest use, or take-up figures, of only 41% and 64%.

The study drew attention to another aspect of standardization that architects have rarely faced up to — that although there may be a range of (Turin example) 97 distinct components available from standard drawings, when differences have been identified for the purposes of obtaining system programme quotations this has risen to 247 choices. The manufacturer, to identify these sufficiently for taking off and production, brings the number of choices to 533. 'Variety', says Turin, 'is *generated* as the need for precise identification increases' (Turin 1972b). This precise identification, involving perhaps special

11.2.2 Nominal and actual starts of projects on site for 4 client sponsored systems. CLASP, SCOLA, SEAC, CMB, 1969-70. A nominal or 'technical' start involves merely the formality of starting work by fencing or placing of a site hut, whereas the actual start is when construction work commences. This shows the peak of nominal starts during March, just before the end of the financial year in April. It also shows the peaking of actual starts during March, April and May which creates material and labour problems

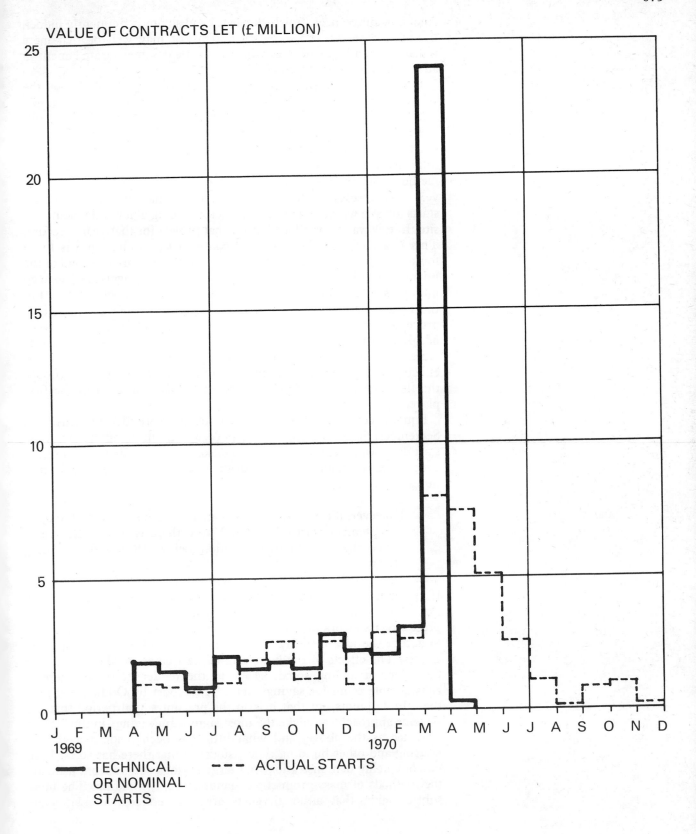

VALUE OF CONTRACTS LET (£ MILLION)

—— TECHNICAL OR NOMINAL STARTS   - - - ACTUAL STARTS

cuts, hole differences, finish alternates and so on, is frequently hidden from the designer and more importantly it may well be that mere reduction in a range of sizes will bring little commensurate benefit in an overall reduction of types.

Turin's study also identified quite clearly, the problem of the 'peaking' of the start and finish dates of school projects. For example in the four systems under study, CLASP, SCOLA, SEAC and CMB, 45% of all primary and secondary starts in the programme years 1968/69–1970/71 occurred in March, just prior to the end of the financial year (**11.2.2**). This is a result of the government fiscal method whereby monies allocated for a given year must be used in that year — the result is that authorities often make merely 'technical starts' on site with the general contractor placing a hut and fencing the site. In this way their allocation for that project for that year is secure. This is a method, of course, not confined to building and is to be found in any government spending department towards the end of the financial year where, often, considerable effort and ingenuity has to go into the spending of the available finance, so that it does not revert to the Treasury and allocations consequently cut the following year. This procedure, in itself, is one so open to abuse and waste that it is worthy of special study. Clearly with many similar projects starting at the same time contractors and suppliers were found to be concerned at the effects upon component supply, and of course it does put into doubt the flow-line concept which motivated the systems idea in the first place.

Equally, completions for secondary schools were found to cluster in September, the opening of the school year, and this, combined with the start date peak occasioned Turin to make a most pertinent observation concerning the claimed attempt of the systems to reduce construction times:

> 'However, if the majority of secondary schools scheduled to open in September start on site in March there is little incentive to reduce the construction duration below 18 months.' (Turin 1972b)

If indeed it was hoped to speed construction the report suggested that the client would have to plan further ahead with regard to precontractual work like site acquisition, and the designers would have to accept the idea of delay and organize themselves to minimize its effects. The effects of the bunching of start dates in March had also the effect of adding some 3.13% to the tender costs, so there is considerable room for savings here. The report by Oddie (1975) for the OECD points out that it is in the precontractual period that the systems approach can make the most significant savings in time. Such evidence as the above, and what little has been made available in housing and other building types, suggests that there has so far been no convincing demonstration that large savings accrue from utilizing the methods of mass production manufacture in building. The more subtle claim, that although costs are not reduced, standards are maintained or increased, is certainly a difficult one to prove in building — especially so since the art of building appraisal is in its infancy. Bold claims are often made in this direction by most systems, but it has equally frequently been the case that systems have gone out of business because of the expense and need to maintain a flow of units.

Turin's observations bear this out in the client sponsored systems in his study for, as has been shown, he established that their costs equalled the national DOE cost index and therefore authorities were either finding extra money (from say minor works programmes or abnormals) or reducing standards. There is thus some evidence in support of Mumford's observation of 1930 that:

> 'Mass-production, just because it involves the utmost specialisation in labour-saving machinery and the careful interlinkage of chain processes, suffers, as I have pointed out elsewhere, from rigidity, from premature standardisation. When the cheapening of the cost is the main object, mass-production tends to prolong the life of designs which should be refurbished.' (Mumford 1930)

This lesson of 1930 had not been learned by the 1960s for we find the Preface to one of the many catalogues of building systems pointing out that:

> 'System building methods should lead eventually to a more frequent change of the housing stock. If building costs can be reduced still further, a much shorter economic life for buildings can be envisaged and it may become possible to build for about 25–30 years, and to replace the structures when they have been made obsolete by the march of time.' (Diamant 1965)

Such a profligate view of resources also encouraged the wholesale removal of older houses that could, for less cost, have been rehabilitated. In his article, Mumford also pointed out that the application of mass production to objects that wear out may be 'socially valuable' but that when applied to more durable goods such as furniture and houses there is the danger that there will be insufficient replacement to keep the plants running that supply the parts. This has certainly been proved true with regard to most of the heavy concrete systems as we have seen, for it is the guaranteed market that permits such systems to survive. An investigation into the 'Housing Delivery System' in Connecticut by a team from consultants Arthur D. Little made a number of pertinent points in connection with new methods:

> 'As a general rule, the more modest the technological change, the more modest will be the time required to assimilate changes, the volume necessary to justify the change, and the expected savings in time and money. The more revolutionary the system, the

longer will be the acceptance time, the greater the volume necessary to provide financial feasibility and the greater the potential savings.' (Little *c* 1970)

If a given system is not market sensitive (as is timber frame construction) it will require government intervention to stabilize the market and/or subsidize production. This happened with many of the housing and schools systems in Britain. However, this distorts the market in turn, as Turin shows, and further it perforce introduces a degree of government intervention which may have more important political implications on the economy. The effects of such intervention in the field of public housing is already presenting major problems for, apart from financial issues, it raises important questions concerning the self-determination of the individual. The difficult question of how much should be provided centrally and how much choice can or should be left to the individual is a recurring one with regard to public housing. Personalization and flexibility have become two of the key words for architects in such discussions. It is only necessary to look at a council built house that has been sold to see the effect of personal ownership. The house will be painted, often in bright colours, porch and greenhouse attachments will sprout whilst walls, fences, shrubs and trees will appear around the home.

One of the most publicized pieces of work which has attempted to come to grips with this public provision/private choice relation was that carried out by Nicholas Habraken in Holland. In his book: *Supports: An Alternative to Mass Housing* (1972a) Habraken proposes a system to take account of such needs. This consists of a framework into which individual dwelling units will be inserted using either kits of parts provided or by any means available. Within the framework are areas of fixed and flexible zones. This proposal to open public housing to a more responsive use of mass production components has seen only one attempt at application in Britain — by the Greater London Council although further experiments are in progress in Holland.

In the mid-sixties two students at the Architectural Association in London used Habraken's ideas to produce a scheme, called Primary Supports Structures and Housing Assembly Kits (PSSHAK). This received considerable publicity including an exhibition at the Institute of Contemporary Arts. The two students, Nabeel Hamdi and Nick Wilkinson, were employed by the GLC to develop their ideas into a practical result. Their first scheme, after coming to terms with the bureaucracy of the GLC, was a twelve unit project at Stamford Hill Depot not complete until 1976. Commenting on this building's 'brutally monotonous' appearance Rabeneck (1975) saw the result as a 'victim of ruthless British parsimony'. The second PSSHAK scheme at Adelaide Road NW1, whilst retaining some of the basic ideas, has become something other than Habraken's original proposal. The support structure turned out to be, quite sensibly, brick crosswall construction as used by the GLC for years. The external skin was also of brick, which brooked no personalization. The planning still retained the concept of alternative internal permutations which could be worked out with future tenants using a model. Kits of internal fittings were provided, supplied by a Dutch firm and these fittings would be owned by the GLC. Whilst clearly this degree of choice introduces a welcome change into local authority housing, it is far from the ideal proposed by Habraken where mass production is turned to the consumer's advantage on a massive scale.

That the myth of mass production and the implications of the machine have occupied a crucial place in the minds of most architects of the western hemisphere during the twentieth century is beyond doubt. In endeavouring to accommodate to the machine a certain willing suspension of disbelief was necessary. For such a concerted and cohesive act to take place some binding myth was necessary for, as Cohen points out:

> '. . . a myth is a narrative of events: the narrative has a sacred quality; the sacred communication is made in symbolic form; at least some of the events and objects which occur in the myth neither occur nor exist in the world other than that of the myth itself; and the narrative refers in dramatic form to origins or transformations.' (Cohen 1969)

The purpose of a myth then, is to mediate contradictions, to offer some device whereby man can come to terms with certain situations. We have seen the dilemma created by the introduction of industrialization and its effects upon the creative responses of architects as they endeavoured to absorb, control or redirect its effects. Until recently few architects questioned the great myths of standardization and mass production on the assembly line model, although many expressed alarm at the results. Abel in 'Ditching the Dinosaur Sanctuary' (1969) elegantly demonstrated that this view of machine production must be discarded. The myth, often created by designers themselves, around the impact of machine technology has given rise to some enormously creative endeavour as well as a large proportion of disastrous building, for it served, as do all myths, as:

> 'originally devices for blocking off explanation. If they were valued as such, then their place in the cognitive scheme would make them eligible as means of *legitimising social practice.*' (Cohen 1969)

The myths surrounding quantity production have certainly legitimized a large amount of decision making in architectural practice, and for a long period the elevation of this aspect of the design of buildings to an all pervading one has had a serious and limiting effect on both design and designers.

12

OPEN SYSTEMS, WHOLE SYSTEMS

THE PERVASIVE SYSTEMS CONCEPT

'Man has a tropism for order. Keys in one pocket,
change in another. Mandolins are tuned G D A E. The
physical world has a tropism for disorder, entropy. Man
against nature... the battle of the centuries. Keys yearn
to mix with change. Mandolins strive to get out of tune.
Every order has within it the germ of destruction. All
order is doomed, yet the battle is worthwhile.'

Nathaniel West
Miss Lonelyhearts 1961, 1933

The commonly understood notion of a building system centres around
the idea of sets of dimensionally related components, largely factory
produced, that fit together on site in a variety of ways to make
buildings. Most of the systems that have been examined here are in
this category and generally most of them are closed systems in the
sense that they have their own ranges of components and subsystems
and do not take kindly to alien components. We have seen how there
are sharp boundaries in Britain between most of the existing building
systems and that these boundaries are identified both technologically
and administratively. As the pressure on resources has become more
pronounced there has been an attempt to bring the client sponsored
systems together and to eliminate the immense duplications that have
existed. Fortunately it is unlikely, with the systems notion in such
disfavour, that this could now result in the once-suggested, centralized,
client-sponsored system. Equally, however, it seems unlikely that it
will give rise to the more organized open component market envisaged
at one time — a notion which could still bring welcome change to the
British building industry and the design professions.

There are signs that those client sponsored systems that remain are
having to recognize that many of the arguments upon which they were
founded have been eroded. As systems they are having to come to
terms with strong influences from their environment. Central to their
original assumptions was the belief that building could, and would,
follow the rest of the consumer goods industries in showing clear and
obvious benefits by espousing the philosophy of the production line.
Much of the earlier work in the design methods field during the 1960s
had a similar orientation, drawing many of its models from systems
engineering. Accounts of such approaches are included in Jones (1970)
and Broadbent (1973), whilst an early example can be found in

Asimow's *Introduction to Design* of 1962 which sets out a problem solving process which has its roots in the work of Winslow Taylor at the turn of the century as well as in the operations research methods developed during the 1939–45 war. Much of the building system work can be fitted to Asimow's stages:

1. Analysis of the problem situation
2. Synthesis of solutions
3. Evaluation and decision
4. Optimization
5. Revision
6. Implementation

Unfortunately, many such approaches, emanating from pure systems analysis or from systems engineering appeared to develop in ignorance of the more embracing ideas offered by General System Theory itself. By 1968 this turn of events had even caused one of the formulators of General System Theory, von Bertalanffy, to point out:

> 'What may be obscured in these developments — important as they are — is the fact that systems theory is a broad view which transcends technological problems and demands a reorientation that has become necessary in science in general and in the gamut of disciplines from physics and biology to the behavioural and social sciences and to philosophy.' (Bertalanffy 1968)

In hastening to apply what was thought to be the systems concept, most building systems designers took the applications which had been successful in other fields, such as product engineering, and used the techniques developed there for design and building, unaware that these were merely techniques developed from the propositions of General System Theory. An examination of this theory might have suggested that what was appropriate in engineering was not necessarily appropriate in architecture. Rapoport puts it like this:

> '. . . it is important to note, therefore, that what is a system and what is environment is a matter of subjective preference, depending on what interests us most. If we are interested in an individual and what happens to him, everything outside the individual's skin is the environment. If we are interested in a family, a firm, an institution, a nation, or the entire biosphere, we define our environment accordingly.' (Rapoport 1974)

The subject matter of General System Theory is '. . . the formation and derivation of those principles which are valid for "systems" in general' (Bertalanffy 1968). That is a theory which looks for the similarities in the models, principles and laws in different disciplines: a theory which seeks generalized underlying principles.

Whilst it is certainly a useful and valid aspect of systems theory that one can take the operations of one discipline and apply it to another — as Churchman (1968) observed when he suggested that we look at engineering as if it were a religion, and religion as if it were engineering — the application of one narrow field, systems engineering, to architecture has caused it to be limited by that very application. Wholeness is one of the central features of the General Systems idea, and what constitutes wholeness in a given situation is no easy question. It immediately raises issues of perception and identity, as elegantly described by Angyal in 1941. Angyal makes a clear distinction between a relation, as with objects on the basis of colour, size or weight, and a system which is 'a distribution of the members in a dimensional domain' and in which 'the members are, from the holistic viewpoint, not significantly connected with each other except by reference to the whole' (Angyal 1967, 1941). This view introduces an overall or structural element which gives a new meaning to the elements. Broadbent (1973) discusses the criticisms of the systems view to the effect that such a view is too general and broad to yield to analysis. He also points out, however, that what it does draw attention to is how such an analysis is made. It is how we divide up a problem to deal with it that becomes important and some methods will be more adequate than others. The systems view can posit alternatives by offering totally new ways of looking at things. So the model of the world that we use is of central importance to the way in which we make decisions. For many of the building systems designers their production line model, their viewing of the building process as just another factory has given rise to the emphasis on technological solutions of a specific type, a concentration on new ranges of factory made components to accord with this image. Even the results embody this view, with housing and schools looking like factories or with their 'systems' philosophy worn proudly (**12.1.1**). Even when not using a system as such many designers obey the imperatives of the modern movement and design their building itself as if it were a giant machine (**12.1.2**). It is clear however, that such reductions of the systems idea are exclusive and not inclusive. For many outside the architectural profession, and many inside, important aspects of that whole called architecture have been left out. Most of the building systems, as has been discussed, ignored fundamental issues of environmental control, of attempting to relate to their context, of the use of energy sources, of long term maintenance and, most crucially, of delight. In limiting meanings to those associated with the production and efficiency model the buildings are seen to be mere essays in cost effectiveness: the other meanings which most people tend to attach to buildings often seem to be missing. They lacked continuity of material or of form, they often lacked colour and texture, they had few natural materials. In one system built school the Headmaster insisted that the second stage to be built should have a variety of materials applied inside, brick, stone and timber, rightly pointing out that the lack of these reduced the value of the building as a learning environment. It is those properties that are systemic, of the system as a whole, that are of interest in the true systems approach. Since architecture consists of so many things it is possible to ignore some and concentrate on others. However, there is danger in this if the chosen areas of concentration make the sort of

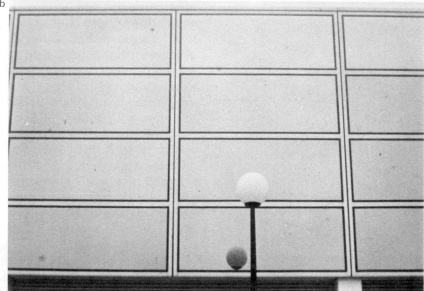

12.1.1 The building as a metaphor for the production line, perhaps an appropriate symbol for that cathedral of consumerism, the covered shopping centre. Milton Keynes shopping centre, Buckinghamshire, England. Derek Walker, Chris Woodward and Syd Green, 1979. (a) The building is on a 3-dimensional grid developed into a seemingly endless identical panel system. (b) The panels in their neoprene gaskets

reductions described in the above examples. For many of the building systems have avowedly set aside much that they thought pretentious and irrelevant in architecture but in so doing they have divided themselves from what constitutes the understood view of what a building should provide. Angyal points out that with:

> '...aggregates it is significant that the parts are added, in a system it is significant that the parts are arranged' (Angyal 1941)

and this is a crucial point for designers who are involved in organizing ideas and a material structure into a particular arrangement to suit a specific set of goals. Much that follows from such an approach is quite different from the normal associations of the idea of a system with things systematic. To be systematic is to work methodically, according to plan, whereas systemic describes those properties that are of the bodily system as a whole. Now clearly things systematic have a large part to play in the making of buildings, most particularly in the way the building is constructed. However, attention to such issues must not cloud the systemic properties of a building, that it must be a meaningful whole. Further, as Alexander indicated in 'Systems Generating Systems' (1968) a good building is one where the building plus the users form the proper whole. Using this definition, those buildings that address themselves to, say, very narrow architectural interests (e.g. to photograph well) or narrow technological interest, must fail. However successful they may be in those terms, looked at as a whole they are unsuccessful. I believe a view that can offer such an explanation for the many aspects of architectural endeavour to be a rich one.

Another important concern central to General System Theory is that of the open and closed system. A closed system is one in which there is no import or export of energy — the common example given is of a number of reactants brought together in a closed vessel. However, truly closed systems are hard to find and it is most useful to think of the terms as relatively descriptive and in this way they have a purpose. For a system to be identifiable at all it has to have a degree of closure and it is again what constitutes that closure that is of interest. As many systems theorists are fond of pointing out the concept of the closed system developed from the physical sciences with their roots in Cartesian rationality, whereas the growth of the human sciences found the models arising from this too mechanistic and reductionist to account for the phenomena in which they were interested.

In this sense then, a system is a set of entities, which may be real or conceptual (material or immaterial) and these interact with one another to form a whole: an open system has a specified environment within which to react whereas a closed system is one that ignores its environment. In the open system energy can pass in and out but the system maintains its homeostatic state — a principle enunciated by Cannon in 1939. The human body is thus an example of such a system, constantly changing, importing and exporting, but remaining essentially the same. A fountain or a candle offer simple examples of an open system. Thus a system, to use Boulding's (1956) term, has an image and the nature of the system, its pattern, its particular order is evidence of that image. The outline, or boundary of a system is its most interesting area, for it is here that change takes place. If within a system there is structure and order as Douglas points out:

'Its outlines contain power to reward conformity and repulse attack, there is energy in its margins and unstructured areas.' (Douglas 1966)

The boundary of a system may be hard, soft or just fuzzy: it may be very rigid in what it lets in and out, or very relaxed: however with the

a

12.1.2 Centre Pompidou, Paris.
Renzo Piano and Richard Rogers,
1976. An exciting, if anachronistic,
vision of the building as an efficient
machine, with its expression of
structure, services, and movement
systems. (a) View across Place
Beauborg. (b) The modern
movement's equation of efficiency and
marine technology here assumes
Oldenburg-like proportions.
(c) & (d) The escalator arrives at a
minuscule viewing platform from
which access is gained to the rooftop
restaurant. (Photos 1978)

b

c

d

SYSTEM AND ENVIRONMENT

A CLEAR HIGHLY
DEFINED SYSTEM
HIGHLIGHTS ITS
ENVIRONMENT.
THE 'HARD' BOUNDARY
HIGHLY STRUCTURES
COMMUNICATION
WITH THAT
ENVIRONMENT.

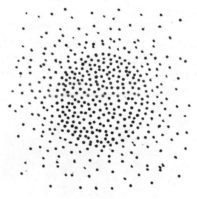

MANY IDENTIFIABLE
SYSTEMS HAVE A
'SOFT' OR OPEN
BOUNDARY ACROSS
WHICH ENERGY
MATERIAL OR
IMMATERIAL (MATTER
OR INFORMATION)
MAY PASS.

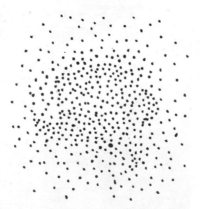

AS THE BOUNDARY
BECOMES INCREASINGLY
'FUZZY' THE SYSTEM
SETTLES INTO
EQUILIBRIUM WITH ITS
ENVIRONMENT UNTIL
ULTIMATELY ANY
SPECIFIC SYSTEM
CHARACTERISTICS
DISAPPEAR.

12.1.3 System and environment

latter the definition between the system and its environment becomes vague (**12.1.3**). The essential point about this transport of energy, or information, is that it has to be in the correct form or the system will reject it (**12.1.4**). In the case of the components of a closed building system, it can tolerate no alien information or components: whereas an open one can, perhaps, include certain sorts of non-system components depending upon their characteristics. Equally one can apply this to organizations, including the complex administrative structures created to deal with most building systems. The very administrative structures that have been set up to achieve the goals set limits to the things that can then be done. In the case of the institutional client sponsored system such issues are compounded by government involvement and the relative insulation this has afforded from the environment: hence change is slow and often reluctant. This is not because individuals are unwilling (although many take on the image of the system) but because the system has a character of its own and operates to its own set of rules.

This is not the place to delve more deeply into General System Theory itself but the foregoing has illustrated, I hope, that there is an important difference between what a theoretical stance can offer and those views which have so far passed for a system view in architectural usage. What it comes to is that architects have seen the system idea only in their own terms, and through their own discipline. Primarily their frame of reference comes from the making of buildings and is formed by the education and the skills that surround that. It is

SYSTEM AND ENVIRONMENT COMMUNICATION

THE FORM OF THE SYSTEM AND ITS COMMUNICATION CHANNELS CONTROL THE IMPORT AND EXPORT OF MATTER AND INFORMATION. A SYSTEM NORMALLY ONLY RECOGNISES AND ACCEPTS COMMUNICATIONS IN THE RIGHT FORM — IN SPECIFIC CODES

12.1.4 System and environment communication. The form of the system and its communication channels control the import and export of matter and information. A system normally only recognizes and accepts communications in the right form — in specific codes

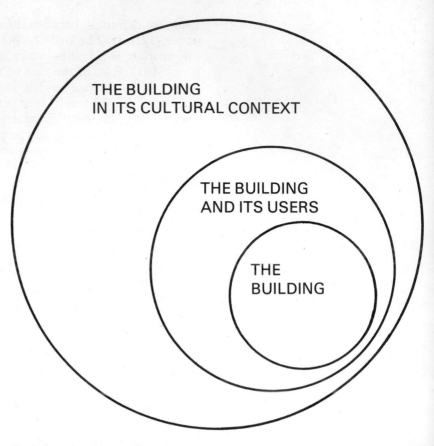

12.1.5 Three levels of system

therefore not surprising that the view of a 'system' they have projected on to their work and the world at large involves the hardware of their stock in trade: it concerns building materials, components and subsystems and the way these relate together. However, this has gone on during an historical period when it has been fashionable to denigrate both the influences of art and of history and for the architect to see himself as something of a journeyman. Much of the time such a philosophy has merely served to describe him to the world at large as a technician — and, when this is compounded by his involvement with bureaucracies, as a state functionary. In this context it may be helpful to look at three levels of system (**12.1.5**): the first resting within the second, both of these within the third:

1. That of the building: a collection of materials and interlocking subsystems of technological bits.
2. The building and the people who use it: the interaction of the forms of the building, its use and administrative patterns.
3. The cultural context: Here we are concerned with the social place of the building and its users: together with the cultural and symbolic meanings.

At the first level, that of the building itself, there are a set of subsystems involved which have been given differing emphasis. For example most histories of architecture describe the form, the structure and the material fabric of the building, with a heavy emphasis on its visual qualities. Now although these are important they are only a part of the experience of any building — our recollections of buildings embody a complex set of sensory reactions, which have achieved little attention. Who can summon up their image of a visit to one of the great cathedrals without bringing to mind that musty, cool, still quality of the interior? Or of their school without the smell? Further, few histories say anything of the complex servicing systems that have gone into buildings of the past one hundred years, even though these may have in large part created the forms of the building, as happened with Wright's Larkin building or with Foster's Cosham IBM building. In many cases the majority of the costs of a building is in such environmental control systems. This concentration on the structure and fabric of construction itself has been endemic in the building systems, the result of identifying the systems to be dealt with in one specific manner.

Moving to the second level, we see the building in relation to its users, again something that features little in histories of architecture. How do people really use buildings, how do they perceive them and what happens over time? These are questions of importance to which there are not easy answers. We all have experience of using buildings and so it is somewhat surprising that such studies have taken so long to develop. In recent years the emergence of interest amongst architects, sociologists, psychologists and others in such issues is gradually developing a body of useful material, although so far it has had little effect on the making of buildings. Nevertheless the public have been vocal in their response to environmental issues and this has partly led to the growth of interest in conservation. The clean sweep planning once beloved of architects and planners has been seriously brought into question, and the building systems themselves have been very slow to respond to such pressures. A whole new field concerns building appraisal studies, such as those pioneered by Markus *et al.* (1970) and Canter at Strathclyde and Surrey Universities and this could serve to identify those issues which surround the use and management of systems in relation to buildings. After all the way in which an organization commissions and makes use of space is both functionally and symbolically important, which makes it even more surprising that, although it goes on daily all over the world, the management of space is not considered a function for which one should have proper skills. Oddie (1975) points out that those in the front line of (school) teaching, for example, may be very aware of the deficiencies of their existing building but are:

> 'often wholly unaware of the wealth of opportunity that a good building can open up: and (he) will only awaken to these opportunities if he is prompted to observe the inter-action of people and buildings more attentively than he is normally expected to. He will only be prompted in this way if he is required to play a creative and active role in developing designs with architects.' (Oddie 1975)

Thirdly, at the cultural level, another set of issues is involved. A building is not only of concern to those who use it directly, but to those who see it and for whom it forms part of a total environment. The forces that bring a building into being are cultural forces, and what a given society decides to build and where, is as important as how it decides to build. That the centres of most large western cities are filled with multi-storey office blocks is as culturally significant as was the Gothic cathedral rising above the complex of mediaeval housing which surrounded it (Russell 1956). No designer can escape from such forces and, although he might choose to ignore them, his work will inexorably have some cultural meaning, whether he sees himself as journeyman, a technician or a grand prima donna. There is little guarantee that a humble attitude in such things produces a more appropriate cultural object. Meaning is attached to buildings and places by people, and this is an important part of the cultural process.

The many new studies which are emerging in this area are important in reassessing much that has been ignored. There are attempts to re-establish more explicitly the links between architecture and the whole cultural process. In contradistinction to much of the argument underlying the modern movement such work is also drawing attention to the continuity of architectural ideas. It was Venturi (1966) who pointed out that the three requirements of architecture defined by Vitruvius — Firmness, Commodity and Delight — had been redefined by Gropius and the Bauhaus as Firmness PLUS Commodity EQUALS delight. It is clear that delight is not the automatic result of dealing with Firmness and Commodity, nor is it necessarily the natural outcome of dealing with so called functional requirements. The importance of the symbolic world is one constantly emphasized by social anthropologists like Douglas (1970) Leach (1976) and Thompson (1979), whilst systems theorists such as von Bertalanffy (1968) Boulding (1956) and Rapoport (1974) have also stressed the necessity to take the symbolic world into consideration.

Architecture, in common with the other activities of man, can be seen as an expressive system and as such will inevitably carry meanings. It is important for architects to understand something of how this works, since it is often the confusion over the meanings of acts that limits communication. Those who have ventured into this difficult area include Norberg-Schulz (1963; 1971), Jencks and Baird (1969), Bonta (1975; 1979), Broadbent (1969), and Balfour (1978). Semiotics, or the science of signs and symbols, has flourished internationally and, whilst its applications to architecture can be revealing (see Broadbent *et al.* 1980) it often appears to be little more than another form of functionalism. This new form of rationalism demonstrates a deep need in some commentators to be explicit in all things, to be able to explain everything in words. Whilst clearly there is a part of architectural endeavour that can be dealt with in this manner there is also a part which can only be understood directly in its own terms, much as with painting, sculpture or music. Whilst there will always be attempts to try and translate this internal language of architecture, the nature of its existence may reside precisely in its impenetrability. It is complex, and often ambiguous. An architecture that is transparent would hold little interest for society. What sort of society would produce such an architecture?

TOWARDS OPEN SYSTEMS

'Humanism has two enemies — chaos and inhuman order.'

Geoffrey Scott
The Architecture of Humanism
1924, 1914

Those ideas surrounding the building systems notion have offered great promise but, sadly, have often become merely a panacea. The impact of industrialization upon architecture has brought forth a wide range of responses, many of them highly innovative. During the formative period of the modern movement many architects and designers looked upon industrialization as a cleansing force, which would sweep away the past and offer us a bright new world with a bright new architecture. The building systems idea became the leading edge of a weapon which would cut through the gordian knot of tradition. It would draw upon science and technology and set aside art, the traditions of which had come to seem unbearably academic and restrictive. The Constructivists in 1920 had tied all this together in their six slogans:

'1. Down with art
   Long live technic
2. Religion is a lie
   Art is a lie
3. Kill human thinking's last remains tying it to art
4. Down with guarding the traditions of art
   Long live the Constructivist technician
5. Down with art, which only camouflages humanity's impotence
6. The collective art of the present is constructive life.' (in Bann 1974)

The building systems idea has also operated at many levels — from the single potent building which has been designed in such a way that it implies a philosophy, through various programmes which bring together a series of buildings using the same approach, to proposals to transform the building industry of a region or country. The argument put forward has often been that, if architects will not get to grips with industrialization and the forces of the market place they will, as a profession, disappear. It is, however, a little naive to assume that if this is the direction in which society is moving, a handful of architects can have much effect upon it. In the end society will decide whether architects disappear or change their role. Nevertheless architects have frequently seen themselves as central in a reshaping of society's attitudes to work. However, Tafuri has acutely pointed out that there has been no avant-garde movement which did not aim at a 'liberation from work'. The Constructivists, he says, gave:

> 'a full affirmation of the ideology of work. This contradiction is permissible because the "new work" they proposed was presented as collective work, and what counts most, as planned.' (Tafuri 1976)

Planning, looking ahead, prediction and control have been underlying forces in all the attempts to create viable building systems. This in turn reflects western man's recurring desire to predict and control events with much of the strength of modern science and technology deriving from this pursuit. Descartes marks the break with that view of the cosmos which has its roots in sense perception and in which the forms, forces and qualities of nature are seen in clear relation to man. The generation of which Descartes was a part, Sutcliffe (1968) points out, was characterized by:

> 'the will to dominate, to control events, to eliminate chance and the irrational. This attitude is present in every field: the political, the military, the scientific. But how can one control phenomena if one cannot foresee the way in which phenomena will behave?'

Clearly prediction and control have formed an important driving force in the building systems argument and have served to give it some of the more sinister overtones it has acquired. But what is the nature of the control sought and why? In part it can be explained by the need for credibility. For politicians and the accountants of society, there are obvious attractions to the many claims made on behalf of industrialized building systems to build faster and quicker than ever before, whilst much of the public credibility of architects rests on their producing appropriate buildings within the predicted time and cost. The wayward ways of the building industry in Britain have made such prediction and control a chancy business and so it should, perhaps, be no surprise that architects have frequently been attracted by the seeming rationale of the building systems idea. However, the attempts both by politicians and architects to bend the industry to such neat and rational models of progress have been beset by problems. The anomalies created by the gap between the symbols of industrialized building and their application is frequently apparent in Britain. In societies with a less technological tradition than Britain this gap becomes increased to a critical point. As Maitland (1979) points out in connection with the massive Polish use of precast concrete systems:

> 'It is as if, in the tradition of the early modern movement, the principle of industrialised building could be separated from the fact. These buildings "stand for" industrialised construction as an organising principle, rather than as a method.'

In other, less industrially developed countries we find this gap even more painfully demonstrated with attempts to use the order, prediction and operational controls associated with industrialized technology. In building terms the introduction of such technology is often reduced to the visual order associated with such approaches in western Europe or North America. Such an approach immediately precludes a more intelligent use of the systems idea, one that may integrate the existing and the new in a subtle fashion without the aesthetic baggage associated with highly industrialized societies. However, this latter may not prove easy since the idea of design exhibiting a transparent total order is deeply embedded in our thought processes. Descartes himself points out that 'buildings undertaken and completed by a single architect are usually more beautiful and better ordered than those that several architects have tried to put into shape, making use of old walls which were built for other purposes', and that those old towns that have grown organically are 'so badly proportioned' compared to those designed of a piece 'at will on some plain'. They are, he says, 'more the product of chance than of a human will operating according to reason' (Descartes 1968, 1637). Reason, order, prediction and control have played a central part in the systems argument. As in other fields this has yielded dividends but it has also gradually brought about the realization that although the phenomena of architecture may be redefined to suit such a desire for control and prediction there are areas that cannot be accounted for in this way and the end results do not automatically produce buildings or environments of quality. Indeed the powerful rationale created to justify system approaches may itself be instrumental in limiting some of the very benefits intended. All buildings are at least buildings, even if it be decided that they are not architecture. As buildings they cannot escape carrying culturally loaded information about where they came from. Further, much of this information will not be explicit but implicit. Thus it will defy analysis in terms of the argument used by most building systems designers.

In one sense the growth of building systems can be seen as a struggle for control, with industry needing to produce more and more goods and to match demand or create new markets, whilst architects have on occasions been attempting to direct such efforts in socially purposeful ways. In promoting building systems, architects have been engaging in a process well known in other industries and characterized by John Galbraith as the management of demand. Here the development of technological production procedures necessitates the producers organizing the market. In this way producers and the state come to work in close association and 'planning extensively replaces the market as the force determining what the economy does' (Galbraith 1966).

In drawing on the common but fallacious analogy between the consumer goods and the building industries it is easy to see that many of the building systems in Britain, and especially the schools and housing consortia, have been primarily in the business of the management of demand, the key to which, says Galbraith, 'is effective control of, or sufficient influence on, the purchases of final consumers'. Whilst the promoters of building systems often genuinely strive to offer the benefits of efficiency, cost effectiveness and speed of construction, they become enmeshed in the control mechanisms made necessary by such an argument. In Britain what started as an attempt by architects to control their own sphere of operations, developed into a device by which local and central government could control and direct large sections of the building industry. Architects had been reduced, as Gropius himself had indicated, to little more than 'organizers of the building economy' (Wilhelm 1979) constantly trying to interpret the requirements of consumers who had become increasingly remote. With the rise of the modern movement the traditional role of the architect became reinterpreted within the framework of a new ideology. Watkin describes this as:

> 'the morally insinuating and widely disseminated argument that modern architecture exercises some special unassailable claim over us since it is not a "style" which we are free to like or dislike as we choose, but is the expression of some unchallengeable "need" or requirement inherent in the twentieth century with which we must conform.' (Watkin 1977)

Many of the apologists for building systems have made such a case as they have argued for more prediction, more control. In its embrace of the imperatives of industrialization the argument is indeed a moral one. At the present time such arguments are found at their most strident in the so called developing countries struggling to industrialize. It is in such countries that we often find proposals to use the most sophisticated and complex technologies to alleviate housing and other problems. If similar proposals have met with such implementation difficulties in highly industrialized western Europe what hope of success can such methods expect in those countries without such a technological background? There is behind it all, the drive to create new markets for products and the symbolic demand for progress. In the very centralized economy of Britain we have seen how the building system approaches have come to grips with such issues, gradually building up their own markets, endeavouring to predict and control within an overall monetary context of some instability. In their study of the British Treasury, published in 1974, Heclo and Wildavsky point out that the 'entire Treasury sector overseeing public spending comprises only 28 people above the level of principal' and they clearly show just how closely controlled is the delegation of decision:

> 'It may be astonishing to learn that Treasury and Parliamentary approval of a department's estimates does not entitle the department to spend a single pound. But that is the case. In one of the few statutory references to Treasury power, the law states that audited expenditure "shall, unless sanctioned by the Treasury, be regarded as not properly chargeable to a

Parliamentary grant, and shall be so reported to the House of Commons". Despite the fact that the Treasury has already approved the estimate and Parliament voted the money, every department spending commitment, except where Treasury chooses to delegate its authority, must be, approved in advance.' (Heclo and Wildavsky 1974)

Subsequent to 1950 the Treasury had begun to look at the rolling programme idea with which broad financial agreements could be suggested over a five or ten year period. However such programmes were no buffer against the stop-go policy which has been the norm for many years in Britain. One of the major aims of the client sponsored building systems was to demonstrate to the Treasury that they could predict and control their area of operations and could, with sophisticated cost control, create a situation of confidence in the building industry. In this way a good case could be argued with the Treasury each year for continuing to provide the finance for the work of a given department. The Reports of two House of Commons Estimates Committees, in 1961 (on school building) and 1968 (on the Public Building Programme), gave encouragement to the consortia and to the development of standardization and industrialized methods, whilst the Treasury was wielding its traditional, short term, control over actual expenditure. One explanation for this seemingly schizophrenic approach is offered by Heclo and Wildavsky with regard to those nations with a written constitution which

'means that almost anything can be done (or at least justified), in Britain the unwritten constitution is read to mean that almost nothing can be done. Its principles are ethereal bodies unable to offer any positive guidance but always ready to descend on any change as a violation of their spirit. To summon these Harpies, you need only suggest something different.' (Heclo and Wildavsky 1974)

Whilst these comments do explain a great deal they also indicate some fundamental aspect of contemporary British culture and the ambiguous attitudes we hold to rationalization. In this way rationalization has been encouraged in the building industry, but at the same time kept at arms length and yet using the mechanisms of very centralized control at the point of application.

What are the effects of such contradictory attitudes to the use of public money? With justification it might be thought that such a level of control would lead to very direct accountability. That this was not the case was revealed in December 1975 in a report of the House of Commons Expenditure Committee which stated that:

'The Treasury's present methods of controlling public expenditure are inadequate in the sense that money can be spent on a scale which was not contemplated when the relevant policies were decided upon' (*Guardian* 1975)

The committee pointed out that in 1974–5 a sum equivalent to 5% of the total United Kingdom gross domestic product had been spent outside announced government policy changes. The measures to control expenditure at national and local government level which followed this revelation of fiscal confusion caused the *Times* leader writer to point out that 1976 was 'the first year in which it has been possible to gain any reliable inkling of whether or not local budgets are being exceeded in the current year' (*Times* 1976). In such a climate it is not difficult to see how such large sums of money could be put into developing and promoting building systems in the 1960s without any realistic appraisal of the claimed advantages.

In this study we have seen how, in Britain, during the last 35 years building system ideas have developed. From the early postwar proposals concerning component building, government was quick to encourage and absorb those aspects that would offer savings, with the Ministry of Education (now DES), the Ministry of Public Buildings and Works (now DOE), and the Department of Health and Social Security, all guiding their member groups along the road to rationalization. Many of the early innovators found themselves in senior positions able to encourage large public building programmes into using industrialized building systems.

Compared to most architectural endeavour, the theory that surrounded the system building idea was simple, clear and direct. After all everybody in the land was experiencing the consumer revolution with regard to household appliances, cars and electronics. The idea also had the attraction of being comprehensible to politicians and easily became translated into slogans. It seemed 'common sense' to assume that the same logic could be applied to building and with the same success. I have tried to indicate throughout the study that on almost every occasion the major attempts were less than successful — and precisely because building involves other factors than those with which 'mobile' consumables concern themselves. Furthermore to remake the world in the image of an industrialized consumer society is, at the very least, questionable. It is here that architects do carry a responsibility since they are involved with developing ideas about building, putting these into the market place, and attempting to convince people to carry them out. There are many models available to choose from but by the wholesale backing of a single set of ideas surrounding building systems architects have made themselves responsible for demonstrating a specific point of view. The pursuit of these ideas has led, as we have seen, to a vast range of building types and a large number of buildings being made in one particular way.

Although many architects involved in this process were against the expression of the will of any one individual (and against the 'prima donnas' of architecture), it is clear that the creation of many of the building systems was precisely on this basis. Indeed working within the ideology of system building such architects were able to be very much more influential than they would have been in single buildings. There is a simple relationship between such a set of ideas that prizes anonymity and an architecture that exhibits similar characteristics. It is perhaps, the architecture of bureaucracy.

Through all of this we still find the idea of 'total design' present. For most architects those buildings which have a consistency and coherence running through from the idea to the smallest detail are those to be admired, and indeed there is something impressive about such an act of will. This desire for a single gestalt is a characteristic of architectural education, and of most architects, with the result that they tend to seek the application of monolithic ideas of total design. The vision of a total architecture of standardization, of industrialized parts is one of these models. Systems terminology would characterize this as the creation of systems with a considerable degree of closure. Within the boundaries of such systems there is a degree of control, the whole can be understood. Similarly the philosophy can be seen at work in single buildings. Often those buildings that are particularly good at integrating all the technological aspects of building design also exhibit a large degree of control over their internal environment. Many of the buildings of Foster, Rogers and the followers of this school show these characteristics: there is a thin sealed envelope which is penetrated at a few clearly defined points. The insides of buildings conceived in this way become sanctuaries where there is total order, and indeed total architecture. Outside is the disorder of the environment at large.

Equally with the building systems, a large measure of order and control is sought within, and there is also a heavy control over inputs and outputs from the environment (Russell 1973; 1977). Whilst there are very clear practical reasons for proceeding in the way described, there is also an important symbolic component. Leach points out that 'when we use symbols (either verbal or non-verbal) to distinguish one class of things or actions from another we are creating artificial boundaries in a field which is "naturally" continuous' (Leach 1976). Where and how the boundaries are drawn is of vital significance since it is by this means that contrasts are made and meanings implied. In addition to this, when we draw boundaries we tend to value that which is inside more than that which is outside. Such a principle can be seen most clearly at work with regard to property boundaries. The result, then, of creating building systems with boundaries that have a high degree of closure is that only those ideas and things within the boundary are thought important because that is what the system has decided to concentrate its attention upon.

It is easy to see how system sanctuaries of this sort can imply that activities that go on outside their boundary are less relevant, or even totally valueless. In terms of the Rubbish Theory of the social anthropologist Michael Thompson, all that outside a given boundary is rubbish to those inside (Thompson 1977; 1979). It is only with the impact of very powerful forces from their environments that such systems change. The dramatic collapse of a Ronan Point, the energy problem, or the impact of many fires on system built schools are all examples of such forces. Under such pressures the role of these closed systems is being re-evaluated. The continuing demands on resources also suggests that even for Britain such monolithic approaches tied to complex programming may not be the answer. Much more responsive, less centralized mechanisms are required in a rapidly changing situation. In his Reith lectures of 1970 Donald Schon pointed out that the stable state was an ideal long since unattainable and what we had to recognize was the inevitability of rapid change (Schon 1971). In such a situation cumbersome organizations tied into complex bureaucratic procedures may not be the best type of institution. Indeed it has been found in recent years that, the more sophisticated such institutions become, the less able they seem to deliver the goods. In this context Connery (1966) makes an interesting observation on a country that has gone further along this road than many others:

'The Swedish state has been involved in housing construction and regulation ever since World War 1. The government insists that it has had to exercise housing controls because the high cost of land, materials and labor would mean high rents in a free market, and because the manpower shortage forces it to try to allocate labor on some priority basis. What it boils down to, however, is a surplus of ˚state regulation and a shortage of housing.'

In Britain, most of the housing systems of the sixties have fallen by the wayside. The schools consortia with much diminished programmes have had a struggle, some like SEAC and MACE have been disbanded. The time is ripe to reassess the situation. Architects have been held by a specific notion of system. Such a notion has grown with the modern movement itself and become deeply embedded in the fabric of architectural ideas. This notion of system has been inextricably bound up with industrialization and the responses of those theorists of the early part of the twentieth century — the conventions emerging from this are now the common currency of much architectural practice. The single building seen as a set of integrated technological systems is the one still most understood by architects, and the one which has offered the most successful results. Where architects have attempted to apply their notion of system more widely it has often had its root in such a view, which has at best met difficulties and at worst hostility and physical failure. However, it has been shown that the notion of a system as merely a technological tool-kit is exceedingly limited and that the real value of the systems idea is in making explicit the nature of all the interacting systems in a given situation. It is clear then, that if architects are making larger claims for the idea of system than they understand, it will founder. Indeed it can be seen that such has frequently been the case, with ideas generated within the systems context being used and misused when placed in a social and political context.

The manner in which technical systems interact with social systems has already given rise to an extensive area of studies dubbed 'sociotech' (Emery and Trist 1960) although architecture has remained remarkably aloof from such developments. Nevertheless the attempts over recent years to establish user response before, during and after design is a recognition of the implications of such interacting systems. Many of the best architects of the past and present understand intuitively the nature of such interacting forces and have little need for the explicitness that seems to obsess the systems designers. However, intuition is but 'tuition inside' and the need to explain how such success may be achieved is one of the demands made by democracies and by the education systems set up by such societies. Nevertheless, intuitive or explicit, there has to be a recognition by architects of the interacting forces in an open system. Using this open system in a responsive and responsible way can offer new and original results as well as drawing sensitively on tradition.

We have seen how the work of, say, Foster Associates, has used the idea of surveying the market and integrating new and developing technologies. Other examples may be found in the Maximum Space House of John Hix who after examining what was industrially available proposed a modified commercial greenhouse as an economic space and efficient space enclosure (Hix 1970; 1974). Walter Segal has also indicated another possible approach, this time using largely timber. Segal has brought together some of the traditions of the modern movement with the pragmatic approach of the timber frame. Unlike most system builders in Britain he has accepted the sizes of materials as they are and designed his approach around them. His lifelong interest in history, building and people has allowed him to develop an approach that is responsive and controlled. He also emphasizes that he takes personal responsibility for his actions and demonstrated this by carrying no indemnity insurance. The few commercial systems that have survived have done so by being very responsive to what people actually want. Guildway, who use a timber-frame house system, do not seem to be hampered by the truth to materials morality of the early modern movement, with the result that the external appearance does not look system built and can also be adapted to a given locale or requirement. Kingsworthy Construction on the other hand has developed from the early systems ideas into a frame with a dry envelope and a variable exterior. However Conder, who market Kingsworthy, have long since realized that the components and subsystems are a minor part of the open system and they have developed a very strong management contracting approach to attempt to solve some of the difficulties inherent in the traditional split between design and construction. The CLASP Research into Site Management Programme, discussed elsewhere, is a similar attempt in the local government field to cut through the traditional divisions. In a different way Alexander has offered a number of interesting propositions concerning an open systems approach. The competition entry by Alexander and his Centre for Environmental Structure for the Lima housing competition of 1969 demonstrated a way of drawing out existing cultural and technological criteria and using them in an innovative design around a set of 'patterns'. A paper of 1970 by

Alexander 'Changes in Form', showed how the early idea of a pattern, which has three parts, works.

The first states the pattern itself, the second is a brief summary of the problem solved by the pattern. The third is a collection of brief refutable hypotheses which can offer some test of the pattern (Alexander 1970). This idea ultimately developed into the ambitious work of Alexander and his colleagues at Berkeley published in 1977, *A Pattern Language* and its two associated volumes *The Timeless Way of Building* (1979) and *The Oregon Experiment* (1975). Alexander's view, put forward in the introduction is that

> 'Towns and buildings will not be able to become alive, unless they are made by all the people in society, and unless these people share a common pattern language, within which to make these buildings, and unless this common pattern language is alive itself.' (Alexander 1977)

It is perhaps significant that Christopher Alexander, who wrote in 1959 a paper entitled 'Perception and Modular Co-ordination' should, in 1977, produce a book *A Pattern Language* in which he sets out a series of patterns which attempt to describe the qualities of regions, towns, buildings and construction. Each of the patterns, says Alexander, describes a recurring problem in our environment and the solutions offered can be used again and again without being identical. Each of the patterns shows a picture and gives the context of the problem. Then follows a brief summarizing statement, for example:

> 'Everybody loves window seats, bay windows and big windows with low sills and comfortable chairs drawn up to them' (Pattern 180, Window Place)

The background of the pattern is then described, followed by the solution, 'the heart of the pattern' which, in the example given is:

> 'In every room where you spend any length of time during the day, make at least one window into a window place.'

Alexander describes how the patterns relate to other patterns, on a hierarchical basis, in sum giving us *A Pattern Language*. Some commentators (Ward 1979) see the work as showing an important new direction for all people, not only trained designers, to design their own environments. Others, like Rabeneck (1979) and Broadbent (1980, 1979) have been more critical of its all embracing claims. The appearance of the book — a cross between a works manual, the Bible and the I Ching shows a welcome recognition that many specialist texts limit the audience by their form. Such an attitude permeates the work, where desirable qualities (to Alexander and his team) are clearly described by the pattern titles: 18, Network of learning; 61, Small public squares; 121, Path shape; 204, Secret place; 225, Frames as thickened edges; 251, Different chairs. If, as we are led to believe, language is central to our perception of the world, such an approach can show an alternative to the bleak, mechanistic prose which has characterized many of the documents and books which attempt to improve design, and in so doing it can cause us to rethink and reform how we make buildings and places.

The strengths of the work are also weaknesses for, in drawing on particular aspects of traditional building it suggests a somewhat romanticized notion of the world whilst at the same time carrying a suggestion of utopianism, the one way to proceed. However, such an approach does indicate a more human based attitude, and one which embodies a much more open set of possibilities than many of the prescriptions offered in recent years.

The cult of efficiency has permeated attempts to create building systems and, as we have seen, it is usually a very machine based efficiency that is being sought — efficiency in construction and efficiency in appearance. The way in which this has influenced building systems has been made evident. The way in which it has effected the planning of buildings is neatly summarized by Banham:

> 'when Gropius was thinking about the Bauhaus teaching programme he thought of it in terms of neat rectangular rooms or drew rectangles and circles connected by long straight lines, like the circulation diagram of a Hertfordshire school. Once you started to think about the programme of the building you were committed to a set of symbolic forms.' (quoted in Watkin 1977)

The appropriately named 'International Style' became the only available model, and with it a particular attitude to the new technologies. Hannes Meyer had, in 1926, stated the argument clearly:

> 'Constructive form is not peculiar to any country; it is cosmopolitan and the expression of an international philosophy of building.' (in Schnaidt 1965)

Fuller also frequently drew attention to the changed circumstances that had been brought about by the increasing facility and speed of transportation. This saw its clearest statement in 'Designing a New Industry' published in 1946:

> 'Even at 300 cubic feet we can get eight packaged houses into a freight car and we can ship by rail to the seaboard — the farthermost point in the United States from Wichita — for $75 a house. We can ship economically to any place in the world, because when we get to seaboard the ocean rates are so cheap we can ship to any place in the world for a few hundred dollars total from Wichita.' (Meller 1972)

Thus the worldwide industrialized vernacular became the common currency of much architectural language, and a key part of the philosophy of most building systems. But do building systems have to be thought of in this way? The systems notion has been examined sufficiently in previous chapters to see that looked at properly it suggests no such thing. An appropriate response would be to look at the whole system of what was available. This would include an examination of existing methods and possible new ones; traditional tried and trusted craft solutions and new organizational configurations. In short it would involve an amalgam of subsystems appropriate to a particular time and place. Such approaches are usually ignored with the international model, often with very destructive results, for the enormous strength and possibilities of industrialization have been allied to specific solutions. These solutions pay homage to repetition and standardization in their appearance, yet this appearance has been created by an idea, not a process: the metaphor of industrialization.

Yet industrialization does not have to be seen in this way, as is evident from some of the examples already discussed, notably the timber-frame house. A more potent example, and one that uses steel, is offered by Robert Bruno, a sculpture (and one time Assistant Professor of Architecture at Texas Tech University) who has created a structure entirely fashioned from the offcuts from steel mills (**12.2.1**).

12.2.1 House at Ransom Canyon, Lubbock, Texas, USA. Robert Bruno, commenced 1974. View of house from Canyon floor. (Photo 1979)

12.2.2 House at Ransom Canyon, 12 m high, 27 m long and 18 m wide there is some 200 sq metres of interior space, largely on one floor. Access is from the road on the right and, ultimately from the area beneath, by means of an elevator up one of the legs. Finishes to be glassfibre and hardwood with stained glass windows. The crane was installed in 1978. Bruno in centre. (Photo 1979)

(Russell 1980). Singlehanded he has created an elegant house which, he believes, demonstrates that the use of industrialized materials can offer enormous creative possibilities if we are prepared to take them (**12.2.2**). Bruno (1980) points out that 'some buildings that cost millions of dollars and thousands of man hours still look as if no-one has made them'. His personal experiment was to allow the building to grow and change as he worked on it: the result is a beautiful expression of technology and human skill, with the welded steel plates moving in subtle ways and having the character of a human hand (**12.2.3**). About his approach Bruno says:

'Spending not just months but years on the site, feeling the changes from hour to hour, season to season, becoming familiar with the mood of this place, is in no way similar to studying site plans and climatic charts. Being constantly affected by this environment, one responds accordingly. I do all the physical work myself which has several advantages (**12.2.4**). In separating the designer from the builder, we deprive the designer of what the building teaches the builder — both as structure and as process. There is another more subtle problem in this separation. It is that when we decide what work is worth while for someone else to do, we are apt to be more careless than if we had to do the

710

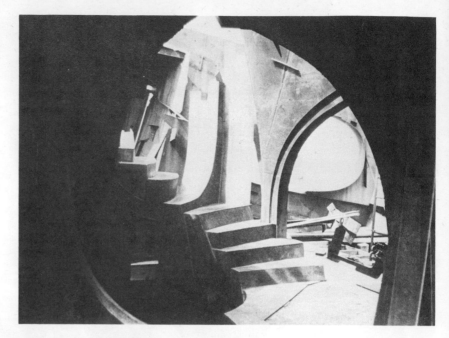

12.2.3 House at Ransom Canyon. The 6 mm offcuts are tack welded and flexed to create undulating convex and concave shapes to form an intricate surface. (Photo 1979)

12.2.4 House at Ransom Canyon. The hand tools used by Robert Bruno. (Photo 1979)

work ourselves; we are apt to delegate trivial work and degrade the craftsman's participation.' (Bruno 1980)

It is in the nature of Bruno's house that it is not universally applicable, and in presenting his work in Mexico City (1979) he was concerned that it may seem 'extravagant and selfish'. The enthusiastic response it received suggested that it had clearly struck an important chord which has to do with man's aspirations. His attitudes certainly have a great deal to teach us concerning the use of the products of industrialization in an organic, not mechanistic manner — the latter being the approach we have come to expect from the modern movement. Even so it must be recognized that many of the valuable lessons of the modern movement should be drawn on, in common with any other body of useful ideas. A simple view which describes architects as either manipulators or as the manipulated is no useful contribution to the discussion. Watkin refers to Taut in whose work we see 'a shift from the belief in architecture as the tail wagged by the ideological dog, to architecture as a central agent in the renewal of society' (Watkin 1977).

The model offered to large parts of the world by the international industrialized vernacular is surely very inappropriate. If we set out the two extremes of a continuum with regional or local approaches at one end and international at the other the argument may become clearer (Russell 1979). Put this simply, we can all recognize some of the characteristics. Architecture that is at the regional end will be place dependent, rooted in the culture, related to the site and place, responsive to climate and custom, and will draw on local technologies. Architecture which is at the international end of the spectrum will be place independent, independent of local culture, free of the site (including surrounding buildings), independent of climate and will employ an international version of high technology — or try to look as if it is (**12.2.5**). Although regional characteristics are frequently absorbed to the international end, they become transformed in the process. Many good buildings fall in the mid area of the spectrum since they attempt to reconcile the demands of context with the larger world of architectural ideas.

Rather than accept the glib dictates of their professional subcultural conventions, architects can make a more appropriate response by a judicious blend of their broad knowledge as professionals, and the

12.2.5 The Regional-International spectrum

REGIONAL   INTERNATIONAL

←   →

ROOTED IN LOCAL CULTURE   INDEPENDENT OF LOCAL CULTURE
PLACE DEPENDENT   PLACE INDEPENDENT
RELATED TO SITE   FREE OF THE SITE
RESPONSIVE TO CLIMATE   IGNORES CLIMATE
USES LOCAL TECHNOLOGY   USES HIGH TECHNOLOGY
                                 (OR MADE TO LOOK AS IF IT DOES)

relevant regional characteristics. Clearly, long obsolete methods cannot be recreated but a true systems approach would surely expect a close examination of the available systems whatever their type, and the options for growth, improvement and change in an evolutionary way. The easiest way to describe the places that we like is by their qualities, for these comprise both the architecture and our experience of it. It is necessary, at this point in time, to draw to the attention of many involved in the making and commissioning of buildings that the creation of such qualities is both possible and necessary. Qualities cannot be created merely by the use of the functional mechanics of planning, nor by the blind use of the machine aesthetic, and certainly not by the gentle coercion of institutionalized central government. They are questions about which individuals must make responsible decisions if we are to create acceptable places in which to carry on our lives.

Building systems have been seen by many as a useful device for bringing order into an unruly industry, and by some as a means by which dirty commerce could be eliminated. In this sense the implementation of many of the ideas were covertly political. Some spoke of a national building system, some of the way it related to the nationalization of the building industry. For those obsessed with rationalization at all costs, such an outcome seemed logical as indeed, in some quarters, it still does. However, serious questions have to be asked about how far down such a road a nation should travel. In the end such questions are of cultural significance and are at the same time intensely personal. 'Humanism', said Geoffrey Scott in 1914, 'has two enemies — chaos and inhuman order'.

Seen only as a panacea, whether by architect, administrator or politician, the building systems idea is of little value. Indeed it may, like many another system, be dangerous in its propensity for removing the sense of responsibility from the individual. Identifying appropriate systems may promise new design possibilities. Such possibilities must be tolerant of human behaviour. For too long architects have sought an architecture of 'fit'. What is required is an architecture of tolerance.

BIBLIOGRAPHY

The Bibliography combines references to books, government reports, papers, journal articles and journals. For material generally the policy has been to cite the author where named; where no author is cited the reference is listed under the journal title or organization. Where two dates are bracketed together, the first refers to the edition consulted and the second to the date when the work was first published. In the text the following abbreviations have been used to simplify referencing:

AD	*Architectural Design*
ABN	*The Architect and Building News*
AF	*Architectural Forum*
AJ	*The Architects' Journal*
AR	*The Architectural Review*
BAC	British Aluminium Company
BD	*Building Design*
BG	*Building*
BRE	Building Research Establishment
BRS	Building Research Station
BSI	British Standards Institution
CLASP	Consortium of Local Authorities Special Programme
DES	Department of Education and Science
EFL	Educational Facilities Laboratory
EPA	European Productivity Agency
GLC	Greater London Council
MHLG	Ministry of Housing and Local Government
MOE	Ministry of Education
MOH	Ministry of Health
MPBW	Ministry of Public Buildings and Works
MQ	*Modular Quarterly*

RIBAJ *The Journal of the Royal Institute of British Architects*
SCOLA Second Consortium of Local Authorities
TDA Timber Development Association
UGC University Grants Committee

Abel, C. (1969). Ditching the Dinosaur Sanctuary, in *Architectural Design,* Aug., pp. 419-24.

Ackerman, J.S. (1977, 1966). *Palladio,* Penguin, Harmondsworth.

Adams, G. (1972). Fast Growing Market for Solar Control Glass, in *Building Specification,* July, p. 47.

Alexander, C. (1959). Perception and Modular Co-ordination, in *RIBAJ,* Oct., pp. 425-29.

Alexander, C. (1964). *Notes on the Synthesis of Form,* Harvard University Press, Harvard.

Alexander, C. (1968). Systems Generating Systems, in *Architectural Design,* December, pp. 605-10.

Alexander, C. (1970). Changes in Form, in *Architectural Design,* pp. 122-25.

Alexander, C., Silverstein, M., Angel, S., Ishikawa, S. and Abrams, A. (1975). *The Oregon Experiment,* Oxford University Press, New York.

Alexander, C., Ishikawa, S., Silverstein, M., with Jacobson, M., Fiksdahl-King, I. and Angel, S. (1977). *A Pattern Language,* Oxford University Press, New York.

Alexander, C. (1979). *The Timeless Way of Building,* Oxford University Press, New York.

Angyal, A. (1967, 1941). *Foundations for a Science of Personalities,* Harvard University Press, Harvard.

Anthony, H. (1945). *Houses: Permanence and Prefabrication,* Pleiades Books, London.

Apollonio, U. (ed.) (1973, 1970). *Futurist Manifestos,* Thames and Hudson, London.

Architect and Building News (1949). User Requirements in School Design, 30 September, pp. 319-26.

Architects' Journal (1954). Housing Scheme in Bentham Road, Hackney, 3 June, pp. 676-81.

Architects' Journal (1958). Privately Built Housing at Ham Common, Surrey, 17 April, pp. 577-82.

Architects' Journal (1959). CLASP: Consortium of Local Authorities' Special Programme, 30 April, pp. 645-50.

Architects' Journal (1961a). Crosswall School, 27 April, p. 610.

Architects' Journal (1961b). The Way Forward, 29 June, pp. 941-60.

Architects' Journal (1961c). School Building: Estimates Committee Urges Extension of CLASP, 30 August, p. 296.

Architects' Journal (1962). The Bison Wall Frame System, 1 August, pp. 262-64.

Architects' Journal (1964). A Smear On Consortia (editorial), p. 661, and subsequent letters on 2.12.64, p. 1290, 16.12.64, p. 1419, 3.2.65, p. 280.

Architects Journal (1966a). House at Highgate, 23 March, pp. 763-69.

Architects' Journal (1966b). The System Built Environment, IBSAC Exhibition Issue, 4 May, pp. 1110-11.

Architects' Journal (1967). Disciple Becomes a Prophet, 22 November, p. 1285.

Architects' Journal (1969a). Fires in CLASP, 4 April, p. 787.

Architects' Journal (1969b). Performance Specification for Building Components, 25 June, pp. 1705-14.

Architects' Journal (1970). Walter Segal's Houses, 30 September, pp. 769-80.

Architects' Journal (1971). Thamesmead: Window and Roof Failures, 4 August, p. 234.

Architects' Journal (1972). Letter, 12 January.

Architectural Design (1949). Prefabricated Unit System of School Construction, designed by Arcon, January.

Architectural Design (1965). SCSD Project USA: School Construction Systems Development, July, pp. 324-39.

Architectural Design (1966). Eames Celebration, September.

Architectural Design (1967). School Construction Systems Development, November, pp. 495-506.

Architectural Forum (1949). The factory built house is here, but not the answer to the $33 million question: How to get it to market? No. 90, May, pp. 107-14.

Architectural Review (1956). 12-Foot Frontage Terrace Houses in Hampstead, November, pp. 288-94.

Arcon (1948). Prefabricated Unit System of School Construction, in *The Builder,* 23 July.

Arcon (1949). User Requirements in School Design, in *The Architect and Building News,* 30 September, pp. 319-26.

Arnold, C. (1970). The 'Classic' Systems Model, in *Building Research,* July/December, pp. 6-16.

Arnold, C. (1977). The Apple Pie Building System: A Parable for the 1970's, in *Beyond the Performance Concept,* National Bureau of Standards (US).

Arnold, C., Rabeneck, A. and Brindle, D. (1971). Building Systems Development, in *Architectural Design,* November, pp. 679-702.

Asimow, M. (1962). *Introduction to Design,* Prentice Hall, Englewood Cliffs.

Association of Building Technicians (1946). *Homes for the People,* Paul Elek, London.

Balency-Bearn, M.A. (1964). *The Evolution of Industrialisation in Building: A Case Study of Prefabrication with Large Units,* Balency Schuhl, Paris.

Balfour, A. (1970). *Portsmouth,* Studio Vista, London.

Balfour, A. (1978). *Rockefeller Center: Architecture as Theatre,* McGraw Hill, New York.

Baljeu, J. (1974). *Theo Van Doesburg,* Studio Vista, London.

Banham, R. (1957). Futurism and Modern Architecture. Paper read at the RIBA on 8 January 1957, in *RIBA Journal,* February, pp. 129-39.

Banham, R. (1960). *Theory and Design in the First Machine Age,* Architectural Press, London.

Banham, R. (1962a). The Thin, Bent Detail, in *Architectural Review,* Vol. 131, No. 782, April, pp. 249-52.

Banham, R. (1962b). Ill-Met by Clip-Joint, in *Architectural Review,* Vol. 131, No. 783, May, pp. 349-52.

Banham, R. (1966). Consortium Scheme, in *New Society,* 6 January. Banham, R. (1969). *The Architecture of the Well-Tempered Environment,* Architectural Press, London.

Banham, R. (1971a). Bennett's Leviathan, in *New Society,* 8 April, pp. 594-95.

Banham, R. (1971b). The Master Builders, *The Sunday Times,* 8 August, London.

Banham, R. (1971c). Rudolph Schindler: Pioneering Without Tears, in

Architectural Design, December, pp. 577-79.

Bann, S. (ed.) (1974). *The Documents of 20th Century Art: The Tradition of Constructivism,* Thames and Hudson, London.

Bannister, T. (1950). The First Iron-Framed Buildings, in *Architectural Review,* April, pp. 231-46.

Barr, C. and Carter, J. (1965). How to Appraise Systems, in *RIBAJ,* November, pp. 534-39.

Beecher, C.E. and Beecher Stowe, H. (1971, 1869). *The American Woman's Home, or Principles of Domestic Science,* Reprint Arno Press, and The New York Times.

Beer, S. (1965). The World, The Flesh and The Metal, in *Nature,* 16 January, pp. 223-31.

Beishon, J. and Peters, G. (1976, 1972). *Systems Behaviour.* The Open University Press and Harper and Row, London.

Bemis, A.F. and Burchard, J. (1933). *The Evolving House, Vol. I: A History of the Home,* The Technology Press, MIT, Cambridge, Mass. This volume largely the work of Burchard.

Bemis, A.F. (1934). *The Evolving House, Vol. II,* The Technology Press, MIT, Cambridge, Mass.

Bemis, A. F. (1936). *The Evolving House, Vol. III: Rational Design,* The Technology Press, MIT, Cambridge, Mass.

Bender, R. (1973). *A Crack in the Rear-View Mirror: A View of Industrialised Building,* Van Nostrand Reinhold, New York.

Benson, A.C., Jones, F.M. and Vaughan, J.E. (1963). Early Example of Prefabrication. Letter in *RIBAJ,* June, p. 251.

Bentley, E. (1955, 1946). *The Playwright as Thinker,* Meridian, New York.

Benton, T., Benton, C. and Sharp, D. (1975). *Form and Function,* Crosby Lockwood Staples, London.

Berry, F. (1974). *Housing: The Great British Failure,* Charles Knight, London.

Bertalanffy, L. von (1956). General System Theory, in *General Systems Yearbook* Vol. 1., pp. 1-10.

Bertalanffy, L. von (1968). *General System Theory,* Braziller, New York.

Blake, P. (1964). *God's Own Junkyard,* Holt, Rinehart and Winston, New York.

Boccioni, Balla, Carra, Russolo and Severini (1972, 1910). Technical Manifesto of Futurist Painting, 11 April 1910, in *Futurismo, 1909-1919,* Catalogue to Exhibition at Royal Academy of Arts, London, Northern Arts and Scottish Arts Council, UK (1972).

Boccioni (1972, 1912). Manifesto of Futurist Sculpture, 11 April 1912, in *Futurismo, 1909-1919,* Catalogue to Exhibition at Royal Academy of Arts, London, Northern Arts and Scottish Arts Council, UK (1972).

Boesiger, W. (1960). *Le Corbusier: 1910-60,* Girsberger, Zurich.

Boesiger, W. and Stonorow, O. (1964). Le Corbusier and Pierre Jeanneret, *The Complete Architectural Works, Vol. I, 1910-29,* Thames and Hudson, London.

Boice, J.R. (n.d. *c.*1970?). *A History and Evaluation of the SCSD Project 1961-67,* Building Systems Information Clearing House and Educational Facilities Laboratory, Menlo Park, Cal.

Bonta, J.P. (1975). *Mies van der Rohe, Barcelona 1929,* Gili, Barcelona.

Bonta, J.P. (1979). *Architecture and Its Interpretation,* Lund Humphries, London.

Borges, J.L. (1971, 1964). *Labyrinths,* Penguin, Harmondsworth.

Boudon, P. (1969). *Lived-in Architecture,* Lund Humphries, London.

Boulding, K.E. (1956). *The Image*, University of Michigan Press, Michigan.

Braun, E. (ed.) (1969). *Meyerhold on Theatre*, Methuen, London.

Brawne, M. (1966). The Wit of Technology, in *Architectural Design*, September, pp. 449-57.

Bristol Aeroplane Company (BAC) (1948). Weston at Work in Bristol, in *Bristol Review, Christmas*, BAC, Bristol.

Bristol Aeroplane Company (BAC) (1951). The Largest Aluminium School Yet Built, in *Bristol Review, Spring*, BAC, Bristol.

British Standards Institution (1966). *Recommendations for the Co-ordination of Dimensions in Building - Basic Sizes for Building Components and Assemblies, BS 4011*, Her Majesty's Stationery Office, London.

British Standards Institution (1968). *Recommendations for the Co-ordination of Dimensions in Building: Controlling Dimensions (Metric Units)*, Her Majesty's Stationery Office, London.

Broadbent, G. (1969). Meaning Into Architecture, in *Meaning in Architecture*, Jencks and Baird, 1969.

Broadbent, G. (1973). *Design in Architecture*, John Wiley and Sons, Ltd, London.

Broadbent, G. (1980, 1979). Review of 'A Pattern Language' in *Design Studies*, Vol. I, No. 4, April, pp. 252-3. First published AAQ Vol. II, No. 4, 1979.

Broadbent, G., Bunt, R. and Jencks, C. (1980). *Signs, Symbols and Architecture*, Wiley, Chichester.

Brooks, H.A. (1972). *The Prairie School: Frank Lloyd Wright and his Mid-West Contemporaries*, University of Toronto Press, Toronto.

Bruno, B. (1980). Unpublished communication to author.

Brutton, M. (1971). Thamesmead Costs Soar as Delays Cause Chaos, in *Building Design*, 29 October.

Brutton, M. (1973). Missing the Point, in *Building Design*, 19 January.

Bryan, E.R. (1973). *Stressed Skin Roof Decks for SEAC and CLASP Building Systems*, CONSTRADO (Constructional Steel Research and Development Organisation), March.

Builder (1943). Experimental Houses at Coventry, 1 October, pp. 271-75.

Builder (1945). Experimental Houses at Coventry, 2 November, pp. 346-47.

Building (1969). School System: Sophisticated Package within DES Cost Limits. 31 October, pp. 83-8.

Building Design (1976). The Frightening Cost of Failures, 26 March.

Building Design (1978). Thamesmead Housing to be Given a £400,000 Face-lift, 15 September.

Building Design (1979a). Troubled Systems, 5 January, p. 7.

Building Design (1979b). 5M Discussions, 12 January.

Building Design (1979c). Repair Shock for Troubled Systems, 6 June.

Building Research Station (1967). Joints Between Concrete Wall Panels: Open Drained Joints, *Digest 85* (second series), August, HMSO, London.

Building Specification (1971). Eastergate Primary School, near Chichester, Sussex, June, pp. 34-6.

Building Systems Information Clearing House. Educational Facilities Laboratory (Staff Boice, J.R. and Burns, J.A.) (1972). Research Report No. 2 Evaluation: Two Studies of SCSD Schools, September.

Burt, G. (1944). *House Construction: Post-War Building Studies No. 1*, Her Majesty's Stationery Office, London.

Cannon, W.B. (1939). *The Wisdom of the Body*, W.W. Norton, New York.

Carter, H. (1924). *The New Theatre and Cinema of Soviet Russia*, Chapman and Dodd, London.

Carter, J. (1968). Nottinghamshire Builds: Project Research into Site Management, in *Architects' Journal*, 11 December, pp. 1389-96.

Cartmell, G.W. (1965). SCSD: Californian Schools Development Project, in *RIBAJ*, August, pp. 409-15.

Caudill, W.W. (1971). *Architecture by Team*, Van Nostrand Reinhold, New York.

Chadwick, G. (1961). *The Works of Sir Joseph Paxton 1803-1865*, Architectural Press, London.

Chipp, H.B. (1968). *Theories of Modern Art*, University of California Press, Berkeley.

Churchman, C.W. (1968). *Challenge to Reason*, McGraw Hill, New York.

CLASP (1966). *Annual Report*, Consortium of Local Authorities' Special Programme, Nottinghamshire County Council.

CLASP (1972). *Annual Report*, Consortium of Local Authorities' Special Programme, Nottinghamshire County Council.

CLASP (1975). *18th Annual Report*, Consortium of Local Authorities' Special Programme, Nottinghamshire County Council.

Cohen, P.S. (1969). Theories of Myth, London School of Economics, Malinovski Lecture, 8 May, in *Man*, September, pp. 337-53.

Collins, P. (1959). *Concrete: The Vision of a New Architecture*, Faber and Faber, London.

Collins, P. (1965). *Changing Ideals in Modern Architecture*, Faber and Faber, London.

Connery, D. (1966). *The Scandinavians*, Eyre and Spottiswood, London.

Corker, E. and Diprose, A. (1963). *Modular Primer*, The Modular Society, London. First published as a Technical Study in *Architects' Journal*, 1 August 1962.

Crawford, D. (ed.) (1975). *A Decade of British Housing 1963-73*, Architectural Press, London and Halsted Press, New York.

Crossman, R. (1975). The Crossman Diaries, *The Sunday Times*, 23 March and 2 February.

Cust, E. (1967). Theories Go Into Practice: Site Management Experiments with CLASP Mark IV, in *Architects' Journal*, 4 October.

Cutler, L.S. and Cutler, S.S. (1974). *Handbook of Housing Systems for Designers and Developers*, Van Nostrand Reinhold, New York.

Daley, J. (1973). The Legacy of Futurism, in *RIBAJ*, March.

Daltry, C.D. and Crawshaw, D.T. (1973). *Working Drawings in Use, June*. Building Research Station, Garston.

Davidson, J.W. (1967). S.F.I., in *Architectural Design*, March, pp. 138-40.

Davies, P. (1972). *The Battle of Trafalgar*, Pan Books, London.

Davies, R.L. and Weeks, J.R. (1952). The Hertfordshire Achievement, in *Architectural Review*, April, pp. 366-87.

Davison, P.L. *et al.* (1970). Research in Action: An Integrated Design Study Applied to Schools Development, in *Architectural Research and Teaching*, Vol. 1., No. 2., November.

Davison, P.L. *et al.* (1971). *Integrated Environmental Design*, December, Gloucester County Council.

Deeson, A.F. (ed.) (1964). *The Comprehensive Industrialised Building Systems Annual 1965*, House Publications, London.

Department of Education and Science (1964). *Controlling Dimensions for Educational Building*, Her Majesty's Stationery Office, London.

Department of Education and Science (1969). *Architects and Building Branch Design Note 3*. Demonstration Rig: Component Fixing Conventions, Her Majesty's Stationery Office, London.

Department of Education and Science and University Grants Committee (1970). *CLASP/JDP: The Development of a Building System for Higher Education,* Building Bulletin No. 45, Her Majesty's Stationery Office, London.

Department of Education and Science and Local Education Authority Consortia (1974?a). *Local Education Authority Consortia: Code of Practice for Obtaining and Administering Consortium Programme Nominations,* Her Majesty's Stationery Office, London (undated).

Department of Education and Science (1974?b). *Local Education Authority Consortia: Common Performance Standards for Building Components: Structure, External Envelope, Internal Subdivision,* Her Majesty's Stationery Office, London (undated).

Department of Education and Science (1975). *Fire and the Design of Schools,* Fifth Edition 1975, Building Bulletin No. 7, Her Majesty's Stationery Office, London.

Department of Education and Science (1976). *The Consortia,* Building Bulletin No. 54, Her Majesty's Stationery Office, London.

Department of Education and Science (1977). *Energy Conservation in Educational Buildings,* Building Bulletin No. 55, Her Majesty's Stationery Office, London.

Department of Health and Social Security (1965-6). *Industrialised Hospital Building Programme Compendium.* Her Majesty's Stationery Office, London.

Descartes, R. (1968, 1637). *Discourse on Method and Other Writings,* ed. by F.E. Sutcliffe, Penguin, Harmondsworth. The Discourse first published 1637.

Diamant, R.M.E. (1965). *Industrialised Building,* Iliffe, London.

Dickinson, Pound, Rostron, Stephens, Stuart and Moizer (1963). The NENK Method of Building: Dimensional Organisation, in *The Modular Quarterly,* Summer.

Diprose, A. (1966). Consortia and Development Groups I: Consortium for Method Building, in *Architects' Journal,* 13 July.

Diprose, A. (1970). First MACE, in *Architects' Journal,* 10 June, pp. 1418-21.

Donnison, D.V. (1967). *The Government of Housing,* Penguin, Harmondsworth.

Douglas, M. (1966). *Purity and Danger,* Routledge and Kegan Paul, London.

Douglas, M. (1970). *Natural Symbols,* Barrie and Rockliff, London.

Duncan, P. and Barr, A.W. (1955). Crosswalls, in *Architects' Journal,* 17 March, pp. 356-74.

Duval, D. (1972). An Appreciation of Walter Segal, unpublished Dissertation, School of Architecture, Portsmouth Polytechnic.

Eames, C. (1966). Annual Discourse, Royal Institute of British Architects 1959, in *Architectural Design,* September, p. 461.

Eaton, L.K. (1969). *Two Chicago Architects and Their Clients: Frank Lloyd Wright and Howard Van Doren Shaw,* The MIT Press, Cambridge, Mass.

Eden, J.F. (1967). Metrology and the Module, in *Architectural Design,* March, p. 148. Also published as a BRE Current Paper, March 1968.

Educational Facilities Laboratory (1965). *SCSD: An Interim Report,* EFL, New York.

Educational Facilities Laboratory (1967). *The Project and the Schools,* EFL, New York.

Educational Facilities Laboratory (1972). *Evaluation: Two Studies of SCSD Schools,* Research Report No.2, September, Building Systems Information

Clearing House and EFL, Menlo Park, Cal.

Ehrenkrantz, E. (1956). *The Modular Number Pattern,* Tiranti, London.

Ehrenkrantz, E., Arnold, C., Rabeneck, A., Meyer, W., Allen, W. and Habraken, J. (1977). *Beyond the Performance Concept,* Institute for Applied Technology, National Bureau of Standards, Washington.

Emery, F.E. (ed.) (1969). *Systems Thinking,* Penguin, Harmondsworth.

Emery, F.E. and Trist, E.L. (1960). Socio-Technical Systems, in Emery (1969), pp. 281-96.

Engel, H. (1964). *The Japanese House,* Charles E. Tuttle Co. Inc., Rutland, Vermont and Tokyo.

English, M. (Chairman) (1975). *Expenditure Committee Report,* General Sub-Committee of House of Commons General Expenditure Committee, December, Her Majesty's Stationery Office, London.

Estimates Committee (1968). *The Public Building Programme.* Fourth Report from the Estimates Committee Session 1967/68 of the House of Commons, 17 July, Her Majesty's Stationery Office, London.

European Productivity Agency (EPA) (1956). *Modular Co-ordination in Building,* Project No. 174, European Productivity Agency of Organisation for European Economic Co-operation, Paris.

European Productivity Agency (EPA) (1961). *Modular Co-ordination,* Second Report of EPA Project No. 174, European Productivity Agency of Organisation for European Economic Co-operation, Paris.

Farrell, T. and Grimshaw, N. (1978). Timber Frame Infill Housing, *RIBAJ,* October, pp. 413-20.

Feldman, P. (1973). Thamesmead House Costs are Eighty Percent Interest, in *Building Design,* 1 June.

Fleetwood, M. (1977). Alton Estate Revisited, in *Architects' Journal,* 30 March, pp. 593-603.

Ford, H. in collaboration with Crowther, S. (1928, 1922). *My Life and Work,* Heinemann, London.

Forssman, E. (1973). *Visible Harmony: Palladio's Villa Foscari at Malcontenta,* Sveriges Arkiteckturmuseum and Konsthögskolans Arkitecksturskole.

Foster Associates (1970). Foster Associates: Recent Work, in *Architectural Design,* May, pp. 235-58.

Foster Associates (1972). Foster Associates: Recent Work, in *Architectural Design,* November, pp. 686-701.

Foster, N. (1970). Exploring the Client's Range of Options, in *RIBAJ,* June, pp. 246-53.

Frampton, K. (1971). Notes on a Lost Avant-Garde, in *Catalogue to Exhibition Art in Revolution,* Hayward Gallery, 26 February to 18 April, Arts Council, London.

Frank, D. (1970). The Greatest Happiness, in *Architects' Journal,* 27 May.

Fraser, I. (1967). Methods of Construction — Concrete, Conference Papers *Industrialised Housing and the Architect,* 12/13 January, RIBA and National Building Agency.

Fuller, R.B. (1946). Designing a New Industry, in *The Buckminster Fuller Reader,* ed. Meller, J. (1970), Cape, London. Booklet originally published by Fuller Research Institute, Wichita, Kansas.

Fuller, R.B. (1958). Experimental Probing of Architectural Initiative, The RIBA Discourse 1958, in *RIBAJ,* October.

Fuller, R.B. (1962). Philosophy and Structure, in *Architects' Journal,* 5 December, pp. 1262-65.

Fuller, R.B. (1969). *Operating Manual for Spaceship Earth,* S. Illinois University Press. Carbondale.

Fuller, R.B. (1972). Interview with Michael Ben-Eli, in *Architectural Design,* December.

Gabo, N. (with Pevsner, A.) (1920). *The Realistic Manifesto,* reprinted in Chipp, 1969.

Galbraith, J.K. (1966). The New Industrial State: Control of Prices and People, The Third Reith Lecture, in *The Listener,* 1 December, pp. 793-812.

Galloway, P. (1980). Phase II Extensions to Frogmore School, in *Architects' Journal,* 12 March, pp. 524-27.

Gans, H.J. (1967). *The Levittowners,* Free Press, New York.

Gay, P. (1968). *Weimar Culture,* Secker and Warburg, London.

Gebhard, D. (1971). *Schindler,* Thames and Hudson, London.

Gebhard, D. and Von Breton, H. (1971). *Lloyd Wright, Architect: 20th Century Architecture in an Organic Exhibition,* California University Press, Santa Barbara.

George, C.S. (1968). *The History of Management Thought,* Prentice-Hall, Englewood Cliffs.

Giedion, S. (1967, 1941). *Space, Time and Architecture,* Harvard University Press.

Giedion, S. (1969, 1948). *Mechanisation Takes Command,* W.W. Norton, New York.

Gilbert, K.R. (1965). *The Portsmouth Block-Making Machinery,* Her Majesty's Stationery Office, London.

Giles, B., Wilson, B. and Mallon, R. (1974a). *Report on MACE,* Greater London Council Architecture Club.

Giles, B., Wilson, B. and Mallon, R. (1974b). Too Much MACE Bashing? Letter in *Architects' Journal,* 3 July.

Gloag, J. and Wornum, G. (1946). *House out of Factory,* George Allen and Unwin, London.

Goldstein, B. (1978). Foster Associates: Designing the Means to Social Ends, in *RIBAJ,* January, pp. 7-22.

Goodman, R. (1972). *After the Planners,* Penguin, Harmondsworth.

Goody, J.E. (1965). *New Architecture in Boston,* MIT Press.

Gordon, A. (1972). Long Life, Loose Fit, Low Energy, in *RIBAJ,* September, pp. 374-75.

Gordon, A. (1978). Self Motivation in Career Development, in *Training and Professional Career Development in the Building Professions. Post Conference Report,* 1-3 November, Institute of Advanced Architectural Studies, University of York.

Gray, C. (1962). *The Great Experiment: Russian Art 1863-1922,* Thames and Hudson, London.

Greater London Council (1973). *Eighty-five Years of Housing by L.C.C. and G.L.C.,* Architects' (undated) Brochure with exhibition of same name at Institute of Contemporary Arts, September, London.

Greater London Council Architecture Club (n.d. *c.* 1974?). Unsigned article in *Acid* 8, The Architecture Clubs News-sheet.

Greater London Council (1976). *Home Sweet Home,* Academy Editions and GLC.

Griffin, C.W. (1971). *Systems: An Approach to School Construction,* Educational Facilities Laboratory, New York.

Griffiths, H., Pugsley, A. and Saunders, O. (1968). *Report of the Inquiry into the Collapse of Flats at Ronan Point, Canning Town,* presented to MHLG, Her Majesty's Stationery Office, London.

Grinberg, D.I. (1977). *Housing in the Netherlands,* Delft University Press.

Gropius, W. (1934). The Formal and Technical Problems of Modern Architecture and Planning, in *RIBAJ*, May, pp. 679-94.

Gropius, W. (1955). *The Scope of Total Architecture*, Harper and Brothers, New York.

Gropius, W. (1961, 1910). *Programme for the Establishment of a Company for the Provision of Housing on Aesthetically Consistent Principles*, Memorandum to AEG, March 1910. First published in *Architectural Review*, July 1961.

Gropius, W. (1965, 1935). *The New Architecture and the Bauhaus*, Faber and Faber, London.

Guardian (1974). School Fire Danger Denied by CLASP, 4 January.

Guardian (1975). Methods are 'Inadequate' says Report, 20 December.

Gutheim, F. (1975). *Frank Lloyd Wright: In The Cause of Architecture*, Architectural Record Books, New York.

Guy, F. (1976). The Lessons of Cannock Lawn, in *Building Design*, 7 November.

Haber, S. (1964). *Efficiency and Uplift: Scientific Management in the Progressive Era 1890-1920*, Chicago University Press.

Habraken, N. (1972a). *Supports: An Alternative to Mass Housing*, Architectural Press, London.

Habraken, N. (1972b). Involving People in the Housing Process, in *RIBAJ*, November, pp. 469-79.

Hacker, M. (1967). R and D takes a Sea Change, in *Architects' Journal*, 22 November.

Hamdi, N., Wilkinson, N. and Evans, J. (1971). PSSHAK: Primary Support Structures and Housing Assembly Kits, in *RIBAJ*, October, pp. 434-35.

Handler, A.B. (1970). *Systems Approach to Architecture*, American Elsevier, New York.

Handlin, D. P. (1979). *The American Home: Architecture and Society, 1815-1915*, Little, Brown and Co., Boston and Toronto.

Hardy, A.C. and O'Sullivan, P.E. (1967a). *Building a Climate*, Electricity Council, London.

Hardy, A.C. and O'Sullivan, P.E. (1967b). *Insolation and Fenestration*, Oriel Press Ltd, Newcastle-upon-Tyne.

Hardy, A.C. and O'Sullivan, P.E. (1967c). *The Building: A Climatic Modifier, Heating and Ventilation for a Human Environment*, Institute of Mechanical Engineers, November, London.

Harnden, K. (1973). Joseph Paxton, Unpublished Mimeo, Portsmouth Polytechnic.

Harrison, H.W. (1969). *Performance Specifications for Building Components*, Building Research Station.

Harvey, J. (1972). *The Mediaeval Architect*, Wayland, London.

Heclo, H. and Wildavsky, A. (1974). *The Private Government of Public Money: Communities and Policy inside British Politics*, Macmillan, London.

Herbert, G. (1978). *Pioneers of Prefabrication: The British Contribution in the Nineteenth Century*, John Hopkins University Press, Baltimore and London.

Herrey, H. (1943). "At Last we have a Prefabrication System which enables Architects to Design any Type of Building with 3-Dimensional Modules", Wachsmann and Gropius, for General Panel Corporation, in *Pencil Points*, April.

Hertfordshire Architectural Department (1948). Planning the New Schools, in *The Architect and Building News*, 16 January, pp. 47-51.

Hills (West Bromwich) Ltd (n.d. *c.*1950s). *Hills Dry Building System Manual.*

Hislop, P. and Walker, C. (1970). *School Construction Systems Development (SCSD)*. Development of System Building Components by Performance Specifications in the USA, The Building Centre Trust.

Hix, J. (1970). Maximum Space House, in *Architectural Design*, March, pp. 148-49.

Hix, J. (1974). *The Glass House*, Phaidon, London.

Holroyd, G. (1966). Architecture Creating Relaxed Intensity, in *Architectural Design*, September 1966.

Howell, W.G. (1970). Vertebrate Buildings: The Architecture of Structural Space, in *RIBAJ*, March.

Howson, B. (1977). Quarry Hill Bites the Dust, in *Building Design*, 13 May.

Huber, G. and Steinegger, J-.C. (eds.) (1971). *Jean Prouvé*, Pall Mall Press, London.

Hughes, Q. (1964). *Seaport: Architecture and Townscape in Liverpool*, Lund Humphries, London.

IBIS — Industrialised Building in Steel (*c*.1964/5). Catalogue, undated.

Illich, I.D. (1973). *Tools for Conviviality*, Calder and Boyars, London.

Iredale, R. (1963). Prefabricated Building: The NENK Method in Operation, in *Architects' Journal*, 13 March, pp. 569-76.

Jackson, A. (1970). The Politics of Architecture, Architectural Press, London.

Jencks, C. (1968/9). Pop-Non Pop, in *Architectural Association Quarterly*, Winter, pp. 49-64. Reprinted in *Modern Movements in Architecture*.

Jencks, C. (1973). *Modern Movements in Architecture*, Penguin, Harmondsworth.

Jencks, C. (1977). *The Language of Post-Modern Architecture*, Academy Editions, London. Revised edition 1978.

Jencks, C. and Baird, G. (1969). *Meaning in Architecture*, Barrie and Rockcliff, London.

Johnson, H.R. and Skempton, A.W. (1955/7). William Strutt's Cotton Mills 1793-1812, *Transactions of the Newcomen Society*, Volume XXX, pp. 179-211.

Jones, J.C. (1970). *Design Methods*, Wiley, London and New York.

Jordan, J. (1964). The NENK Method: First Project — Maidstone, in *Architects' Journal*, 25 March, pp. 697-706.

Jordan, J. (1969). CLASP 1969: Speculative Criticism, in *Architects' Journal*, 19 November, pp. 1283-4.

Jordy, W.H. (1976, 1972). *American Buildings and Their Architects*, Doubleday, Garden City, New York (Anchor Press).

Jupp, K., Hanson, J., Robinson, P. and Tomlinson, P. (1975). *Report of the Committee of Inquiry into the Fire at Fairfield Home, Edwalton, Nottinghamshire on 15 December 1974*. Presented to Department of Health and Social Security, Cmnd 6149, Her Majesty's Stationery Office, London.

Kainrath, W. (1970). Segal's Significance, in *Architects' Journal*, 30 September, pp. 775-80.

Katz, D. and Kahn, R.L. (1969, 1966). Common Characteristics of Open Systems, in Emery (1969).

Kaufmann, E. Jnr. (1970). *The Rise of an American Architecture*, Pall Mall, London, in association with the Metropolitan Museum of Art, New York.

Kaufmann, E. and Raeburn, B. (1960). *Frank Lloyd Wright: Writings and Buildings*, Meridian, New York.

Kelly, B. (ed.) (1959). *The Prefabrication of Houses, Design and the Production of Houses*, McGraw Hill, New York.

Kidder Smith, G.E. (1962, 1961). *The New Architecture of Europe,* Penguin, Harmondsworth.

Killeen, J. (1968). Component Building for Schools, in *Official Architecture and Planning,* Volume 31, No. 10, October, pp. 1289-1304.

King, H. and Everett, A. (1971). *Components and Finishes,* Mitchell's Building Construction, Batsford, London. Revised Edition, ed. King and Osbourn, 1979.

Koch, C., with Lewis A. (1958). *At Home With Tomorrow,* Rinehart, New York.

Kropotkin, P. (1974, 1899). *Fields, Factories and Workshops Tomorrow,* Allen and Unwin, London. First published 1899.

Lacey, W.D. and Swain, H.T. (1955). Hertfordshire Schools Development, in *Architects' Journal,* 12 May, pp. 643-52, 26 May, pp. 717-23, 11 August, 22 December, pp. 849-55.

Lambert, G. (1963, 1959). *The Slide Area,* Penguin, Harmondsworth. First published Hamish Hamilton, 1959.

Landau, R. (ed.) (1969). Despite Popular Demand, Architectural Design is Thinking about Architecture and Planning, in *Architectural Design,* September, Special Issue.

Landau, R. (ed.) (1972). Complexity (or How to See the Wood in Spite of the Trees), in *Architectural Design,* Volume XII, October, Special Issue.

Larsen, C.T. (1965). *The Effect of Windowless Classrooms on Elementary School Children: An Environmental Case Study,* Architectural Research Laboratory, Department of Architecture, University of Michigan.

Leach, E. (1976). *Culture and Communication: The Logic by which Symbols are Connected,* Cambridge University Press.

Le Corbusier (1923). *Vers une Architecture,* Editions Crès, Paris.

Le Corbusier (1927). *Towards a New Architecture,* Architectural Press, London.

Le Corbusier (1929). Les Techniques sont l'Assiette même du Lyrisme: Elles ouvrient un nouveau Cycle de l'Architecture, given at second conference *Amis des Arts,* 5 October 1929. Reprinted in *Précisions,* Le Corbusier, 1960.

Le Corbusier (1960, 1930). *Précisions sur un Etat Présent de l'Architecture et de l'Urbanisme,* Vincent, Freal and Co., Paris.

Le Corbusier (1961). *The Modulor,* Faber and Faber, London.

Le Corbusier (1965, 1927). *Towards a New Architecture,* Lund Humphries, London.

Lewicki, B. (1966). *Building with Large Prefabricates,* Institute for Building Research, Warsaw, Elsevier Publishing Company, Amsterdam, London, New York.

Little, A.D., Inc. (1970?). *Technology in Connecticut's Housing Delivery System,* Arthur D. Little, Inc., Cambridge, Mass.

Loweth, S.H. (1949). Post-War Schools: New Materials and Methods of Construction, in *Architect and Building News,* 15 April, pp. 4450-53.

Lyall, S. (1977a). Thamesmead: IB's Death-Rattle? in *Building Design,* 10 June, pp. 12-13.

Lyall, S. (1977b). PSSHAKed up at County Hall, in *Building Design,* 17 June, pp. 18-19.

Lyons, E. (1968). 'Too Often We Justify Our Ineptitudes by Our Moral Postures', in *RIBAJ,* May, pp. 214-22.

MacCormac, R.C. (1968). The Anatomy of Wright's Aesthetic, in *Architectural Review,* March, pp. 143-6.

MACE (n.d. 1968?). *MACE Mock-up and Field Laboratory: Architectural and*

Educational Performance Standards.

MACE (1972). *Annual Report,* Publisher uncredited, possibly Surrey County Council, for MACE.

Maitland, B. (1979). When the Duck Speaks for the State, in *Architectural Design,* Volume 49, No. 3-4, pp. 96-97.

Malpass, P. (1975). Professionalism and the Role of Architects in Local Authority's Housing, in *RIBAJ,* June, pp. 6-29.

Manning, P. (ed.) (1967). *The Primary School: An Environment for Education,* Pilkington Research Unit, Liverpool.

Manson, G. (1953). Wright in the Nursery, in *Architectural Review,* June, pp. 349-51.

March, L. (1970). Imperial City of the Boundless West: Chicago's Impact on Frank Lloyd Wright, in *The Listener,* 30 April, Vol. 83, No. 2144, pp. 581-4.

March, L. and Steadman, P. (1974, 1971). *The Geometry of Environment,* Methuen and Co. Ltd, London.

Marks, R.W. (1960). *The Dymaxion World of Buckminster Fuller,* Southern Illinois University Press, Carbondale and Edwardsville.

Markus, T.A. (1967). The Function of Windows: A Reappraisal, in *Building Science,* Vol. 2, pp. 97-121, Pergamon.

Markus, T.A. *et al.* (1970). St. Michael's Academy, Kilwinning, in *Architects' Journal,* 7 January, pp. 9-50.

Marsh, P. (1976). GLC Get £45,000 Bill for Thamesmead Walls Mistake, in *Building Design,* 6 August.

Martin, B. (ed.) (1965). *The Co-ordination of Dimensions for Building,* RIBA, London.

Martin, B. (1971). *Standards and Building,* RIBA Publications, London.

Martin, J.L., Nicholson, B. and Gabo, N. (eds.) (1937). *Circle: International Survey of Constructive Art,* Faber and Faber, London.

Martin, M.W. (1968). *Futurist Art and Theory 1909-1915,* Clarendon Press, Oxford.

Marx, K. (1946, 1867). *Capital,* Allen and Unwin, London, 1946. Edition being a reprint of 1889 Edition. Volume I of 'Capital' first published in German in 1867. First English Edition 1887.

Marx, K. and Engels, F. (1948, 1848). *The Communist Manifesto,* Lawrence and Wishart, London. First German and French Editions 1848. First English Translation 1850.

Maur, K. von (1972). *Oskar Schlemmer,* Thames and Hudson, London.

Maybeck, B.R. (1862-1957). Selections from the Writings of this Year's Gold Medallist, *Journal of the American Institute of Architects,* May 1951.

McCallum, I. (1959). *Architecture U.S.A.,* Reinhold, New York and Architectural Press, London.

McGuire, P. (1978). Fire Risks in 'Airey' Homes, in *Building Design,* 24 March.

McHale, J. (1956). Buckminster Fuller, in *The Architectural Review,* July, pp. 13-20.

McKean, J.M. (1976a). Walter Segal, in *Architectural Design,* May, pp. 288-95.

McKean, J.M. (1976b). Walter Segal — Pioneer, in *Building Design,* 20 and 27 February.

McLuhan, M. (1962). The Gutenberg Galaxy: the Making of Typographic Man, Routledge and Kegan Paul, London.

McLuhan, M. (1964). *Understanding Media,* Routledge and Kegan Paul, London.

Medd, D. (1972). Responding to Change, in *RIBAJ*, December. pp. 522-3.

Meller, J. (ed.) (1972). *The Buckminster Fuller Reader*, Penguin, Harmondsworth.

Merriman, J.R. (1978). Development of Client Sponsored School Building Systems since 1945, unpublished Dissertation, Portsmouth Polytechnic, School of Architecture.

Ministry of Education (GB) (1952). *Development Projects: Wokingham School*, Building Bulletin No. 8, Her Majesty's Stationery Office, London.

Ministry of Education (1961). *The Story of CLASP*, Building Bulletin No. 19, Her Majesty's Stationery Office, London.

Ministry of Health (GB) (1920). Report of Standardisation and New Methods of Construction, Committee Report.

Ministry of Health (GB) (1925). Third Interim Report of Committee on New Methods of House Construction, 29 January.

Ministry of Health (1964). *Dimensional Co-ordination and Industrialised Building*, Hospital Design Note No. 1, Her Majesty's Stationery Office, London.

Ministry of Housing and Local Government (1963a). *Dimensions and Components for Housing*, Design Bulletin No. 8, Her Majesty's Stationery Office, London.

Ministry of Housing and Local Government (1963b). *Space in the Home (Imperial)*, Her Majesty's Stationery Office, London.

Ministry of Housing and Local Government (1966a). *Housebuilding in the USA: A Study of Rationalisation and its Implications*, Her Majesty's Stationery Office, London.

Ministry of Housing and Local Government (Research and Development Group) (1966b). *5-Minute Guide to Economic Design in 5M System Housing*, July.

Ministry of Housing and Local Government (1967). *12M Jespersen Design Guide*, Research and Development Group, January.

Ministry of Housing and Local Government (1968a). *Space in the Home (Metric)*, Her Majesty's Stationery Office, London.

Ministry of Housing and Local Government (1968b). *Co-ordination of Components in Housing: Metric Dimensional Framework*, Design Bulletin No. 16, Her Majesty's Stationery Office, London.

Ministry of Housing and Local Government (1970). *Designing a Low-Rise Housing System*, Her Majesty's Stationery Office, London.

Ministry of Public Building and Works (1963a). *Dimensional Co-ordination for Industrialised Building (DCI)*, February, Her Majesty's Stationery Office, London.

Ministry of Public Building and Works (1963b). *Dimensional Co-ordination for Industrialised Building: Preferred Dimensions for Housing (DC2)*, September, Her Majesty's Stationery Office, London.

Ministry of Public Building and Works (1964a). *Dimensional Co-ordination for Industrialised Building: Preferred Dimensions for Educational, Health and Crown Office Buildings (DC3)*, July, Her Majesty's Stationery Office, London.

Ministry of Public Building and Works (1964b). *Dimensional Co-ordination for Crown Office Buildings*, Her Majesty's Stationery Office, London.

Ministry of Public Building and Works (1965). 5M Housing at Sheffield, in *Architects' Journal*, 12 May.

Ministry of Public Building and Works (1967a). *Dimensional Co-ordination for Building: Recommended Vertical Dimensions for Educational, Health, Housing, Office and Single-Storey general purpose Industrial Buildings*

(DC4), January, Her Majesty's Stationery Office, London.

Ministry of Public Building and Works (1967b). *Dimensional Co-ordination for Building: Recommended Horizontal Dimensions for Educational, Health, Housing, Office and Single-storey general purpose Industrial Buildings (DC5)*, May, Her Majesty's Stationery Office, London.

Ministry of Public Building and Works (1967c). *Dimensional Co-ordination for Building (DC6): Guidance on the Application of Recommended Vertical and Horizontal Dimensions for Educational, Health, Housing, Office and Single-storey general purpose Industrial Buildings,* Her Majesty's Stationery Office, London.

Ministry of Public Building and Works (1968a). *Dimensional Co-ordination for Building: Recommended intermediate vertical controlling Dimensions for Educational, Health, Housing and Office Buildings, and Guidance on their Application (DC7),* Her Majesty's Stationery Office, London.

Ministry of Public Building and Works (1968b). *Dimensional Co-ordination for Building: Recommended Dimensions of Spaces allocated for selected Components and Assemblies used in Educational, Health, Housing and Office Buildings,* June, Her Majesty's Stationery Office, London.

Modular Quarterly (MQ) (1963). The NENK Method of Building: Dimensional Organisation, Summer, pp. 10-17.

Modular Quarterly (MQ) (1963/64). Dimensional Co-ordination for Industrialised Housing, Winter.

Moffett, N. (1955). Architect/Manufacturer Co-operation, in *Architectural Review,* September 1955, Vol. 118, No. 705, pp. 201-4, and December 1955, Vol. 118, No. 707, pp. 412-16.

Morgan, B.G. (1961). *Canonic Design in English Mediaeval Architecture,* Liverpool University Press.

Morris, H. (1966). Architects' Approach to Architecture, in *RIBAJ,* April, pp. 154-63.

Morris, P. (1961). *Homes for Today and Tomorrow* (known as the Parker Morris Report), Ministry of Housing and Local Government, Her Majesty's Stationery Office, London.

Morris, W. (1944, 1884). Art and Socialism, in *William Morris: Selected Writings* edited by G.D.H. Cole, Nonesuch Press 1944. First published in *Architecture, Industry and Wealth,* Longmans, Green, 1902.

Morse, E.S. (1972, 1885). *Japanese Homes and their Surroundings,* Charles E. Tuttle Co., Rutland, Vermont and Tokyo, Japan.

Mumford, L. (1930). *Mass Production and the Modern House:* 1. The Limits of Mechanisation; 2. The Role of Community Planning. Reprinted in *Architecture* as 'A Home for Man'. Edited by J.M. Davern, 1975.

Mumford, L. (1946, 1930). Mass Production and Housing, The Architectural Record, January/February 1930. Reprinted in *City Development,* Secker and Warburg, London, 1946.

Mumford, L. (1970, 1964). *The Myth of the Machine: The Pentagon of Power,* Secker and Warburg, London.

Mumford, L. (1972, 1952). *Roots of Contemporary American Architecture,* Dover, New York, containing Montgomery Schuyler's A Critique of the Works of Adler and Sullivan in *Architectural Record,* December 1895.

Mumford, L. (1975). *Architecture as a Home for Man: Essays for Architectural Record,* edited by J.M. Davern, Architectural Record Books.

Münz, L. and Künstler, G. (1966, 1964). *Adolf Loos, Pioneer of Modern Architecture,* Thames and Hudson, London.

Murray, P. (1977). It's About the Slowest and Relatively Most Expensive Package the GLC has Ever Been Sold, in *Building Design,* 25 March.

Naylor, G. (1971). *The Arts and Crafts Movement: A Study of its Sources, Ideals and Influence on Design Theory,* Studio Vista, London.

Nelson, G. (1980). Energy Comment (on Frogmore School Phase II), in *Architects' Journal,* 12 March, pp. 528-9.

Nissen, H. (1972). *Industrialised Building and Modular Design,* Cement and Concrete Association, London.

Nomitch, R. (1970). Operation Breakthrough, in *Architectural Design,* August, pp. 403-4.

Norberg-Schulz, C. (1963). *Intentions in Architecture,* Allen and Unwin, London.

Norberg-Schulz, C. (1971). *Existence, Space and Architecture,* Praeger, New York.

Northern Arts and The Scottish Arts Council (1972). *Futurismo 1909-1919.* Exhibition of Italian Futurism, Royal Academy of Arts, London, Catalogue.

Oddie, G. (1975). *Industrialised Building for Schools,* Organisation for Economic Co-operation and Development, Paris.

Official Architecture and Planning (1966). Schools Issue, Vol. 29, No. 9, September.

Official Architecture and Planning (1968). Ministry of Public Buildings and Works: Building Development, June.

O'Sullivan, P.E. and Cole, R.J. (1974). The Thermal Performance of School Buildings, in *Journal of Architectural Research,* May.

Padovan, R. (1974). The MACE System, in *Architects' Journal,* 7 July, pp. 101-14.

Panofsky, E. (1970, 1955). *Meaning in the Visual Arts,* Penguin, Harmondsworth.

Parkinson, C.N. (1958). *Parkinson's Law: Or the Pursuit of Progress,* Murray, London.

Pawley, M. (1970). Mass Housing: The Desperate Effort of Pre-Industrial Thought to Achieve the Equivalent of Machine Production, in *Architectural Design,* January, pp. 32-8.

Pawley, M. (1971a). *Architecture Versus Housing,* Studio Vista, London.

Pawley, M. (1971b). The Philosopher's Stone, in *New Society,* 29 April.

Pearce, D. (1977). No Room for Achievement, in *Building Design,* 9 September.

Pearman, H. (1978). Fears Mount Over Housing Blocks using Bison Frame System, in *Building Design,* 20 October.

Pehnt, W. (1973). *Expressionist Architecture,* Thames and Hudson, London.

Peisch, M.L. (1964). *The Chicago School of Architecture,* Phaidon Press, London.

Perkins, M. (1980). Frogmore Comprehensive School, Phase 1, in *Architects' Journal,* 12 March, pp. 521-23.

Pevsner, N. (1951, 1943). *An Outline of European Architecture,* Penguin, Harmondsworth.

Pevsner, N. (1959). Roehampton: LCC Housing and the Picturesque Tradition, in *The Architectural Review,* Vol. CXXVII, No. 750, July, pp. 21-35.

Pevsner, N. (1968). Architecture and the Bauhaus, in *Bauhaus 50 Years:* Exhibition Catalogue, Royal Academy of Arts, and Federal German Republic.

Piacentini, M. (1922). Le Corbusier's 'The Engineer's Aesthetic: Mass-Production Houses', in *Architectura Arti Decorative* **11,** pp. 220-23, in Serenyi, 1975.

Pickett, W.V. (1845). *A New System of Architecture Founded on the Forms of Nature and Developing the Properties of Metals*, Longman & Co., London.

Pike, A. (1968). Techcrete: An American Industrialised Building System, in *Architectural Design*, June, pp. 278-89.

Pike, A. (1969). Thamesmead Report, in *Architectural Design*, November, pp. 602-13.

Placzek, A.K. (ed.) (1970). *The Origins of Cast Iron Architecture in America*, Badger and Bogardus, Da Capo Press, New York.

Powell, J.A. and Matthews, R. (1976). *Flexibility and Appropriation of Space in Schools*. Paper presented at 3rd International Architectural Psychology Conference, Strasbourg, June.

Pritchard, J. (1969). Gropius, The Bauhaus, and the Future: A Paper read to the Royal Society of Arts, 13 November 1968 in *Journal of the RSA*, January 1969.

Rabeneck, A. (1970). Five Systems, in *Architectural Design*, August, pp. 404-7.

Rabeneck, A. (1971). IBM Head Office, in *Architectural Design*, August, pp. 474-8.

Rabeneck, A. (1973). PSA Method of Building, in *Architects' Journal*, 14 November, pp. 1169-79.

Rabeneck, A. (1975). The New PSSHAK, in *Architectural Design*, October, pp. 629-33.

Rabeneck, A. (1976a). Self-Fulfilling Prophets, in *Architectural Design*, March, pp. 162-65.

Rabeneck, A. (1976b). Research into Site Management, in *Architectural Design*, May, pp. 275-80.

Rabeneck, A. (1976c). Whatever Happened to the Systems Approach, in *Architectural Design*, May, Vol. XLVI, pp. 298-303.

Rabeneck, A. (1979). A Pattern Language, in *Architectural Design*, January, pp. 19-20.

Rapoport, A. (1974). *Conflict in Man-Made Environment*, Penguin, Harmondsworth.

Ravetz, A. (1971). The History of a Housing Estate, in *New Society*, 27 May, pp. 907-10.

Ravetz, A. (1974). *Model Estate: Planned Housing at Quarry Hill, Leeds*, Croom Helm, London and Rowntree Memorial Trust.

Ray-Jones, A. (ed.) (1968). Construction Indexing Manual. RIBA, London.

Ray-Jones, A. and Clegg, D. (eds.) (1976). C1/SfB Construction Indexing Manual. RIBA Publications, London.

Rees, A. (1819-20). *The Cyclopedia or Universal Dictionary of Arts, Science and Literature*, Volume 22. (of 45 volumes)

Richards, J.M. (1971, 1937). The Condition of Architecture and the Principle of Anonymity, in *Circle*, edited by J.L. Martin, B. Nicholson, and N. Gabo, Faber and Faber, London.

Richards, J.M. (1972). The Hollow Victory: 1932-72, in *RIBAJ*, May, pp. 192-97.

Robbie, R. (1970a). The Turning Point: Report of a Meeting in the 'Technology and the Profession' series, in *RIBAJ*, June, pp. 254-61.

Robbie, R. (1970b). Tomorrow's World: Planned or Damned by Man or Machine, in *Building Design*, 17 April, pp. 6-7.

Robertson, E.G. and Robertson, J. (1977). *Cast Iron Decoration: A World Survey*, Thames and Hudson, London.

Robinson, K.J. (1961). Worktown Mementos, in *The Observer*, London, 28 May.

Roskill, O.W. (1964). *Consortia of Local Authorities: Advantages and Disadvantages.* The Builder.

Rostron, R.M. (1964). *Light Cladding of Buildings,* Architectural Press, London.

Rowe, C. (1947). The Mathematics of the Ideal Villa: Palladio and Le Corbusier Compared, in *Architectural Review,* March, pp. 101-4.

Rowe, C. (1956). Chicago Frame — Chicago's Place in the Modern Movement, in *Architectural Review,* November, pp. 285-9.

Royal Academy of Arts and German Federal Republic (1968). *50 Years Bauhaus Exhibition Catalogue,* Royal Academy of Arts, London and Federal German Republic. Gebr. Rasch & Co., Braunsche bei Osnabrück.

Royal Institute of British Architects Journal (1955). LCC Technical Development and Research, October, pp. 471-79.

Rudofsky, B. (1965). *The Kimono Mind,* Gollancz, London.

Ruskin (1907, 1849). *The Seven Lamps of Architecture,* Bernhard Tauchnitz, Leipzig.

Russell, B. (1956). The Foci of Cities, unpublished Dissertation.

Russell, B. (1973). Barbarians in the Living Room, or Systems Patterns and Design, in *Architectural Design,* December, pp. 792-95.

Russell, B. (1977). Frames of Reference (Is Architecture What Architects do?) in *Design Methods and Theories,* January/March, pp. 2-10.

Russell, B. (1979). The Vernacular, The Industrialised Vernacular and other Convenient Myths, in *Journal of the Faculty of Architecture,* Middle East Technical University, Ankara, Vol. 5, No. 1, Spring, pp. 101-7.

Russell, B. (1980). Bruno's Dream, in *Architectural Review,* December, pp. 331-32.

Safdie, M. (1970). *Beyond Habitat,* MIT Press, Cambridge, Mass.

Sahlins, M. (1978, 1976). *Culture and Practical Reason,* Chicago University Press.

Salisbury, F. (1973). Home Truths from the Building Research Establishment, in *RIBAJ,* November, pp. 562-63.

Sant'Elia, A. (1972, 1914). Manifesto of Futurist Architecture, in *Futurismo 1909-1919.* Catalogue to Exhibition at Royal Academy of Arts, London, Northern Arts and Scottish Arts Council, United Kingdom (1972).

Scamozzi, O.B. (1786). *Les Bâtiments et les Desseins de André Palladio,* Second Edition, chez Jean Rossi.

Schindler, R.M. (1947). The Schindler Frame, *Architectural Record,* May.

Schlemmer, O., Moholy-Nagy, L. and Molnar, F. (1961). *The Theater of the Bauhaus,* Wesleyan University Press.

Schmid, T. and Testa, C. (1969). *Systems Building: An International Survey of Methods,* Pall Mall, London.

Schnaidt, C. (1965). *Hannes Meyer: Buildings, Projects and Writings,* Tiranti, London.

Schon, D. (1971). *Beyond the Stable State,* Smith, London.

Scott, G. (1924, 1914). *The Architecture of Humanism,* Second Edition, Constable, London.

Scott, G. (1976). *Building Disasters and Failures - A Practical Report,* Construction Press, London.

Sebestyén, G. (1965). *Large Panel Buildings,* Akadémiai Kaidó, Publishing House of the Hungarian Academy of Sciences, Budapest.

Second Consortium of Local Authorities (1962?). *Second Consortium of Local Authorities:* with contributions from member authorities, Hampshire County Council.

Second Consortium of Local Authorities (1966). *Report to Elected Represen-*

tatives, published by Shropshire County Council, November.

Segal, W. (1948). *Home and Environment,* Leonard Hill, London.

Segal, W. (1974). Into the 20's, in *The Architectural Review.* Vol. CLV, No. 923, pp. 31-8.

Segal, W. (1977). Timber Framed Housing, in *RIBAJ,* July. pp. 284-95.

Serenyi, P. (1975). *Le Corbusier in Perspective,* Prentice Hall, Englewood Cliffs.

Sergeant, J. (1976). *Frank Lloyd Wright's Usonian Houses: The Case for Organic Architecture,* Watson Guptill, New York.

Shankland, H. (1972). Futurism in Literature and in the Theatre, in *Futurismo 1909-1919,* Catalogue to Exhibition at Royal Academy of Arts, London, Northern Arts and Scottish Arts Council, United Kingdom.

Shepherd, C. (1971). Eastergate: The 'Integrated Design' School, in *RIBAJ,* July, pp. 282-86.

Silcock, A. and Tucker, D.M. (1976). *Fires in Schools: an Investigation of actual Fire Development and Building Performance,* Fire Research Station and Building Research Establishment, Garston, January.

Smithson, P. (1966). Just a Few Chairs and a House: An Essay on the Eames-Aesthetic, in *Architectural Design,* September, pp. 443-6.

Stamp, G. (1976). Stirling's Worth, The History Faculty Building, in *The Cambridge Review,* January, pp. 77-82.

Stephenson, K. (1970). All-carpeted, air-conditioned, all-flexible, all new school, *Building Design,* December 11th, pp. 13-14.

Stern, E.G. (1968). *Nails in Residential Buildings and Farm Structures,* Wood Research and Wood Construction Laboratory, Virginia Polytechnic Institute.

Stevens, E. (1976). Spending Cuts Force Closure of Consortium, in *Building Design,* 10 December.

Stevens, E. (1977). Stop the Home Fires Burning, in *Building Design,* 6 May.

Stevens, E. (1978). French School Fire — UK Architect Appeals, in *Building Design,* 7 April.

Stillman, C.G. (1949). Prototype Classroom for the Middlesex County Council, in *Architects' Journal,* 1 September, pp. 223-5.

Stillman, C.G. and Cleary, R.C. (1949). *The Modern School,* The Architectural Press, London.

Stirling, J. (1965). An Architect's Approach to Architecture, Article based on a Paper given at the RIBA, in *RIBAJ,* May, pp. 231-40.

Stirling, J. (1974). Exhibition Catalogue. RIBA Drawings Collection. London.

Stirling, J. (with Leon Krier) (1975). *James Stirling: Buildings and Projects, 1950-1974,* Thames and Hudson, London.

Stirling, J. and Gowan, J. (1961). Rehousing at Preston, in *Architects' Journal,* 8 June, pp. 845-50.

Storrer, W.A. (1974). *The Architecture of Frank Lloyd Wright,* MIT Press, Cambridge, Mass. and London.

Sturges, W.K. (1970). *The Origins of Cast Iron Architecture in America,* De Capo Press, New York.

Sutcliffe, F.E. (1968). *Introduction to Descartes: Discourse on Method and other Writings,* Penguin, Harmondsworth.

Swain, H. (1960). The Mass Production Spirit, in *Architects' Journal,* 21 January, pp. 127-30.

Swain, H. (1972). Notts Builds: Project RSM, in *Architects' Journal,* 12 January, pp. 75-96.

Swain, H. (1974). Successful Design for Mining Subsidence, in *Architects'*

Journal, 8 May, pp. 1047-54.

Tafuri, M. (1976). *Architecture and Utopia: Design and Capitalist Development,* MIT, Cambridge, Mass.

Tatton-Brown, W.G. (1957). Some Aspects of Modular Co-ordination in USA, in *The Modular Quarterly,* Spring, pp. 16-23.

Taut, B. (1929). *Modern Architecture,* The Studio, London.

Taylor, B.B. (1972). *Le Corbusier et Pessac 1914-1928,* Volume 1 and 2, Fondation Le Corbusier en collaboration avec Harvard University.

Taylor, B.B. (1975). Le Corbusier at Pessac, in Serenyi, 1975.

Taylor, B.B. (1977). Le Corbusier at Pessac: Professional and Client Responsibilities, in Walden, 1977.

Taylor, F.W. (1911). Principles of Scientific Management, 1911, Reprinted in Taylor (1964) *Scientific Management,* Harper Row, Ltd, New York, 1964.

Taylor, F.W. (1912). Hearings before Special Committee of the House of Representatives to Investigate the Taylor and Other Systems of Shop Management under Authority of House Resolution, Reprinted in Taylor (1964) *Scientific Managament,* Harper and Row, New York, 1964.

Taylor, F.W. (1964). *Scientific Management,* Harper Row, Ltd, New York.

Terkel, L. (1975). *Working,* Avon, New York.

Thomas, M.H. (1947). *Building Is Your Business,* Wingate, London.

Thompson, M. (1977). *Rat-Infested Slum or Glorious Heritage: An Introduction to Rubbish Theory,* International Institute for Environment and Society, Berlin.

Thompson, M. (1979). *Rubbish Theory: The Destruction and Creation of Value,* Oxford University Press.

Thompson, W. d'Arcy (1961, 1917). *On Growth and Form,* Cambridge University Press.

Timber Development Association Ltd (1947, 1944). *Prefabricated Timber Houses,* TDA, London.

Times, The (1976). *Putting the Squeeze on the Big Spenders, Times* Leading Article, 7 July.

Tindale, P. and O'Toole, T. (1969). Post Mortem on 5M: Report of a Lecture given at the RIBA on 15 October 1968, in *RIBAJ,* March, pp. 112-14.

Trump, J.L. (1959). *Images of the Future: The Commission on the Experimental Study of the Utilisation of the Staff in the Secondary School,* National Association of Secondary School Principals, Washington D.C.

Trump, J.L. and Baynham, D. (1970). *Focus on Change: Guide to Better Schools,* Rand McNally, Chicago.

Turin, D.A. (1972a). *The Mechanism of Response to Effective Demand,* Building Economics Research Unit of University College Environmental Research Group, London, January.

Turin, D.A. (1972b). *Recent Work of the Building Economics Research Unit,* University College Environmental Research Group, London, March.

Twist, K.C., Redpath, J.T. and Evans, K.C. (1956). Hertfordshire Schools Development, in *Architects' Journal,* 19 April, pp. 379-84, 2 August, pp. 156-62.

Venturi, R. (1966). *Complexity and Contradiction in Architecture,* The Museum of Modern Art, New York, and the Graham Foundation, Chicago.

Wachsmann, K. (1930). *Holzhausbau,* Berlin.

Wachsmann, K. (1957). Building in Our Time, in *Architectural Association Journal,* London, April, pp. 224-33.

Wachsmann, K. (1961). *The Turning Point of Building: Structure and Design*, Reinhold Publishing Corporation, New York.

Walden, R. (ed.) (1977). *The Open Hand: Essays on Le Corbusier*, MIT Press, Cambridge, Mass.

Wallis, M. (1966). Hertford: A System, Yesterday, Today and Tomorrow, in *Official Architecture and Planning*, September, pp. 1286-89.

Walters, R. and Iredale, R. (1964). The NENK Method of Building, in *RIBAJ*, June, pp. 259-74.

Wang, J.C. (1970). Participant Observation: An Experience in Field Design Research. Study of Two Form Entry, Hampshire Junior School and Ilford Jewish Primary School, unpublished Study, Portsmouth School of Architecture.

Wanzel, J.G. (1969). Systems and Society, in *Architectural Design*, October, p. 565.

Ward, A. (1969). Right and Wrong, in *Architectural Design*, July, pp. 384-89.

Ward, A. (1979). A Pattern Language, in *Architectural Design*, January, pp. 15-17.

Watkin, D. (1977). *Morality and Architecture*, Oxford University Press.

Webb, S. (1969a). Ronan Point: The Blind Risk, in *Architectural Design*, January, pp. 3-4.

Webb, S. (1969b). In Need of Support, in *Architectural Design*, August.

Webb, S. (1980). The Ramifications of Ronan Point, in *RIBAJ*, March, p. 15.

Weisman, W. (1970). A New View of Skyscraper History, in *The Rise of an American Architecture*, Kaufmann E. (1970) Pall Mall, London.

West, N. (1961, 1933). *Miss Lonelyhearts*, Penguin, Harmondsworth.

White, R.B. (1965). *Prefabrication: A History of its Development in Great Britain*, Her Majesty's Stationery Office, London.

Wiks, R.M. (1972). *Henry Ford and Grass Roots America*, University of Michigan Press.

Wilbur, D.E. (1971). *Housing: Expectations and Realities*, Gryphon House, Washington.

Wilhelm, K. (1979). From the Fantastic to Fantasy, in *Architectural Association Quarterly*, Vol. II, No. 1, pp. 4-15.

Wilson, C.B., Facey, P., Giles, R.W. and Mallon, R. (1974). *Schools Division: Participation Movement Report of Working Party on the MACE School Building System*, GLC Department of Architecture and Civic Design.

Wingler, H.M. (1969, 1962) *The Bauhaus: Weimar, Dessau, Berlin, Chicago*. MIT Press, Cambridge, Mass., and London. (Original German publication 1962).

Wittkower, R. (1973, 1949). *Architectural Principles in the Age of Humanism*, Academy Editions, London.

Wood, R. (1944). *Standard Construction for Schools: Post-War Building Studies No. 2.*, Her Majesty's Stationery Office, London.

Wright, F.L. (1945, 1932). *An Autobiography*, Faber and Faber, and The Hyperion Press, London.

Wright, F.L. (1975). In the Cause of Architecture: I. The Architect and the Machine, in *In the Cause of Architecture*, edited by F. Gutheim, Architectural Record Books, New York.

Wright, J.A. (1967). Schools: Design Criteria, unpublished thesis for Portsmouth School of Architecture.

Wright, J.A. (1969). *Ilford Jewish Primary School: A New Environment for Learning* (mimeo).

Wright, L. (1972). IBM Offices, Cosham, Hants., in *Architectural Review*, January, Vol. CLI No. 899, pp. 5-24.

Yorke, F.R.S. (1943, 1934). *The Modern House,* The Architectural Press, Fourth Edition, London.

Yorke, F.R.S. and Gibberd, F. (1937). *The Modern Flat,* The Architectural Press, London.

ILLUSTRATIONS: SOURCES AND ACKNOWLEDGEMENTS

Wherever possible the source of an illustration has been given, together with the appropriate credit. I am grateful to all those individuals and organisations who have given their permission for the use of material. In a small number of cases it has not been possible to trace the copyright owner, and to these apologies are extended.

Although often heavily reliant upon photographs the majority of books on architectural topics fail to give the date that these were taken, often a crucial piece of information for subsequent studies. Wherever possible, therefore, the date that a photograph was taken is given in the caption.

1. 'A ROSE BY ANY OTHER NAME'

1.2. THE JAPANESE HOUSE

1.2.1. and **1.2.2.** *The Japanese House* by Heinrich Engel. Charles E. Tuttle Co., Inc., Tokyo, 1964

1.2.3, 1.2.4. and **1.2.5.** Drawings by the Author

1.2.6. *The Japanese House* by Heinrich Engel. Charles E. Tuttle Co., Inc., Tokyo, 1964.

1.3 A MATTER OF PROPORTION

1.3.1. Drawing by the Author

1.3.2. Drawing by Leonardo da Vinci

1.3.3. Drawing by the Author after Cesariano, 1521

1.3.4. **a,b,c.** *Les Bâtiments et les Desseins de André Palladio*, Octave Bertotti Scamozzi, Chez Jean Rossi, 1786

    d. Photo by the Author

1.3.5. Photo by the Author

2. INDUSTRIALIZED CORNUCOPIA

2.1. STANDARDS AND ICONS

2.1.1. a,b,c,d,e,f. and **2.1.2.a,b.** City of Portsmouth Museums. Photos by Karen Goodwin

2.1.3. *London Illustrated News,* 31 May 1851, Southampton Public Library. Photos by Karen Goodwin

2.1.4. *London Illustrated News,* Southampton Public Library. Photos by Karen Goodwin.
a. 3 May 1851; **b.** 25 January 1851

2.1.5. *London Illustrated News,* Southampton Public Library. Photos by Karen Goodwin.
a. 23 November 1850; **b.** 16 November 1850; **c.** 7 December 1850; **d.** 4 January 1851; **e.** 30 November 1850

2.1.6. a,b. *London Illustrated News,* 16 November 1850, Southampton Public Library. Photos by Karen Goodwin

2.1.7. *London Illustrated News,* Southampton Public Library. Photos by Karen Goodwin
a. 7 December 1850; **b.** 4 January 1851

2.1.8. Drawing by the Author after *London Illustrated News,* 18 February 1851

2.1.9. a,b. and **2.1.10.a,b.** Photos by Richard Ford

2.2. CAST IRON

2.2.1. *The Cyclopædia or Universal Dictionary of Arts, Sciences and Literature,* Abraham Rees, 1819-20. Volume 22, plate XIV. Birmingham Public Library

2.2.2. and **2.2.3.** Photos by the Author

2.2.4. *London Illustrated News,* 12 April 1851, Southampton Public Library. Photo by Karen Goodwin

2.3. IRON AND STEEL FRAMES

2.3.1. and **2.3.2.** Collection of the Art Institute of Chicago. Photos by J. W. Taylor

2.3.3. Photo by the Author

3 RESPONSES TO MACHINES

3.2. MORAL MACHINES

3.2.1. The Anatomy of Wrights Aesthetic by Richard MacCormac, *Architectural Review,* March 1968

3.2.2. and **3.2.3.** Drawings by the Author

3.2.4, 3.2.5. and **3.2.6.** *The Architecture of the Well-tempered Environment* by Reyner Banham. Architectural Press 1969

3.3 EFFICIENT MACHINES

3.3.1. By permission of Volkswagen and VAG (UK) Ltd
3.3.2. *Mechanization Takes Command* by Siegfried Giedion, Oxford University Press, 1969, 1948

4 EUROPE AND THE IDEAL MACHINE STYLE

4.2. MACHINE LOVE

4.2.1. and **4.2.2.** *Futurismo 1909-1919.* Exhibition catalogue, Northern Arts and Scottish Arts Council, 1973

4.3 THE FORMALITIES OF EFFICIENCY

4.3.1, 4.3.2. and **4.3.3.** *The New Theatre and Cinema of Soviet Russia* by Huntly Carter. Chapman and Dodd, London, 1924
4.3.4. a. *Art in Revolution,* Exhibition catalogue, Arts Council, 1971
b. Photo by the Author
4.3.5. Peter Dickens

4.4. THE ICONOGRAPHY OF THE MACHINE

4.4.1. *Le Corbusier 1910-1960* by W. Boesiger, Artemis Verlag, Zürich, 1960
4.4.2. Mass Housing: Frameworks by Nicolas Habraken, *Architectural Design,* January 1970
4.4.3. Photo by the Author
4.4.4. *Le Corbusier and Pierre Jeanneret: The Complete Architectural Works,* Vol. 1: 1910-1929, by W. Boesiger and O. Stonorow. Thames and Hudson, London. By permission of Artemis Verlag, Zürich
4.4.5. a. *Les Bâtiments et les Desseins de André Palladio,* Octave Bertotti Scamozzi, Chez Jean Rossi, 1786
b. Photo by the Author
4.4.6, 4.4.7. and **4.4.8.** *Le Corbusier 1910-1960* by W. Boesiger, Artemis Verlag, Zürich
4.4.9. a. *Le Corbusier and Pierre Jeanneret: The Complete Architectural Works,* Vol. 1: 1910-1929, by W. Boesiger and O. Stonorow. Thames and Hudson, London. By permission of Artemis Verlag, Zürich
b. *Le Corbusier 1910-1960* by W. Boesiger, Artemis Verlag, Zürich, 1960
4.4.10. and **4.4.11.** *Le Corbusier 1910-1960* by W. Boesiger, Artemis Verlag, Zürich, 1960
4.4.12. *Bauhaus.* Exhibition catalogue, Royal Academy of Arts, London, 1968. By permission of H. M. Wingler

4.4.13. *Theo van Doesburg* by Joost Baljeu, Studio Vista, London, 1974

4.4.14. *Oskar Schlemmer* by Karin von Maur, Thames and Hudson, London, 1972. By permission of Verlag Gerd Hatje, Stuttgart

4.4.15, 4.4.16, 4.4.17, 4.4.18, 4.4.19. and **4.4.20.** *Bauhaus,* Exhibition catalogue, Royal Academy of Arts, London, 1968. By permission of H. M. Wingler

4.4.21.a. *Bauhaus.* Exhibition catalogue, Royal Academy of Arts, London, 1968. By permission of H. M. Wingler. **b.** and **c.** *The Modern House* by F. R. S. Yorke. Architectural Press, London, 1943, 1934.

4.4.22. and **4.4.23.** *Bauhaus.* Exhibition catalogue, Royal Academy of Arts, London, 1968. By permission of H. M. Wingler

4.4.24. *Pencil Points,* April 1943

4.4.25. *The Architecture of the Well-tempered Environment,* by Reyner Banham, Architectural Press, 1969

4.4.26. Photo by Nick Wolfenden

4.4.27. a. *Jean Prouvé* by B. Huber and J-.C. Steinegger, Pall Mall Press, London, 1971 **b.** and **c.** Photos by Nick Wolfenden

4.4.28. *Jean Prouvé* by B. Huber and J-.C. Steinegger, Pall Mall Press, London, 1971

4.4.29. Photo by Michael McKee

4.4.30, 4.4.31. and **4.4.32** *Jean Prouvé* by B. Huber and J-.C. Steinegger, Pall Mall Press, London, 1971

4.4.33. *9 × 10* Portsmouth Polytechnic, 1971

4.4.34. *The Modern House* by F. R. S. Yorke, Architectural Press, London, 1943, 1934

4.4.35. Reprinted by permission of Faber and Faber Ltd. from *The New Architecture and the Bauhaus* by Walter Gropius, 1965, 1935.

5 TRANSATLANTIC TONIC

5.1. ACHIEVEMENTS AND DREAMS

5.1.1. Drawings by the Author

5.1.2, 5.1.3, 5.1.4. and **5.1.5.** *The Modern House* by F. R. S. Yorke, Architectural Press, London, 1943, 1934

5.2. RUDOLPH SCHINDLER

5.2.1. *Schindler* by David Gebhard, Thames and Hudson, London, 1971

5.2.2. Photo by John Griffin

5.2.3. *Schindler* by David Gebhard, Thames and Hudson, London, 1971

5.3. RICHARD BUCKMINSTER FULLER

5.3.1, 5.3.2, 5.3.3, 5.3.4, 5.3.5, 5.3.6. and **5.3.7.** *The Dymaxion World of Buckminster Fuller* by R. W. Marks, Southern Illinois University Press, 1960. By permission of R. Buckminster Fuller

5.4. BEMIS AND MODULAR DESIGN

5.4.1, 5.4.2. and **5.4.3.** *The Evolving House,* Vol. III, *Rational Design* by Albert Farwell Bemis, The Technology Press, MIT, 1936. Copyright 1963 Alan Cogswell Bemis

6   BRITAIN FLIRTS WITH MECHANIZATION

6.1. THEY CALLED IT PREFABRICATION

6.1.1. Photo by the Author

6.2. LIGHT AND DRY

6.2.1, 6.2.2, 6.2.3, 6.2.4, 6.2.5, 6.2.6. and **6.2.7.** *House Construction,* Postwar Building Study No. 1, 1944. Reproduced with the permission of the Controller of Her Majesty's Stationery Office
6.2.8. Leeds City Libraries
6.2.9. *Building Equipment News,* reprinted in *Prefabrication: A History of its Development in Britain,* National Building Studies Report No. 36, R. B. White, 1965
6.2.10. Leeds City Libraries
6.2.11, 6.2.12, 6.2.13. and **6.2.14.** Photos by the Author

6.3. COMMERCE AND STANDARDS

6.3.1. Photo by the Author
6.3.2. *The Modern School,* by C. G. Stillman and R. C. Cleary, The Architectural Press, London, 1949
6.3.3. *Hills Dry Building System* catalogue, 1957
6.3.4. Building Bulletin No. 8. *Development Projects: Wokingham School,* Ministry of Education, 1952. Reproduced with the permission of the Controller of Her Majesty's Stationery Office

6.4. HOMES FIT FOR HEROES, AGAIN

6.4.1. *Homes for the People* by the Association of Building Technicians, Paul Elek, 1945. By permission of Granada Publishing Ltd.
6.4.2. and **6.4.3.** Architect/Manufacturer Co-operation (III) by Noel Moffett, *Architectural Review,* September 1955

7.1.10. *Architecture USA,* by Ian McCallum, Architectural Press, London, 1959

7.1.11. Architecture creating relaxed intensity, by Geoffrey Holroyd, *Architectural Design,* September, 1966

7.1.12. Composition with Red, Yellow and Blue by P. Mondrian, SPADEM, Paris, and the Tate Gallery, London

7.1.13. Photos by the Author

7.1.14. *Architecture USA,* by Ian McCallum, Architectural Press, London, 1959

7.1.15. *The Evolving House,* Vol. III, *Rational Design,* by Albert Farwell Bemis, The Technology Press, MIT, 1936. Copyright 1963 by Alan Cogswell Bemis

7.1.16. Drawing by the Author

7.1.17. and **7.1.18.** *Modular Co-ordination,* OEEC 1961. By permission of OECD

7.1.19, 7.1.20, 7.1.21, 7.1.22, 7.1.23, 7.1.24. and **7.1.25** Redrawn by the Author from *The Modular Primer* by E. Corker and A. Diprose, The Modular Society, London, 1963. By permission of Eric Corker

7.1.26.a. *Pencil Points,* April 1943
   b. Unknown

7.1.27, 7.1.28. and **7.1.29.** *The Turning Point of Building* by Konrad Wachsmann, Reinhold Publishing Corp., New York, 1961

7.1.30. Redrawn by the Author from *The Turning Point of Building,* by Konrad Wachsmann, Reinhold Publishing Corp., New York, 1961

7.1.31, 7.1.32, 7.1.33, 7.1.34. and **7.1.35.** *The Turning Point of Building,* by Konrad Wachsmann, Reinhold Publishing Corp., New York, 1961

7.1.36. *SCSD: The Project and the Schools,* Educational Facilities Laboratory, 1967

7.1.37. *The Turning Point of Building* by Konrad Wachsmann, Reinhold Publishing Corp., New York, 1961

7.2. SYSTEMATIZATION WITHOUT A SYSTEM

7.2.1. a. Photo by the Author
   b. Drawing by the Author

7.2.2. Photos by the Author

7.2.3. *The Architecture of the Well-tempered Environment* by Reyner Banham, Architectural Press, 1969

7.2.4. *The Dymaxion World of Buckminster Fuller,* by R. W. Marks, Southern Illinois University Press, 1960. By permission of R. Buckminster Fuller

7.2.5. a,b,c. Photos by the Author
   d. *Housebuilding in the USA,* Ministry of Housing and Local Government, 1966. Reproduced with the permission of the Controller of Her Majesty's Stationery Office

7.2.6. a,b,c,d. *God's Own Junkyard,* by Peter Blake, Holt Rinehart and Winston, New York, 1964. Photos by Wiliam Garnett
e,f. Photos by the Author

7.2.7. Trade literature, Council of Forest Industries of British Columbia

7.2.8. a. Redrawn by the Author from *Housebuilding in the USA,* Ministry of Housing and Local Government, 1966.
b. Photo by the Author

7.2.9. a. *Housebuilding in the USA,* Ministry of Housing and Local Government, 1966. Reproduced with the permission of the Controller of Her Majesty's Stationery Office
b. Photo by the Author

7.2.10. and **7.2.11.** *Housebuilding the USA,* Ministry of Housing and Local Government, 1966. Reproduced with the permission of the Controller of Her Majesty's Stationery Office

7.2.12. Drawing by the Author

7.2.13. *Housebuilding in the USA,* Ministry of Housing and Local Government, 1966. Reproduced by permission of the Controller of Her Majesty's Stationery Office

7.2.14. a. Housebuilding in the USA, Ministry of Housing and Local Government, 1966. Reproduced by permission of the Controller of Her Majesty's Stationery Office
b. National Homes, Lafayette, Indiana, USA

7.2.15. Timber Frame Infill Housing, *RIBA Journal,* October 1978. By permission of Farrell/Grimshaw Partnership

7.2.16. Photos by Nigel Grundy

7.2.17. *Architectural Design,* May 1976. By permission of Walter Segal

7.3. COMPONENTS IN CONTEXT

7.3.1. Photo by the Author

7.3.2. Bauhaus, Exhibition catalogue, Royal Academy of Arts, London, 1968. By permission of H.M. Wingler

7.3.3, 7.3.4. and **7.3.5.** Photos by the Author

7.3.6. Drawing by the Author

7.3.7, 7.3.8, 7.3.9. and **7.3.10.** Crosswalls, *The Architects' Journal,* 17 March 1955

7.3.11, 7.3.12, 7.3.13. and **7.3.14.** Photos by the Author

7.3.15. Photo by Charles Fripp

7.3.16. House in the Isle of Wight, *The Architectural Review,* April 1958. By permission of James Stirling

7.3.17. *Mechanization Takes Command,* by Siefgried Giedion, Oxford University Press, 1969, 1948

7.3.18. Photo by Charles Fripp

7.3.19. Photo by the Author

7.3.20. *Art in Revolution.* Exhibition catalogue, Art Council, 1971

7.3.21.a. James Stirling
     b. Photo by the Author
7.3.22. *Hills Dry Building System* catalogue, 1957
7.3.23. Building is Your Business, by M. H. Thomas, Wingate, 1947. By permission of W. H. Allen & Co. Ltd (Allan Wingate)
7.3.24. House in the Isle of Wight, The Architectural Review, April 1958. By permission of James Stirling
7.3.25. *SCSD: The Project and the Schools,* Educational Facilities Laboratory, 1967

8 'SOME SORT OF ARCHITECTURE'

8.2. THE ADVENT OF CLASP

8.2.1. Successful design for mining subsidence, by Henry Swain, *The Architects' Journal,* 8 May 1974
8.2.2. and **8.2.3.** *The Story of CLASP,* Ministry of Education, 1961. Reproduced with the permission of the Controller of Her Majesty's Stationery Office
8.2.4. Photo by the Author
8.2.5. and **8.2.6.** Drawings by the Author
8.2.7. Photos by the Author
8.2.8. Reprinted by permission of Faber and Faber Ltd from *The New Architecture and the Bauhaus,* by Walter Gropius, 1965, 1935
8.2.9. *The Modern School,* by C. G. Stillman and R. C. Cleary, The Architectural Press, London, 1949
8.2.10. Drawing by the Author
8.2.11. Courtesy The Tate Gallery, London
8.2.12. Photos by the Author

8.3. GO FORTH AND MULTIPLY

8.3.1. Unknown
8.3.2. The NENK Method of Building, by R. Iredale and R. Walters, *RIBA Journal,* June 1964. Reproduced by permission of the Controller of Her Majesty's Stationery Office
8.3.3. Drawing by the Author
8.3.4. The Turning Point of Building by Konrad Wachsmann, Reinhold Publishing Corp, New York, 1961
8.3.5. MPBW Systems Development, *Official Architecture and Planning,* June 1968. Reproduced with the permission of the Controller of Her Majesty's Stationery Office

8.3.6. The NENK Method of Building, by R. Iredale and R. Walters. RIBA Journal, June 1964. Reproduced with the permission of the Controller of Her Majesty's Stationery Office

8.3.7. Drawing by the Author

8.3.8. Unknown

8.3.9. a. 5M Construction System, *The Architects' Journal*, 8 January 1964
b. Post-mortem on 5M, by P. Tindale and T. O'Toole, *RIBA Journal*, March 1969. Reproduced with the permission of the Controller of Her Majesty's Stationery Office

8.3.10, 8.3.11. and **8.3.12.** *5 Minute Guide to Economic Design in 5M System Housing*, MHLG, July 1966. Reproduced with the permission of the Controller of Her Majesty's Stationery Office

8.3.13, 8.3.14. Photos by the Author

8.3.15. Socea Balency (Sobea)

8.3.16. Drawings by the Author

8.3.17. Keystone Press Agency Ltd.

8.3.18. *Bauhaus*, Exhibition catalogue, Royal Academy of Arts, London, 1968. By permission of H. M. Wingler

8.3.19. Photo by John Griffin

8.3.20.a,b. *Housing in the Netherlands 1900/40*, by D. I. Grinberg, Delft University Press and Sijthoff and Noordhoff International Publishers, 1977
c,d,e,f. Photos by the Author

8.3.21.a. In need of support, by Sam Webb, *Architectural Design*, August 1969. By permission of Sam Webb, RIBA
b. Photo by the Author

8.3.22. Photo by the Author

8.3.23. The Author

8.3.24. Photos by the Author

8.3.25, 8.3.26. and **8.3.27.** Drawings by the Author

8.3.28, 8.3.29. and **8.3.30.** Photos by the Author

8.3.31.a,b. *12M Jespersen Design Guide*, MHLG, 1966
c. Photo by the Author

8.3.32. Drawing by the Author

8.3.33. Photos by the Author

8.4. AND THEY WENT FORTH

8.4.1. The Consortia, *DES Building Bulletin No. 54*, 1971. Reproduced with the permission of the Controller of Her Majesty's Stationery Office

8.4.2. *First SCOLA Report*, 1963. Second Consortium of Local Authorities

8.4.3, 8.4.4, 8.4.5, 8.4.6, 8.4.7. and **8.4.8.** Photos by the Author

8.4.9. The Author

8.4.10. Dimensional Co-ordination for Building, DC6. Ministry of Public Building and Works, 1967

8.4.11. Drawing by the Author

8.4.12. SCOLA Mark 2 Documentation. Second Consortium of Local Authorities

8.4.13. Drawings by the Author

8.4.14.a,b. SCOLA Mark 2 Documentation. Second Consortium of Local Authorities

c,d,e. Photos by the Author

8.4.15. SCOLA Mark 2 Documentation. Second Consortium of Local Authorities

8.4.16. Drawing by the Author

8.4.17. and **8.4.18.** Consortium for method building. *The Architects' Journal,* 13 July 1966

8.4.19. Reprinted by permission of Faber and Faber Ltd. from *The New Architecture and the Bauhaus,* by Walter Gropius, 1965, 1935

8.4.20. The Consortia; *DES Building Bulletin No. 54,* 1976. Reproduced with the permission of the Controller of Her Majesty's Stationery Office

8.4.21. South East Architects' Collaboration Mark 3 System Documentation, 1972

8.4.22. Photo by Allan Wood

9   THUNDERBIRD AND MODEL 'T'

9.2.   THE EHRENKRANTZ ICON

9.2.1. *SCSD: The Project and the Schools,* Educational Facilities Laboratory, 1967

9.2.2, 9.2.3, 9.2.4. and **9.2.5.** *Images of the Future,* by J. Lloyd Trump, Educational Facilities Laboratory and National Association of Secondary-School Principals, 1959

9.2.6. Drawing by the Author

9.2.7, 9.2.8, 9.2.9, 9.2.10. and **9.2.11.** SCSD: The Project and the Schools, Educational Facilities Laboratory, 1967

9.3.   TORONTO'S SEF

9.3.1. Redrawn by the Author from The Turning Point, by R. Robbie, *RIBA Journal,* June 1970. Copyright Metro Toronto School Board

9.3.2. and **9.3.3.** The Turning Point, by R. Robbie, *RIBA Journal,* June 1970. Copyright by Metro Toronto School Board

9.4. METROPOLITAN ARCHITECTURAL CONSORTIUM FOR EDUCATION

9.4.1. and **9.4.2.** MACE Brochure, *c.*1968. Metropolitan Architectural Consortium for Education

9.4.3. MACE documentation, 1970. Metropolitan Architectural Consortium for Education

9.4.4. MACE documentation, 1969. Metropolitan Architectural Consortium for Education

9.4.5. **a.** MACE documentation, 1969
b,c,d. MACE brochure, *c.* 1968

9.4.6. Photos by the Author

9.4.7. **a,b.** Photos by the Author
c. Photo courtesy of John Anthony Wright

9.4.8. MACE documentation, 1972

9.4.9. Drawing by the Author

9.5. SYSTEMS AT LARGE

9.5.1. and **9.5.2.** PSA method of building, by A. Rabeneck, *The Architects' Journal,* 14 November, 1973

9.5.3. *Architecture USA,* by I. McCallum, The Architectural Press, London, 1959. By permission of Carl Koch

9.5.4. **a.** Photo courtesy of New England Components and Techbuilt
b. Drawing by Carl Koch

9.5.5. Photos by the Author

9.5.6. Techcrete brochure, *c.*1965. By permisison of Carl Koch

10 INTEGRATED SYSTEMS

10.2. BIG BOXES, DISAPPEARING WALLS

10.2.1. John Anthony Wright

10.2.2.a. Photo by the Author
b. Photo by John Anthony Wright

10.2.3.a. Photo by the Author
b. John Anthony Wright

10.2.4. Photos courtesy of John Anthony Wright

10.2.5, 10.2.6. and **10.2.7.** Foster Associates

10.2.8. and **10.2.9.** Photos by the Author

10.2.10. IBM Concept Report, Foster Associates

10.2.11. Foster Associates

10.2.12. and **10.2.13.** Photos by the Author

10.2.14. Drawing by the Author

10.2.15. Photo by Graham Dugan

11 ENTROPIC DRIFT

11.1. FEEDBACK, FEEDFORWARD

11.1.1. and **11.1.2.** Photos by the Author
11.1.3.a,b,c. Photos by the Author
     d. Drawings by the Author
11.1.4. Photo by the Author
11.1.5. Keystone Press Agency Ltd.
11.1.6. Photos by the Author

11.2. MASS PRODUCTION AND ITS MYTHS

11.2.1. The Story of CLASP. *MOE Building Bulletin B No. 19.* Reproduced with the permission of the Controller of Her Majesty's Stationery Office

11.2.2. Drawing by the Author

12 OPEN SYSTEMS, WHOLE SYSTEMS

12.1. THE PERVASIVE SYSTEMS CONCEPT

12.1.1. and **12.1.2.** Photos by the Author
12.1.3, 12.1.4. and **12.1.5.** Drawings by the Author

12.2. TOWARDS OPEN SYSTEMS

12.2.1. Photo by Robert Bruno
12.2.2, 12.2.3. and **12.2.4.** Photos by the Author
12.2.5. Drawing by the Author